Oleksandr Savateev, Markus Antonietti, Xinchen Wang (Eds.)
Carbon Nitrides

Also of interest

Active Materials
(Open Access)
Fratzl, Friedman, Krauthausen, Schäffner (Eds.) 2022
ISBN 978-3-11-056181-4, e-ISBN (PDF) 978-3-11-056206-4,
e-ISBN (EPUB) 978-3-11-056220-0

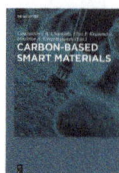

Carbon-Based Smart Materials
(Open Access)
Charitidis, Koumoulos, Dragatogiannis, 2020
ISBN 978-3-11-047774-0, e-ISBN (PDF) 978-3-11-047913-3,
e-ISBN (EPUB) 978-3-11-047775-7

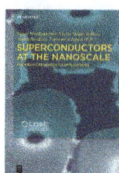

Superconductors at the Nanoscale.
From Basic Research to Applications
Wördenweber, Moshchalkov, Bending, Tafuri (Eds.), 2017
ISBN 978-3-11-045620-2, e-ISBN (PDF) 978-3-11-045680-6,
e-ISBN (EPUB) 978-3-11-045624-0

Materials Science.
Vol. 1: Structure
Hu, Cai, Rong, 2021
ISBN 978-3-11-049512-6, e-ISBN (PDF) 978-3-11-049534-8,
e-ISBN (EPUB) 978-3-11-049272-9

Materials Science.
Vol. 2: Phase Transformation and Properties
Hu, Cai, Rong, 2020
ISBN 978-3-11-049515-7, e-ISBN (PDF) 978-3-11-049537-9,
e-ISBN (EPUB) 978-3-11-049270-5

Carbon Nitrides

Structure, Properties and Applications in Science
and Technology

Edited by
Oleksandr Savateev, Markus Antonietti and Xinchen Wang

DE GRUYTER

Editors

Dr. Oleksandr Savateev
Max-Planck-Institut für Kolloid- und
Grenzflächenforschung
Abteilung Kolloidchemie
Forschungscampus Golm
14424 Potsdam
Germany
Oleksandr.Savatieiev@mpikg.mpg.de

Prof. Dr. Markus Antonietti
Max-Planck-Institut für Kolloid- und
Grenzflächenforschung
Abteilung Kolloidchemie
Forschungscampus Golm
14424 Potsdam
Germany
Markus.Antonietti@mpikg.mpg.de

Prof. Xinchen Wang
Fuzhou University
2 Xue Yuan Road
350116 Fuzhou
China
xcwang@fzu.edu.cn

ISBN 978-3-11-074696-9
e-ISBN (PDF) 978-3-11-074697-6
e-ISBN (EPUB) 978-3-11-074709-6

Library of Congress Control Number: 2023932363

Bibliographic information published by the Deutsche Nationalbibliothek
The Deutsche Nationalbibliothek lists this publication in the Deutsche Nationalbibliografie;
detailed bibliographic data are available on the internet at http://dnb.dnb.de.

© 2023 Walter de Gruyter GmbH, Berlin/Boston
Cover image: "Crucibles with carbon nitride powders"/Dr. Oleksandr Savateev
Typesetting: Integra Software Services Pvt. Ltd.
Printing and binding: CPI books GmbH, Leck

www.degruyter.com

Preface

Carbon nitride is a two-dimensional, covalent structure like graphene, and it is a recent "supermolecule," which makes the difference in many fields of chemistry. Unlike graphene, it is a semiconductor and is therefore used in a wide range of applications. In this book, experts with more than 20 years of experience in carbon nitride research summarize the vision of the community on structure and unique properties of carbon nitride materials, including actual applications in artificial photosynthesis, organic photoredox catalysis and super-optics, through electronics and sensing to next-technology devices, as artificial ion pumps and self-propelled microrobots.

https://doi.org/10.1515/9783110746976-202

Contents

Contributing authors

Markus Antonietti
Colloid Chemistry Department, Max Planck
Institute of Colloids and Interfaces, Am
Muehlenberg 1, 14476 Potsdam, Germany
markus.antonietti@mpikg.mpg.de

Bettina V. Lotsch
Nanochemistry Department, Max Planck
Institute for Solid State Research,
Heisenbergstrasse 1, 70569 Stuttgart, Germany

Department of Chemistry, University of Munich
(LMU), Butenandtstrasse 5-13, 81377 München,
Germany
b.lotsch@fkf.mpg.de

Xinchen Wang
State Key Laboratory of Photocatalysis on Energy
and Environment, College of Chemistry, Fuzhou
University, Fuzhou 350116, P.R. China.
xcwang@fzu.edu.cn

Shaowen Cao
State Key Laboratory of Advanced Technology
for Materials Synthesis and Processing, Wuhan
University of Technology, Wuhan 430070,
P.R. China
swcao@whut.edu.cn

Jingsan Xu
School of Chemistry and Physics, Science and
Engineering Faculty, Queensland University of
Technology, Brisbane, QLD 4000, Australia
jingsan.xu@qut.edu.au

Gabriel Ali Atta Diab
Department of Chemistry, Federal University of
São Carlo, 13565-905, São Carlos, São Paulo,
Brazil

Shi-Nan Zhang
School of Chemistry and Chemical Engineering,
Shanghai Jiao Tong University, Shanghai 200240,
P.R. China

Lu-Han Sun
School of Chemistry and Chemical Engineering,
Shanghai Jiao Tong University, Shanghai 200240,
P.R. China

Peng Gao
School of Chemistry and Chemical Engineering,
Shanghai Jiao Tong University, Shanghai 200240,
P.R. China

Stefano Mazzanti
Colloid Chemistry Department, Max Planck
Institute of Colloids and Interfaces, Am
Muehlenberg 1, 14476 Potsdam, Germany

Vincent W.-H. Lau
Department of Chemistry, National Cheng Kung
University, Tainan 701, Taiwan

Takuzo Aida
RIKEN Center for Emergent Matter Science,
Saitama, Japan
Department of Chemistry and Biotechnology,
The University of Tokyo, Tokyo, Japan
aida@macro.t.u-tokyo.ac.jp

Jian Liu
Qingdao Institute of Bioenergy and Bioprocess
Technology, Chinese Academy of Sciences,
Shandong Energy Institute, Qingdao 266101,
P.R. China
liujian@qibebt.ac.cn

Kai Xiao
Department of Biomedical Engineering
Southern University of Science and Technology
(SUSTech), 518055 Shenzhen, P.R. China
xiaok3@sustech.edu.cn

Oleksandr Savateev
Colloid Chemistry Department, Max Planck
Institute of Colloids and Interfaces, Am
Muehlenberg 1, 14476 Potsdam, Germany
oleksandr.savatieiev@mpikg.mpg.de

https://doi.org/10.1515/9783110746976-204

Yuanxing Fang
State Key Laboratory of Photocatalysis on Energy
and Environment, College of Chemistry, Fuzhou
University, Fuzhou 350116, P.R. China.

Han Li
State Key Laboratory of Advanced Technology
for Materials Synthesis and Processing, Wuhan
University of Technology, Wuhan 430070,
P.R. China

Qi Xiao
State Key Laboratory for Modification of
Chemical Fibers and Polymer Materials, College
of Materials Science and Engineering, Donghua
University, Shanghai 201620, China

Ivo Freitas Teixeira
Department of Chemistry, Federal University of
São Carlo, 13565-905, São Carlos, São Paulo,
Brazil
ivo@ufscar.br

Xiu Lin
School of Chemistry and Chemical Engineering,
Shanghai Jiao Tong University, Shanghai 200240,
P.R. China

Dong Xu
School of Chemistry and Chemical Engineering,
Shanghai Jiao Tong University, Shanghai 200240,
P.R. China

Guang-Yao Zhai
School of Chemistry and Chemical Engineering,
Shanghai Jiao Tong University, Shanghai 200240,
P.R. China

Xin-Hao Li
School of Chemistry and Chemical Engineering,
Shanghai Jiao Tong University,
Shanghai 200240, P.R. China
xinhaoli@sjtu.edu.cn

Filip Podjaski
Nanochemistry Department, Max Planck
Institute for Solid State Research,
Heisenbergstrasse 1, 70569 Stuttgart,
Germany
Department of Chemistry,
Imperial College London, 82 Wood Lane,
W12 0BZ London, United Kingdom

Nobuhiko Mitoma
RIKEN Center for Emergent Matter Science,
Saitama, Japan
Department of Chemistry and Biotechnology,
The University of Tokyo, Tokyo, Japan
nobuhiko.mitoma@riken.jp

Yuanyuan Zhang
Qingdao Institute of Bioenergy and Bioprocess
Technology, Chinese Academy of Sciences,
Shandong Energy Institute, Qingdao 266101,
P.R. China

Paolo Giusto
Colloid Chemistry Department, Max Planck
Institute of Colloids and Interfaces,
Am Muehlenberg 1, 14476 Potsdam, Germany
paolo.giusto@mpikg.mpg.de

Lei Jiang
Key Laboratory of Bio-inspired Smart Interfacial
Science and Technology of Ministry of Education,
School of Chemistry, Beihang University, 100191
Beijing, P.R. China

Markus Antonietti*, Oleksandr Savateev and Bettina V. Lotsch

Chapter 1
Celebrating 200 years of carbon nitride

Carbon nitride has a remarkably long history, and it is indeed one of the first artificial polymer structures and its synthesis was described in the scientific literature.

Reports on carbon nitride in the modern scientific sense are usually believed to begin with the observation of Jöns Jakob Berzelius, who identified a yellow polymer produced by the ignition of mercury thiocyanate. Justus von Liebig was inspired to work on carbon nitride by a longer letter exchange with Berzelius, arbitrarily naming the product of his synthesis "Melon" [1]. He also gave a composition of C:N = 3:4 [2] and called the product odd ("merkwürdig"), as it has no similarity to any other product except "molecules from animals" (indicating its organic character, while the concept of organic chemistry was only in its infancy those days). Interestingly, Liebig did synthesize the product using a variety of ways, for instance, by the oxidative coupling of KSCN to thiocyanogen $(SCN)_2$, which thermally decomposes to the "very oxidation-stable" Melon. This is an *essentially* hydrogen-free synthesis of carbon nitride, as it starts from *hydrogen-free precursors*, and minimizing the amount of hydrogen has been a quest in the carbon nitride field ever since.

There is, however, a second, more entertaining version of the story, attributed to another outstanding chemist of the nineteenth century, Friedrich Wöhler. In the first postmortem Wöhler Biography by A. F. Hofmann [3], we find the story of the young 20-year-old Wöhler (in 1820) who – still a student of medicine and without chemistry courses – ignited mercury thiocyanate in his student flat and observed the curiosity of "Pharaoh's serpent" for the first time. Needless to say, because of that, he ran into serious trouble with his mentor and landlord, and inorganic chemistry would look different today, if Wöhler would have been allowed to continue with such slow suicide! He published his observations as a rather young student in a smaller letter journal, where we still can observe his passion for thiocyanogen and the foaming serpent described [4]. The original German sentence is pure joy: "Erhitzt man es (the mercury thiocyanate) gelinde, so schwillt es plötzlich, sich gleichsam aus sich selbst in wurmartigen Gestalten windend, um das Vielfache seines vorigen Umfangs auf" (*If it (the mercury thiocyanate)*

*Corresponding author: Markus Antonietti, Colloid Chemistry Department, Max Planck Institute of Colloids and Interfaces, Am Muehlenberg 1, 14476 Potsdam, Germany,
e-mail: markus.antonietti@mpikg.mpg.de
Oleksandr Savateev, Colloid Chemistry Department, Max Planck Institute of Colloids and Interfaces, Am Muehlenberg 1, 14476 Potsdam, Germany
Bettina V. Lotsch, Nanochemistry Department, Max Planck Institute for Solid State Research, Heisenbergstrasse 1, 70569 Stuttgart, Germany; Department of Chemistry, University of Munich (LMU), Butenandtstrasse 5–13, 81377 München, Germany

https://doi.org/10.1515/9783110746976-001

is heated gently, it suddenly swells, as it were writhing out of itself in worm-like shapes, to many times its previous size). Three years later, Wöhler decided to join the labs of Jöns Jakob Berzelius for one year, which would nowadays be considered as an internship, and the two remained good colleagues for the rest of their lives. Other retrospective sources confirm that Wöhler is indeed the first to report the Pharaoh's serpent [5], while Berzelius in 1822 reported the first synthesis of pure HgSCN. We think it is safe to assume that the two quickly exchanged information about the Pharaoh's serpent and the yellow material it is made of, and that is how it became known to Liebig. In any case, this makes us able to celebrate "200 years of carbon nitride" these days.

The challenge with this family of materials is that they are inert, insoluble and thermally rather stable (up to 550 °C without even melting), so much so that deducing their structures has been difficult ever since their inception. In the 1920s, chemists proposed several structures for Melon, including a series of tri-*s*-triazines [6]. It was, then, Linus Pauling who, in 1937, showed with crystallography that the molecules are made up of a heptazine heterocycle (the "cyameluric nucleus") with a planar structure consisting of three fused *s*-triazine rings [7]. Pauling obviously retained his interest in these fancy structures. After his death in 1994, he left behind a structure others already called "mysterious" on his blackboard: a heptazine substituted with two hydroxyls and an azide (Fig. 1.1) [8]. It is essentially this fascination, also felt by others much later, which made this book possible!

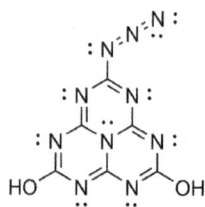

Fig. 1.1: The "mysterious" structure of a heptazine derivative found on Linus Pauling's blackboard after his death.

Now, after overall 200 years, this strange and mysterious family of nitrogen-rich compounds experiences an unexpected renaissance, thanks to countless applications and the disclosed potential of carbon nitride materials as 2D semiconductors. It indeed started by resolving the structure of the crystalline, molecular intermediates, such as Melem [10], and both the groups of Tamikuni Komatsu [9] and Wolfgang Schnick [10] reported strongest evidence that graphitic carbon nitride and especially 1D polymer Melon [11] are built up from heptazine units.

Admittedly, focusing on the functional electronic properties rather than structural questions, Markus Antonietti's group first studied synthesis of carbon nitride quantum dots and delaminated nanodisks [12]. Interestingly, analysis of differently sized nanoparticles already indicated a quantum size effect and, thereby, the semiconducting character of the particles. The following work was devoted to synthesize the inverse hard-templated mesoporous carbon nitrides and their use in heterogeneous organocatalysis, first for efficient Friedel–Crafts reactions [13]. Carbon nitride performance left the expected range immediately afterwards by catalyzing an unusual redox reaction, where benzene was oxidized to phenol, while CO_2 was reduced to CO [14]. From a contemporary view, this was presumably the first (photo)redox reaction with carbon nitride (in photochemistry, both half reactions are very effective and now well documented), just that it was not identified as such, back then. Shortly thereafter, Xinchen Wang et al. identified carbon nitride as an effective heterogeneous photocatalytic semiconductor and reported, in their seminal paper, the photochemical splitting of water into the elements, running both half reactions independently, using sacrificial agents [15].

The way to improve crystallinity and better-defined structures was less straightforward. Salt melt synthesis turned out to be an enabling solution. The first work by Michael J. Bojdys et al. [16] described the highly improved crystallinity, but assigned the structure to the traditional structure model. This was quickly corrected by Eva Wirnhier et al. [17], who identified an unexpected poly(triazine imide) structure as the actual product. Highly crystalline poly(heptazine imide) (PHI) derivatives made by high temperature "template" synthesis were known before [18], but the templates stayed in the structure. Condensation of triazoles in salt melt [19] allowed for improved crystallinity, while offering an opportunity for ion exchange [20] in the cavities of PHIs, which are four times larger compared to that in fully condensed heptazine-based g-CN (see Fig. 2.15). The improved crystallinity, the possibility of water and proton transport through the channels, but also ion exchange and the altered band positions extended the potential of PHI in photochemical reactions enormously. It was then first shown by Bettina V. Lotsch and her group that such PHIs enable, for instance, the trapping and storage of the photogenerated electrons. This feature was applied in a solar battery to design solar battery microswimmers, photomemristors and to promote "dark photocatalysis" [21, 22]. Thus, Liebig's Melon turned into an advanced, complex, functional chemical platform to inspire further progress.

What is clear from these last lines is that progress in carbon nitride chemistry became incredibly fast and efficient, and if one combs through an electronic library with the term "carbon nitride," one will find no less than 38,000 publications on the subject. From a mainly structural curiosity, the topic transformed into one of the most valuable assets in functional nanochemistry in only 20 years: even within the globalization of science, this is phenomenal, and, therefore, serves as inspiration for this book.

References

[1]   Fun Fact: Liebig: "Ich enthalte mich, Gründe für die Namen anzuführen, denen man in dieser Abhandlung begegnet; sie sind, wenn man will, aus der Luft gegriffen, was den Zweck genau so gut erfüllt, als wären sie von der Farbe oder einer Eigenschaft abgeleitet", i.e. Liebig gave arbitrary names to Melon (and other carbon nitrides) and did not consider it necessary to give reasons for his choice of names.
[2]   Liebig J. Analyse der Harnsäure. Ann Pharm 1834, 10(1), 47–8.

[3]   Hofmann AW. Zur Erinnerung an Friedrich Wöhler. Berlin, Ferd. Dümmlers Verlagsbuchhandlung (Harrwitz & Grossmann), 1883. 164.

[4]   Wöhler F. Ueber die eigenthümliche Säure, welche entsteht, wenn Cyan (Blaustoff) von Alkalien aufgenommen wird. Ann Phys 1822, 71(5), 95–103.

[5]   Irving H. An historical account of Pharaoh's serpents. Sci Prog (1933-) 1935, 30(117), 62–6.

[6]   Wilson EK. Old molecules, new chemistry. Chem Eng News 2004, 82(22), 34–5.

[7]   Pauling L, Sturdivant JH. The structure of cyameluric acid, hydromelonic acid and related substances. Proc Natl Acad Sci 1937, 23(12), 615–20.

[8]   Wilson EK. A prized collection: Pauling memorabilia. Chem Eng News 2000, 78(32), 62–3.

[9]   Komatsu T. The first synthesis and characterization of cyameluric high polymers. Macromol Chem Phys 2001, 202(1), 19–25.

[10]   Jürgens B, Irran E, Senker J, Kroll P, Müller H, Schnick W. Melem (2,5,8-triamino-tri-s-triazine), an important intermediate during condensation of melamine rings to graphitic carbon nitride: Synthesis, structure determination by X-ray powder diffractometry, solid-state NMR, and theoretical studies. J Am Chem Soc 2003, 125(34), 10288–300.

[11]   Lotsch BV, Döblinger M, Sehnert J, Seyfarth L, Senker J, Oeckler O, et al. Unmasking Melon by a complementary approach employing electron diffraction, solid-state NMR spectroscopy, and theoretical calculations – structural characterization of a carbon nitride polymer. Chem Eur J 2007, 13(17), 4969–80.

[12]   Groenewolt M, Antonietti M. Synthesis of g-C_3N_4 nanoparticles in mesoporous silica host matrices. Adv Mater 2005, 17(14), 1789–92.

[13]   Goettmann F, Fischer A, Antonietti M, Thomas A. Chemical synthesis of mesoporous carbon nitrides using hard templates and their use as a metal-free catalyst for Friedel–Crafts reaction of benzene. Angew Chem Int Ed 2006, 45(27), 4467–71.

[14]   Goettmann F, Thomas A, Antonietti M. Metal-free activation of CO_2 by mesoporous graphitic carbon nitride. Angew Chem Int Ed 2007, 46(15), 2717–20.

[15]   Wang X, Maeda K, Thomas A, Takanabe K, Xin G, Carlsson JM, et al. A metal-free polymeric photocatalyst for hydrogen production from water under visible light. Nat Mater 2009, 8(1), 76–80.

[16]   Bojdys MJ, Müller J-O, Antonietti M, Thomas A. Ionothermal synthesis of crystalline, condensed, graphitic carbon nitride. Chem Eur J 2008, 14(27), 8177–82.

[17]   Wirnhier E, Döblinger M, Gunzelmann D, Senker J, Lotsch BV, Schnick W. Poly(triazine imide) with intercalation of lithium and chloride ions [$(C_3N_3)_2(NH_xLi_{1-x})_3\cdot LiCl$]: A crystalline 2D carbon nitride network. Chem – Eur J 2011, 17(11), 3213–21.

[18]   Döblinger M, Lotsch BV, Wack J, Thun J, Senker J, Schnick W. Structure elucidation of polyheptazine imide by electron diffraction – a templated 2D carbon nitride network. Chem Commun 2009, (12), 1541–3.

[19]   Dontsova D, Pronkin S, Wehle M, Chen Z, Fettkenhauer C, Clavel G, et al. Triazoles: A new class of precursors for the synthesis of negatively charged carbon nitride derivatives. Chem Mater 2015, 27 (15), 5170–9.

[20]   Savateev A, Pronkin S, Willinger MG, Antonietti M, Dontsova D. Towards organic zeolites and inclusion catalysts: Heptazine imide salts can exchange metal cations in the solid state. Chem – Asian J 2017, 12(13), 1517–22.

[21]   Podjaski F, Kröger J, Lotsch BV. Toward an aqueous solar battery: Direct electrochemical storage of solar energy in carbon nitrides. Adv Mater 2018, 30(9), 1705477.

[22]   Lau VW-H, Klose D, Kasap H, Podjaski F, Pignié M-C, Reisner E, et al. Dark photocatalysis: Storage of solar energy in carbon nitride for time-delayed hydrogen generation. Angew Chem Int Ed 2017, 56 (2), 510–4.

Oleksandr Savateev* and Bettina V. Lotsch

Chapter 2
Classification, synthesis and structure of carbon nitrides

2.1 Introduction

In a very broad context, carbon nitrides are materials made of carbon and nitrogen. The stoichiometry between C and N may be different. Nitrogen-doped carbon that contains few atomic percent of nitrogen is at one end of the scale. Binary materials with the stoichiometry C_3N_4 and nitrogen content of 57 atomic percent are at the other end of the spectrum. In this chapter, we limit our discussion to carbon nitrides with a close-to -3C:4N ratio. There is more than one way, as defined by valency rules and hybridization, how carbon and nitrogen may be connected and spatially arranged to create a periodic structure composed of strictly alternating C–N (or C =N) bonds. It includes, for example, β-C_3N_4 composed of sp^3-carbon (tetrahedral) and sp^2-nitrogen (trigonal planar, to a first approximation), as well as graphitic carbon nitride in which both carbon and nitrogen are trigonal planar. We additionally narrow down our discussion to structures related to graphitic carbon nitride, which, in modern research, are abbreviated as g-C_3N_4 or g-CN. We use the term graphitic carbon nitride to denote an ensemble of materials that satisfy the following set of rules: (1) the material is composed of trigonal planar carbon and nitrogen; (2) the ratio of carbon and nitrogen is close to 3:4; hydrogen or other heteroatoms can be present, as well, in small amounts; (3) carbon and nitrogen are assembled into 1,3,5-triazine or tri-s-triazine units; (4) the units are interconnected covalently via sp^2-hybridized tertiary nitrogen atoms or NH-groups to form layers; and (5) the layers are stacked via weak van der Waals interactions to form structures resembling that of graphite. Graphitic carbon nitrides are often called "polymeric carbon nitrides," which

Acknowledgments: Authors are grateful to Dr. Lihua Lin for providing an SEM image of PTI/Li⁺Cl⁻ (Fig. 2.23b).

*Corresponding author: Oleksandr Savateev, Colloid Chemistry Department, Max Planck Institute of Colloids and Interfaces, Am Muehlenberg 1, 14476 Potsdam, Germany,
e-mail: oleksandr.savatieiev@mpikg.mpg.de
Bettina V. Lotsch, Nanochemistry Department, Max Planck Institute for Solid State Research, Heisenbergstrasse 1, 70569 Stuttgart, Germany; Department of Chemistry, University of Munich (LMU), Butenandtstrasse 5-13, 81377 München, Germany

https://doi.org/10.1515/9783110746976-002

refers to the nanocrystalline or (semi-)amorphous nature of these materials and also the fact that one of the oldest carbon nitrides, Liebig's Melon, is composed of threads held together via hydrogen bonding, which in turn form layers. Therefore, on the one hand, it is a quasi-2D material, and on the other hand, it is a polymer. Although the local chemical structure of carbon nitrides, i.e., their molecular building blocks, is well-defined, weak van der Waals interactions almost always result in a certain degree of stacking disorder, i.e., shift or rotation of the neighboring layers with respect to each other. Such planar defects create substantial difficulties in the characterization of the crystal structure of carbon nitrides. It is remarkable that although graphitic carbon nitrides have been known for 200 years, the structure of some of them, such as Liebig's Melon, has been solved only in the beginning of the twenty-first century, using advanced spectroscopic techniques! As the area of carbon nitride research is very broad and ranges from organic synthesis and polymer science to solid-state physics, in the following chapters, the term "graphitic carbon nitride," "polymeric carbon nitride," "polymeric graphitic carbon nitride" or the very general term "carbon nitride" will be used, depending on the most relevant context. In this and subsequent chapters, labels of carbon nitride materials are taken from the publications without significant modification, to preserve original terminology.

The objective of this chapter is to summarize synthesis, structures and a few general properties of carbon nitrides that are the most relevant to their applications in the subsequent chapters. The chapter also links heptazine derivatives, which are carbon nitride precursor molecules, with the carbon nitride materials.

2.2 From molecules to materials

By using a pencil and a sheet of paper, chemists can generate a variety of structures that satisfy Hückel's aromaticity rule (a cyclic conjugated system with [4n +2] π electrons is aromatic and, thus, particularly stable) [1, 2]. Figure 2.1 shows a series of structures created by fusion of several 1,3,5-triazine rings. All-carbon counterparts are provided for comparison to illustrate how the combination of two types of atoms with different valence in an alternating fashion generates a scope of possible structures saturated at their termini by hydrogen. The first cyclic representative from a series is 1,3,5-triazine (or s-triazine). The molecule is thermodynamically stable but reactive. It may be considered as a trimer of hydrogen cyanide. Substitution of residual hydrogen atoms with amino groups gives melamine – a major precursor used to synthesize carbon nitride materials.

Derivatives of s-triazine

All-carbon analogues

1,3,5-triazine
(s-triazine)

melamine

benzene

naphthalene

anthracene

phenanthrene

tri-s-triazine
(heptazine)

Melem

phenalene

phenalenyl radical

triangulene

trimesityltriangulene

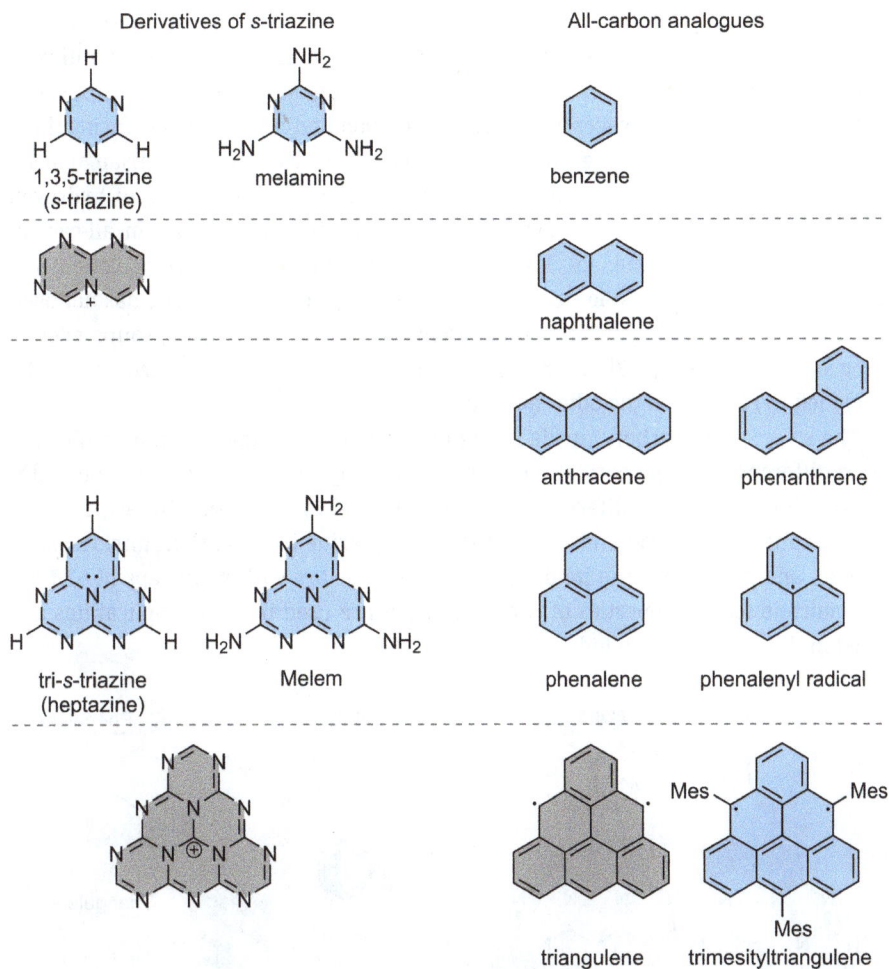

Fig. 2.1: A scope of azines created by fusion of several 1,3,5-triazine units and all-carbon counterparts. Structures filled with blue color have been either isolated as individual compounds or observed spectroscopically; gray structures have not been reported.

Fusion of two phenyl rings gives naphthalene, an aromatic thermodynamically stable 10π-electron compound. On the other hand, fusion of two 1,3,5-triazine molecules gives a structure in which the bridging nitrogen atom must be positively charged. The variety of aromatic structures increases, while the structures themselves become more interesting once we fuse three hexagons. For all-carbon polyaromatic hydrocarbons, three structures are possible – anthracene, phenanthrene and phenalene. The phenalenyl core is the smallest polycyclic odd alternant hydrocarbon with 13 carbon atoms. While many polyaromatic hydrocarbons, such as phenanthrene, can adapt only a Kekulé structure with all π-electrons being fully paired and conjugated, the phenalenyl radical can adapt only a non-Kekulé structure – an open-shell structure with an unpaired π-electron. We will not discuss counterparts of anthracene and phenanthrene with alternating C–N bonds, as such structures, at first glance, would appear unstable at ambient conditions. However, the valence of carbon and nitrogen allow fusing three 1,3,5-triazine units into a stable closed-shell 14π-electron structure with a triangular topology similar to phenalene – tri-s-triazine or heptazine [3]. Substitution of hydrogen atoms with NH₂-groups gives Melem. Expansion of a triangular motif in all-carbon structures gives triangulene, also known as Clar's hydrocarbon [4]. Triangulene is another example of a non-Kekulé structure. Although the parent triangulene has not been isolated, bulky *tert*-butyl, mesityl substitutents as well as three keto-groups exert a strong stabilizing effect [5–7]. Extended structures of tri-s-triazine (gray structure at the bottom left in Fig. 2.1), on the other hand, are yet to be discovered.

The structure of carbon nitride layers featuring large triangular "holes" (Fig. 2.2, left) is different from the continuous structure of graphene (Fig. 2.2, right). Such differences should result in different chemical and physical properties. In fact, they do. While graphene is a zero-band gap semimetal, heptazine-based carbon nitrides, such as Melon and poly(heptazine imides), are semiconductors with band gaps of ~2.7 eV. The semiconducting properties of carbon nitrides are used in many applications discussed in this book and beyond.

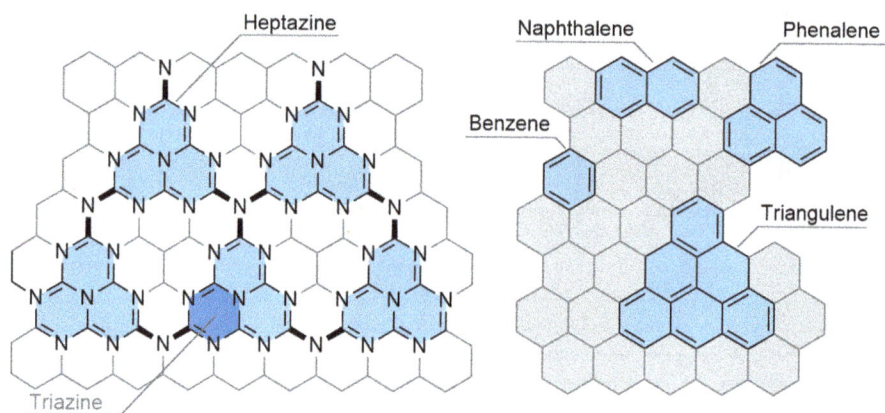

Fig. 2.2: Schematic image of a fully condensed layer of heptazine-based g-CN (left) and graphene (right) with molecular building blocks marked with blue color.

The following section discusses the synthesis and properties of heptazine derivatives as models of heptazine-based g-CN materials.

2.3 Derivatives of heptazine

Without doubt, the carbon nitride field is mainly associated with material science and, hence, extended solids. At the same time, the molecular compound tri-s-triazine forms the elementary building block postulated for most graphitic carbon nitrides in modern research. This is where molecular chemistry meets the solid state. Tri-s-triazine possesses a triangular topology and is used as a scaffold for subsequent functionalization. It is a nitrogen-rich, electron-deficient and thermally stable heterocycle, while the scope of compounds based on this structural motif is abundant.

Cyclazines 3a–d are molecules composed of a fused conjugate ring system held planar by three covalent bonds to the internal nitrogen atom. These compounds are prepared via a two-step procedure from diaminotriazine 1a, diaminopyrimidine 1b, diaminopyridine 1c and diaminopyrazine 1d, using flash vacuum pyrolysis in 42–67% isolated yield (Fig. 2.3) [8].

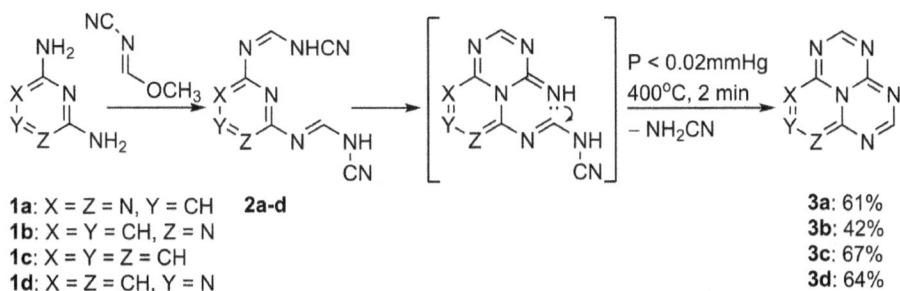

1a: X = Z = N, Y = CH   2a-d

1b: X = Y = CH, Z = N

1c: X = Y = Z = CH

1d: X = Z = CH, Y = N

3a: 61%
3b: 42%
3c: 67%
3d: 64%

Fig. 2.3: A general pathway of cyclazines synthesis. The legend for X, Y and Z in structures **1**, **2** and **3** is the same.

Bare tri-s-triazine **3a** is synthesized by vacuum flash pyrolysis of substituted 1,3,5-triazine at 400° [9]. X-ray single crystal analysis of **3a** revealed that it is a planar molecule – the lone pair of the central nitrogen atom is indeed in conjugation with the π-aromatic system. Electron deficiency of cyclazines is supported by a progressively increasing ionization potential, which correlates with the number of pyridinic-type nitrogen atoms in the cycle. Thus, the ionization potential of **3c** that contains four pyridinic-type nitrogen atoms and one nitrogen atom in the center is 7.94 eV, while the ionization potential of tri-s-triazine **3a** having seven nitrogen atoms in total is 9.16 eV. Calculations show that the lone electron pairs are delocalized over the aromatic heterocycle to a greater extent than in pyridine. It explains the aromatic nature and poor reactivity of **3a**. Tri-s-triazine **3a** does not react with acids, alkylating agents such as CH_3I at room temperature and oxidants such as m-chloroperbenzoic acid both at room temperature and reflux [3]. Poor reactivity is also illustrated by the fact that it does not form charge transfer complexes with electron donors or with electron acceptors. Fluorescence lifetime of tri-s-triazine **3a** in acetonitrile is 19 ns, while the quantum yield is 1% at 283 K [10].

2,5,8-triamino-tri-s-triazine **4**, or Melem, is a molecule with a triangular topology similar to melamine (Fig. 2.4) [11]. It is synthesized by calcination of melamine. Despite its relatively compact size, due to multiple hydrogen bonds, Melem is not soluble in organic solvents. Therefore, we will discuss it from the materials chemistry perspective in the next section.

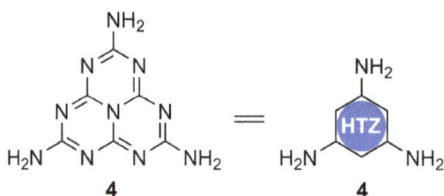

Fig. 2.4: Chemical structure of Melem.

Due to the high chemical stability of the heptazine moiety, functionalization in the positions 2,5 and 8 may be conducted conveniently and selectively using various and, often, quite aggressive reagents. Melem is a precursor for a series of derivatives, such as 2,5,8-trihydrazino-s-heptazine **5**, [12] 2,5,8-triazido-s-heptazine **6** and tris(tri-n-butylphosphinimino)-s-heptazine **7** (Fig. 2.5).

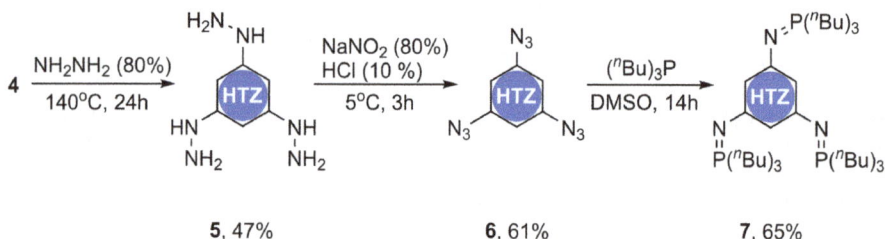

Fig. 2.5: Synthesis of heptazine derivative **7**.

Hydrolysis of Melem **4** in concentrated alkaline solution at elevated temperature gives water-soluble potassium cyamelurate **8** with ~24% isolated yield (Fig. 2.6) [13]. Treatment of **8** with PCl_5 at 130 °C in the solid state gives trichloroheptazine **9** – a yellow solid that is purified by sublimation in vacuum and soluble in tetrahydrofuran (THF) and acetone.

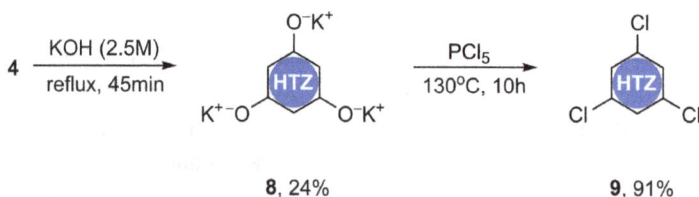

Fig. 2.6: Synthesis of trichloroheptazine **9**.

Similar to **3a**, the molecule of **9** is planar. In the solid state, similar to cyanuric chloride, **9** forms extensive donor–acceptor networks. Chlorine atoms in **9** can be substituted in a reaction with different nucleophiles, such as secondary phosphines and phosphineoxides (Fig. 2.7) [14].

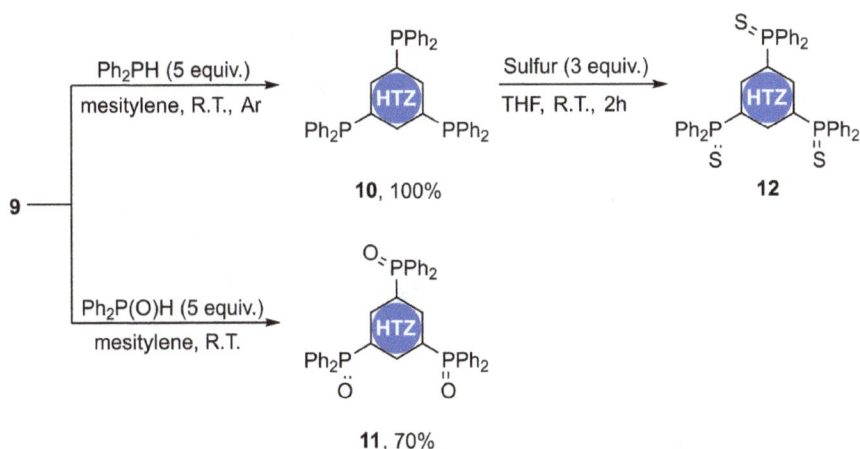

Fig. 2.7: Phosphorylation of trichloroheptazine **9**.

An alternative pathway for derivatization of the heptazine core starts from 2,5,8-tris (3,5-diethyl-pyrazolyl)-heptazine **13**, which is synthesized from trihydrazine **5** via mechanochemical treatment with heptanedione-3,5 (Fig. 2.8) [15]. In the solid state, pyrazolyl groups in **13** are coplanar with the tri-s-triazine unit. Pyrazolyl groups in **13** undergo nucleophilic substitution in a reaction with secondary amines and thiols at moderate heating, affording a series of derivatives of tri-s-triazine **14a-d** and **15a-c** with 22–75% isolated yield, which are soluble in dichloromethane.

Compounds **13** and **15a–c** possess reduction potentials of –1.3 to –1.0 V versus SCE. The molar extinction coefficients of **13**, **14a–d** and **15a–c** measured in dichloromethane are in the range $8-9 \times 10^4$ M^{-1} cm^{-1} [15]. Compounds **14** show an onset of absorption at ~300 nm. Due to stronger donor–acceptor character of compounds **15** (heptazine is electron deficient, while RS moieties are electron rich) the onset of absorption is shifted to ~350 nm. Compound **13** demonstrates temperature-activated delayed fluorescence (TADF) with a lifetime of 66.8 ns, which is slightly quenched by O_2 to 62.0 ns. The quantum yield of compounds **13–15** fluorescence in solution is 3–19%. Compounds **13–15** also fluoresce in solid state. The results of time-dependent density functional theory (TD DFT) calculations suggest a singlet-triplet energy gap of ~0.15 eV. Together with the spatially separated HOMO and LUMO orbitals, it explains TADF via reversed intersystem crossing (RISC) [16].

Eliminating extensive hydrogen bonding in Melem by substituting NH_2-groups with $N({}^{t}Bu)_2$-groups affords a series of heptazine derivatives, which are soluble in

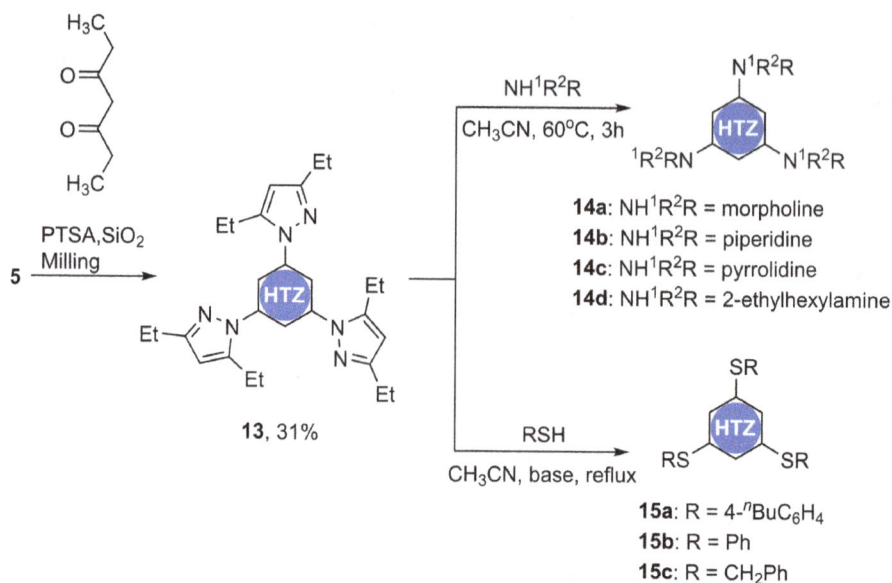

Fig. 2.8: Mechanochemical synthesis of **13** and nucleophilic substitution of pyrazolyl groups.

acetonitrile, tetrahydrofuran, chloroform, and hexane [17]. The first representative is hexa(*iso*-butyl)-alkylated Melem **16** (the monomer, Fig. 2.9).

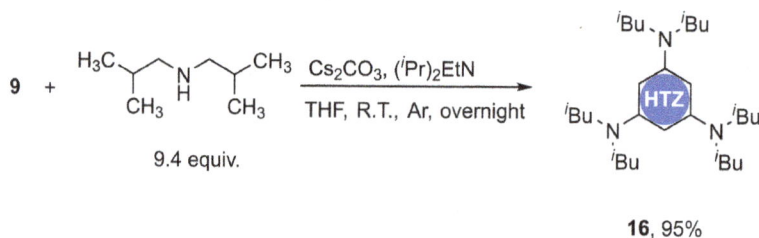

Fig. 2.9: Synthesis of **16** from trichloroheptazine and di(*iso*-butyl)amine.

Unsymmetrically substituted tri-*s*-triazine derivative **17** is obtained with 85% isolated yield upon addition of two equivalents of secondary amine to **9** in THF cooled to −95 °C (Fig. 2.10).

Unsymmetrical tri-substituted heptazine **18** is synthesized in quantitative yield upon addition of excess of butylamine to **17** at room temperature (Fig. 2.11).

The dimer **19** is synthesized with 85% yield by coupling **18** and **17** (Fig. 2.12). In the solid state, the angle between two heptazine rings is 66.7°, but in solution, rotation is not restricted [17].

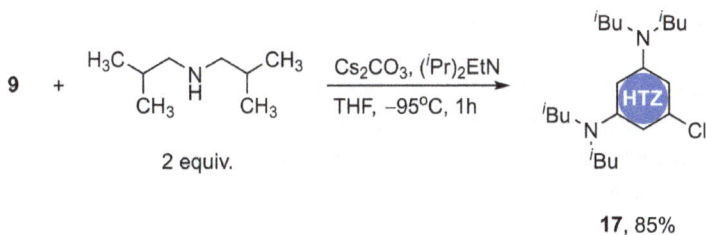

Fig. 2.10: Synthesis of **17** from trichloroheptazine **9** and di(*iso*-butyl)amine.

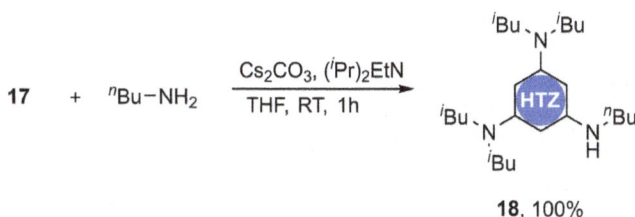

Fig. 2.11: Synthesis of unsymmetrically substituted heptazine derivative **18**.

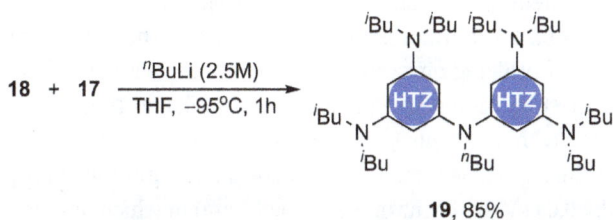

Fig. 2.12: Synthesis of heptazine dimer **19**.

As shown in Fig. 2.13, the trimer **20** is synthesized with 17% yield by coupling two equivalents of **17** with one equivalent of **21** (synthesis of this compound is similar to **18** and not discussed here).

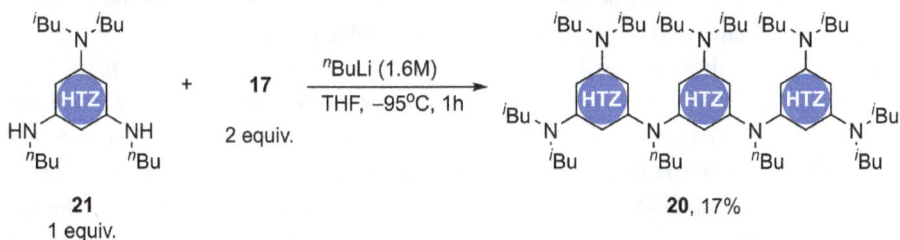

Fig. 2.13: Synthesis of heptazine trimer **20**.

Derivatives of Melem – monomer **16**, dimer **19** and trimer **20**, which are soluble in organic solvents, are excellent models to correlate the degree of conjugation in heptazine oligomers with their photo-physical properties. Thus, onset of absorption shifts bathochromically with the increase of the number of heptazines in the conjugation (Fig. 2.14a). Extinction coefficients of the bands that may be assigned to π–π^* transitions (8.94–$11.13 \cdot 10^4$ M^{-1} cm^{-1}) and n–π^* transitions ($2,000$–$4,900$ M^{-1} cm^{-1}) also increase along this series (Tab. 2.1).

Tab. 2.1: Molar extinction coefficients of heptazine derivative solution in MeCN.

Compound	ε (M^{-1} cm^{-1})[a]	ε (M^{-1} cm^{-1})[a]
Monomer **16**	89,400 (273 nm)	2,000 (315 nm)
Dimer **19**	90,200 (276 nm)	2,400 (368 nm)
Trimer **20**	111,300 (272 nm)	4,900 (370 nm)

[a]Wavelength is given in parentheses.

Results of DFT modeling confirmed that HOMO and LUMO are primarily localized at the planar heptazine units. The low-intensity tail transitions were assigned to Ext. N/Cyc. N and Bdg. C/Cyc. C atoms (Fig. 2.14b). Furthermore, with the increase of the size of the linear polymer, the contribution of bridging nitrogens is expected to increase and n–π^* transitions show a bathochromic shift. In dimelem (a compound similar to **19** but with all alkyl groups being substituted with hydrogens), DFT modeling indicates that increase of the dihedral angle between the heptazine units from 0 to 90 °C decreases the energy of low-energy tail transitions by 0.69 eV (bathochromic shift by 65 nm) in which bridging carbon and nitrogen atoms are involved. In carbon nitride materials, in which free rotation of heptazine units is restricted as they are in the solid state, deviation from perfectly planar arrangement induces evolution of a strong absorption band in the visible range of the spectrum. Such a phenomenon is the basis for chromoselective photocatalysis.

In the PL spectra, the maximum is bathochromically shifted when the oligomer size increases, while the Stokes shift changes from ~109 to 171 nm (Fig. 2.14d). For the same compound, decrease of the excitation wavelength does not change the PL spectrum, which points at relaxation of the same excited state. Quantum yield of fluorescence progressively increases from 0.4% (λ_{exc} = 315 nm) for the monomer **16**, to 3.4% (λ_{exc} = 353 nm) for the dimer **19** and to 14.7% (λ_{exc} = 356 nm) for the trimer **20**. When dissolved in DMSO, compounds **16**, **19** and **20** show an irreversible oxidation potential at +2.11 to +2.12 V versus NHE. The reduction potential progressively shifts from –2.17 V for the monomer **16**, to –1.60 V for the dimer **19** and to –1.30 V versus NHE for the trimer **20**. Transition from the monomer to trimer increases the size of the LUMO. In the trimer, upon one-electron reduction, the excess negative charge is delocalized over the larger conjugated system, compared to the dimer and monomer. As a result, reduction becomes more energetically favorable and occurs at a less negative potential. Overall,

a)

b)

c)

d)

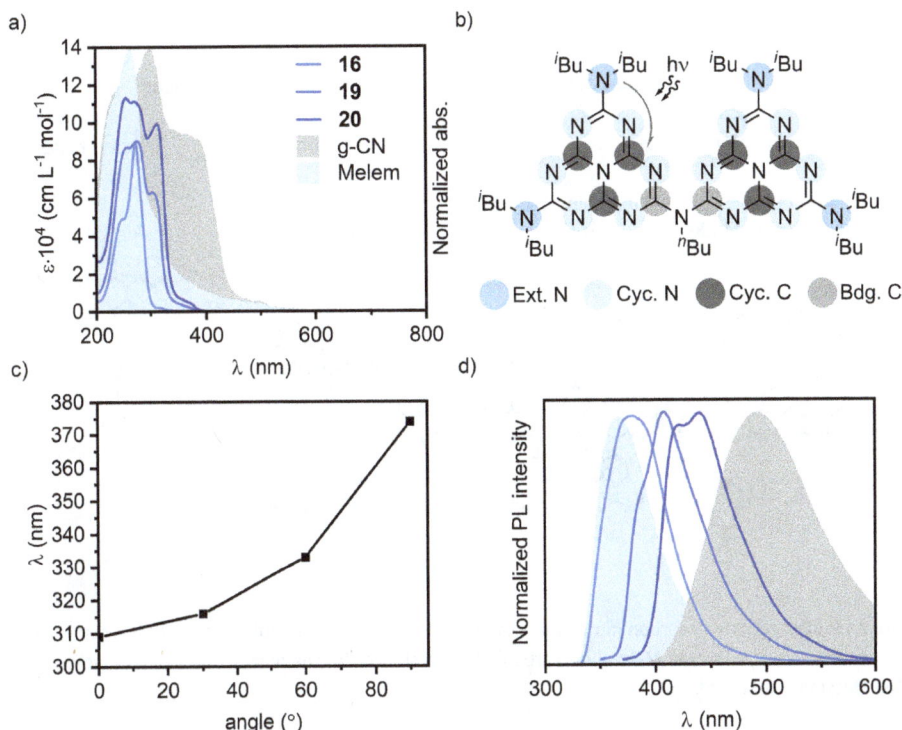

Fig. 2.14: Photophysical properties of heptazine monomer **16**, dimer **19**, trimer **20**, Melem and g-CN. (a) UV–vis absorption spectra of **16**, **19** and **20** in MeCN; Melem and g-CN in solid state (normalized absorption). (b) Atoms in dimer **19** involved in tail transitions. (c) Maximum of tail-transition absorption band as a function of dihedral angle between the heptazine units in dimelem. (d) Steady-state photoluminescence spectra of **16**, **19** and **20** MeCN solutions, g-CN and Melem in solid state at room temperature. Data shown in panels (a), (c) and (d) are reproduced from [17] (Copyright 2016 the Royal Society of Chemistry) and adapted with permission from[11] (Copyright 2003 American Chemical Society).

methods and reactions commonly used in organic chemistry are also applicable for synthesis of heptazine derivatives, which, in turn, are valuable precursors of porous polymeric frameworks and OLEDs [18, 19].

2.4 Classification and structure of carbon nitrides

Although carbon nitrides are primarily built of s-triazine and/or tri-s-triazine units, the arrangement of these building blocks gives a variety of structures with peculiar photophysical and chemical properties, as illustrated in Fig. 2.15 [20, 21].

Typically, for applications discussed in this book and beyond, carbon nitrides are prepared according to the "open synthesis" method – i.e., gases, primarily NH_3, are allowed

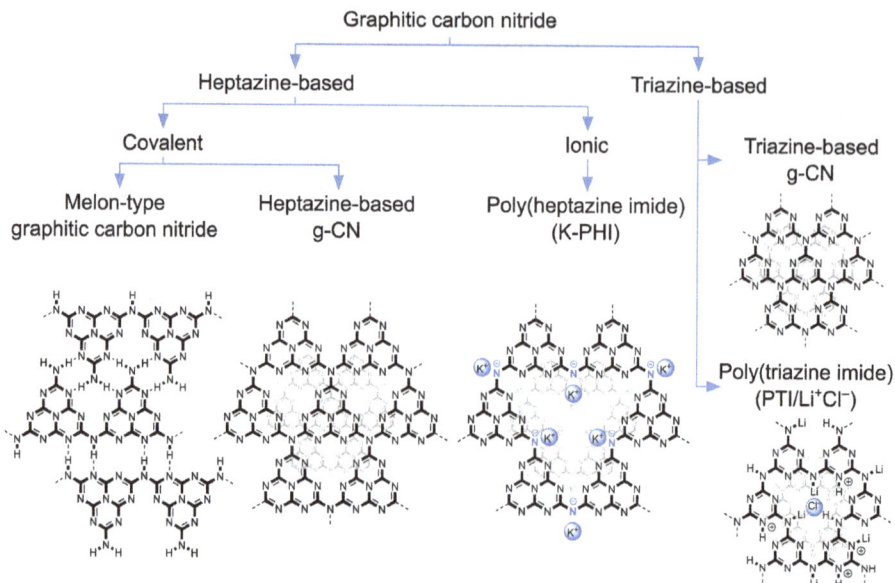

Fig. 2.15: Classification of carbon nitride materials with a (close-to) 3C:4N stoichiometry that is discussed in the book. A single layer of Melon-type carbon nitride is shown. For the other g-CN materials, eclipsed AAA stacking is assumed.

to escape from the vessel in which the precursors are calcined. This results in a substantial number of structural defects. Therefore, "open synthesis" produces amorphous or nanocrystalline materials. When precursors are calcined in a sealed glass tube ("closed synthesis") under autogenous NH_3 pressure, equilibrium between polymerization of the precursor and depolymerization of carbon nitride is established. It allows for repairing the defects in the structure of carbon nitrides and, as a result, affords more crystalline material. Combination of "closed synthesis" and electron microscopy that became widely accessible at the beginning of the twenty-first century allowed resolving the crystal structure of many carbon nitride materials. In this section, we summarize carbon nitride materials whose crystal structure was determined by combining powder X-ray diffraction, electron diffraction and electron microscopy.

2.4.1 Covalent heptazine-based carbon nitrides

2.4.1.1 Melem or triamino-tri-s-triazine

Melem suitable for crystal structure determination is synthesized by calcining a small (80 mg) amount of cyanamide, ammonium dicyanamide, dicyandiamide or melamine in a sealed glass ampoule (outer diameter 16 mm, inner diameter 12 mm, length 120 mm) at

450 °C for 5 h [11]. The crystal structure was solved and refined based on conventional powder diffraction data (Fig. 2.16a). The individual $C_6N_7(NH_2)_3$ molecules are packed in the crystal with the interlayer distance of 3.27 Å (Fig. 2.16b), which is comparable to the respective distance in melamine (3.2–3.4 Å) and pure graphite (3.34 Å). The ring system is almost planar. There are eight hydrogen bonds per $C_6N_7(NH_2)_3$ molecule (Fig. 2.16c), which anchor individual molecules within the crystal and explain insolubility of Melem in solvents. Melem absorbs in the UV and on the edge of the visible spectrum (Fig. 2.14a), while tailing in the visible range could be explained by defects of different kinds, as concluded in an earlier study [11]; or taking into account a more recent study of heptazine oligomers, this feature could be intrinsic to certain heterocycles and is assigned to $n–\pi^*$ transitions [17]. The photoluminescence maximum is observed at 466 nm (Fig. 2.14d), while quantum yield of fluorescence is 40%.

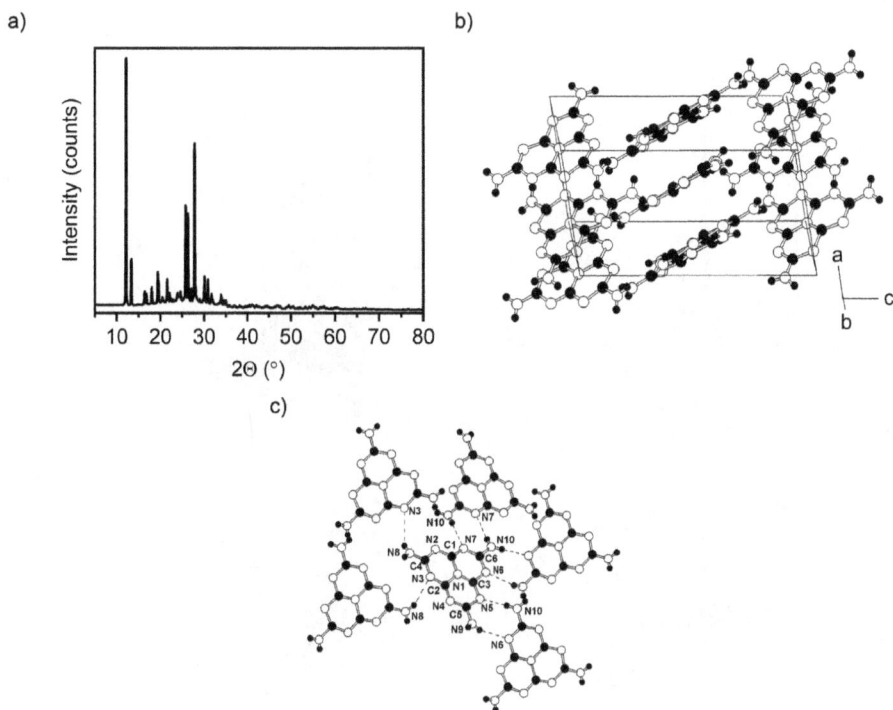

a)

b)

c)

Fig. 2.16: Structure of Melem. (a) Rietveld refinement of the PXRD pattern. (b) Crystal structure of Melem. (c) Hydrogen bonds between Melem molecules (reproduced with permission from [11]. Copyright 2003 American Chemical Society).

2.4.1.2 Melem hydrate

Unlike the above described synthesis of crystalline Melem in a sealed glass tube, preparation of the material on a larger scale following the "open synthesis" method gives a mixture of Melem and unreacted melamine. On purification of the product, under hydrothermal conditions at 200°C in an autoclave, single crystals of Melem hydrate with a composition $C_6N_7(NH_2)_2 \cdot 1.98\ H_2O$ are obtained (Fig. 2.17) [22]. At the time of publication, it was the first example of a heptazine-based structure with a hydrogen-bonded rosette-like pattern that is similar to a complex of cyanuric acid and melamine [23]. Boiling water breaks the intermolecular hydrogen bonds in Melem, which leads to the rearrangement of molecules into a hydrogen-bonded rosette-like network. Water molecules serve as a template that result in the formation of hexagonal channels with a diameter of 8.9 Å.

a)

b)

c)

d)

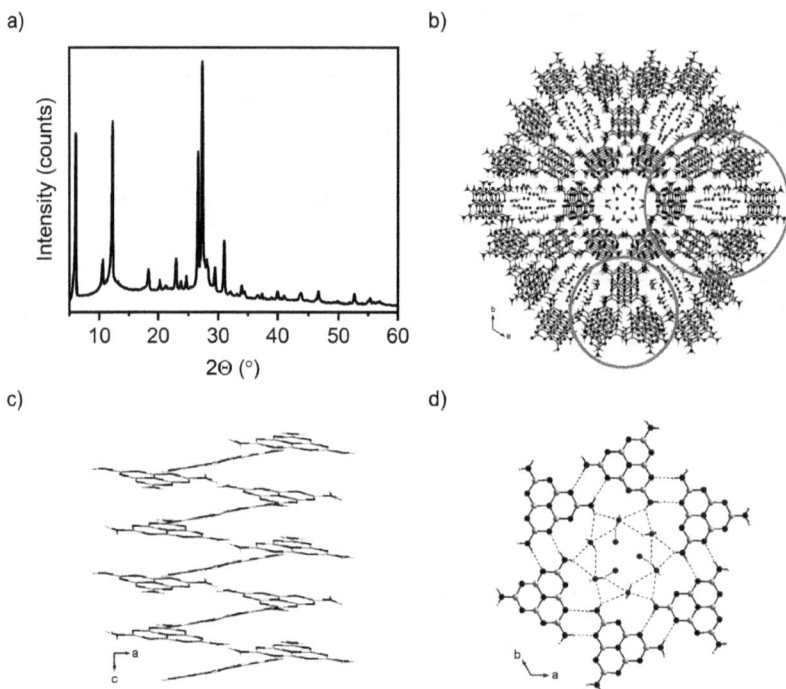

Fig. 2.17: A structure of Melem hydrate. (a) PXRD pattern. (b) Porous structure with hexagonal channels along c direction, filled with water molecules. (c) Arrangement of Melem molecules in helices along c direction. (d) Rosette formed by six Melem molecules and filled with water. Dashed lines represent hydrogen bonds (reproduced with permission from [22]. Copyright 2012 John Wiley & Sons, Inc.).

The Melem units are not coplanar but tilted out of the *ab* plane and are arranged in a helix motif along the crystallographic *c* axis (Fig. 2.17c) resulting in a corrugated layered structure perpendicular to the channels. The interlayer distance is 3.37 Å, which is slightly larger than in anhydrous Melem. Dehydration of the material at 150 °C in vacuum results in a decrease in crystallinity. Despite a relatively large diameter of the pores, the Brunauer-Emmett-Teller (BET) model applied to nitrogen physisorption at 77 K revealed a specific surface area in the dehydrated Melem of only 29 m^2 g^{-1}. Decrease in crystallinity and low surface area indicate that, upon water removal, the Melem layers are shifted perpendicular to the direction of the channels.

2.4.1.3 Melon-type carbon nitride

The material is typically prepared by polycondensation of melamine, cyanamide, dicyanamide or other nitrogen-rich compounds at 500–600 °C using the "open synthesis" route. The crystallite size of Melon-type carbon nitride prepared in this fashion is limited to tens of nm in the vertical direction [24]. On the contrary, calcination in a closed vessel under the autogenous pressure of ammonia released upon polycondensation of nitrogen-rich precursors at 630 °C facilitates the formation of samples with larger crystallites (Fig. 2.18a) [25].

The structure of Melon-type carbon nitride was solved in projection using electron diffraction, while the layer spacing of 3.2 Å was determined from X-ray powder diffraction. Melon-type carbon nitride is composed of linear chains of heptazine units linked via NH-groups. The chains are arranged in a zigzag fashion that supports a tight hydrogen bonding network between the adjacent threads into quasi 2D layers with a distance of 3.2–3.3 Å. The layers are stacked with an offset with respect to each other in ABA configuration, which minimizes repulsive π–π interactions that would be maximal in the perfect AAA arrangement [26]. Lateral shifts in *a* and *b* directions of one Melon layer with respect to another are characterized by the shift vector **S** = [Δ*a*Δ*b*]. Thus, **S** = [0% 0%] = [0.0 Å 0.0 Å] denotes eclipsed layer stacking – adjacent Melon layers are located perfectly above each other. Such stacking is energetically disfavored. When both Δ*a* and Δ*b* are in the range 0–50%, it corresponds to a partially staggered arrangement. Combination of cross polarization 2D NMR spectroscopy and the results of force field modeling revealed that there are two partially staggered arrangements of adjacent layers that have similar energy, giving rise to a rather flat stacking energy landscape [27, 26]. These are characterized by **S** = [0% 40%] = [0.0 Å 6.7 Å] and **S** = [22% 5%] = [3.7 Å 0.6 Å]. The perfectly staggered configuration, **S** = [50% 50%], is still by about 12 kcal mol^{-1} higher than the global minimum. Such relatively small energy difference is the origin of the nanocrystalline or amorphous nature of most carbon nitrides in the lateral direction.

a)

b)

c)

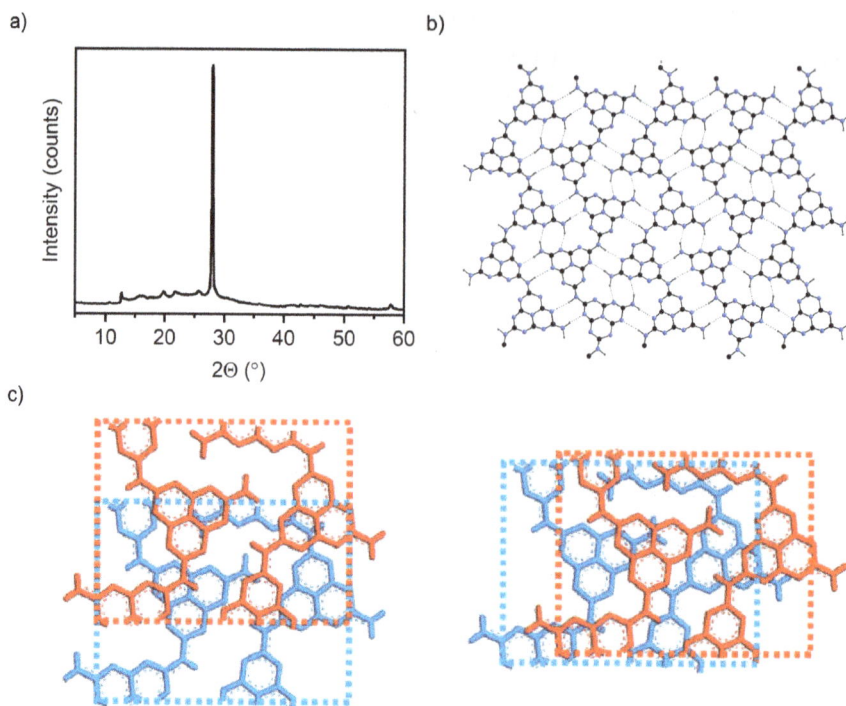

Fig. 2.18: Structure of Melon-type carbon nitride. (a) PXRD pattern. (b) Structure of a single layer of the material. Dashed lines represent hydrogen bonds between the adjacent Melon strands. (c) Structures of two favored conformations of Melon layers. Panels (a) and (b) are reproduced with permission from [25] (Copyright 2007 John Wiley & Sons, Inc.). Panel (c) is reproduced with permission from [27] (Copyright 2010 Royal Society of Chemistry).

2.4.1.4 Melamine-templated poly(heptazine imide)

Condensation of melamine is a multistep process and involves formation of at least three noncovalent crystalline adducts of melamine and Melem with molar ratios between the components of 2:1, 1:1 and 1:3 [28]. Although polycondensation of melamine under the "closed synthesis" conditions at 630 °C yields predominantly Melon-type carbon nitride [25], careful inspection of the pyrolysis products revealed another carbon nitride material – poly(heptazine imide) with isolated melamine molecules encapsulated in the trigonal pores (Fig. 2.19a) [29].

Using selected area electron diffraction (SAED), the structure was solved in [001] projection in the plane group *p31m*. Information on the third dimension was added by taking the interlayer distance of 3.2 Å, a typical value for graphitic carbon nitride materials. The hexagonal symmetry of SAED patterns supports an eclipsed AAA-type stacking sequence of the layers. The similarity of the crystal structure of this material

a)

b)

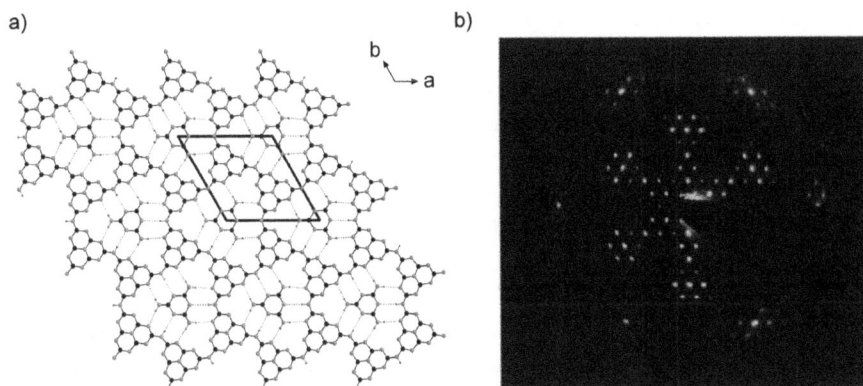

Fig. 2.19: Structure of melamine-templated poly(heptazine imide). (a) A single layer of the material. (b) Selected area electron diffraction (SAED) pattern demonstrates hexagonal symmetry (reproduced with permission from [29]. Copyright 2009 Royal Society of Chemistry).

and Melon-type carbon nitride become more obvious when considering the [100] direction – the heptazine units are connected in a zig-zag fashion. The structure can also be viewed as composed of an infinite array of parallel Melon strands connected to their mirror images via C–NH–C bridges. Such interconnectivity of heptazine units yields a degree of condensation higher (2D) than that in Melon (1D). The above connectivity scheme introduces large triangular voids into the planar network, each hosting a single melamine molecule. In this material, melamine serves as a templating agent, which likely directs the formation of 2D structure. This conclusion is supported by theoretical calculations – incorporation of melamine stabilizes the network by 136.36 kJ mol^{-1}. Theoretical calculations also predict the existence of poly(heptazine imide) – a material free of templating melamine molecules, which, in fact, was synthesized several years later [30, 31].

2.4.1.5 Fully condensed heptazine-based graphitic carbon nitride

Large triangular pores in poly(heptazine imide) that are free of templating melamine molecules provide exact space to accommodate a heptazine unit and to create a fully condensed heptazine-based graphitic carbon nitride (Fig. 2.15). Except for the edges, which are rich in NH and NH$_2$ groups due to termination of the periodic structure, such a material must be free of hydrogen. However, conventional "open synthesis" always gives a material with at least 1–2 wt% of hydrogen, which indicates the presence of a substantial number of NH- and NH$_2$-groups. In agreement with earlier results (Fig. 2.18), a combination of quantitative X-ray photoelectron spectroscopy and X-ray powder diffraction techniques suggests that the structure of carbon nitrides prepared by the thermal condensation of dicyandiamide is closer to Melon-type carbon nitride than to fully condensed heptazine-based g-CN [24]. While a large number of publications, especially those focused on application of carbon nitrides rather than rigorous characterization, propose a structure of a fully condensed heptazine-based g-CN, infinitely large sheets of such material are yet to be synthesized.

2.4.2 Ionic metal poly(heptazine imides)

Until now, we considered carbon nitride materials composed of carbon, nitrogen and hydrogen, which are required to terminate the periodic structure. Ionic metal poly(heptazine imides), denoted as M-PHI with M being a metal cation, contain up to 15 wt% of alkali metals, which are introduced during ionothermal synthesis. Condensation of a melamine:NaCl mixture in a 1:10 mass ratio at 600 °C for 4 h under N_2 gives sodium poly(heptazine imide), Na-PHI [30]. Potassium poly(heptazine imide), K-PHI, is prepared by calcining a mixture of dicyandiamide, KSCN and potassium melonate pentahydrate ($8 \cdot 5H_2O$) under argon atmosphere in a sealed tube at 500 °C for 6 h [31]. Alternatively, K-PHI is prepared by calcining 5-aminotetrazole monohydrate in a deep eutectic mixture of LiCl:KCl (9:11 weight ratio, m.p. 352 °C) with the precursor to salt mixture mass ratio of 1:4 at 550 °C for 4 h under N_2 [32]. Ca-PHI is prepared by calcining a Melon-type carbon nitride with $CaCl_2$ in a 1:2 mass ratio at 700 °C under Ar [33]. Note that calcination of melamine with LiCl in a 1:10 mass ratio at 600 °C for 8 h does not give Li-PHI, but PTI/Li$^+$Cl$^-$ (see also next section) [30]. Similar to the melamine molecule in melamine-templated poly(heptazine imide), alkali metal ions direct the polycondensation of the nitrogen rich precursor and assembly of heptazine units into a 2D structure with large triangular pores. Na$^+$ and K$^+$ are mobile and may be conveniently replaced by other metal ions: alkali (Li$^+$, Cs$^+$), alkaline earth (Mg^{2+}, Ca^{2+}, Ba^{2+}) and transition metals (Zn^{2+}, Co^{2+}, Ni^{2+}, Au^{3+}, Ru^{3+}, Fe^{3+}) [34–37]. Protonated PHI (H-PHI), in which K$^+$ are replaced by H$^+$, is obtained by dispersing K-PHI in an aqueous solution of HCl or other inorganic acids [34, 31]. In the case of H-PHI, K-PHI and Mg-PHI, ion exchange is complete and reversible – it is possible to convert Mg-PHI into K-PHI by dispersing the material in excess of KCl. At the same time, K$^+$ ions are inserted back into the network upon treatment of H-PHI with 0.1 M aqueous KOH solution [38]. On the other hand, Fe-PHI does not exchange iron back to K$^+$, which is due to formation of strong coordinative bonds between Fe and four nitrogen atoms – two from each of the adjacent layers, as evident from X-ray absorption spectroscopic data [36].

One of the signatures of crystalline M-PHIs is a diffraction peak at ~8° (CuKα1 radiation λ = 1.540598 Å), which is absent in Melon-type carbon nitride (Fig. 2.20a). Decrease in K-PHI lateral crystallite size from ~120 to ~20 nm, disordered pore content and the presence of significant stacking faults broaden the diffraction peaks and especially lead to a decrease in the intensity of the diagnostic 010 reflection at 8° 2theta [31]. For amorphous K-PHI samples, the peak at 8° might be absent, while the overall PXRD pattern exhibits only a broad stacking peak at ~27°, which could lead to a false

a)

b)

c)

d)

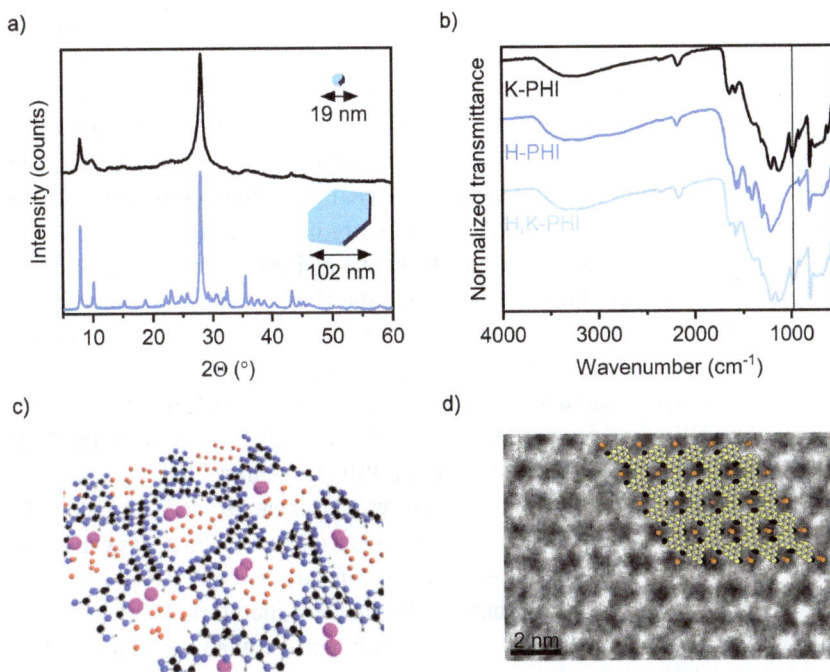

Fig. 2.20: Structure of *M*-PHIs. (a) PXRD patterns of K-PHI with variable lateral size, but nearly the same vertical size. (b) FTIR spectra of K-PHI, H-PHI and K-PHI with reinstalled K$^+$ upon H-PHI treatment with KOH. (c) Crystal structure of K-PHI as obtained by Rietveld refinement of X-ray powder data. (d) High-resolution transmission electron microscopy (TEM) image with superimposed structure of Na-PHI. Panels (a) and (c) are reproduced with permission from [31]. License CC-BY. Copyright 2019 American Chemical Society. Panel (b) is reproduced with permission from [34] (Copyright 2017 John Wiley & Sons, Inc.). Panel (d) is reproduced with permission from [30] (Copyright 2017 John Wiley & Sons, Inc.).

conclusion that Melon-type carbon nitride is observed. However, regardless of crystallinity, in Fourier transform infrared (FT-IR) spectra of K-PHI, a peak at 980 cm^{-1} is present (Fig. 2.20b). This peak is assigned to the symmetric vibration in K-NC$_2$ groups [34]. The peak disappears upon removal of K$^+$ ions from the structure by washing with acid and is recovered upon treatment of H-PHI with KOH aqueous solution. Considering the availability of powder X-ray diffraction, elemental analysis and FT-IR analytical techniques in most laboratories, the diffraction peak at ~8°, high content of alkali metals, and a vibrational peak at 980 cm^{-1} may collectively be considered as fingerprints of the *M*-PHI structure. Although single crystal X-ray diffraction data is not yet available for K-PHI, crystallites as large as 100 nm are readily available via the ionothermal "closed synthesis" approach. Thus, the combination of SAED, TEM, PXRD patterns and pair distribution function (PDF) analysis allows solving the structure of *M*-PHIs [30–32]. Unlike Melon-type carbon nitride, in which the layers are arranged in partially staggered configuration to minimize the repulsion (Fig. 2.18), in K-PHI the layers are stacked in a slip-stacked fashion that still allows for continuous channels to form along the c direction. The ratio between K^+ ions and heptazine units likely depends on the synthesis conditions, which, in turn, influences the amount of NCN-groups terminating the periodic structure, proton content and crystallization water content, and ranges from 1:1 (as shown in Fig. 2.15) [32] to 1:3 (Fig. 2.20c) [31]. Variable alkali metal content is supported by facile ion exchange in aqueous dispersion and direct ionothermal synthesis of NaK-PHI in a mixture of NaCl/KCl salts [39]. K^+ ions are solvated by 6–7 water molecules, which form one hydration shell with the radius of 3.3 Å. Solvated K^+ ions fit almost perfectly into the channel. From this standpoint, solvated Na^+ ions have a slightly larger radius of 3.6 Å, which is in perfect agreement with the pore diameter and explains the roughly one order of magnitude higher ionic conductivity of Na-PHI, compared to K-PHI at 42% relative humidity [37]. K-PHI crystals are terminated by $NH_2–$ and cyanamide groups. The latter are observed in the FT-IR spectrum as a band at ~2,200 cm^{-1}. The degree of K-PHI crystallinity affects the physico-chemical properties of the material in several ways. For example, more crystalline K-PHI is more active in oxidation of benzylalcohol to benzaldehyde, which was explained by a greater exposure of the lateral dimension [40]. Ion transport along the c direction facilitates electron transfer from the bulk to the surface. Therefore, K-PHI particles with greater fraction of lateral facets exhibit faster kinetics in the bespoken reaction. On the other hand, in the hydrogen evolution reaction, K-PHI with smaller crystallite sizes (~20 nm) shows three times higher activity compared to K-PHI with the crystallite size of 100 nm [31]. Such increase in activity was assigned to a higher density of cyana-mide terminations that can bind the Pt cocatalyst more efficiently, rather than improving bulk properties of K-PHI. Nevertheless, higher crystallinity might be also beneficial for increasing the mobility of charge carriers within the material.

2.4.3 Triazine-based graphitic carbon nitride

At the end of the twentieth century and until around the end of the first decade of the twenty-first century, efforts were concentrated toward the synthesis of triazine-based g-CN as the closest structural analogue of graphite [41]. In the vast majority of publications around that time, the periodic structure shown in Fig. 2.21a was labeled with the general term, "graphitic carbon nitride." The situation changed around 2007 with the discovery that Liebig's Melon is composed of tri-s-triazine units [25]. Nowadays, the term, "graphitic carbon nitride" is used almost exclusively to denote materials composed of heptazine units. However, with the discovery of crystalline films of triazine-based g-CN [42], this area of materials science has been revived.

Earlier strategies for synthesis of triazine-based g-CN are based on condensation of various s-triazine derivatives, ideally those that do not contain any strong C–H and N–H bonds in their structure. Thus, films of the thickness 120–500 nm were prepared by decomposing $C_3N_3(Hal)_2N(XMe_3)_2$ precursors at 300–500 °C and N_2 pressure of

a)

X = Si, Sn
Hal = F, Cl

b)

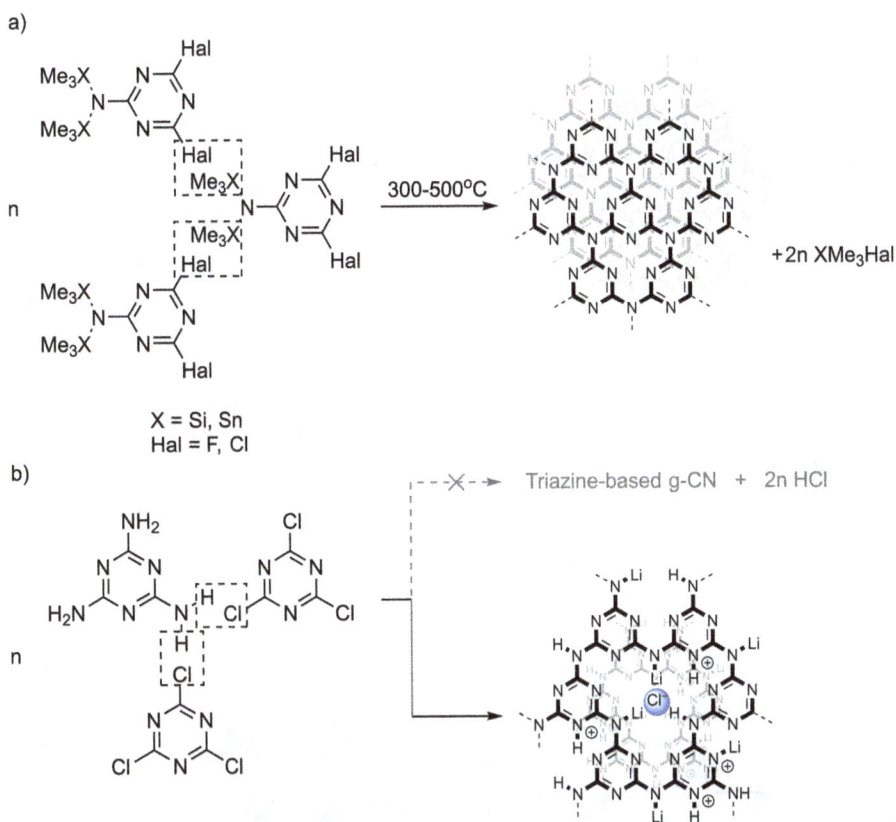

Fig. 2.21: Synthetic approaches toward triazine-based g-CN. (a) Synthesis of triazine-based g-CN possessing ABC stacking by polycondensation of $C_3N_3(Hal)_2N(XMe_3)_2$. (b) Synthesis of PTI/H$^+$Cl$^-$ by polycondensation of a mixture of melamine and cyanuric chloride in 1:2 ratio.

0.01–0.5 Torr, followed by deposition on SiO_2 in a chamber of the chemical vapor deposition (CVD) reactor setup [43, 44]. Quantitative formation of XMe$_3$Hal was observed in all cases. Elemental analysis revealed chemical composition of films ranging from C_3N_4 to $C_{3.2}N_4$, trace amounts of Si, F and O and, importantly, absence of hydrogen. Compared to the Si-analogues, $C_3N_3(Hal)_2N(SnMe_3)_2$ precursors do not require a carrier gas and reactions are carried out at lower pressure, 0.01 versus 0.5 Torr, and lower temperature, 350–400 versus 500 °C. $C_3N_3(Hal)_2N(SnMe_3)_2$ also gave a material with an elemental composition similar to that of thin films, when the reaction was conducted in a sealed tube.

Polycondensation of a mixture of melamine and cyanuric chloride (or metathesis polymerization of 2-amino-4,6-dichloro-1,3,5-triazine) at 500–600 °C and a pressure of 10,000 bar is accompanied by evolution of 2 equivalents of HCl [45]. Although in the original publication, this route was envisioned to prepare triazine-based g-CN, in fact, a material with the stoichiometry $C_6N_9H_3\cdot HCl$, poly(triazine imide) with intercalated

HCl was obtained, as indicated by the authors. HCl does not leave the reaction mixture and serves as a templating agent, similar to the already mentioned role of melamine in melamine-templated PHI (Fig. 2.19) and alkali metal ions in the synthesis of K-PHI and Na-PHI, and directs the formation of PTI/H$^+$Cl$^-$ instead of triazine-based g-CN.

Similarly, calcination of dicyandiamide in LiBr:KBr (52:48 wt%) gives PTI/Li$^+$Br$^-$ (see next section) as the major phase suspended in the liquid eutectic. Triazine-based g-CN is formed as a continuous film at the gas–liquid and solid–liquid interface in the same reactor [42]. The interlayer spacing is 3.28(8) Å. Comparison of calculated (Fig. 2.22b) and experimental (Fig. 2.22c) TEM images revealed that ABC stacking in which each triazine unit is superimposed on a bridging nitrogen followed by a void is the most plausible, but due to the low difference in energy, other stacking configurations – eclipsed AAA and staggered ABA – are also possible. The UV-vis diffuse reflectance spectrum in combination with the modeling of the band structure of the material single layer revealed a band gap of 1.6–2.0 eV (Fig. 2.22d).

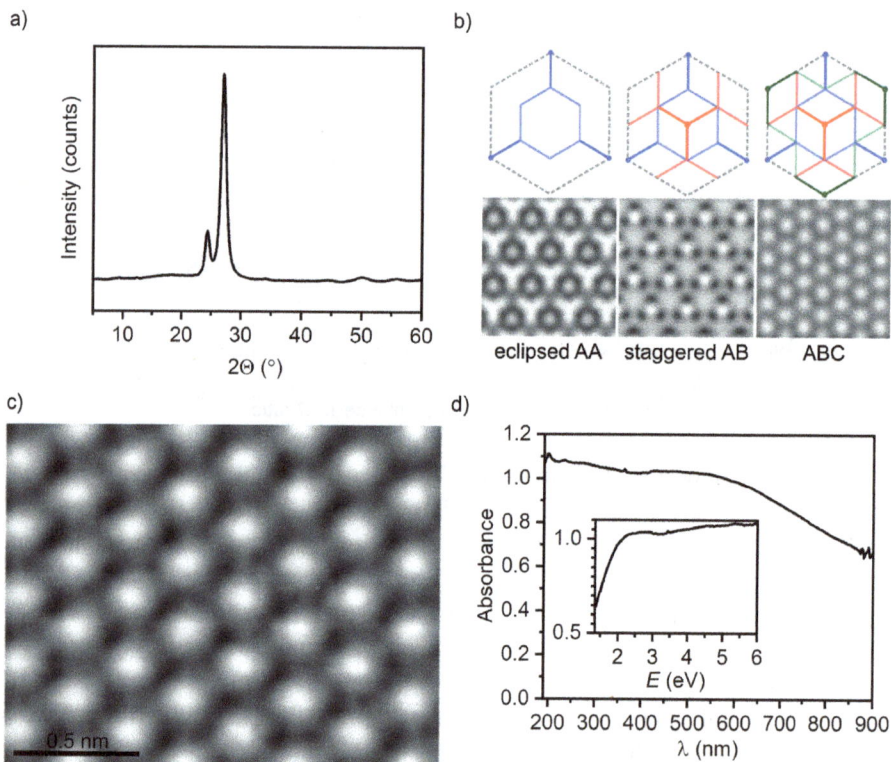

Fig. 2.22: Structure and characterization of triazine-based g-CN. (a) PXRD pattern. (b) Calculated images of three stacking arrangements of layers within the crystal. (c) High-resolution TEM image. (d) UV–vis diffuse reflectance spectrum of triazine-based g-CN and Kubelka-Munk plot in inset (reproduced with permission from [42]. Copyright 2014 John Wiley & Sons, Inc.).

2.4.3.1 Poly(triazine imide) intercalated with LiCl and HCl

PTI/Li$^+$Cl$^-$ is a material with the formula [(C$_3$N$_3$)$_2$(NH$_x$Li$_{1-x}$)$_3$·LiCl] (x is typically 2/3), in which the imide groups are partially protonated. Variable content of LiCl explains the (partial) disorder in the system. PTI/Li$^+$Cl$^-$ is prepared via a two-step procedure [46]. In the first step, a mixture of dicyandiamide and an eutectic mixture of LiCl:KCl (45:55 wt%, m.p. 352 °C) in a 1:5 mass ratio is calcined in an opened crucible at 400 °C for 6 h. In the second step, the homogeneous anhydrous block of LiCl-KCl and dicyandiamide condensation intermediates are transferred into a quartz glass tube, sealed and heated at 600 °C for 12 h, followed by washing with water and drying in vacuum. Alternatively, a mixture of dicyandiamide and LiCl:KCl (45:55 wt%) in 1:10 mass ratio is first calcined in an open glass tube at 400 °C for 4 h [47]. After cooling to room temperature and sealing the tube, the mixture is further calcined at 500, 550 or 600 °C for 24 h, followed by washing with boiling water. PTI with intercalated LiBr is prepared by calcining a mixture of dicyandiamide with LiBr:KBr (52:48 wt%, m.p. 348 °C in a 1:5 mass ratio at 600 °C for 48 h, followed by washing with boiling water and drying in vacuum [48]. In a fourth example, a mixture of thoroughly ground melamine (1.0 g), LiCl (0.93 g) and KCl (1.31 g) is placed into a reaction vessel (diameter 2.5 cm × height 25 cm) and sealed under vacuum [49]. Heating at 470 °C for 36 h, followed by cooling to 350 °C with the rate of 2 °C h^{-1} and spontaneous cooling to room temperature afforded PTI/Li$^+$Cl$^-$ upon washing with deionized water. From these three examples, it is evident that despite similar chemical composition of the initial reaction mixture, conditions that afford PTI/Li$^+$Cl$^-$ rather than K-PHI are: 1) calcination at 400 °C (preheating step) that is used to create higher molecular weight intermediates by eliminating NH$_3$. A similar effect is achieved by conducting the synthesis in a closed vessel under reduced pressure at 470 °C and 2) extended calcination time, 12–96 h, in case of PTI/Li$^+$X$^-$ versus 4 h in case of K-PHI. Depending on the preparation method and aqueous workup, Li$^+$ can be completely removed from the structure, but substantial amounts of HCl and HBr remain bound to the triazine units [50]. The ionic character of this g-CN is translated in crystallite sizes exceeding 200 nm [46]. Therefore, formation of a crystalline PTI/Li$^+$Cl$^-$ phase can be concluded from powder X-ray diffraction patterns (Fig. 2.23a) [51]. In scanning electron microscope (SEM), PTI/Li$^+$Cl$^-$ is seen as hexagonal prisms (Fig. 2.23b) [47]. Both PTI/Li$^+$Cl$^-$ and PTI/Li$^+$Br$^-$ are composed of 1,3,5-triazine units linked through sp^2-hybridized, partially protonated imide nitrogen atoms, forming C$_{12}$N$_{12}$ ring voids. The layers are stacked in an ABA fashion and are almost planar, which is due to the conjugation within the layers and intercalation of salt ions (Fig. 2.23c). Due to their relatively large diameter, halide ions are located between the layers (Fig. 2.23d). It should be mentioned that in contrast to M-PHI, PTI is intercalated by "neutral" LiCl, rather than Li$^+$ only. In PTI/Li$^+$Cl$^-$, this leads to an extension of the interlayer distance to 3.37 Å, in comparison to Melon-type carbon nitride (3.2 Å) and PTI/H$^+$Cl$^-$ (3.21 Å, Fig. 2.21b). Compared to PTI/Li$^+$Cl$^-$, the distance between the layers in PTI/Li$^+$Br$^-$ is extended even further to 3.52 Å [48]. Exchange of bromide ions in PTI/Li$^+$Br$^-$ by fluorides upon material dispersion in 8 M NH$_4$F solution leads to the contraction of the interlayer distance to 3.32 Å and is accompanied by a shift of the neighboring layers with respect to each other [48].

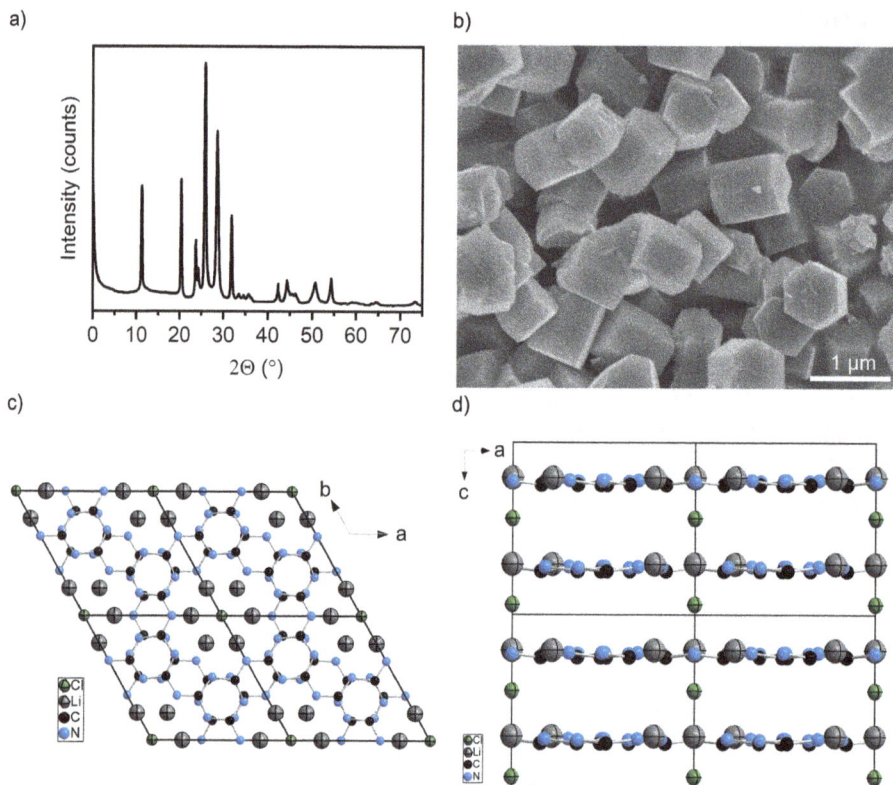

a)

b)

c)

d)

Fig. 2.23: Structure of PTI/Li$^+$Cl$^-$. (a) PXRD pattern. (b) SEM image. (c) Projection of layers with ABA stacking. (d) Stacking of the layers in ABA fashion along the c direction. Panels (a), (c) and (d) are reproduced from [51] (Copyright 2011 John Wiley & Sons, Inc.).

2.5 Comparative summary of synthesis, structure and properties of graphitic carbon nitrides

Synthesis conditions affect the structure of carbon nitrides as well as the number and type of defects, which is reflected in their properties. Table 2.2 summarizes the synthesis conditions and crystal structures, and provides typical values of some parameters obtained for five types of carbon nitride materials discussed in this chapter.

The structure of the precursor is transferred into a structure of the carbon nitride material. Molecules rich in nitrogen and composed of only alternating C–N and C=N bonds are used to synthesize carbon nitrides with the approximate 3C:4N stoichiometry. Although precursors are the same, Melem (0D) is typically synthesized at lower temperatures, 450 °C, compared to 1D and 2D carbon nitrides with a higher degree of condensation, 550–630 °C. Furthermore, calcination under ionothermal conditions for a relatively short time, 4 h, gives M-PHIs, while extended reaction time, 12–96 h, is essential to obtain PTI/M^+X^-. Melon-type carbon nitride may be converted into PTI/Li^+Cl^- upon calcination in LiCl:KCl eutectics [52]. It should be noted that 5–10-fold excess of alkali metal salt versus organic precursor produces phase-pure M-PHIs or PTI/M^+X^-, while addition of only few wt% of alkali metal salt results in "alkali metal doped carbon nitrides," likely without substantial change of its bulk chemical structure, i.e., Melon-type carbon nitride [53].

In Melem, the size of the conjugated system is limited to 13 atoms. Therefore, it has a wide band gap (E_g) that is comparable to heptazine derivatives (Fig. 2.14a). Increase of the conjugation narrows the band gap to ~2.7 eV in Melon-type carbon nitride and M-PHIs. The UV-vis electronic spectrum of PTI/Li^+Cl^- shows an onset of absorption at 400 nm (Tab. 2.2) [54]. Ionothermal copolymerization of dicyandiamide with 4-amino-2,6-dihydroxypyrimidine, a compound possessing C-C and C=C bonds, gives amorphous PTI/Li^+Cl^- with a shift of the absorption edge up to 600 nm, which, in turn, affords higher visible light quantum efficiency in proton reduction to H_2 [55]. The copolymerization strategy is also applicable for introducing heteroelements into other carbon nitride materials. It will be discussed in the following chapters and linked to a specific application.

Fluorescence lifetime (τ_{FL}) and, in general, photophysics of carbon nitrides, depend not only on synthesis conditions, but also excitation wavelength and excitation fluence [56]. Thus, for Melem, τ_{FL} as long as 360 ns and fluorescence quantum yield (FLQY) as high as 56.9% were reported [57]. On the other hand, τ_{FL} of Melon-type carbon nitrides is 3.11 ± 1.23 ns (data from 5 references acquired in 5 laboratories) [30, 56, 58–60]. PTI/Li^+Cl^- and M-PHIs (M = Li, Na, K, Cs, Ba) demonstrate overall shorter τ_{FL}, 0.49 and 0.58 ± 0.19 ns, respectively (data from four references acquired in two laboratories) [30, 32, 37, 58].

Several research labs demonstrated that dynamics of g-CN excited states is complex. Via intersystem crossing (ISC), singlet excited states are converted into triplet states. Radiative relaxation of triplet back to singlet ground state is known as phosphorescence. Phosphorescence lifetime (τ_{PH}), which is an important parameter to characterize the lifetime of a triplet excited state, in Melon-type g-CN is 124 [59] or 126 ± 4 µs [60]. In K-PHI, τ_{PH} is 2.07 µs [32]. One of the parameters that define how facile the ISC is, is the energy gap between singlet and triplet states (ΔE_{ST}). ΔE_{ST} values for Melon-type carbon nitride are 0.243 and 0.248 eV [59, 60], while for K-PHI it is 0.2 eV [32]. Such relatively low ΔE_{ST} values facilitate the ISC and have implications in TADF and photocatalysis (Chapter 8).

Absorption of a photon with energy greater than the optical gap of the semiconductor produces an exciton – a quasi particle composed of two oppositely charged species, the hole and the electron. The electron and the hole can dissociate into free charges or remain bound as exciton. The chemical structure of a semiconductor defines the exciton binding energy (E_b) – the amount of energy required to split excitons into free charges. E_b determined from the temperature-dependent steady-state fluorescence measurements is the highest for Melon-type carbon nitride, 64 meV [52]. In PTI/Li$^+$Cl$^-$, the E_b is reduced to 50.4 meV [47]. On the other hand, as predicted by theory, in Melon-type carbon nitride E_b is 840 meV, while in heptazine-based g-CN, E_b is 1,350 meV [61].

Although mobility of charge carriers and conductivity are two important parameters that characterize semiconductor materials, the exact values have been largely elusive for carbon nitrides, mainly linked to the fact that the materials are prepared in the form of powders. Therefore, macroscopic conductivity (σ) measured by electrochemical impedance spectroscopy by pressing the powder into a pellet is strongly affected by grain boundaries between the particles. However, in 306 ± 30 nm crystalline thin films of triazine-based g-CN grown on top of quartz glass, conductivity in vertical (out-of-plane, measured using four-point-probe method) direction is $101 \pm 15 \times 10^{-6}$ S m^{-1}, which is 65 times higher than in the lateral (in-plane, measured by two-probe method) direction [62]. These values are comparable to some conjugated organic frameworks, but two orders of magnitude lower than for 2D semiconducting MoS$_2$, reflecting the more localized character of carbon nitrides. Mobility (μ) in such films was determined to be 2.43 cm^2 V^{-1} s^{-1} assuming 1D diffusion of charges and 0.81 cm^2 V^{-1} s^{-1} for 3D diffusion.

Tab. 2.2: Summary of synthesis conditions, crystal structure and some of the physicochemical properties of a few representative carbon nitride materials.

	Melem	Melon-type g-CN	*M*-PHI	PTI/Li$^+$Cl$^-$	Triazine-based g-CN
Synthesis conditions	Melamine, 450 °C, 5 h	Melamine, 550–630 °C	Precursor and KCl (NaCl) or LiCl:KCl (9:11), 550–600 °C, 4 h	Precursor and LiCl: KCl (9:11), 470–600 °C, 12–96 h	Similar to PTI/Li$^+$Cl$^-$
Space/plane group	$P2_1/c$	$p2gg$ $P2,1212$	$P31m$	$P6_3cm$	$P\bar{6}m2$
ρ_{calc} (g cm^{-3})^a	1.717	–	K-PHI 1.88^b	–	–
d_{max} (nm)^c	–	7–10 (vertical)	102 ± 10 (lateral); 25 ± 2 (vertical)	>200	–
λ_{ons} (nm)^d	~350	460	450 (Na-PHI) 460 (K-PHI) 440 (H-PHI)	400	>900

Tab. 2.2 (continued)

	Melem	Melon-type g-CN	*M*-PHI	PTI/Li$^+$Cl$^-$	Triazine-based g-CN
E_g (eV)	3.5	2.7	2.7–2.8	3.1	1.6–2.0
τ_{FL} (ns)	360	3.1 ± 1.2	0.58 ± 0.19	0.49	–
FLQY (%)	56.9	–	<0.1 (K-PHI)	–	–
τ_{PH} (μs)	–	124, 126 ± 4	2.07	–	–
ΔE_{ST} (eV)	–	0.243, 0.248	0.2	–	–
τ_{ISC}	–	Few ps	470 ps	–	–
E_b (meV)	–	64	–	50.4	–
σ (10^{-6} S m^{-1})	–	–	–	–	1.55 ± 0.19 (lateral); 101 ± 15 (vertical)
μ (cm^2 V^{-1} s^{-1})	–	–	–	–	2.43

[a]ρ_{calc}, calculated gravimetric density.
[b]Crystallization water content in the sample 3 wt%.
[c]d_{max}, largest crystallite size reported.
[d]Steep onset of absorption due to π–π* transitions.

Beyond the carbon nitride materials summarized in this chapter are carbon nitrides "doped" either with elemental or molecular substituents. The question of whether doping with carbon or heteroelements is random or produces, for instance, an alternating Melon-type heptazine-cyclazine block-copolymer (see Fig. 2.3 for structures of cyclazines), remains open.

Of course, Fig. 2.15 summarizes carbon nitride structures that are built of only one type of heterocycle – either heptazines or *s*-triazines. Combination of these two units in one material alters their properties. For example, in comparison to Melon-type g-CN, E_b in triazine–heptazine-based g-CN is lower, 48 meV [52]. In conjugated covalent heptazine frameworks constructed from heptazine and diphenyldiacetylene linkers, which may not be considered as carbon nitrides as they contain substantial amount of C =C bonds, it is only 24 meV [63]. Nevertheless, such E_b value is comparable to the magnitude of thermal excitation at room temperature.

In this chapter, we have seen that materials, such as triazine-based g-CN (Figs. 2.21a, 2.22) and melamine-templated PHI (Fig. 2.19), are formed as minor components in the reaction. Careful inspection of the reaction mixture allowed isolating and characterizing them by means of electron microscopy. Progress in the development of new unconventional synthesis methods as well as improvements in the sensitivity of analytical methods are bound to reveal new phases of graphitic carbon nitrides in the future.

2.6 Summary

Compared to 1,3,5-triazine, whose derivatives were used in industrial, pharmaceutical and biological applications, tri-*s*-triazine or heptazine is far less studied. The scientific community has resumed research into this molecule at the beginning of the twenty-first century as a response to the discovery that photocatalytic water splitting catalyzed by its condensation product, Melon, is indeed feasible [64]. Heptazine is electron-deficient and a remarkably stable aromatic heterocycle. Its chemistry in solution has been explored and dozens of derivatives, including oligomers, have been prepared.

1,3,5-Triazine and heptazine are building blocks of graphitic carbon nitrides. In the majority of publications focused on studying the semiconducting properties of g-CN for certain uses, the materials are conveniently prepared following the "open synthesis," route where nitrogen-rich precursors, such as melamine, are calcined at 500–600 °C in air. The as-obtained materials are (semi-)amorphous or nanocrystalline, which created substantial difficulties in determining their crystal structures until recently. "Closed synthesis," on the other hand, under autogeneous pressure of ammonia released upon polycondensation of the precursors, can afford materials with crystallites large enough to unambiguously confirm their structure. Structural elucidation is typically based on a combination of SAED, powder X-ray diffraction, electron microscopy and PDF analyses.

The amount of data accumulated in the past 200 years and, in particular, at the beginning of the twenty-first century, allows us to conclude that the family of g-CN materials consists of at least five classes: Melem and its derivatives; Melon-type carbon nitride; metal poly(heptazine imides); poly(triazine imides) with intercalated alkali metal salts; and triazine-based graphitic carbon nitride. Nevertheless, there are more carbon nitrides that are yet to be discovered.

References

[1]   Hückel E. Quantentheoretische Beiträge zum Benzolproblem. Z Physik 1931, 70(3), 204–86.
[2]   Hückel E. Quanstentheoretische Beiträge zum Benzolproblem. Z Physik 1931, 72(5), 310–37.
[3]   Shahbaz M, Urano S, LeBreton PR, Rossman MA, Hosmane RS, Leonard NJ. Tri-*s*-triazine: Synthesis, chemical behavior, and spectroscopic and theoretical probes of valence orbital structure. J Am Chem Soc 1984, 106(10), 2805–11.
[4]   Clar E, Stewart DG. Aromatic hydrocarbons. LXV. Triangulene derivatives. J Am Chem Soc 1953, 75(11), 2667–72.
[5]   Allinson G, Bushby RJ, Paillaud J-L, Thornton-Pett M. Synthesis of a derivative of triangulene; the first non-Kekulé polynuclear aromatic. J Chem Soc, Perkin Trans 1995, 1(4), 385–90.
[6]   Valenta L, Mayländer M, Kappeler P, Blacque O, Šolomek T, Richert S, et al. Trimesityltriangulene: A persistent derivative of Clar's hydrocarbon. Chem Commun 2022, 58(18), 3019–22.
[7]   Arikawa S, Shimizu A, Shiomi D, Sato K, Shintani R. Synthesis and isolation of a kinetically stabilized crystalline triangulene. J Am Chem Soc 2021, 143(46), 19599–605.

[8]   Rossman MA, Leonard NJ, Urano S, LeBreton PR. Synthesis and valence orbital structures of azacycl [3.3.3]azines in a systematic series. J Am Chem Soc 1985, 107(13), 3884–90.

[9]   Hosmane RS, Rossman MA, Leonard NJ. Synthesis and structure of tri-s-triazine. J Am Chem Soc 1982, 104(20), 5497–9.

[10] Halpern AM, Rossman MA, Hosmane RS, Leonard NJ. Photophysics of the $S_1 \leftrightarrow S_0$ transition in tri-s-triazine. J Phys Chem 1984, 88(19), 4324–6.

[11] Jürgens B, Irran E, Senker J, Kroll P, Müller H, Schnick W. Melem (2,5,8-Triamino-tri-s-triazine), an important intermediate during condensation of melamine rings to graphitic carbon nitride: Synthesis, structure determination by X-ray powder diffractometry, solid-state NMR, and theoretical studies. J Am Chem Soc 2003, 125(34), 10288–300.

[12] Saplinova T, Bakumov V, Gmeiner T, Wagler J, Schwarz M, Kroke E. 2,5,8-Trihydrazino-s-heptazine: A precursor for heptazine-based iminophosphoranes. Z Anorg Allg Chem 2009, 635(15), 2480–7.

[13] Kroke E, Schwarz M, Horath-Bordon E, Kroll P, Noll B, Norman AD. Tri-s-triazine derivatives. Part I. From trichloro-tri-s-triazine to graphitic C_3N_4 structures. New J Chem 2002, 26(5), 508–12.

[14] Posern C, Böhme U, Kroke E. The reactivity of cyameluric chloride $C_6N_7Cl_3$ towards phosphines and phosphine oxides. Z Anorg Allg Chem 2018, 644(2), 121–6.

[15] Galmiche L, Allain C, Le T, Guillot R, Audebert P. Renewing accessible heptazine chemistry: 2,5,8-tris (3,5-diethyl-pyrazolyl)-heptazine, a new highly soluble heptazine derivative with exchangeable groups, and examples of newly derived heptazines and their physical chemistry. Chem Sci 2019, 10(21), 5513–8.

[16] Ehrmaier J, Rabe EJ, Pristash SR, Corp KL, Schlenker CW, Sobolewski AL, et al. Singlet–triplet inversion in heptazine and in polymeric carbon nitrides. J Phys Chem A 2019, 123(38), 8099–108.

[17] Zambon A, Mouesca JM, Gheorghiu C, Bayle PA, Pécaut J, Claeys-Bruno M, et al. s-Heptazine oligomers: Promising structural models for graphitic carbon nitride. Chem Sci 2016, 7(2), 945–50.

[18] Bala I, Ming L, Yadav RAK, De J, Dubey DK, Kumar S, et al. Deep-blue OLED fabrication from heptazine columnar liquid crystal based AIE-active sky-blue emitter. ChemistrySelect 2018, 3(27), 7771–7.

[19] Kumar S, Sharma N, Kailasam K. Emergence of s-heptazines: From trichloro-s-heptazine building blocks to functional materials. J Mater Chem A 2018, 6(44), 21719–28.

[20] Miller TS, Jorge AB, Suter TM, Sella A, Corà F, McMillan PF. Carbon nitrides: Synthesis and characterization of a new class of functional materials. Phys Chem Chem Phys 2017, 19(24), 15613–38.

[21] Lau VW-h, Lotsch BV. A tour-guide through carbon nitride-land: Structure- and dimensionality-dependent properties for photo(electro)chemical energy conversion and storage. Adv Energy Mater 2022, 12(4), 2101078.

[22] Makowski SJ, Köstler P, Schnick W. Formation of a hydrogen-bonded heptazine framework by self-assembly of Melem into a hexagonal channel structure. Chem – Eur J 2012, 18(11), 3248–57.

[23] Wang Y, Wei B, Wang Q. Crystal structure of melamine cyanuric acid complex (1:1) trihydrochloride, MCA·3HCl. J Crystallogr Spectrosc Res 1990, 20(1), 79–84.

[24] Alwin E, Nowicki W, Wojcieszak R, Zieliński M, Pietrowski M. Elucidating the structure of the graphitic carbon nitride nanomaterials via X-ray photoelectron spectroscopy and X-ray powder diffraction techniques. Dalton Trans 2020, 49(36), 12805–13.

[25] Lotsch BV, Döblinger M, Sehnert J, Seyfarth L, Senker J, Oeckler O, et al. Unmasking Melon by a complementary approach employing electron diffraction, solid-state NMR spectroscopy, and theoretical calculations – Structural characterization of a carbon nitride polymer. Chem – Eur J 2007, 13(17), 4969–80.

[26] Fina F, Callear SK, Carins GM, Irvine JTS. Structural investigation of graphitic carbon nitride via XRD and neutron diffraction. Chem Mater 2015, 27(7), 2612–8.

[27] Seyfarth L, Seyfarth J, Lotsch BV, Schnick W, Senker J. Tackling the stacking disorder of Melon – Structure elucidation in a semicrystalline material. Phys Chem Chem Phys 2010, 12(9), 2227–37.

[28] Sattler A, Pagano S, Zeuner M, Zurawski A, Gunzelmann D, Senker J, et al. Melamine–Melem adduct phases: Investigating the thermal condensation of melamine. Chem – Eur J 2009, 15(47), 13161–70.

[29] Döblinger M, Lotsch BV, Wack J, Thun J, Senker J, Schnick W. Structure elucidation of polyheptazine imide by electron diffraction – A templated 2D carbon nitride network. Chem Commun 2009(12), 1541–3. https://pubs.rsc.org/en/content/articlelanding/2009/cc/b820032g.

[30] Chen Z, Savateev A, Pronkin S, Papaefthimiou V, Wolff C, Willinger MG, et al. "The easier the better" preparation of efficient photocatalysts – Metastable poly(heptazine imide) salts. Adv Mater 2017, 29(32), 1700555.

[31] Schlomberg H, Kröger J, Savasci G, Terban MW, Bette S, Moudrakovski I, et al. Structural insights into poly(heptazine imides): A light-storing carbon nitride material for dark photocatalysis. Chem Mater 2019, 31(18), 7478–86.

[32] Savateev A, Tarakina NV, Strauss V, Hussain T, ten Brummelhuis K, Sánchez Vadillo JM, et al. Potassium poly(heptazine imide): Transition metal-free solid-state triplet sensitizer in cascade energy transfer and [3+2]-cycloadditions. Angew Chem Int Ed 2020, 59(35), 15061–8.

[33] Burrow JN, Ciufo RA, Smith LA, Wang Y, Calabro DC, Henkelman G, et al. Calcium poly(heptazine imide): A covalent heptazine framework for selective CO_2 adsorption. ACS Nano 2022, 16(4), 5393–403.

[34] Savateev A, Pronkin S, Willinger MG, Antonietti M, Dontsova D. Towards organic zeolites and inclusion catalysts: Heptazine imide salts can exchange metal cations in the solid state. Chem – Asian J 2017, 12(13), 1517–22.

[35] Sahoo SK, Teixeira IF, Naik A, Heske J, Cruz D, Antonietti M, et al. Photocatalytic water splitting reaction catalyzed by ion-exchanged salts of potassium poly(heptazine imide) 2D materials. J Phys Chem C 2021, 125(25), 13749–58.

[36] da Silva MAR, Silva IF, Xue Q, Lo BTW, Tarakina NV, Nunes BN, et al. Sustainable oxidation catalysis supported by light: Fe-poly (heptazine imide) as a heterogeneous single-atom photocatalyst. Appl Catal: B, 2022, 304, 120965.

[37] Kröger J, Podjaski F, Savasci G, Moudrakovski I, Jiménez-Solano A, Terban MW, et al. Conductivity mechanism in ionic 2D carbon nitrides: From hydrated ion motion to enhanced photocatalysis. Adv Mater 2022, 34(7), 2107061.

[38] Markushyna Y, Teutloff C, Kurpil B, Cruz D, Lauermann I, Zhao Y, et al. Halogenation of aromatic hydrocarbons by halide anion oxidation with poly(heptazine imide) photocatalyst. Appl Catal: B, 2019, 248, 211–7.

[39] Zhang G, Liu M, Heil T, Zafeiratos S, Savateev A, Antonietti M, et al. Electron deficient monomers that optimize nucleation and enhance the photocatalytic redox activity of carbon nitrides. Angew Chem Int Ed 2019, 58(42), 14950–4.

[40] Savateev A, Dontsova D, Kurpil B, Antonietti M. Highly crystalline poly(heptazine imides) by mechanochemical synthesis for photooxidation of various organic substrates using an intriguing electron acceptor – Elemental sulfur. J Catal 2017, 350, 203–11.

[41] Kawaguchi M, Nozaki K. Synthesis, structure, and characteristics of the new host material $[(C_3N_3)_2 (NH)_3]_n$. Chem Mater 1995, 7(2), 257–64.

[42] Algara-Siller G, Severin N, Chong SY, Björkman T, Palgrave RG, Laybourn A, et al. Triazine-based graphitic carbon nitride: A two-dimensional semiconductor. Angew Chem Int Ed 2014, 53(29), 7450–5.

[43] Kouvetakis J, Todd M, Wilkens B, Bandari A, Cave N. Novel synthetic routes to carbon-nitrogen thin films. Chem Mater 1994, 6(6), 811–4.

[44] Todd M, Kouvetakis J, Groy TL, Chandrasekhar D, Smith DJ, Deal PW. Novel synthetic routes to carbon nitride. Chem Mater 1995, 7(7), 1422–6.

[45] Zhang Z, Leinenweber K, Bauer M, Garvie LAJ, McMillan PF, Wolf GH. High-pressure bulk synthesis of crystalline $C_6N_9H_3 \cdot HCl$: A novel C_3N_4 graphitic derivative. J Am Chem Soc 2001, 123(32), 7788–96.

[46] Bojdys MJ, Müller J-O, Antonietti M, Thomas A. Ionothermal synthesis of crystalline, condensed, graphitic carbon nitride. Chem – Eur J 2008, 14(27), 8177–82.

[47] Lin L, Lin Z, Zhang J, Cai X, Lin W, Yu Z, et al. Molecular-level insights on the reactive facet of carbon nitride single crystals photocatalysing overall water splitting. Nat Catal 2020, 3(8), 649–55.

[48] Chong SY, Jones JTA, Khimyak YZ, Cooper AI, Thomas A, Antonietti M, et al. Tuning of gallery heights in a crystalline 2D carbon nitride network. J Mater Chem A 2013, 1(4), 1102–7.

[49] Pauly M, Kröger J, Duppel V, Murphey C, Cahoon J, Lotsch BV, et al. Unveiling the complex configurational landscape of the intralayer cavities in a crystalline carbon nitride. Chem Sci 2022, 13(11), 3187–93.

[50] Ham Y, Maeda K, Cha D, Takanabe K, Domen K. Synthesis and photocatalytic activity of poly(triazine imide). Chem – Asian J 2013, 8(1), 218–24.

[51] Wirnhier E, Döblinger M, Gunzelmann D, Senker J, Lotsch BV, Schnick W. Poly(triazine imide) with intercalation of lithium and chloride ions $[(C_3N_3)_2(NH_xLi_{1-x})_3 LiCl]$: A crystalline 2D carbon nitride network. Chem Eur J 2011, 17(11), 3213–21.

[52] Pan Z, Liu M, Zhang G, Zhuzhang H, Wang X. Molecular triazine–heptazine junctions promoting exciton dissociation for overall water splitting with visible light. J Phys Chem C 2021, 125(18), 9818–26.

[53] Savateev A, Antonietti M. Ionic carbon nitrides in solar hydrogen production and organic synthesis: Exciting chemistry and economic advantages. ChemCatChem 2019, 11(24), 6166–76.

[54] Lin L, Wang C, Ren W, Ou H, Zhang Y, Wang X. Photocatalytic overall water splitting by conjugated semiconductors with crystalline poly(triazine imide) frameworks. Chem Sci 2017, 8(8), 5506–11.

[55] Schwinghammer K, Tuffy B, Mesch MB, Wirnhier E, Martineau C, Taulelle F, et al. Triazine-based carbon nitrides for visible-light-driven hydrogen evolution. Angew Chem Int Ed 2013, 52(9), 2435–9.

[56] Zhang H, Yu A. Photophysics and photocatalysis of carbon nitride synthesized at different temperatures. J Phys Chem C 2014, 118(22), 11628–35.

[57] Zheng HB, Chen W, Gao H, Wang YY, Guo HY, Guo SQ, et al. Melem: An efficient metal-free luminescent material. J Mater Chem C 2017, 5(41), 10746–53.

[58] Savateev A, Pronkin S, Epping JD, Willinger MG, Wolff C, Neher D, et al. Potassium poly(heptazine imides) from aminotetrazoles: Shifting band gaps of carbon nitride-like materials for more efficient solar hydrogen and oxygen evolution. ChemCatChem 2017, 9(1), 167–74.

[59] Wang H, Jiang S, Chen S, Zhang X, Shao W, Sun X, et al. Insights into the excitonic processes in polymeric photocatalysts. Chem Sci 2017, 8(5), 4087–92.

[60] Wang H, Jiang S, Chen S, Li D, Zhang X, Shao W, et al. Enhanced singlet oxygen generation in oxidized graphitic carbon nitride for organic synthesis. Adv Mater 2016, 28(32), 6940–5.

[61] Melissen S, Le Bahers T, Steinmann SN, Sautet P. Relationship between carbon nitride structure and exciton binding energies: A DFT perspective. J Phys Chem C 2015, 119(45), 25188–96.

[62] Noda Y, Merschjann C, Tarábek J, Amsalem P, Koch N, Bojdys MJ. Directional charge transport in layered two-dimensional triazine-based graphitic carbon nitride. Angew Chem Int Ed 2019, 58(28), 9394–8.

[63] Cheng H, Lv H, Cheng J, Wang L, Wu X, Xu H. Rational design of covalent heptazine frameworks with spatially separated redox centers for high-efficiency photocatalytic hydrogen peroxide production. Adv Mater 2022, 34(7), 2107480.

[64] Wang X, Maeda K, Thomas A, Takanabe K, Xin G, Carlsson JM, et al. A metal-free polymeric photocatalyst for hydrogen production from water under visible light. Nat Mater 2009, 8(1), 76–80.

Yuanxing Fang and Xinchen Wang*

Chapter 3
Photocatalytic water splitting by carbon nitride polymers

3.1 Introduction

Over the past 10 years, global primary energy consumption has been increasing by 1.7% annually on average and reached 19.7 TW in 2017. This value will be ca. 22 TW by 2030 [1]. To date, around 80% of the consumed energy is derived from fossil fuels, dominating the energy market, making it important to find alternatives to fulfill the demand. Sunlight is an inexhaustible, renewable and decentralized natural energy resource that is independent of geographical location [2]. Photocatalysis mimics photosynthesis by driving redox reaction under light illumination. It directly converts solar energy and stores it in the reactive chemicals – compounds that have weakly negative or zero enthalpy of formation [3–5]. In 1972, Fujishima and Honda first reported photoelectrochemical water oxidation on a titanium oxide single crystal photo anode [6]. This phenomenon, known today as the Honda–Fujishima effect, showed the potential of semiconducting materials for the conversion of light energy into chemical energy, and its storage in fuels via the uphill water splitting reaction. By splitting water, the solar energy can be stored in the form of hydrogen fuel. Water is one of the most abundant hydrogen sources available on the Earth and its supply is unlimited. Importantly, hydrogen gas is a fuel with a higher energy density than general fossil fuels – typically, 286 kJ mol^{-1} of energy (which is equivalent to standard enthalpy of water formation in liquid state) is released by reacting one mole of hydrogen with a half a mole of oxygen in gas phase [7, 8]. Hydrogen fuel is excluded from the carbon cycle, and thus, its consumption results in low impact on environment [9–11]. Hydrogen gas is also a vital raw material for electronics, metallurgy, aerospace, fine chemical and other industries [12–14]. The hydrogen market has been increasing extensively in the recent years and will continue growing until at least the end of this decade. Typically, the revenue of H_2 market was ca. 121.6 billion USD and is estimated to be ca. 228.9 billion USD in 2027. Assuming solar-to-hydrogen conversion efficiency (STH) values of

*Corresponding author: Xinchen Wang State Key Laboratory of Photocatalysis on Energy and Environment, College of Chemistry, Fuzhou University, Fuzhou 350116, P. R. China,
e-mail: xcwang@fzu.edu.cn
Yuanxing Fang, State Key Laboratory of Photocatalysis on Energy and Environment, College of Chemistry, Fuzhou University, Fuzhou 350116, P. R. China

https://doi.org/10.1515/9783110746976-003

5–10% and a system lifetime of 5 years, the estimated average cost of the hydrogen produced by photocatalysis devices is in the range of 1.6–3.5 USD kg^{-1}. Therefore, the price range meets the goal of a cost of 2.00–4.00 USD kg^{-1} hydrogen, which is, thus, economically competitive with fossil fuels.

It is obvious that the photocatalyst is the key element in realizing water splitting. In 2009, polymeric carbon nitride (PCN) was reported to mediate photocatalytic hydrogen evolution reaction and oxygen evolution reaction under visible light illumination [15]. This sparked development of a plethora of PCN materials as well as theoretical studies in the field of photocatalysis, which, at present, has resulted in more than 17,000 scientific articles. In this chapter, the thermodynamics and kinetics of water splitting by photocatalysis are discussed. Special attention is given to explain the role of fundamental properties of PCN in water splitting. The strategies for the modifications of PCN structure that are the most relevant to water splitting reaction are presented and include synthesis of copolymerized PCN and crystalline PCN. Additionally, application of PCN in photoelectrochemical systems is discussed.

3.2 Thermodynamics and kinetics of photocatalytic water splitting

Photocatalytic water splitting by semiconducting photocatalysts to generate hydrogen and oxygen generally consists of three processes (Fig. 3.1): (i) the absorption of a photon having energy greater than the semiconductor band gap, which excites the electron from the valence band to the conduction band and creates electron–hole pair, (ii) the separation of the photoexcited carriers into free carriers and subsequent migration and accumulation at the active sites on the particle surface and (iii) the initiation of the redox reactions involving these charges to generate hydrogen and oxygen with the assistance of the co-catalysts. Among them, surface reaction is generally considered as the rate-determining step that defines the overall reaction rate, as it is the slowest among the three stages.

The water splitting reaction is summarized further [16–18].

Hydrogen evolution half-reaction:

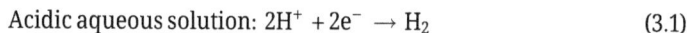

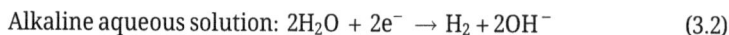

$$\text{Acidic aqueous solution: } 2H^+ + 2e^- \rightarrow H_2 \tag{3.1}$$

$$\text{Alkaline aqueous solution: } 2H_2O + 2e^- \rightarrow H_2 + 2OH^- \tag{3.2}$$

Oxygen evolution half-reaction:

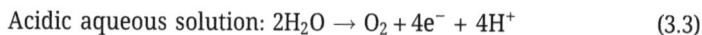

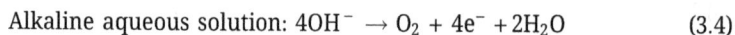

$$\text{Acidic aqueous solution: } 2H_2O \rightarrow O_2 + 4e^- + 4H^+ \tag{3.3}$$

$$\text{Alkaline aqueous solution: } 4OH^- \rightarrow O_2 + 4e^- + 2H_2O \tag{3.4}$$

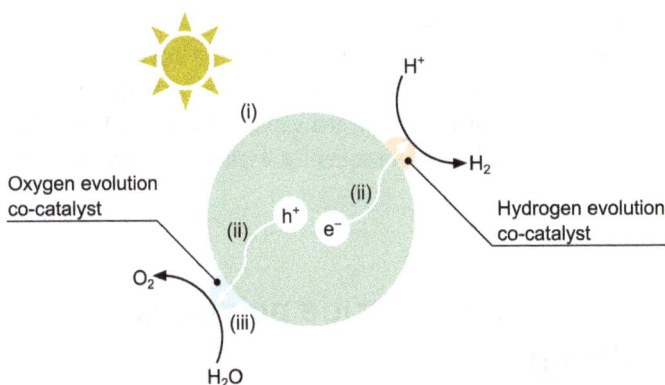

Fig. 3.1: Schematic diagram to present the process of water splitting on a semiconductor photocatalyst: (i) light absorption, (ii) charge separation and transport and (iii) redox reactions.

The total efficiency (η_{total}) of overall solar energy conversion is associated with three parameters that are determined by the photocatalyst, namely, efficiency of light absorption ($\eta_{absorption}$), charge separation ($\eta_{separation}$) and surface reaction ($\eta_{surface}$), as shown in the equation 3.5:

$$\eta_{absorption} \times \eta_{separation} \times \eta_{surface} = \eta_{total} \tag{3.5}$$

A large positive change in the Gibbs free energy ($\Delta G = 237$ kJ mol^{-1}, equivalent to 2.46 eV) is required to split a water molecule into the elements, H_2 and $\frac{1}{2}$ O_2. For the full dioxygen molecule, this sums up to four photons of 1.23 eV minimal each. The semiconductor band gap must be sufficiently wide to accumulate this quantum of energy required to overcome this positive Gibbs free energy change. The band gap is considered as the primary factor that determines the capability of a photocatalyst to split water. Typically, for transfer issues, the valence band potential must be more positive than +1.23 V versus a reversible hydrogen electrode (vs. RHE, pH 0), while the conduction band should be more negative than 0 V (vs. RHE, pH 0). Beyond the thermodynamics, an additional kinetic overpotential is needed to drive the water splitting reaction. Overpotential is needed to bring the reaction along higher energy intermediates; therefore, the band gap of the semiconductor photocatalyst in practical cases has to be at least 1.6–1.8 eV, corresponding to a maximal useful photon wavelength of ca. 775–690 nm. In addition to providing sufficient energy, cocatalysts are needed to form H_2 and O_2. As examples, oxygen evolution cocatalysts are represented by RuO_2, Co $(OH)_2$, $Ni(OH)_2$ and Mn, while Ru [19], Rh [20], Pt [21], Au [22] or MoS_2 [23] are used as hydrogen evolution co-catalysts.

Between the two half-reactions, oxygen evolution is kinetically sluggish and normally considered as the rate-limiting step that determines the overall reaction rate of water splitting. The oxygen evolution reaction consists of multiple steps, including the dissociation of relatively strong O–H bonds and the formation of weak O–O bonds. Naturally, the oxygen evolution reaction is an important metabolic process in photosynthesis, which has been occurring for billions of years and has led to the oxygen-abundant atmosphere on the Earth. Specifically, the solar-driven oxygen evolution reaction frees electrons and protons that accumulate and are subsequently used to convert CO_2 into carbohydrates.

3.3 Fundamentals of carbon nitride polymers for water splitting

PCN generally possesses an optical band gap of 2.7 eV. The potentials of the conduction band and valence band are −1.4 V and +1.3 V (vs. RHE at pH 7), respectively [24]. As such, from the thermodynamic standpoint, PCN allows driving the overall water splitting. In addition, owing to the polymeric nature, the band gap of PCN can be readily optimized by incorporating electron donor or acceptor to tune the physicochemical properties. In the past years, research on the optimizations of the PCN band gap to maximize light absorption has been witnessed [25]. For instance, both organic comonomers and metal elements can be introduced into the framework of the PCN to manipulate the band structure and improve the light absorption [25–27].

On the contrary, poor charge separation and transfer are considered as the main obstacles that restrict performance of PCN materials in water splitting. Unlike PCN materials, crystallinity of metal oxides can be readily improved by thermal posttreatment. Therefore, the charge carrier mobility can be sufficiently increased. In addition, the introduction of shallow donors and acceptors in the semiconducting polymer could efficiently promote charge carrier density. The situation is different for PCN materials. PCN is composed of organic units, and thus, a large Coulomb force normally restricts the dissociation of excitons into free charge carriers. This force, or in other words, the so-called exciton binding energy, can be compensated by the introduction of other forms of energy. For instance, PCN can be integrated into a photoelectrode by forming a film. In such a device, charge separation is facilitated by supplying electric energy.

PCN can be synthesized by thermal treatment of carbon- and nitrogen-abundant precursors. Typically, such molecules evaporate at temperatures lower than the temperature at which formation of the extended PCN structure takes place. For instance, urea is a typical PCN precursor. The polycondensation of urea into PCN takes place at a temperature of 500 °C, while the sublimation temperature is only ca. 135 °C at atmospheric pressure. Therefore, instead of improving crystallinity of PCN, in this case, increasing the annealing temperature normally leads to the formation of amorphous PCN in low yields. Recent studies have shown that ionothermal approaches can afford single crystal PCN as large as several hundred nanometers with significantly improved conductivity.

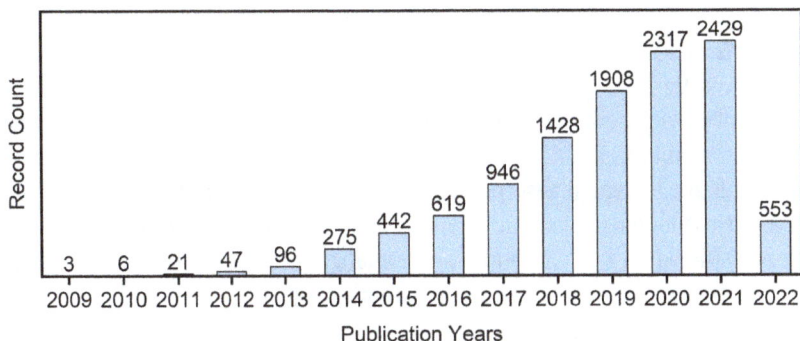

Fig. 3.2: The publication number of carbon nitride polymer for photocatalysis (data source: Web of Science, and the searching words: carbon nitride or g-C$_3$N$_4$, and photocatalysis).

On the basis of the research data meta-analysis, to date, there are more than 10,000 SCI articles that have been published in the field of carbon nitride photocatalysis, with an increasing trend, as illustrated in Fig. 3.2. This is presumably motivated by the fact that by using the PCN material for photocatalytic water splitting, you can meet sustainability goals while substituting fossil fuels with renewable, solar-light based alternatives in the future.

3.4 Synthesis of polymeric carbon nitrides via copolymerization of monomers

The physicochemical properties of PCN can be regulated by the copolymerization of nitrogen-rich precursor with additional monomers to modulate the photochemical and physical properties. Therefore, the band gap width and positions of band edges can be readily controlled. The PCN semiconductor is based on a highly delocalized π-conjugated system formed by the sp^2-hybridized carbon and nitrogen atoms. Small organic molecules, acting either as electron acceptors or donors, could be introduced to modify the π-conjugated system of Melon-based PCN.

As a typical example, barbituric acid (BA) was used as a monomer to be incorporated into the PCN matrix through thermal polycondensation with dicyandiamide, as shown in Fig. 3.3a [28]. The C/N molar ratio was determined by elemental analysis, which unambiguously confirmed incorporation of BA into the PCN matrix, since the weight percentage increases from 0.01 to 0.5 with the quantity of BA taken for materials synthesis. The chemical composition of the modified materials can be studied by ^{13}C solid-state nuclear magnetic resonance (NMR), but not by the conventional powder X-ray diffraction (XRD) or X-ray photoelectron spectroscopy (XPS), since the difference in crystallinity of the materials and binding energies of all elements are rather similar. More importantly, the potential of the conduction band was identified to shift to less negative values, while valence band potential becomes less positive, which results in a shrinkage of the band gap. In this research, the band gap of the copolymerized PCN can be effectively controlled between 1.58 and 2.67 eV by adjusting the concentration of BA. As a result, photocatalytic activity of the PCN modified in this fashion in hydrogen evolution reaction is significantly improved, owing to the increased absorption of visible light. On the basis of this development, other kinds of monomers were attempted for the synthesis of PCN by copolymerization strategy, as shown in Fig. 3.3b, and correspondingly, the hydrogen evolution rates are presented below the monomer structure in parentheses [28].

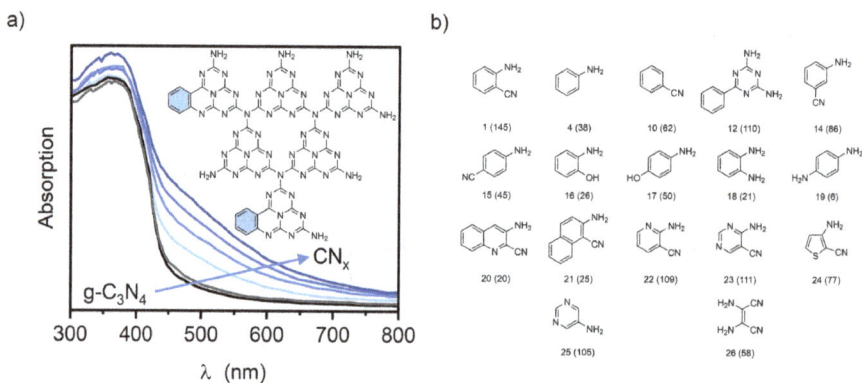

Fig. 3.3: (a) UV–Vis absorption spectra of the PCN materials (inset). (b) List of the monomer structures that have been attempted for copolymerization; the number in parentheses are the hydrogen evolution rate (μmol h^{-1}) of the copolymerized PCN with optimal quantities (reproduced with permission from ref. [28]. Copyright 2012, Wiley-VCH Verlag GmbH & Co. KGaA, Weinheim).

On the basis of this example, intense research has been conducted on applying copolymerization to modify PCN photocatalysts. For instance, benzamide was employed as a new, low-cost comonomer together with urea to prepare a phenyl-modified PCN photocatalyst [29]. Introducing a phenyl group into the PCN extends the original π-conjugation system, leading to the improved utilization of visible light and increases the rate of separation of the electron–hole pairs. As a result, under visible light irradiation, the copolymerized PCN exhibits superior catalytic activity in hydrogen evolution reaction, namely 2.5 mmol h^{-1} g^{-1}, which is ca. 3.2 times higher, compared to the pristine PCN that is synthesized from urea. 4,4′-Sulfonyldiphenol (BPS) was applied as a comonomer to modify the physiochemical properties of PCN [30]. In this research, by inserting the BPS into the framework of the PCN, both light absorption and photoexcitation charge separation and transport are promoted, as evidenced by the experimental and theoretical studies. The material exhibits enhanced photocatalytic hydrogen evolution efficiency with a rate of 12.3 mmol h^{-1} g^{-1}, which is almost 12 times higher than that of pristine PCN, and the quantum efficiency reached 17.7% at 420 nm.

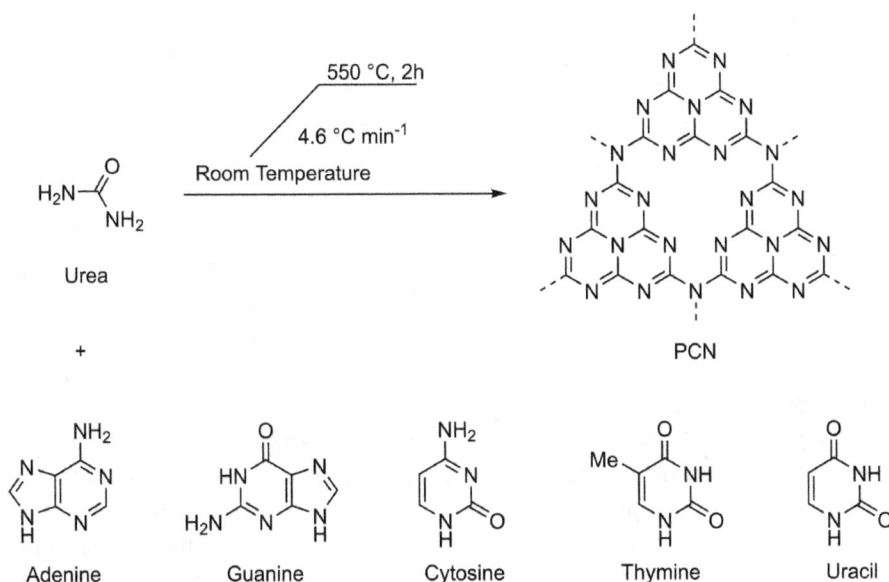

Fig. 3.4: Schematic illustration of the synthesized PCN using nucleobases and urea (reproduced with permission from ref. [31]. Copyright 2017, Wiley-VCH Verlag GmbH & Co. KGaA, Weinheim).

Biotic compounds can be used to construct PCN semiconductors, as shown in Fig. 3.4 [31]. Nucleobases and urea were used as the comonomers and precursor, respectively, to form PCN with tunable physicochemical properties. Thermally induced condensation using such biological precursors represents the extreme terrestrial and thalassic thermal/hydrothermal conditions of primordial Earth. For instance, the temperature in hydrothermal vents close to the volcanic edifices ranges from 60 to 500 °C [31]. Incorporation of a biotic compound in PCN by this preparation method, on one hand, narrows the band gap and results in improved visible-light absorption. On the other hand, under different synthetic conditions, the layered structure of PCN is preserved, and porosity and functional groups in the PCN can be installed to fit certain photocatalytic reactions.

Modification of the PCN semiconductors by adding comonomer to the precursor during the synthesis leads to broadening of physical and chemical properties compared to pristine PCN, with the stoichiometry matching closely to C_3N_4. To date, the main area where the optimized PCN semiconductors are applied is photoredox catalysis [32].

3.5 Metal-doped polymeric carbon nitride

Doping of the PCN framework with metal elements also modifies the physicochemical properties of the material. Particularly, as photocatalyst for water splitting, it may act as an active site for the surface transfer reaction. Noble metals, including Pt and Pd, are typical examples, and they are introduced into the PCN structure to promote the surface photoredox reaction. PCN can be applied as a ligand, and Pt^{2+} or Cu^{+} can be introduced into the PCN structure by a bottom-up route from dicyandiamide and H_2PtCl_6 [33]. By virtue of this structure, it is demonstrated that the metal-to-ligand charge transfer (MLCT) between M^{n+} (M = Pt or Cu) and PCN plays an essential role in improving photo-catalytic performance, as shown in Fig. 3.5. Isolated single Pt atoms may be also an-chored on PCN as cocatalysts by a solution processing reaction of PCN and H_2PtCl_6, followed by annealing at low temperature [21]. The single-atom cocatalyst traps elec-trons and provides proton reduction sites. As such, the lifetime of the photogenerated electrons can be prolonged and, therefore, it dramatically enhances the photocatalytic performance of the hydrogen evolution reaction. In addition, compared with the physi-cal loading of Pt on PCN photocatalysts, the photocatalytic activity in hydrogen evolution reaction improved 8.6 times, when Pt was deposited onto holey PCN nanotubes prepared using the supramolecular precursor strategy. A strong reactive metal-support interaction between Pt single atom and electron-deficient PCN was demonstrated by the in situ pho-tocatalytic reduction method. It was found that the rich N vacancies are important to anchor Pt single atoms at PCN. Nitrogen vacancies not only stabilize the Pt single atoms by a strong chemical bonding, they also optimize the electronic and geometric structures of Pt single-atom for capturing photogenerated electrons and, thus, facilitate the evolu-tion of hydrogen.

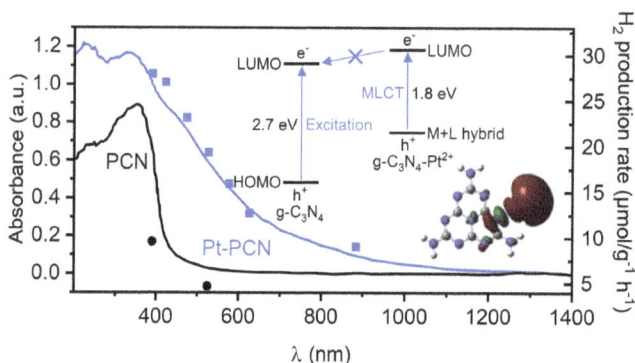

Fig. 3.5: The light absorption spectrum of pristine PCN and PCN with anchored Pt single atoms (reproduced with permission from ref. [33]. Copyright 2016, Wiley-VCH Verlag GmbH & Co. KGaA, Weinheim).

The formation of a built-in electric gradient into a PCN-metal superstructure provides a driving force for directional charge transfer in both vertical and in-plane directions. The vertical channels formed by bridged Pd atoms in interlayers facilitate the transfer of photogenerated electrons from the bulk to the outermost surface, while the Pd atoms anchored on the surface serve as active sites to capture electrons for proton reduction. The synergistic effect between the accelerated charge transport and higher affinity toward reactant molecules greatly enhances the photocatalytic activity of PCN.

Attention has been paid to binding the transition metals and post-transition metals to PCN material. Pt-based transition metal alloys were prepared and in situ anchored on PCN to construct Pt–M (M = Co, Ni and Fe)/PCN photocatalysts. For instance, partially oxidized Ni single-atom sites can be embedded into the framework of PCN to modulate the electron configuration of Ni atoms [34]. When the oxidation state of the Ni single-atom sites is controlled to a precise Ni^{2+}/Ni^0 ratio of 2, the material brings the most abundant unpaired d-electrons into the photocatalyst and optimizes the electronic structure, which elevates the light response, conductivity, charge separation and transfer concurrently and thus, remarkably enhances the photocatalytic performance to 354.9 $\mu mol\ h^{-1}\ g^{-1}$, which is ca. 30 times higher compared to the pristine PCN. Meanwhile, control experiments were carried out with different metal species (Fe, Co and Cu), whose valence states and unpaired electron configurations were also probed. It was proven that the presence of unpaired d-electrons was an essential factor for improved photocatalytic performance.

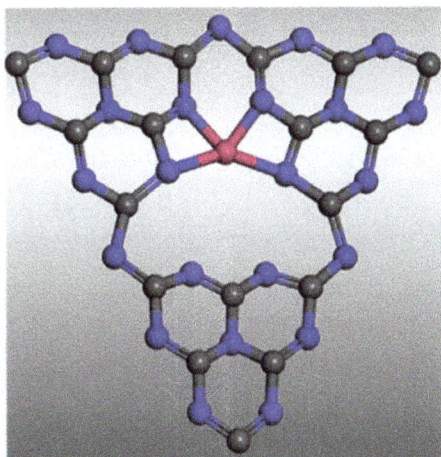

Fig. 3.6: Illustration of the structure of the GaN_4 site-embedded PCN photocatalyst (reproduced with permission from ref. [35]. Copyright 2021, Wiley-VCH Verlag GmbH & Co. KGaA, Weinheim).

An active GaN_4 site was integrated into PCN as demonstrated in Fig. 3.6, displaying a high performance in photocatalytic hydrogen evolution reaction [35]. Theoretically, localizations of the conduction and valence bands are found at the GaN_4 site and aromatic ring, respectively. Owing to the electron-deficient nature of Ga, photogenerated electrons are transferred and stored at the GaN_4 site, which contributes to charge separation. In the excited state, the negatively charged GaN_4 also promotes proton adsorption. After the electron was transferred to a proton, the GaN_4 site returns to the ground state where the adsorption energy of hydrogen is positive, promoting hydrogen desorption. This process is cycled at the GaN_4 site, so that rapid hydrogen evolution with a maximal rate of ca. 9,904 $\mu mol\ h^{-1}\ g^{-1}$ is achieved.

In the other example, single-atom Ni could be considered as a terminating moiety that coordinates with the sp^2 or sp^3 N atoms in the PCN, which results in the formation of extra hybrid orbitals [36]. Consequently the visible light absorption is increased via a metal-to-ligand charge-transfer process and accelerates the separation and migration of photoexcited charge carriers, synergistically. In the obtained single-atom Ni-terminated PCN, the single-atom Ni and the neighboring C atom acted as active sites for oxidation and reduction for H_2O_2 production and H_2 evolution, respectively, with matching rates of 24.0 and 26.6 $\mu mol\ g^{-1}\ h^{-1}$.

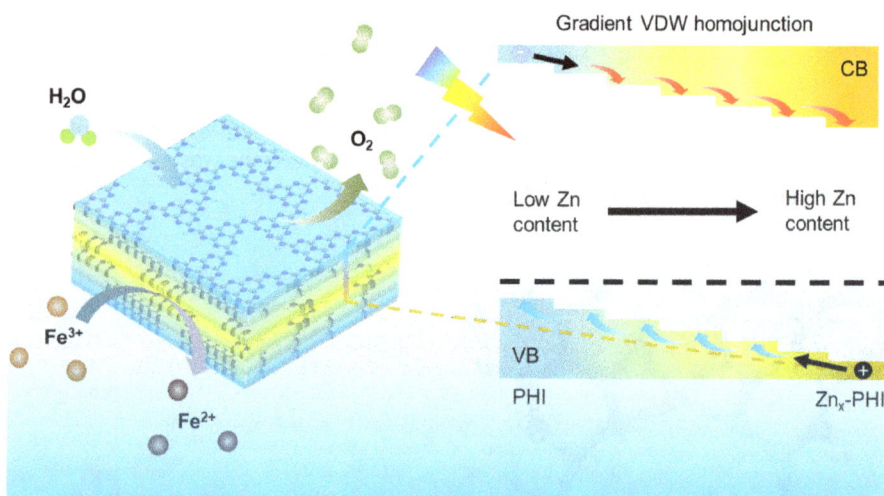

Fig. 3.7: Zn-doped PHI with gradient shift of the band structure for improved photocatalytic oxygen evolution reaction (reproduced with permission from ref. [37]. Copyright 2021, American Chemical Society).

Ionothermal synthesis of carbon nitride in a ternary salt melt ($ZnCl_2$–LiCl–KCl) promotes the polycondensation process and improves the degree of crystallinity [37]. As Zn^{2+} ions tend to coordinate to edge nitrogen atoms of carbon nitride via Lewis acid–base interactions, a series of crystalline hybrid poly(heptazine imides) (PHI), namely, Zn-PHI/PHI, assembled through van der Waals interactions, were generated as a photocatalyst. Interestingly, the content of the dopant, Zn^{2+} ions, in the as-synthesized Zn-PHI/PHI gradually decreases from the bulk interior to surface, due to the evaporation of $ZnCl_2$ during thermal heating processes. Accordingly, a built-in electric field gradient is created in the catalyst, which greatly accelerates the separation of the photogenerated charge carriers at the interfaces of the band structure, as illustrated in Fig. 3.7. Owing to the improved charge carrier separation and transfer, this hybrid polymer is active in photocatalytic water oxidation, with an apparent quantum yield of up to 3.6% at 420 nm.

3.6 Crystalline polymeric carbon nitrides

The structure of PCN prepared by conventional thermal polymerization is usually described as Melon-based. In recent years, ionothermal approach has been optimized to synthesize crystalline carbon nitride derivatives, such as PHI and poly(triazine imide) (PTI). Bojdys and coworkers obtained PTIs with tunable gallery heights by intercalating different halide ions between the layers [38]. Similar to graphite, it is possible to exfoliate layered PTI into thin sheets and scroll them by mechanical and chemical methods. PTI/Br is an ideal candidate for exfoliation, owing to its relatively large interlayer distance. By exfoliation with Scotch tape, Bojdys and coworkers obtained thin sheets of PTI/Br [39]. The thickness was analyzed by atomic force microscopy, and steps composed of 10–100 layers were found in the molecularly flat terraces of PTI. PTI/Br can also be exfoliated by treatment with vapor-phase potassium metal. When subjected to water, the K(PTI/Br) material was exfoliated into thin sheets; this process was accompanied by hydrogen evolution reaction due to the reaction of the intercalated K with water.

PTI thin sheets can be prepared by exfoliation, and thus, an improved active site can be provided to increase the mass transfer efficiency [40] Lotsch and coworkers obtained highly crystalline nanosheets (Fig. 3.8a) by liquid exfoliation of PTI/Li$^+$Cl$^-$, followed by centrifugation at different rotation speed in water for photocatalytic applications [41]. Atomic force microscopy revealed that the lateral size and the height of the as-prepared nanosheets were less than 100 nm and 1–2 nm, respectively, indicating that the nanosheets consisted of only a few layers. Transmission electron microscopy (TEM) analysis further confirmed that the crystal structure of the PTI/Li$^+$Cl$^-$ was retained after exfoliation (Fig. 3.8b). Under visible-light irradiation, bulk PTI exhibited a hydrogen evolution rate of 4.3 μmol h^{-1}. After sonication, the activity decreased by approximately 43%. However, a precipitate II, which contained partially exfoliated PTI, showed improved photocatalytic activity. When using approximately equal amounts, the PTI nanosheets showed 18 times enhanced photocatalytic activity compared to the bulk PTI (Fig. 3.8c). The photocatalytic experiment conducted for 130 h confirmed the high stability of the PTI nanosheets.

Combination of the copolymerization with ionothermal synthesis is a highly appealing method to tune the photocatalytic performance of PTI. Lotsch and coworkers obtained amorphous PTI (aPTI) modified with 4-amino-2,6-dihydroxypyrimidine (TAP) as the molecular dopant [41]. Unlike the sample synthesized in a sealed glass tube, in open environment, the as-prepared samples adopted an amorphous structure. Compared with PTI/Li$^+$Cl$^-$, the organo-doped aPTI exhibits a red-shift in the optical absorption spectrum, suggesting that the visible-light absorption is widened, leading to the absorption edge at ca. 560 nm. The photocatalytic test demonstrates that optimal performance for hydrogen evolution reaction is ca. 204 μmol h^{-1} by 25 mg photocatalyst, and the apparent quantum efficiency was determined as ca. 15% at the incident wavelength of 400 nm [42].

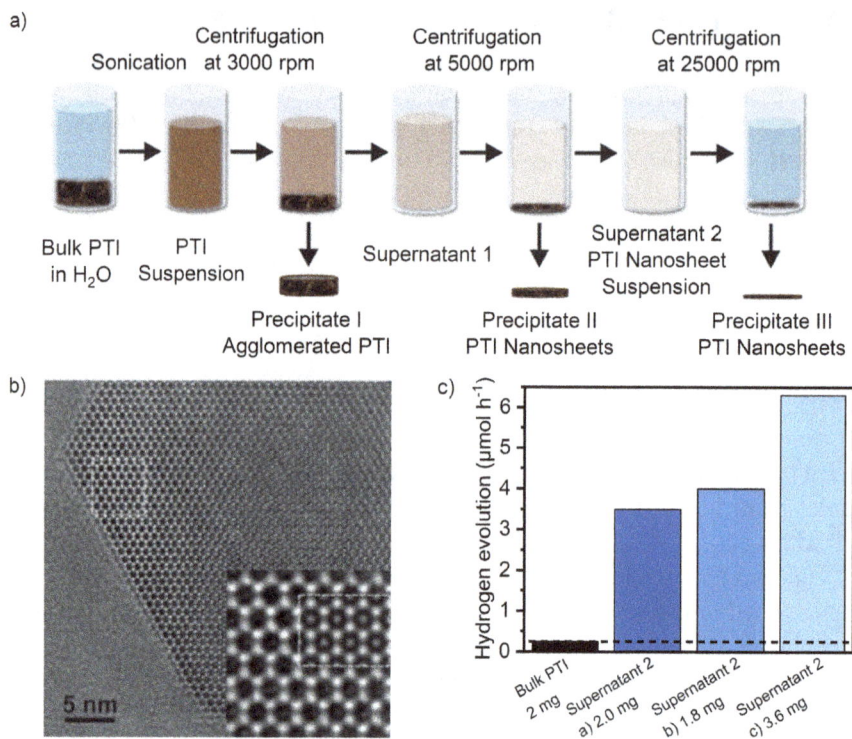

Fig. 3.8: (a) Exfoliation process and product labeling. (b) Magnified image of a PTI nanosheet edge viewed along the [001] direction and simulation (inset). (c) Photocatalytic activity towards H_2 production by loading Pt on the PTI nanosheet and bulk PTI (reproduced with permission from ref. [41]. Copyright 2014, American Chemical Society).

Although the abovementioned studies demonstrated the photoactivity of crystalline PTI in hydrogen production, it is still a great challenge to achieve photocatalytic overall water splitting with PTI, owing to the high reaction barrier of the oxygen evolution reaction via a four-electron process. Recently, Wang and coworkers realized photocatalytic overall water splitting by using crystalline PTI modified with suitable cocatalysts [43]. As revealed by density functional theory calculations, the band gap of the PTI straddles over the redox potential of water, which guarantees that overall water splitting is thermodynamically feasible (Fig. 3.9a). Owing to the absence of Li and K ions, the sample was termed as PTI·HCl. Photocatalytic water splitting was performed under the full spectrum of a xenon lamp. There was no detectable hydrogen or oxygen evolution without any cocatalyst. With 1 wt% Pt as the cocatalyst, PTI·HCl produced hydrogen and oxygen simultaneously, albeit in a nonstoichiometric ratio. This fact is likely due to the high energy barrier of the water oxidation, leading to a low oxygen evolution rate. In the presence of both Pt and Co cocatalysts, the water splitting was accelerated. The ratio between H_2 and O_2 can be maintained at 2:1 by

a)

b)

c)

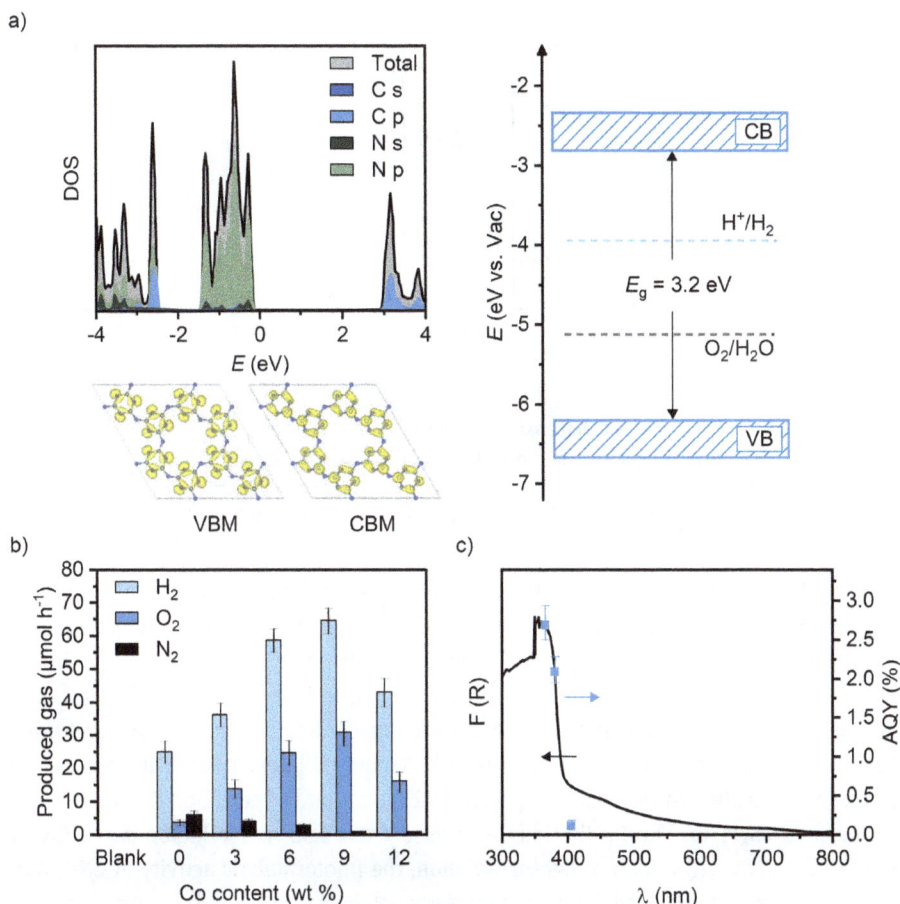

Fig. 3.9: (a) The electronic and band structure of PTI. (b) Photocatalytic overall water splitting using PTI·HCl upon irradiation with a 300 W Xe lamp. All samples were co-loaded with 1 wt% Pt, except for the blank. (c) Wavelength-dependent apparent quantum yield of PTI·HCl with 1 wt% Pt and 9 wt% Co (reproduced with permission from ref. [43]. Copyright 2017, Royal Society of Chemistry).

varying the amount of the Co cocatalyst (Fig. 3.9b). The apparent quantum yield was measured to be 2.1% at 380 nm. PTI·HCl was inactive under irradiation with light of >400 nm owing to its relatively wide band gap, which is consistent with the UV-Vis diffuse reflectance spectroscopy result (Fig. 3.9c).

Yan and coworkers obtained a PCN intercalation compound (CNIC) with heptazine subunits by heating melamine in KCl/LiCl/NaCl molten salts in a semi-closed environment [44]. Interestingly, the XRD pattern of the as-obtained sample was different from that of PTI/Li$^+$Cl$^-$ (Fig. 3.10a). The composition of the CNIC was analyzed by XPS, in which all ions from the salt mixture were found. The as-prepared samples demonstrated high activity in photocatalytic hydrogen production under visible-light irradiation, reaching

a)

b)

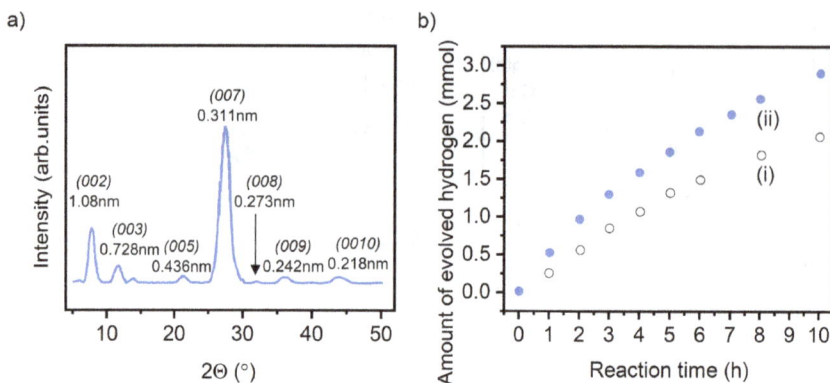

Fig. 3.10: (a) XRD pattern of the CNIC prepared by heating melamine in molten salts. (b) A typical time course of H_2 production under visible-light irradiation over 3.0 wt% Pt-deposited CNIC photocatalyst in aqueous solution of triethanolamine (10 vol. %) (i) and in saturated aqueous solution of KCl, NaCl and LiCl and triethanolamine (10 vol. %) (reproduced with permission from ref. [44]. Copyright 2013, PCCP Owner Societies).

346 µmol h^{-1} in the first hour, which is 34.6 times higher than that of Melon-based PCN (Fig. 3.10b). One important reason for the enhanced photocatalytic performance of CNIC was noted to be the significantly increased interlayer distance, which is due to the incorporation of alkali metal and chloride ions, changing the crystal field and prompting charge separation. These conclusions were further confirmed by photoluminescence (PL) spectroscopy. In contrast to Melon-based PCN, the PL intensity of CNIC was greatly decreased, indicating the efficient suppression of the radiative recombination of the photogenerated carriers. The PL lifetime decreased from 4.99 ns for Melon-based CN to 0.083 ns for CNIC. After 20 h of light irradiation, the photocatalytic activity of CNIC had decreased by about 50%, and the alkali ions were detected in solution by inductively coupled plasma mass spectrometry. These results further supported the conclusion that the incorporation of alkali ions could accelerate the separation of photogenerated carriers.

Heptazine subunits can also be formed in molten salt environment from traditional precursors. By using preheated melamine, Wang and coworkers obtained a heptazine-based crystalline PCN by the ionothermal method, which is referred to as g-CN-1 [45]. The XRD pattern was different from that of bulk PCN and PTI/Li$^+$Cl$^-$, indicating that a new phase was obtained (Fig. 3.11a). The PL spectra show that the emission peak intensity of g-CN-1 is lower than those of reference samples (Fig. 3.11b). Furthermore, different from the amorphous bulk PCN, the TEM image shows clear lattice fringes of g-CN-1, indicating the further polymerization of Melon in a molten salt (Fig. 3.11c). Photocatalytic tests demonstrate that crystalline g-CN-1 possesses higher activity in hydrogen evolution reaction under visible-light irradiation (Fig. 3.11d). Remarkably, upon adding phosphates to the reaction solution, the AQY of g-CN-1 reached 50.7% at 405 nm with triethanolamine as the sacrificial agent. In addition, an oxygen evolution test also confirmed the higher oxidation capability of g-CN-1 compared with bulk PCN.

Inspired by the exfoliation of PTI/Li$^+$Cl$^-$, Wang and coworkers obtained heptazine-based crystalline PCN nanosheets by liquid-phase exfoliation in isopropanol solvent [46]. The as-prepared nanosheets were further characterized by TEM, and the d-spacing was measured to be 3–4 nm; this value corresponds to the stacking of ten layers, confirming the successful exfoliation of bulk crystalline PCN. This was corroborated by atomic force microscopy analysis, in which the height of the nanosheets was determined to be 3.6 nm. Compared to bulk PCN and bulk crystalline PCN, the crystalline nanosheets showed significantly increased photoactivity in hydrogen evolution. The AQY was measured to be 8.57% at 420 nm using methanol as the sacrificial agent.

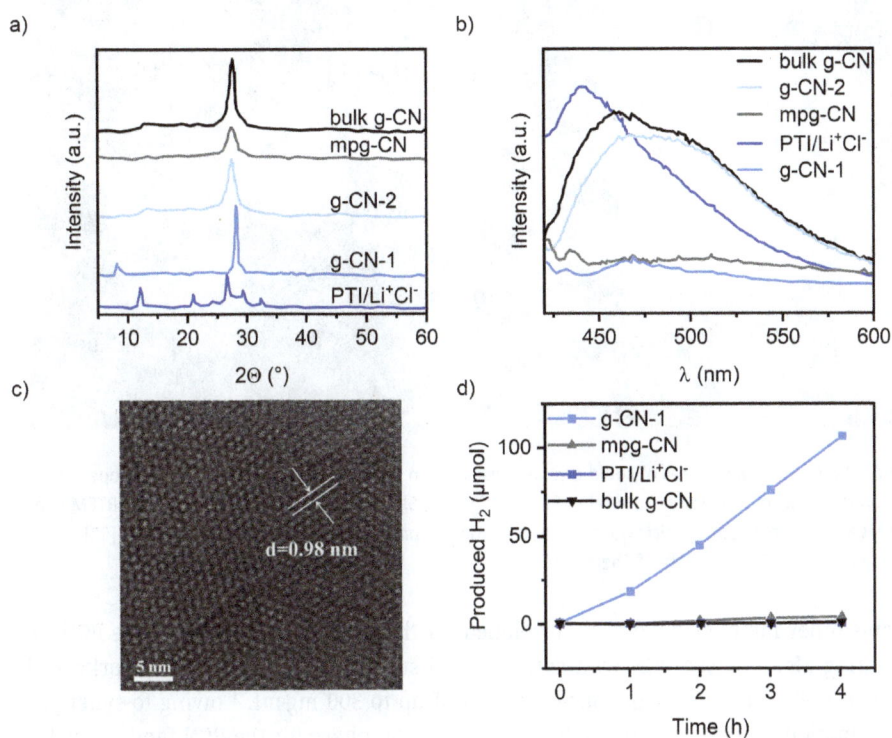

Fig. 3.11: (a) XRD patterns of the samples. (b) PL spectra of the samples under 400 nm excitation. (c) TEM image of g-CN-1. (d) H$_2$ production by the samples (reproduced with permission from ref. [45]. Copyright 2016, American Chemical Society).

Recently, a carbon-modified crystalline PCN was obtained by co-condensation of urea and oxamide, followed by treatment in molten salts [47]. The light absorption ability of copolymerized crystalline PCN and the reference sample was investigated by diffuse reflectance spectroscopy. Interestingly, the absorption shoulders are observed for oxamide-containing PCN either with or without salt treatment, which were ascribed to the n–π* transition accompanying the major π–π* transition. As a consequence of carbon doping, the onset of absorption is extended from 462 to 700 nm. The copolymerized crystalline PCN shows significantly increased hydrogen evolution rates in comparison with pristine PCN, reaching 84 µmol h^{-1}. Other salts, such as K$_2$HPO$_4$, KBr and Na$_2$SO$_4$, also showed a similar promotional effect. The copolymerized crystalline PCN exhibited superior photoactivity in comparison with other reference samples when irradiated with green LEDs. The AQYs in the presence of NaCl reach 57% and 10% at 420 and 525 nm, respectively.

a)

b)

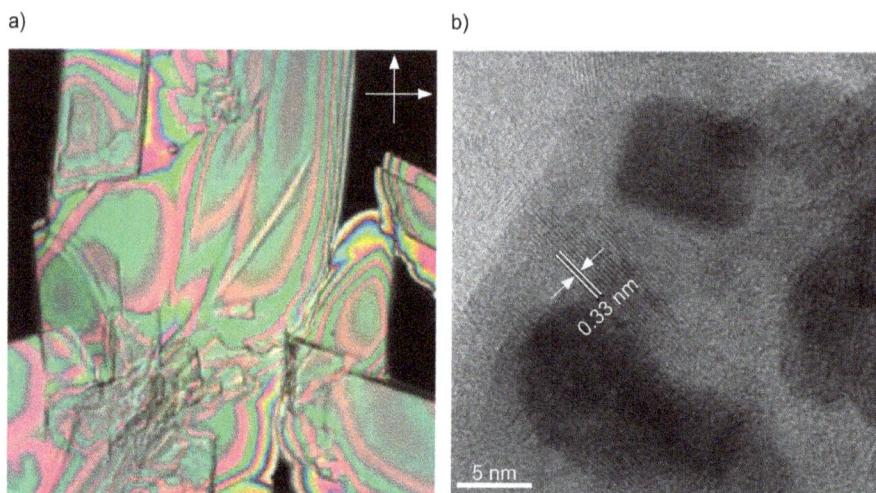

Fig. 3.12: (a) Optical textures of liquid-crystal-phase carbon nitride under a polarizing microscope (reproduced with permission from ref. [48]. Copyright 2015, American Chemical Society). (b) HRTEM image of MCN$_{1000-18}$ prepared by microwave-assisted heating (reproduced with permission from ref. [50]. Copyright 2014, Royal Society of Chemistry).

Some other methods have been developed for the preparation of crystalline PCNs. By treating Melon-based PCN with concentrated sulfuric acid, Zhou and coworkers obtained a PCN solution with concentrations of up to 300 mg mL^{-1} owing to synergistic protonation and intercalation [48]. A liquid-crystal phase for the PCN family was first observed at high concentrations (Fig. 3.12a). Wang and coworkers reported a simple method for the synthesis of crystalline PCN, based on the protonation of specific intermediate species during conventional polymerization processing [49]. The as-obtained crystalline PCN exhibited improved photoactivity that is up to seven times higher than that of bulk PCN. It was found that microwave-assisted synthesis is a faster and a more robust method to prepare PCN with improved crystallinity. By using urea as the precursor, Yuan and coworkers prepared a crystalline PCN by microwave-assisted heating [50]. A very clear HRTEM image with a spacing of 0.36 nm indexed to the (002) facets was found and is shown in Fig. 3.12b.

3.7 Polymeric carbon nitrides for photoelectrochemical water splitting

The research on polymer-based photoanodes prepared by coating metal oxide powder with PCN is inspired from the dye-sensitized photocatalysis or solar cell system [51, 52]. The examples of metal oxides are TiO_2, [53–60] ZnO, [61–63] Fe_2O_3 [64, 65], WO_3 [66, 67] and many more [68–72]. Several different approaches were developed to prepare such hybrid photoanodes. They are mainly based on physical adsorption of PCN powder on metal oxides, which are analyzed in few review papers [73]. In this section, we will focus on the interfaces between metal oxide and polymeric photocatalysts, as it is important for the charge transport, offering in-depth understanding and stimulating further developments in this field.

The first example describes the preparation of hybrid photoanode by anchoring PCN on TiO_2 particles, followed by coating of the powder at conductive substrate [74]. For the synthesis of the hybrid material, TiO_2 powder was placed into a Schlenk tube connected via an adapter to a flask containing urea, which was heated in a muffle oven at 500 °C to form in situ PCN layer on TiO_2 powder. The films were prepared by dip-coating the fluorine doped tin oxide (FTO) glass with the prepared PCN coated TiO_2 powder, followed by hot-pressing the photoanodes to strengthen them physically [74, 75]. In this case, the band gap of the photoanode can be regulated between 3.2 and 2.1 eV, and surprisingly, the value can be even smaller than that for typical PCN. The decrease of the band gap may be attributed to partial reduction of TiO_2 caused by grafting of PCN at TiO_2 via secondary amine group (—NH—). The photoanode generates O_2 with the concentration of ca. 200 ppb in a 90 min test under visible-light illumination (>420 nm) [74].

By integrating PCN with the inorganic photocatalyst that only absorbs light in the UV region, such as TiO_2 and ZnO, the absorption of visible light by the system is significantly improved, which leads to the higher performance [76]. In this case, the electric conductivity in metal oxide is also affected. Typically, the conductivity of metal oxides is improved by introduction of electron donors or electron acceptors into the crystal [77].

The attention was therefore changed to in situ polymerization of PCN films. As an example, a supramolecular precursor, namely cyanuric acid–melamine complex, was coated on TiO_2, while BA was added into the supramolecular complex to adjust the band structure of the resultant PCN [78]. Synthesis of PCN at TiO_2 particles was realized by thermal treatment [79]. The synthesized powder was coated on FTO to form a photoanode by the doctor-blade method, but the performance was still unsatisfactory [79]. Supramolecular complex is formed via hydrogen bonding between cyanuric acid, melamine and 2,4-diamino-6-phenyl-1,3,5-triazine, or a combination thereof.

a)

b)

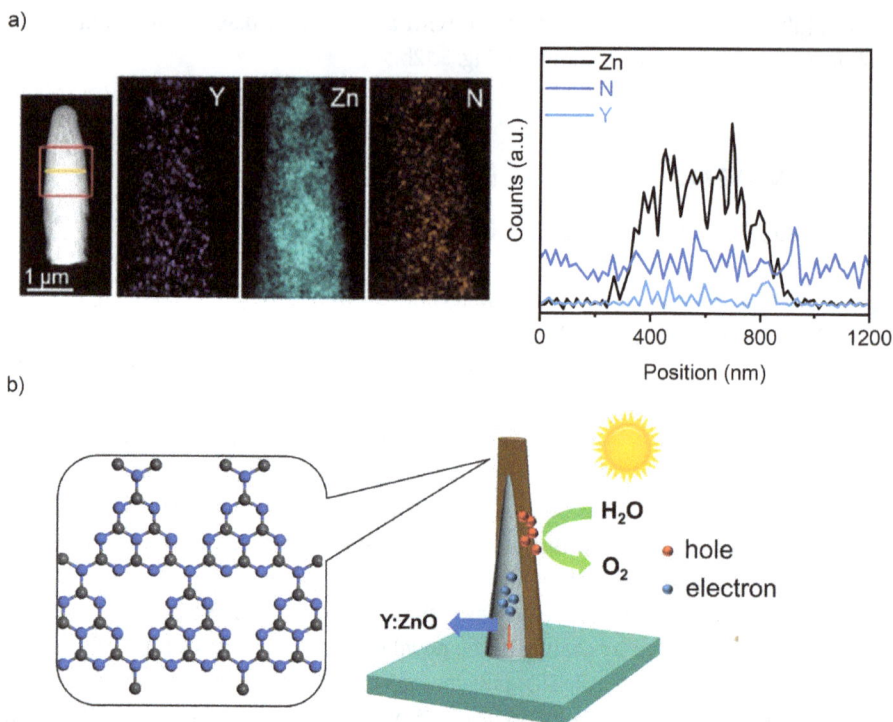

Fig. 3.13: (a) STEM studies of a typical core-shell structured nanorode formed by coating PCN on Y:ZnO. (b) Schematic illustration of the core-shell photoanode formed by PCN on Y:ZnO nanorod (reproduced with permission from ref. [62]. Copyright 2018, Wiley-VCH Verlag GmbH & Co. KGaA, Weinheim).

Thermal vapor condensation of PCN at ZnO can be achieved by using melamine as a precursor to form a photoanode. However, the carrier mobility in ZnO limits the performance of the PEC system [80]. In a notable example, PCN films were coated on yttrium-doped ZnO (Y:ZnO) nanorod arrays to form the photoanode. In this case, Y acts as electron donor in the ZnO crystal to improve the carrier mobility [81]. This change compensates the poor charge transfer in the amorphous PCN caused by low polymerization degree and poor conductivity, and improves the photoanodic current density. When PCN films are deposited at the Y:ZnO nanorods, a core-shell structure is realized, as shown in the TEM image (Fig. 3.13a) [62]. In this analysis, the Y:ZnO core and PCN shell can be clearly distinguished by the linear profile of the mapping image. This heterostructure significantly promotes the charge separation and transport in the PCN films, as illustrated in Fig. 3.13b. As a result, the photocurrent density is ca. 0.4 µA cm^{-2}, which is more than 50-fold with respect to the films that formed by dip-coating of pristine PCN powder. In addition, this photoanode possesses excellent stability – no obvious degradation occurred in a four-hour experiment.

PCN films were also integrated into the traditional metal oxide photocatalysts that are active under visible light, such as WO_3, Fe_2O_3 and $BiVO_4$ [72, 82, 83]. The band gap of WO_3 is ca. 2.65 eV that is similar to PCN [84]. Therefore, the light absorption by the system is insignificantly affected. In addition, this structure leads to a type-II heterojunction, which boosts the separation of photogenerated charges [85].

Among the inorganic metal oxide photocatalysts, $BiVO_4$ is one of the most developed materials for construction of photoanodes, mainly due to its suitable (for water splitting) positions of band edges and excellent stability [86–88]. However, the efficiencies of charge transfer and the rate of surface reaction are, in this case, the bottlenecks that restrict the performance [89, 90]. PCN was loaded on the $BiVO_4$ photoanode to promote surface activity in oxygen evolution reaction [91]. Typically, ultrathin PCN films are prepared by an oxidation and exfoliation process in acidic environment. The prepared PCN is suspended in ethylene glycol as ink, which can be directly used to coat on the surface of $BiVO_4$ films to form the hybrid photoanode. The prepared sample was annealed to impregnate a 2 nm-thick PCN layer into $BiVO_4$. The role of PCN is to promote surface reaction, but it does not act as a conducting photocatalyst. The ultrathin PCN films were used to maximize light delivery to $BiVO_4$. The PCN thin films-coated $BiVO_4$ photoanode exhibits 7 times higher photocurrent density with respect to the pristine $BiVO_4$ with optimal photocurrent density ca. 3.12 mA cm^{-2} at 1.23 V versus RHE.

Since 2009, the photoresponse of PCN materials was studied to apply the knowledge in PEC devices [92]. While dip-coating was generally used to prepare PCN films in these early studies, the low mechanical stability of the films restricts the long-term uses [93, 94]. The PCN powder was converted into a gel to improve stability of the films. The sol–gel PCN was prepared by suspending PCN powder in HNO_3 solution (Fig. 3.14). The strong acid breaks the hydrogen bonds and leads to the formation of a homogeneous sol [95]. Carbon nanotube powder, when blended with the sol, improves the charge transport in the mixed films and translates into improved performance in photoanodic oxygen evolution reaction. Thus, an improved performance of photoanodic water oxidation was detected for photoanode prepared for films composed of conductive carbon nanotubes, PCN gel and TiO_2 due to improved efficiency of charge transfer from the photocatalyst to the conductor [96], which was confirmed by Mott–Schottky measurements and EIS [97].

The PCN suspension can also be coated on an FTO glass by a high-pressure spraying gun [98]. This spraying method could readily achieve homogeneous films with strong adhesion to the substrate. As a result, due to the improved adhesion, the photocurrent density of the photoanode prepared by the electrophoresis and spray-coating can be improved to a few tens of microamps. Although adhesion is improved, the physical contacts are still insufficient for charge transfer and likely even act as recombination centers. Studies revealed the importance of the chemical bonding of PCN to the conductive substrate, which stimulated the development of in situ synthesis [99, 100].

Prior to the developments of PCN films for PEC water splitting, several reports described in situ growth of the PCN films, and they served as a source of inspiration. One of

Fig. 3.14: Preparation of the PCN sol–gel by suspending PCN into HNO_3 solution (reproduced with permission from ref. [95]. Copyright 2015, Wiley-VCH Verlag GmbH & Co. KGaA, Weinheim).

the issues related to the synthesis of PCN materials is due to two competing processes – evaporation of precursors, such as melamine, and its thermal polymerization, which thus induces grain boundary effect and weak adhesion of the bulk PCN to the substrate. The discussed supramolecular complex is rather nonvolatile, when compared to common low molecular weight precursors. Therefore, PCN films of higher quality can be formed.

The supramolecular complex precursor was prepared by blending cyanuric acid, elemental sulfur and 2,4-diamino-6-phenyl-1,3,5-triazine [101]. In this example, sulfur melts first and serves as a reactive dispersion agent to initiate polymerization of PCN material. The reaction progress was monitored using a transparent setup. By increasing sulfur content in the precursor, it is possible to dope it into the PCN. As a consequence, the visible light absorption of the prepared PCN films improved with a red shift from 500 to 525 nm. For this photoanode, an optimal photocurrent density of ca. 60 μA cm^{-2} was achieved at 0 V (vs. Ag/AgCl) under illumination with a sunlight stimulator [101].

In another contribution, a supramolecular complex-based paste is prepared by dispersing precursors in ethylene glycol. The paste is used to coat FTO glass either by doctor blade or electrophoretic deposition techniques [102], followed by PCN photoanode synthesis upon thermal condensation of the paste films [103]. By this approach, modifications of the PCN can be achieved conventionally by adding extra chemicals into the paste. For instance, BA can be used as comonomer to improve visible light absorption, and thereby, the performance for photoanodic oxygen evolution reaction [78]. In addition, reduced graphene oxide was added to the supramolecular paste as a conduction aid, while synthesizing PCN films. The role of the reduced graphene oxide is to compensate for poor charge separation and transfer induced by the defects in the PCN films. As such, photocurrent density of ca. 660 μA cm^{-2} can be achieved, which is ca. 20 times higher than that of the pristine PCN photoanode [104]. Apart from reduced graphene oxide, graphene oxide and graphene were used as conductive bridges to improve charge separation and transfer [105]. Blending of graphene oxide with the supramolecular complex precursor results in gradual reduction of the former during synthesis of PCN films, and thus, formation of covalent bonds with PCN films that further improves charge transport [106]. As a result, a photoanodic current density of 124 µA cm^{-2} was achieved at 1.23 V (vs. RHE) under illumination with a sunlight stimulator, which is higher than using reduced graphene oxide as a dopant.

The adhesion between the PCN films and FTO glass was also optimized by grafting a precursor at the substrate and formation of chemical bonds during thermal condensation. Uniform films of thiourea with thickness of ca. 0.8 mm were formed at FTO upon immersion of the substrate into hot saturated aqueous solution of the former, followed by slow cooling. PCN films were obtained by annealing the as-prepared thiourea films [107]. By this method, the thickness of the PCN films can be readily optimized to control the photocurrent density. In addition, different types of precursors, such as melamine and dicyandiamide, can also be used to synthesize PCN films by this approach [108, 109]. Owing to the different solubility of the precursor, calcination time, thickness and uniformity, the PCN films possess different properties for light harvesting, charge separation and transfer, and surface reaction in oxygen evolution reaction. As a result, thiourea-derived PCN presents the optimal photocurrent density of ca. 266 µA cm^{-2} at 1.23 V (vs. RHE), under illumination with a sunlight stimulator, and the highest incident photon conversion efficiency of ca. 18% at the incident wavelength of 400 nm.

This supramolecular-based approach also allows to coat PCN films on different substrates, including FTO glass, conductive carbon cloth and bare glass [107–109]. When PCN films are formed at conductive carbon cloths, the photocurrent density is generally lower than that of the FTO glass, but the photoanode is flexible and allows extending the scope of applications to, for example, optoelectronics, energy storage and others [110–112]. However, moderate photocurrent density is generally obtained, owing to the poor connection between the carbon substrate and PCN films.

The supramolecular approach has lately inspired us to develop a method for growing PCN films on FTO in solid state. Five different precursors were investigated, including melamine, trithiocyanuric acid, thiosemicarbazide, thiourea and urea. It was found that the PCN films could grow on FTO glasses by combining sulfur-containing and sulfur-free precursors upon thermal condensation [113]. The growth mechanism was explained to proceed in three steps: (1) S-containing precursor initiates the formation of the PCN composite by bonding to Sn with the formation S–Sn bond; (2) the pre-deposited product allows for the sulfur-free precursor to initiate a cascade of polymerization on sulfur-containing layer to form PCN films; and (3) termination of the polymerization yields the PCN films. The structure of the PCN films is illustrated in Fig. 3.15a. Sulfur facilitates transfer of the photogenerated electron from the photoactive films to the FTO and, as such, increases its photocatalytic performance. As a result, a photoanodic current density of 100 µA cm^{-2} was obtained in the optimal photoanode, which was prepared using a mixture of trithiocyanuric acid and melamine in a ratio of 1:2.

Based on photoanodes synthesis in solid state, the physiochemical properties of PCN were further adjusted by using different precursors [114, 115]. For instance, PCN films doped with sulfur were obtained using a mixture of ammonium thiocyanate and melamine. As confirmed by XPS depth profile, the concentration of sulfur among the films gradually decreases from top to bottom of the PCN films. Doping PCN with sulfur not only affects potentials of the valence band and the conduction band (Fig. 3.15b), but also its band gap. By formation of the films with gradient shift of the band gap, the separation of the photoexcited electron and hole is significantly promoted. Also, by the gradient sulfur doping, the top of the PCN films has a larger band gap than the bottom part. Therefore, the top part of the films' harvests light in near-UV region, while visible light with longer wavelength penetrates through the top part and is harvested by the bottom part of the films. As such, absorption of light from the visible range by the PCN films is maximized. As a result, photoanodic current density of ca. 120 $\mu A\ cm^{-2}$ at 1.23 V (vs. RHE) under illumination with a sunlight stimulator was obtained.

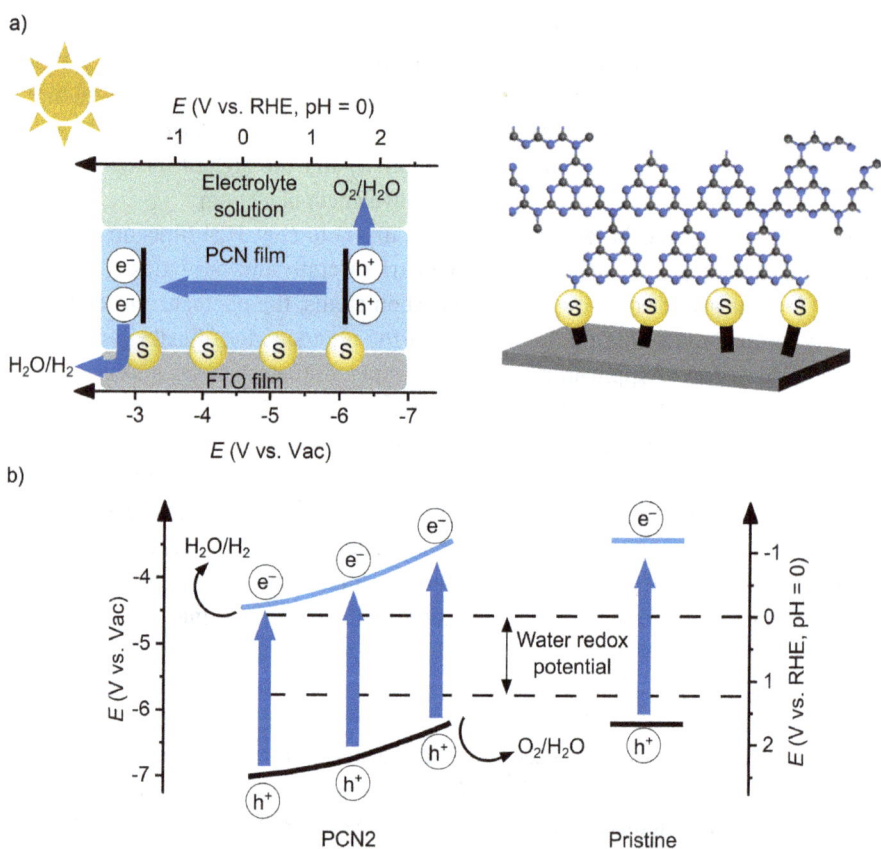

Fig. 3.15: (a) The band gap and chemical composition of the PCN films obtained using sulfur-containing and sulfur-free precursors (reproduced with permission from ref. [113]. Copyright 2018, American Chemical Society). (b) The band gap of PCN films with gradient sulfur doping and pristine PCN films (reproduced with permission from ref. [114]. Copyright 2020, Elsevier B.V.).

Other elements were also introduced into the PCN films by synthesis in solid state. Phosphorus was doped into the PCN films by simply adding hexachlorocyclotriphosphazene into the mixed precursor [115]. It was found that vertical sheets grow on the FTO substrate. In this structure, extra positive charges localized at phosphorus were induced in the π-conjugated heterocyclic rings to increase the photoexcitation charge transport [116]. Despite this, the band gap of PCN films increased from 2.80 to 2.88 eV, and thus, the visible light absorption decreased. However, the performance was improved by a factor of 2 compared to the pristine PCN, giving the optimal photocurrent density ca. 120 $\mu A\ cm^{-2}$. This improvement is mainly attributed to the improved charge transfer and minimized recombination probability [117].

Thermal vapor deposition was developed to synthesize PCN films. The principle behind the technology is similar to chemical vapor deposition [118]. Generally, melamine is vaporized and activated at temperature up to 300 °C, followed by immobilization by reacting with FTO substrate [119]. With further increasing the condensation temperature to 550 °C, polymerization is initialized between the vaporized melamine and the active layer that is immobilized on the substrate. PCN films are formed by thermal condensation for a sufficiently long time. Few precursors can be used to synthesize PCN films by this approach. For instance, poly melamine formaldehyde resin was used to produce modified PCN films [120]. In this case, the dominant direction for growth of the PCN crystal turns from (100), which corresponds to the in-plane packing of the heptazine units to (001) – interlayer stacking of the aromatic heterocyclic segments [121]. The improved arrangement of the stacked PCN layers improves charge transfer and thus promotes photocatalytic activity. As a result, the optimal photocurrent density of 228.2 $\mu A\ cm^{-2}$ was recorded at 1.23 V (vs. RHE) under illumination with a sunlight stimulator [122–125].

Boron-doped PCN (B-PCN) films were synthesized by thermal vapor deposition using a mixture of boric acid and dicyandiamide as precursors [126]. By this approach, boron atoms are likely to substitute carbon in the $CN_2(NH_2)$ environment, and thus improve polymerization and crystallinity of the films. Gradient concentration of boron within the B-PCN films, and therefore, formation of a graded nanojunction that features different potentials of the valence band and conduction band was obtained. This junction facilitates charge separation and minimizes the recombination probability, which improves performance in photoanodic oxygen evolution reaction [127, 128]. Efficiency of charge separation and transfer in optimal B-PCN films are 10 times greater, compared to that in the pristine one owing to the formed nanojunction within the films. As a result, a photocurrent density of 103.2 $\mu A\ cm^{-2}$ was realized at 1.23 V (vs. RHE) under illumination with a sunlight stimulator, while the incident photon-to-current efficiency of ca. 10% was achieved at the incident light wavelength of 400 nm.

The two different developed methods, namely thermal vapor deposition and polymerization of solid precursor possess unique advantages. In thermal vapor deposition, precursor is vaporized and activated at high temperature. As such, a strong bond is formed between PCN films and the substrate. In comparison, synthesis of PCN from the precursor in solid state implies the occurrence of phase transitions throughout the condensation of the material. The binding of the PCN films to the substrate is relatively weak. On the other hand, the general restrictions for choosing precursors are different. For instance, the boiling point of the precursor should be low when opting for thermal vapor deposition method. At the same time, melting point of the precursor deposited over the substrate should be taken into account, when considering direct synthesis of PCN.

Fabrication of multilayered PCN photoanodes is a recent strategy to improve their performance and stability in oxygen evolution reaction. Thermal vapor deposition was demonstrated as a typical approach to form a multilayered PB-PCN photoanode composed of phosphorus-doped PCN (P-PCN), boron-doped PCN (B-PCN) and pristine PCN layer on the top, middle and bottom, with respect to the FTO [129]. The corresponding layers were prepared by thermal condensation of the mixed precursors, namely H_3BO_3–melamine complex and H_3PO_4–dicyandiamide complex. The formation of the layers in a distinct order is explained by different thermodynamics and kinetics associated with the precursor polymerization as shown in Fig. 3.16a. Route c leads to polymerization of melamine first to form PCN films at the bottom of the photoanode. By increasing the temperature, H_3BO_3 is vaporized, reacts with melamine, and thus, leads to B-PCN layer on the top of PCN layer as shown in Route b. Kinetically, polymerization of dicyandiamide is the slowest among the three routes. It reacts with thermally activated H_3PO_4 to give a P-PCN layer at the top of the photo anode, as shown in Route a. It was found that P and B substitute C at the corner and bay sites of heptazine-based framework, respectively. An optimal photocurrent density of ca. 150 $\mu A\ cm^{-2}$ at 1.23 V (vs. RHE) under illumination with a sunlight stimulator was obtained and preserved for over 5 h, which points at superior stability of the multilayered PB-PCN photoanode. In this multilayered structure, the separation and transfer of the photogenerated charge is the dominant factor that explains improved performance. The mechanism of charge separation and transfer were investigated by intensity modulated photocurrent spectroscopy and time-resolved photoluminescence, as shown in Fig. 3.16b and c, respectively. The constants of charge transfer and recombination were derived from the intensity-modulated photocurrent spectroscopy and are 3.3 and 0.9 s^{-1}, respectively, which are 1.6 times higher and 1.2 times lower than for PCN photoanode. This result suggests that doped layer improves charge transfer. In addition, the lifetime of photogenerated charges was determined from time-resolved photoluminescence. PB-PCN demonstrates the longest lifetime among a series of photoanodes doped with various concentrations of the elements. This observation confirms that the rate of electron–hole radiative recombination is significantly reduced with respect to the other two samples.

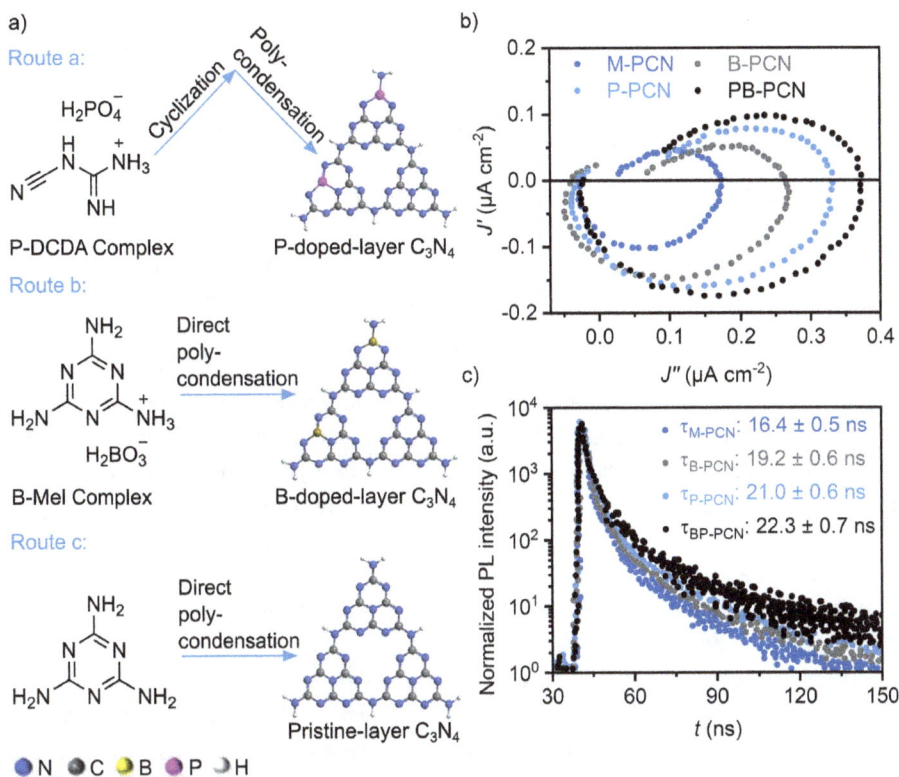

a)

Route a:

Cyclization   Poly-condensation

$H_2PO_4^-$

$N \equiv$ ... $\overset{H}{\underset{NH}{N}} \overset{+}{\underset{}{N}} NH_3$

P-DCDA Complex

P-doped-layer C_3N_4

Route b:

NH_2

$H_2N \overset{N}{\underset{N}{\bigtriangleup}} \overset{+}{N} NH_3$

$H_2BO_3^-$

Direct poly-condensation

B-Mel Complex

B-doped-layer C_3N_4

Route c:

NH_2

$H_2N \overset{N}{\underset{N}{\bigtriangleup}} N NH_2$

Direct poly-condensation

Pristine-layer C_3N_4

● N ● C ● B ● P ● H

b)

J' (µA cm^{-2})

- M-PCN   - B-PCN
- P-PCN   - PB-PCN

J'' (µA cm^{-2})

c)

Normalized PL intensity (a.u.)

- $\tau_{M\text{-}PCN}$: 16.4 ± 0.5 ns
- $\tau_{B\text{-}PCN}$: 19.2 ± 0.6 ns
- $\tau_{P\text{-}PCN}$: 21.0 ± 0.6 ns
- $\tau_{BP\text{-}PCN}$: 22.3 ± 0.7 ns

t (ns)

Fig. 3.16: (a) Synthesis of the multilayered PCN photoanode via several routes with different kinetics. Routes a, b and c demonstrate the formation of the P-PCN, B-PCN and pristine PCN layers. (b) Intensity-modulated photocurrent spectroscopy. (c) Time-resolved photoluminescence used to investigate the excited state dynamics of the photoanodes (reproduced with permission from ref. [129]. Copyright 2020, Wiley-VCH Verlag GmbH & Co. KGaA, Weinheim).

Two-layer PCN films can be synthesized by solid-state approach followed by thermal vapor condensation [130]. Specifically, the first PCN layer was prepared by doctor blade approach using supramolecular complex as a precursor. The bottom films must be sufficiently thin to mediate charge transfer to top layer and yet be mechanically stable. Several supramolecular complexes were attempted to adjust the physicochemical properties of the layer, such as melamine-Melem (MeM) [130] and bismuthiol-melamine [131, 132]. MeM was found to be the best one. The second layer was prepared by thermal vapor deposition. Generally, melamine powder is placed at the bottom of the glass tube, and the as-prepared MeM films are placed at the top of the glass tube. The air in the tube is replaced by purging with N_2 gas to avoid oxidation of the precursors, and the tube is sealed with Al foil for thermal treatment. After polymerization, the PCN-based photoanode composed of two layers is obtained with formation of a type-II heterojunction [130], and thus, improved charge separation. The optimal photocurrent density of this two-layered photoanode is 118 μA cm^{-2} at 1.23 V (vs. RHE) under illumination with a sunlight stimulator. This work also demonstrates that the photocurrent density is boosted by adding a sacrificial agent as it allows overcoming sluggish surface oxygen evolution reaction. The photocurrent density is increased to 383 μA cm^{-2} by adding triethanolamine in the PEC system, and ca. 30% of photocurrent density retained after 9 h with added triethanolamine, which illustrates long-term stability of the photoanode [130].

Note that multilayered photoanodes generally possess an extended working life with respect to the single-layered one. However, the reduction and oxidation reaction do not reach unity, revealing that a large portion of the harvested light energy is released in the other forms that decrease performance of the films. Two aspects can be considered as the main obstacles, which researchers need to overcome to improve performance of PCN-based photoanodes in oxygen evolution reaction. The first issue is related to the abundant defects of the PCN films, which act as recombination centers for the photogenerated charges. Therefore, improving crystallinity would sufficiently improve performance of the photoanode and its stability. The second issue is the sluggish kinetic of the surface oxygen evolution reaction. An appropriate cocatalyst is required to guide the photogenerated charges to promote oxygen evolution reaction. However, to the best of our knowledge, literature still lacks an example of crystalline PCN films for application in oxygen evolution reaction.

There are several examples of loading a cocatalyst on PCN films. Phosphate ions were loaded on the PCN film, which serves as a photoanode [133]. Due to the abundant vacancies present in a PCN film, phosphate ion anchors to the film by immersing it into phosphate solution [134]. In this system, phosphate ion acts as a cocatalyst to promote water oxidation reaction. The surface properties of the phosphorylated PCN films were investigated by XPS and UPS. By phosphorylation, the valence band is shifted by ca. 1.0 eV, as shown in Fig. 3.17a. In other words, the phosphate ion stabilizes the valence band, as illustrated in the band structure of the phosphorylated PCN photoanode in Fig. 3.17c. When the phosphate ion is removed by sputtering with an Ar$^+$ ion gun, which is observed as disappearance of the XPS P 2p peak (Fig. 3.17b), the valence band potential decreases gradually and eventually reaches the same value as in pristine PCN film. The results reveal that the phosphate ions on the PCN surface increase the oxidative power of the photogenerated holes, and thus compensates for the kinetic difficult oxygen evolution reaction to promote the performance. As a result, the optimal photocurrent density is ca. 120 μA cm^{-2}, which is two times higher than that of the pristine PCN photoanode.

A metal organic framework, Ni/Fe-MIL-53, was deposited at PCN film to form a two-layered photoanode for oxygen evolution reaction [135]. The bottom porous PCN films were prepared by the doctor blade approach using the supramolecular complex as a precursor, followed by thermal polymerization. Ni/Fe-MIL-53 nanoneedles grew on the surface of the PCN film under the solvothermal conditions. By this approach, the Ni/Fe-MIL-53 needles are anchored strongly to the surface of PCN. As a result, the two-layered photoanode exhibits an optimal photocurrent density of 472 $\pm$ 20 μA cm^{-2}

a)

b)

c)

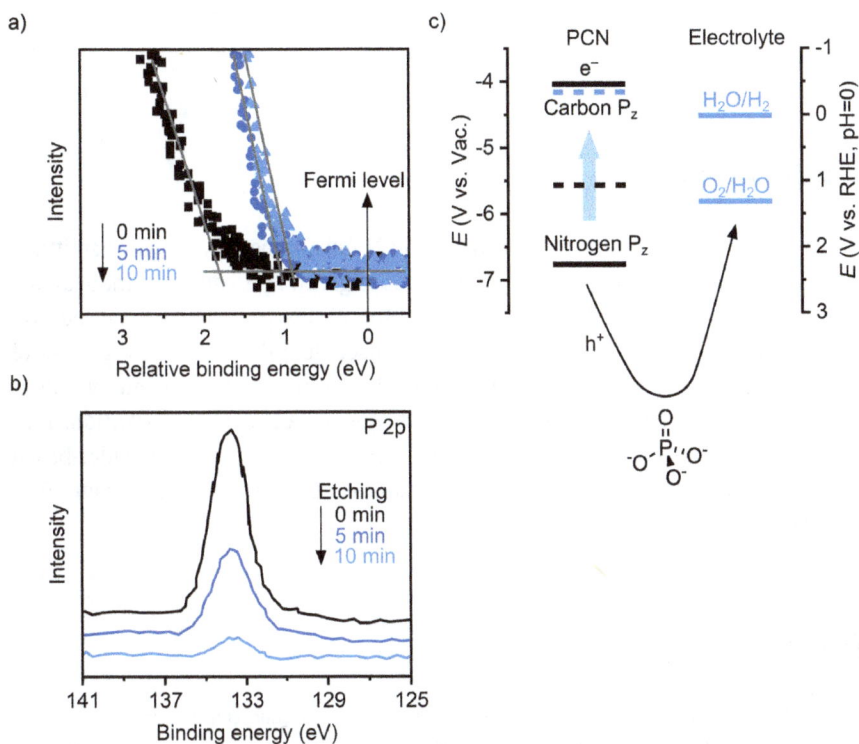

Fig. 3.17: (a) UPS and (b) XPS P2p spectra of the phosphorylated PCN films with different etching time. (c) Illustration of the band structure of the phosphorylated PCN photoanode (reproduced with permission from ref. [133]. Copyright 2019, Wiley-VCH Verlag GmbH & Co. KGaA, Weinheim).

at 1.23 V (vs. RHE) under illumination with a sunlight stimulator with an exceptionally low onset potential of ca. 0.034 V (vs. RHE). The photoanode retains its photocurrent density even after 35 h, without noticeable degradation. The Faradic efficiency for oxygen evolution reaction and hydrogen evolution reaction are $7.8 \pm 3\%$ and $83.0 \pm 5\%$, respectively. In this system, Ni/Fe-MIL-53 nanoneedles act as a cocatalyst to facilitate the oxygen evolution reaction. By overcoming the kinetic difficulty of the surface oxygen evolution reaction, both the conversion performance and stability of PCN are improved. Compared with powder-based oxygen evolution reaction system, the number of studies employing loading of cocatalyst on polymer photoanode for PEC water splitting is still limited, and in these studies, the cocatalyst improved the performance of the polymer photoelectrode only modestly. This result can be attributed to the fact that the applied voltage bias not only promotes separation of the exciton and charge transfer, but also compensates for the kinetic hindrances of the surface reaction. Therefore, the role of cocatalyst is less significant, compared to purely photocatalytic oxygen evolution reaction using catalyst dispersion. Nevertheless, by selecting an appropriate cocatalyst, the loading approach can be expected to improve the performance of polymer photoanode.

3.8 Summary

This chapter has focused on the application of PCN for photocatalytic water splitting. Water splitting is an appealing approach to produce green hydrogen fuel that can act as an ideal replacement for fossil fuels in the future. In this section, the basic requirements of water splitting by photocatalysis were discussed; the general properties of semiconducting PCN were discussed mainly in the context of the thermodynamics and kinetics. Subsequently, the diverse approaches of PCN copolymerization, PCN doped with metals and crystalline PCN were discussed in the context of powder-based photocatalytic water splitting reaction. In addition, water splitting by PEC using PCN was also discussed.

References

[1]   Wu XF, Chen GQ. Global primary energy use associated with production, consumption and international trade. Energy Policy 2017, 111, 85–94.
[2]   Marcus RJ. Chemical conversion of solar energy. Science 1956, 123(3193), 399–405.
[3]   Fang Y, Zheng Y, Fang T, Chen Y, Zhu Y, Liang Q, et al. Photocatalysis: An overview of recent developments and technological advancements. Sci China Chem 2020, 63(2), 149–81.
[4]   Fang Y, Wang X. Metal-free boron-containing heterogeneous catalysts. Angew Chem Int Ed 2017, 56(49), 15506–18.
[5]   Hou Y, Zhu Z, Xu Y, Guo F, Zhang J, Wang X. Efficient photoelectrochemical hydrogen production over p-Si nanowire arrays coupled with molybdenum sulfur clusters. Int J Hydrogen Energy 2017, 42(5), 2832–8.
[6]   Fujishima A, Honda K. Electrochemical photolysis of water at a semiconductor electrode. Nature 1972, 238(5358), 37–8.
[7]   Hisatomi T, Kubota J, Domen K. Recent advances in semiconductors for photocatalytic and photoelectrochemical water splitting. Chem Soc Rev 2014, 43(22), 7520–35.
[8]   Wang Q, Domen K. Particulate photocatalysts for light-driven water splitting: Mechanisms, challenges, and design strategies. Chem Rev 2020, 120(2), 919–85.
[9]   Chen X, Shen S, Guo L, Mao SS. Semiconductor-based photocatalytic hydrogen generation. Chem Rev 2010, 110(11), 6503–70.
[10]   Dahlberg R. Replacement of fossil fuels by hydrogen. Int J Hydrogen Energy 1982, 7(2), 121–42.
[11]   Sobrino FH, Monroy CR, Pérez JLH. Critical analysis on hydrogen as an alternative to fossil fuels and biofuels for vehicles in Europe. Renew Sustain Energy Rev 2010, 14(2), 772–80.
[12]   Zhou B, Song J, Zhou H, Wu T, Han B. Using the hydrogen and oxygen in water directly for hydrogenation reactions and glucose oxidation by photocatalysis. Chem Sci 2016, 7(1), 463–8.
[13]   Li W. Highly selective olefin production from CO_2 hydrogenation on Fe catalysts enabled by a subtle synergy between Mn and Na additives. Acta Phys Chim Sin 2021, 37(5), 2010062.

[14]   Lyu H, Hu B, Liu G, Hong X, Zhuang L. Inverse decoration of ZnO on small-sized Cu/SiO_2 with controllable Cu-ZnO interaction for CO_2 hydrogenation to produce methanol. Acta Phys Chim Sin 2020, 36(11), 1911008.

[15]   Wang X, Maeda K, Chen X, Takanabe K, Domen K, Hou Y, et al. Polymer semiconductors for artificial photosynthesis: Hydrogen evolution by mesoporous graphitic carbon nitride with visible light. J Am Chem Soc 2009, 131(5), 1680–1.

[16]   Fang Y, Hou Y, Fu X, Wang X. Semiconducting polymers for oxygen evolution reaction under light illumination. Chem Rev 2022, 122(3), 4204–56.

[17]   Lan ZA, Fang Y, Zhang Y, Wang X. Photocatalytic oxygen evolution from functional triazine-based polymers with tunable band structures. Angew Chem Int Ed 2018, 57(2), 470–4.

[18]   Teng ZY, Cai W, Liu SX, Wang CY, Zhang QT, Su CL, et al. Bandgap engineering of polymetric carbon nitride copolymerized by 2,5,8-triamino-tri-s-triazine (Melem) and barbituric acid for efficient nonsacrificial photocatalytic H_2O_2 production. Appl Catal: B 2020, 271, 12.

[19]   Kong C, Li Z, Lu G. The dual functional roles of Ru as co-catalyst and stabilizer of dye for photocatalytic hydrogen evolution. Int J Hydrogen Energy 2015, 40(17), 5824–30.

[20]   Kang HW, Lim SN, Song D, Park SB. Organic-inorganic composite of g-C_3N_4–$SrTiO_3$:Rh photocatalyst for improved H_2 evolution under visible light irradiation. Int J Hydrogen Energy 2012, 37(16), 11602–10.

[21]   Li X, Bi W, Zhang L, Tao S, Chu W, Zhang Q, et al. Single-Atom Pt as co-catalyst for enhanced photocatalytic H_2 evolution. Adv Mater 2016, 28(12), 2427–31.

[22]   Shen P, Zhao S, Su D, Li Y, Orlov A. Outstanding activity of sub-nm Au clusters for photocatalytic hydrogen production. Appl Catal: B 2012, 126, 153–60.

[23]   Zang S, Zhang G, Lan Z-A, Zheng D, Wang X. Enhancement of photocatalytic H_2 evolution on pyrene-based polymer promoted by MoS_2 and visible light. Appl Catal: B 2019, 251, 102–11.

[24]   Wang X, Maeda K, Thomas A, Takanabe K, Xin G, Carlsson JM, et al. A metal-free polymeric photocatalyst for hydrogen production from water under visible light. Nat Mater 2009, 8(1), 76–80.

[25]   Zhang G, Wang X. A facile synthesis of covalent carbon nitride photocatalysts by Co-polymerization of urea and phenylurea for hydrogen evolution. J Catal 2013, 307, 246–53.

[26]   Dharmapurikar SS, Arulkashmir A, Mahale RY, Chini MK. Synthesis of amphiphilic isoindigo co-polymers for organic field effect transistors: A comparative study. J Appl Polym Sci 2017, 134(43), 45461.

[27]   Bai Y, Woods DJ, Wilbraham L, Aitchison CM, Zwijnenburg MA, Sprick RS, et al. Hydrogen evolution from water using heteroatom substituted fluorene conjugated co-polymers. J Mater Chem A 2020, 8(17), 8700–5.

[28]   Zhang J, Zhang G, Chen X, Lin S, Möhlmann L, Dołęga G, et al. Co-monomer control of carbon nitride semiconductors to optimize hydrogen evolution with visible light. Angew Chem Int Ed 2012, 51(13), 3183–7.

[29]   Li YF, Jin RX, Li GJ, Liu XC, Yu M, Xing Y, et al. Preparation of phenyl group functionalized g-C_3N_4 nanosheets with extended electron delocalization for enhanced visible-light photocatalytic activity. New J Chem 2018, 42(9), 6756–62.

[30]   Liu J, Yu Y, Qi R, Cao C, Liu X, Zheng Y, et al. Enhanced electron separation on in-plane benzene-ring doped g-C_3N_4 nanosheets for visible light photocatalytic hydrogen evolution. Appl Catal: B 2019, 244, 459–64.

[31]   Yang C, Wang B, Zhang L, Yin L, Wang X. Synthesis of layered carbonitrides from biotic molecules for photoredox transformations. Angew Chem Int Ed 2017, 56(23), 6627–31.

[32]   Ou H, Chen X, Lin L, Fang Y, Wang X. Biomimetic donor–acceptor motifs in conjugated polymers for promoting exciton splitting and charge separation. Angew Chem 2018, 57(28), 8729–33.

[33]   Li Y, Wang Z, Xia T, Ju H, Zhang K, Long R, et al. Implementing metal-to-ligand charge transfer in organic semiconductor for improved visible-near-infrared photocatalysis. Adv Mater 2016, 28(32), 6959–65.

[34]   Jin X, Wang R, Zhang L, Si R, Shen M, Wang M, et al. Electron configuration modulation of nickel single atoms for elevated photocatalytic hydrogen evolution. Angew Chem Int Ed 2020, 59(17), 6827–31.

[35]   Jiang W, Zhao Y, Zong X, Nie H, Niu L, An L, et al. Photocatalyst for high-performance H_2 production: Ga-doped polymeric carbon nitride. Angew Chem Int Ed 2021, 60(11), 6124–9.

[36]   Li Y, Wang Y, Dong C-L, Huang Y-C, Chen J, Zhang Z, et al. Single-atom nickel terminating sp^2 and sp^3 nitride in polymeric carbon nitride for visible-light photocatalytic overall water splitting. Chem Sci 2021, 12(10), 3633–43.

[37]   Pan Z, Zhao M, Zhuzhang H, Zhang G, Anpo M, Wang X. Gradient Zn-doped poly heptazine imides integrated with a van der Waals homojunction boosting visible light-driven water oxidation activities. ACS Catal 2021, 11(21), 13463–71.

[38]   Chong SY, Jones JTA, Khimyak YZ, Cooper AI, Thomas A, Antonietti M, et al. Tuning of gallery heights in a crystalline 2D carbon nitride network. J Mater Chem A 2013, 1(4), 1102–7.

[39]   Bojdys MJ, Severin N, Rabe JP, Cooper AI, Thomas A, Antonietti M. Exfoliation of crystalline 2D carbon nitride: Thin sheets, scrolls and bundles *via* mechanical and chemical routes. Macromol Rapid Commun 2013, 34(10), 850–4.

[40]   Zhang M, Wang X. Two dimensional conjugated polymers with enhanced optical absorption and charge separation for photocatalytic hydrogen evolution. Energy Environ Sci 2014, 7(6), 1902–6.

[41]   Schwinghammer K, Mesch MB, Duppel V, Ziegler C, Senker J, Lotsch BV. Crystalline carbon nitride nanosheets for improved visible-light hydrogen evolution. J Am Chem Soc 2014, 136(5), 1730–3.

[42]   Bhunia MK, Yamauchi K, Takanabe K. Harvesting solar light with crystalline carbon nitrides for efficient photocatalytic hydrogen evolution. Angew Chem Int Ed 2014, 53(41), 11001–5.

[43]   Lin L, Wang C, Ren W, Ou H, Zhang Y, Wang X. Photocatalytic overall water splitting by conjugated semiconductors with crystalline poly(triazine imide) frameworks. Chem Sci 2017, 8(8), 5506–11.

[44]   Gao H, Yan S, Wang J, Huang YA, Wang P, Li Z, et al. Towards efficient solar hydrogen production by intercalated carbon nitride photocatalyst. Phys Chem Chem Phys 2013, 15(41), 18077–84.

[45]   Lin L, Ou H, Zhang Y, Wang X. Tri-*s*-triazine-based crystalline graphitic carbon nitrides for highly efficient hydrogen evolution photocatalysis. ACS Catal 2016, 6(6), 3921–31.

[46]   Ou H, Lin L, Zheng Y, Yang P, Fang Y, Wang X. Tri-*s*-triazine-based crystalline carbon nitride nanosheets for an improved hydrogen evolution. Adv Mater 2017, 29(22), 1700008.

[47]   Zhang G, Li G, Lan Z-A, Lin L, Savateev A, Heil T, et al. Optimizing optical absorption, exciton dissociation, and charge transfer of a polymeric carbon nitride with ultrahigh solar hydrogen production activity. Angew Chem Int Ed 2017, 56(43), 13445–9.

[48]   Zhou Z, Wang J, Yu J, Shen Y, Li Y, Liu A, et al. Dissolution and liquid crystals phase of 2D polymeric carbon nitride. J Am Chem Soc 2015, 137(6), 2179–82.

[49]   Wang J, Shen Y, Li Y, Liu S, Zhang Y. Crystallinity modulation of layered carbon nitride for enhanced photocatalytic activities. Chem – Eur J 2016, 22(35), 12449–54.

[50]   Yuan Y-P, Yin L-S, Cao S-W, Gu L-N, Xu G-S, Du P, et al. Microwave-assisted heating synthesis: A general and rapid strategy for large-scale production of highly crystalline g-C_3N_4 with enhanced photocatalytic H_2 production. Green Chem 2014, 16(11), 4663–8.

[51]   Zhang S, Ye H, Hua J, Tian H. Recent advances in dye-sensitized photoelectrochemical cells for water splitting. EnergyChem 2019, 1(3), 100015.

[52]   Wei Z-Q, Hou S, Lin X, Xu S, Dai X-C, Li Y-H, et al. Unexpected boosted solar water oxidation by nonconjugated polymer-mediated tandem charge transfer. J Am Chem Soc 2020, 142(52), 21899–912.

[53]   Su JY, Geng P, Li XY, Chen GH. Graphene-linked graphitic carbon nitride/TiO_2 nanowire arrays heterojunction for efficient solar-driven water splitting. J Appl Electrochem 2016, 46(8), 807–17.

[54] Yang Y, Wang S, Jiao Y, Wang Z, Xiao M, Du A, et al. An unusual red carbon nitride to boost the photoelectrochemical performance of wide bandgap photoanodes. Adv Funct Mater 2018, 28(47), 1805698.

[55] Zhou X, Jin B, Li L, Peng F, Wang H, Yu H, et al. A carbon nitride/TiO_2 nanotube array heterojunction visible-light photocatalyst: Synthesis, characterization, and photoelectrochemical properties. J Mater Chem 2012, 22(34), 17900–5.

[56] Liu C, Wang F, Zhang J, Wang K, Qiu Y, Liang Q, et al. Efficient photoelectrochemical water splitting by g-C_3N_4/TiO_2 nanotube array heterostructures. Nano-Micro Lett 2018, 10(2), 37.

[57] Xiao L, Liu T, Zhang M, Li Q, Yang J. Interfacial construction of zero-dimensional/one-dimensional g-C_3N_4 nanoparticles/TiO_2 nanotube arrays with Z-scheme heterostructure for improved photoelectrochemical water splitting. ACS Sustainable Chem Eng 2019, 7(2), 2483–91.

[58] Kong W, Zhang X, Chang B, Guo Y, Li Y, Zhang S, et al. TiO_2 nanorods Co-decorated with metal-free carbon materials for boosted photoelectrochemical water oxidation. ChemElectroChem 2020, 7(3), 792–9.

[59] Wang L, Wang R, Feng L, Liu Y. Coupling TiO_2 nanorods with g-CN using modified physical vapor deposition for efficient photoelectrochemical water oxidation. J Am Ceram Soc 2020, 103(11), 6272–9.

[60] Christoforidis KC, Montini T, Fittipaldi M, Jaén JJD, Fornasiero P. Photocatalytic hydrogen production by boron modified TiO_2/carbon nitride heterojunctions. ChemCatChem 2019, 11(24), 6408–16.

[61] Xiao JG, Zhang XL, Li YD. A ternary g-C_3N_4/Pt/ZnO photoanode for efficient photoelectrochemical water splitting. Int J Hydrogen Energy 2015, 40(30), 9080–7.

[62] Fang Y, Xu Y, Li X, Ma Y, Wang X. Coating polymeric carbon nitride photoanodes on conductive Y: ZnO nanorod arrays for overall water splitting. Angew Chem Int Ed 2018, 57(31), 9749–53.

[63] Liu C, Wu P, Wu K, Meng G, Wu J, Hou J, et al. Advanced bi-functional CoPi co-catalyst-decorated g-C_3N_4 nanosheets coupled with ZnO nanorod arrays as integrated photoanodes. Dalton Trans 2018, 47(18), 6605–14.

[64] Yi S-S, Yan J-M, Jiang Q. Carbon quantum dot sensitized integrated Fe_2O_3@g-C_3N_4 core–shell nanoarray photoanode towards highly efficient water oxidation. J Mater Chem A 2018, 6(21), 9839–45.

[65] Arora P, Singh AP, Mehta BR, Basu S. Metal doped tubular carbon nitride (tC_3N_4) based hematite photoanode for enhanced photoelectrochemical performance. Vacuum 2017, 146, 570–7.

[66] Zhan F, Xie R, Li W, Li J, Yang Y, Li Y, et al. In situ synthesis of g-C_3N_4/WO_3 heterojunction plates array films with enhanced photoelectrochemical performance. RSC Adv 2015, 5(85), 69753–60.

[67] Li H, Zhao F, Zhang J, Luo L, Xiao X, Huang Y, et al. A g-C_3N_4/WO_3 photoanode with exceptional ability for photoelectrochemical water splitting. Mater Chem Front 2017, 1(2), 338–42.

[68] Gopalakrishnan S, Bhalerao GM, Jeganathan K. $SrTiO_3$ NPs/g-C_3N_4 NSs coupled Si NWs based hybrid photocathode for visible light driven photoelectrochemical water reduction. ACS Sustainable Chem Eng 2019, 7(16), 13911–19.

[69] Yang F, Kuznietsov V, Lublow M, Merschjann C, Steigert A, Klaer J, et al. Solar hydrogen evolution using metal-free photocatalytic polymeric carbon nitride/$CuInS_2$ composites as photocathodes. J Mater Chem A 2013, 1(21), 6407–15.

[70] Wang L, Si W, Tong Y, Hou F, Pergolesi D, Hou J, et al. Graphitic carbon nitride (g-C_3N_4)-based nanosized heteroarrays: Promising materials for photoelectrochemical water splitting. Carbon Energy 2020, 2(2), 223–50.

[71] Chen S-h, Wang J-j, Huang J, Li Q-x. g-C_3N_4/SnS_2 heterostructure: A promising water splitting photocatalyst. Chin J Chem Phys 2017, 30(1), 36–42.

[72] Wang B, Yu H, Quan X, Chen S. Ultra-thin g-C_3N_4 nanosheets wrapped silicon nanowire array for improved chemical stability and enhanced photoresponse. Mater Res Bull 2014, 59, 179–84.

[73]   Zou X, Sun Z, Hu YH. g-C$_3$N$_4$-based photoelectrodes for photoelectrochemical water splitting: A review. J Mater Chem A 2020, 8(41), 21474–502.

[74]   Bledowski M, Wang LD, Ramakrishnan A, Khavryuchenko OV, Khavryuchenko VD, Ricci PC, et al. Visible-light photocurrent response of TiO$_2$-polyheptazine hybrids: Evidence for interfacial charge-transfer absorption. Phys Chem Chem Phys 2011, 13(48), 21511–9.

[75]   Beranek R, Kisch H. Tuning the optical and photoelectrochemical properties of surface-modified TiO$_2$. Photochem Photobiol Sci 2008, 7(1), 40–8.

[76]   Liu C, Qiu Y, Zhang J, Liang Q, Mitsuzaki N, Chen Z. Construction of CdS quantum dots modified g-C$_3$N$_4$/ZnO heterostructured photoanode for efficient photoelectrochemical water splitting. J Photochem Photobiol A 2019, 371, 109–17.

[77]   Fang Y, Commandeur D, Lee WC, Chen Q. Transparent conductive oxides in photoanodes for solar water oxidation. Nanoscale Adv 2020, 2(2), 626–32.

[78]   Zhang J, Chen X, Takanabe K, Maeda K, Domen K, Epping JD, et al. Synthesis of a carbon nitride structure for visible-light catalysis by copolymerization. Angew Chem Int Ed 2010, 49(2), 441–4.

[79]   Xu J, Herraiz-Cardona I, Yang X, Gimenez S, Antonietti M, Shalom M. The complex role of carbon nitride as a sensitizer in photoelectrochemical cells. Adv Opt Mater 2015, 3(8), 1052–8.

[80]   Wang J, Su F-Y, Zhang W-D. Preparation and enhanced visible light photoelectrochemical activity of g-C$_3$N$_4$/ZnO nanotube arrays. J Solid State Electrochem 2014, 18(10), 2921–9.

[81]   Commandeur D, Brown G, McNulty P, Dadswell C, Spencer J, Chen Q. Yttrium-doped ZnO nanorod arrays for increased charge mobility and carrier density for enhanced solar water splitting. J Phys Chem C 2019, 123(30), 18187–97.

[82]   Safaei J, Ullah H, Mohamed NA, Mohamad Noh MF, Soh MF, Tahir AA, et al. Enhanced photoelectrochemical performance of Z-scheme g-C$_3$N$_4$/BiVO$_4$ photocatalyst. Appl Catal: B 2018, 234, 296–310.

[83]   Seabold JA, Choi K-S. Efficient and stable photo-oxidation of water by a bismuth vanadate photoanode coupled with an iron oxyhydroxide oxygen evolution catalyst. J Am Chem Soc 2012, 134(4), 2186–92.

[84]   Wu H, Xu M, Da P, Li W, Jia D, Zheng G. WO$_3$–reduced graphene oxide composites with enhanced charge transfer for photoelectrochemical conversion. Phys Chem Chem Phys 2013, 15(38), 16138–42.

[85]   Li Y, Wei X, Yan X, Cai J, Zhou A, Yang M, et al. Construction of inorganic–organic 2D/2D WO$_3$/g-C$_3$N$_4$ nanosheet arrays toward efficient photoelectrochemical splitting of natural seawater. Phys Chem Chem Phys 2016, 18(15), 10255–61.

[86]   Kudo A, Ueda K, Kato H, Mikami I. Photocatalytic O$_2$ evolution under visible light irradiation on BiVO$_4$ in aqueous AgNO$_3$ solution. Catal Lett 1998, 53(3), 229–30.

[87]   Kim TW, Choi K-S. Nanoporous BiVO$_4$ photoanodes with dual-layer oxygen evolution catalysts for solar water splitting. Science 2014, 343(6174), 990–4.

[88]   Kim JH, Lee JS. Elaborately modified BiVO$_4$ photoanodes for solar water splitting. Adv Mater 2019, 31(20), 1806938.

[89]   Zhang B, Huang X, Zhang Y, Lu G, Chou L, Bi Y. Unveiling the activity and stability origin of BiVO$_4$ photoanodes with FeNi oxyhydroxides for oxygen evolution. Angew Chem Int Ed 2020, 59(43), 18990–5.

[90]   Cooper JK, Gul S, Toma FM, Chen L, Glans P-A, Guo J, et al. Electronic structure of monoclinic BiVO$_4$. Chem, Mater 2014, 26(18), 5365–73.

[91]   Feng C, Wang Z, Ma Y, Zhang Y, Wang L, Bi Y. Ultrathin graphitic C$_3$N$_4$ nanosheets as highly efficient metal-free cocatalyst for water oxidation. Appl Catal: B 2017, 205, 19–23.

[92]   Chen X, Jun Y-S, Takanabe K, Maeda K, Domen K, Fu X, et al. Ordered mesoporous SBA-15 type graphitic carbon nitride: A semiconductor host structure for photocatalytic hydrogen evolution with visible light. Chem, Mater 2009, 21(18), 4093–5.

[93] Jiang XX, De Hu X, Tarek M, Saravanan P, Alqadhi R, Chin SY, et al. Tailoring the properties of g-C_3N_4 with CuO for enhanced photoelectrocatalytic CO_2 reduction to methanol. J CO2 Util 2020, 40, 101222.

[94] Yang P, Zhuzhang H, Wang R, Lin W, Wang X. Carbon vacancies in a Melon polymeric matrix promote photocatalytic carbon dioxide conversion. Angew Chem Int Ed 2019, 58(4), 1134–7.

[95] Zhang J, Zhang M, Lin L, Wang X. Sol Processing of conjugated carbon nitride powders for thin-film fabrication. Angew Chem Int Ed 2015, 54(21), 6297–301.

[96] Chaudhary D, Kumar S, Khare N. Boosting the visible-light photoelectrochemical performance of C_3N_4 by coupling with TiO_2 and carbon nanotubes: An organic/inorganic hybrid photocatalyst nanocomposite for photoelectrochemical water spitting. Int J Hydrogen Energy 2020, 45(55), 30091–100.

[97] Gelderman K, Lee L, Donne SW. Flat-Band Potential of a Semiconductor: Using the Mott–Schottky Equation. J Chem Educ 2007, 84(4), 685.

[98] Sima M, Vasile E, Sima A, Preda N, Logofatu C. Graphitic carbon nitride based photoanodes prepared by spray coating method. Int J Hydrogen Energy 2019, 44(45), 24430–40.

[99] Guo B, Tian L, Xie W, Batool A, Xie G, Xiang Q, et al. Vertically aligned porous organic semiconductor nanorod array photoanodes for efficient charge utilization. Nano Lett 2018, 18(9), 5954–60.

[100] Liu J, Wang H, Chen ZP, Moehwald H, Fiechter S, van de Krol R, et al. Microcontact-printing-assisted access of graphitic carbon nitride films with favorable textures toward photoelectrochemical application. Adv Mater 2015, 27(4), 712–8.

[101] Xu J, Cao S, Brenner T, Yang X, Yu J, Antonietti M, et al. Supramolecular chemistry in molten sulfur: Preorganization effects leading to marked enhancement of carbon nitride photoelectrochemistry. Adv Funct Mater 2015, 25(39), 6265–71.

[102] Abisdris L, Tzadikov J, Karjule N, Azoulay A, Volokh M, Shalom M. Electrophoretic deposition of supramolecular complexes for the formation of carbon nitride films. Sustainable Energy Fuels 2020, 4(8), 3879–83.

[103] Peng G, Xing L, Barrio J, Volokh M, Shalom M. A General synthesis of porous carbon nitride films with tunable surface area and photophysical properties. Angew Chem Int Ed 2018, 57(5), 1186–92.

[104] Peng G, Volokh M, Tzadikov J, Sun J, Shalom M. Carbon nitride/reduced graphene oxide film with enhanced electron diffusion length: An efficient photo-electrochemical cell for hydrogen generation. Adv Energy Mater 2018, 8(23), 1800566.

[105] Christoforidis KC, Syrgiannis Z, La Parola V, Montini T, Petit C, Stathatos E, et al. Metal-free dual-phase full organic carbon nanotubes/g-C_3N_4 heteroarchitectures for photocatalytic hydrogen production. Nano Energy 2018, 50, 468–78.

[106] Peng G, Qin J, Volokh M, Liu C, Shalom M. Graphene oxide in carbon nitride: From easily processed precursors to a composite material with enhanced photoelectrochemical activity and long-term stability. J Mater Chem A 2019, 7(19), 11718–23.

[107] Qin J, Barrio J, Peng G, Tzadikov J, Abisdris L, Volokh M, et al. Direct growth of uniform carbon nitride layers with extended optical absorption towards efficient water-splitting photoanodes. Nat Commun 2020, 11(1), 4701.

[108] Peng G, Qin J, Volokh M, Shalom M. Freestanding hierarchical carbon nitride/carbon-paper electrode as a photoelectrocatalyst for water splitting and dye degradation. ACS Appl Mater Interfaces 2019, 11(32), 29139–46.

[109] Peng G, Albero J, Garcia H, Shalom M. A Water-splitting carbon nitride photoelectrochemical cell with efficient charge separation and remarkably low onset potential. Angew Chem Int Ed 2018, 57(48), 15807–11.

[110] He Y, Xu X-Q, Lv S, Liao H, Wang Y. Dark ionic liquid for flexible optoelectronics. Langmuir 2019, 35(5), 1192–8.

[111] Lee C-H, Kim Y-J, Hong YJ, Jeon S-R, Bae S, Hong BH, et al. Flexible optoelectronics: Flexible inorganic nanostructure light-emitting diodes fabricated on graphene films Adv. Mater. 40/2011. Adv Mater 2011, 23(40), 4591.

[112] Luo B, Fang Y, Li J, Huang Z, Hu B, Zhou J. Improved stability of metal nanowires *via* electron beam irradiation induced surface passivation. ACS Appl Mater Interfaces 2019, 11(13), 12195–201.

[113] Fang Y, Li X, Wang X. Synthesis of polymeric carbon nitride films with adhesive interfaces for solar water splitting devices. ACS Catal 2018, 8(9), 8774–80.

[114] Fang Y, Li X, Wang Y, Giordano C, Wang X. Gradient sulfur doping along polymeric carbon nitride films as visible light photoanodes for the enhanced water oxidation. Appl Catal: B 2020, 268, 118398.

[115] Li X, Cheng Z, Fang Y, Fu X, Wang X. *In situ* synthesis of phosphorus-doped polymeric carbon nitride sheets for photoelectrochemical water oxidation. Sol RRL 2020, 4(8), 2000168.

[116] Guo S, Tang Y, Xie Y, Tian C, Feng Q, Zhou W, et al. P-doped tubular g-C_3N_4 with surface carbon defects: Universal synthesis and enhanced visible-light photocatalytic hydrogen production. Appl Catal: B 2017, 218, 664–71.

[117] Patel MA, Luo F, Savaram K, Kucheryavy P, Xie Q, Flach C, et al. P and S dual-doped graphitic porous carbon for aerobic oxidation reactions: Enhanced catalytic activity and catalytic sites. Carbon 2017, 114, 383–92.

[118] Bian JC, Li Q, Huang C, Li JF, Guo Y, Zaw M, et al. Thermal vapor condensation of uniform graphitic carbon nitride films with remarkable photocurrent density for photoelectrochemical applications. Nano Energy 2015, 15, 353–61.

[119] Costa L, Camino G. Thermal behaviour of melamine. J Therm Anal 1988, 34(2), 423–9.

[120] Xiong W, Chen S, Huang M, Wang Z, Lu Z, Zhang R-Q. Crystal-face tailored graphitic carbon nitride films for high-performance photoelectrochemical cells. ChemSusChem 2018, 11(15), 2497–501.

[121] Lan H, Li L, An X, Liu F, Chen C, Liu H, et al. Microstructure of carbon nitride affecting synergetic photocatalytic activity: Hydrogen bonds *vs*. structural defects. Appl Catal: B 2017, 204, 49–57.

[122] Wang X, Zhou C, Shi R, Liu Q, Waterhouse GIN, Wu L, et al. Supramolecular precursor strategy for the synthesis of holey graphitic carbon nitride nanotubes with enhanced photocatalytic hydrogen evolution performance. Nano Res 2019, 12(9), 2385–9.

[123] Yu H, Shi R, Zhao Y, Bian T, Zhao Y, Zhou C, et al. Alkali-assisted synthesis of nitrogen deficient graphitic carbon nitride with tunable band structures for efficient visible-light-driven hydrogen evolution. Adv Mater 2017, 29(16), 1605148.

[124] Luo L, Gong Z, Ma J, Wang K, Zhu H, Li K, et al. Ultrathin sulfur-doped holey carbon nitride nanosheets with superior photocatalytic hydrogen production from water. Appl Catal, B 2021, 284, 119742.

[125] Luo L, Wang K, Gong Z, Zhu H, Ma J, Xiong L, et al. Bridging-nitrogen defects modified graphitic carbon nitride nanosheet for boosted photocatalytic hydrogen production. Int J Hydrogen Energy 2021, 46(53), 27014–25.

[126] Ruan QS, Luo WJ, Xie JJ, Wang YO, Liu X, Bai ZM, et al. A nanojunction polymer photoelectrode for efficient charge transport and separation. Angew Chem Int Ed 2017, 56(28), 8221–5.

[127] Ishiguro Y, Inagi S, Fuchigami T. Gradient doping of conducting polymer films by means of bipolar electrochemistry. Langmuir 2011, 27(11), 7158–62.

[128] Luo Z, Li C, Liu S, Wang T, Gong J. Gradient doping of phosphorus in Fe_2O_3 nanoarray photoanodes for enhanced charge separation. Chem Sci 2017, 8(1), 91–100.

[129] Luan P, Meng Q, Wu J, Li Q, Zhang X, Zhang Y, et al. Unique layer-doping-induced regulation of charge behavior in metal-free carbon nitride photoanodes for enhanced performance. ChemSusChem 2020, 13(2), 328–33.

[130] Xia J, Karjule N, Abisdris L, Volokh M, Shalom M. Controllable synthesis of carbon nitride films with type-II heterojunction for efficient photoelectrochemical cells. Chem, Mater 2020, 32(13), 5845–53.

[131] Karjule N, Barrio J, Xing L, Volokh M, Shalom M. Highly efficient polymeric carbon nitride photoanode with excellent electron diffusion length and hole extraction properties. Nano Lett 2020, 20(6), 4618–24.
[132] Karjule N, Barrio J, Tashakory A, Shalom M. Bismuthiol-mediated synthesis of ordered carbon nitride nanosheets with enhanced photocatalytic performance. Sol RRL 2020, 4(8), 2000017.
[133] Fang Y, Li X, Wang X. Phosphorylation of polymeric carbon nitride photoanodes with increased surface valence electrons for solar water splitting. ChemSusChem 2019, 12(12), 2605–8.
[134] Shi L, Chang K, Zhang H, Hai X, Yang L, Wang T, et al. Drastic enhancement of photocatalytic activities over phosphoric acid protonated porous g-C_3N_4 nanosheets under visible light. Small 2016, 12(32), 4431–9.
[135] Karjule N, Singh C, Barrio J, Tzadikov J, Liberman I, Volokh M, et al. Carbon nitride-based photoanode with enhanced photostability and water oxidation kinetics. Adv Funct Mater 2021, 31(25), 2101724.

Han Li and Shaowen Cao*

Chapter 4
CO$_2$ fixation and transformation technology with carbon nitride

4.1 Introduction

The rapid population growth and increased fossil fuel consumption have not only re-sulted in shortage of fossil reserves, but also led to tremendous increase in CO$_2$ emis-sion, causing serious greenhouse effect [1, 2]. Solar energy is recognized as one of the potential energy resources as it is free, clean, abundant and sustainable. Inspired by natural photosynthesis, the conversion of CO$_2$ into chemical fuels, driven by photoca-talytic reaction, is particularly appealing, which could address energy shortage and environmental issues.

In 1978, Halmann reported that the photo-assisted CO$_2$ reduction could be achieved over p-GaAs [3]. In the next year, Inoue et al. studied the photocatalytic CO$_2$ reduction reaction on various materials such as WO$_3$, TiO$_2$, ZnO, CdS, GaP and SiC [4]. Since then, various semiconductors, such as metal oxides (TiO$_2$, ZnO, WO$_3$ and Cu$_2$O), metal chalco-genides (CdS, ZnS and ZnIn$_2$S$_4$), Bi-based semiconductors (Bi$_2$WO$_6$ Bi$_2$MoO$_6$ and BiOCl), metal–organic frameworks and inorganic perovskite have been developed as efficient photocatalysts for CO$_2$ reduction [5–8].

Among the various photocatalysts discussed earlier, polymeric carbon nitride (PCN) is considered a promising photocatalyst because of its excellent stability, suitable band structure, low-cost and facile preparation [9–11]. PCN was first reported as a pho-tocatalyst for CO$_2$ reduction by Zhang and Dong in 2011 [12]. Since then, various efforts have been devoted to improve the photocatalytic performance of PCN. Despite achiev-ing great progress, PCN-based photocatalysis for CO$_2$ reduction still suffers from moder-ate efficiency [13]. We summarize here the recent advancements in using PCN-based photocatalysts for CO$_2$ reduction, which is hoped to provide some understanding and inspire more intriguing ideas on designing highly efficient PCN-based photocatalysts for solar energy conversion.

In this chapter, we will first introduce the fundamentals and possible mecha-nisms of photocatalytic CO$_2$ reduction. Special emphasis is placed on the strategies of PCN-based photocatalysts preparation for improving CO$_2$ reduction performance,

---

***Corresponding author: Shaowen Cao**, State Key Laboratory of Advanced Technology for Materials Synthesis and Processing, Wuhan University of Technology, Wuhan 430070, P. R. China,
e-mail: swcao@whut.edu.cn
Han Li, State Key Laboratory of Advanced Technology for Materials Synthesis and Processing, Wuhan University of Technology, Wuhan 430070, P. R. China

https://doi.org/10.1515/9783110746976-004

such as nanostructure design, defect engineering, crystallinity modulation and hetero-structure construction, followed by the advanced characterization techniques to get insights of the mechanistic aspects of photocatalytic CO_2 reduction over PCN-based photocatalysts. Finally, some future perspectives and challenges to PCN-based photo-catalysis for CO_2 reduction are discussed.

4.2 Theoretical foundation of photocatalytic CO_2 reduction

4.2.1 Band structure

The band structure, with the potentials of conduction band and valence band, is the key factor of semiconductor photocatalysis. First, the band gap is defined as the potential difference between the conduction band and valence band, which determines the light absorption capability of a photocatalyst. The band gap of pristine PCN is around 2.7 eV, corresponding to an absorption edge of 460 nm, which makes the material visible light active [9]. Second, the conduction and valence bands' potential values of a photocatalyst define the possible photoreaction thermodynamics, namely the reduction and oxidation capabilities. Specifically, to realize photocatalytic CO_2 reduction, the bottom of the conduction band must be more negative than the reduction potential of CO_2, while the top of the valence band must be more positive than the oxidation potential of the electron donor, ideally water.

It is well known that CO_2 is a stable linear molecule with high C—O bond energy of 750 kJ mol^{-1}, thus requiring a high energy input to cleave it [14]. Moreover, being in its highest oxidation state, +4, carbon in CO_2 can be reduced to products with various oxidation states, including C_1 products, such as CO (carbon in oxidation state +2), HCOOH (+2), HCHO (0), CH_3OH (–2) and CH_4 (–4), and C_2 products, such as CH_2CH_2 (–2, –2), CH_3CH_2OH (–3, –1) and CH_3COOH (–3, +3). The redox potentials of different CO_2 reduction reactions are listed in Tab. 4.1 [15, 16]. As shown in eq. (4.1), the potential of the direct single electron reduction of CO_2 to $CO_2{}^{\cdot-}$ is –1.9 V, which is, in comparison, a highly negative value. From the perspective of thermodynamics, this reaction is extremely unfavorable. All the candidate photocatalysts have a conduction band potential insufficiently negative to drive such single-electron reaction. To bypass such a large energy barrier, the proton-assisted multiple electron transfer process is an alternative and, consequently, the more advantageous route because of the lower reduction potential (eqs. (4.2)–(4.9)).

Tab. 4.1: Redox potential of different reactions that involve CO$_2$ and H$_2$O.

Equation	Reaction	E^0 (V) versus NHE pH = 7
(4.1)	$CO_2 + e^- \rightarrow CO_2^{\cdot-}$	−1.90
(4.2)	$CO_2 + 2H^+ + 2e^- \rightarrow HCOOH$	−0.61
(4.3)	$CO_2 + 2H^+ + 2e^- \rightarrow CO + H_2O$	−0.53
(4.4)	$CO_2 + 4H^+ + 4e^- \rightarrow HCHO + H_2O$	−0.48
(4.5)	$CO_2 + 6H^+ + 6e^- \rightarrow CH_3OH + H_2O$	−0.38
(4.6)	$CO_2 + 8H^+ + 8e^- \rightarrow CH_4 + 2H_2O$	−0.24
(4.7)	$2CO_2 + 12H^+ + 12e^- \rightarrow C_2H_4 + 4H_2O$	−0.34
(4.8)	$2CO_2 + 12H^+ + 12e^- \rightarrow C_2H_5OH + 3H_2O$	−0.33
(4.9)	$2CO_2 + 14H^+ + 14e^- \rightarrow C_2H_6 + 4H_2O$	−0.27
(4.10)	$2H^+ + 2e^- \rightarrow H_2$	−0.42
(4.11)	$2H_2O + 4h^+ \rightarrow O_2 + 4H^+$	+0.81
(4.12)	$2H_2O + 2h^+ \rightarrow H_2O_2 + 2H^+$	+1.36
(4.13)	$H_2O + h^+ \rightarrow {}^\bullet OH + H^+$	+2.32
(4.14)	$O_2 + 2e^- + 2H^+ \rightarrow H_2O_2$	+0.26

4.2.2 Charge separation and transfer

The charge separation and transfer play a critical role in determining the catalytic efficiency of a semiconductor. The low efficiency of most reported photocatalysts is largely due to the interface and bulk recombination of the photogenerated charges during migration to surface reactive sites. Recombination mostly occurs at undesirable structural defects such as grain boundaries, dangling bonds and vacancies. Thus, the efficiency of charge separation and transfer depend on the structure and electronic properties of PCN. Therefore, many strategies have been developed to improve the utilization of the charge carriers, including nanostructure design, heterojunction construction, molecular structure engineering and co-catalysts engineering.

It is well known that high crystallinity improves the bulk properties of a semiconductor, such as charge migration rate and electronic structure. However, from the perspective of catalysis, the active materials also need appropriate surface defects and edge sites because these surface features could promote the charge transfer to the solution and provide the adsorption sites for the substrate. Moreover, the charge separation and transfer are also affected by the band bending as a product of interfacial electric fields, which drives the electron and hole transfer in opposite directions.

4.2.3 Surface CO_2 reduction

4.2.3.1 Adsorption and activation of CO_2

The adsorption and activation of CO_2 on the surface of photocatalysts is a critical step in photocatalytic activation of this molecule. The adsorption interaction between the CO_2 molecule and the surface atoms of the photocatalyst can change its linear structure [17]. The possible configuration of the adsorbed CO_2 can be classified into three modes: carbon coordination, oxygen coordination and mixed coordination (Fig. 4.1a–c) [18]. Typically, CO_2 adsorption and activation occur simultaneously. The interaction between CO_2 and the surface atom results in a partially charged specie $CO_2^{\delta\cdot-}$ and, as a result, lower the energy barrier of the subsequent reaction due to molecule bending. The interaction mode depends on the state of the exposed atoms on the surface of the photocatalyst. Moreover, the surface atoms also have an influence on the stabilization and binding of the reaction intermediates, thereby affecting the catalytic activity and selectivity. The adsorption and activation of CO_2 could be promoted by various strategies, which include increasing the photocatalyst surface area, enhancing basic sites, creating surface defects and co-catalyst loading [18].

4.2.3.2 Proposed reaction pathways and mechanisms of CO_2 reduction

The reduction of CO_2 is a complex and multiple electron/proton stepwise process, which could generate different intermediates [2]. Depending on the reaction conditions, the various intermediates lead to different reaction pathways and final products. The CO_2 reduction pathway can be affected by many factors such as composition and structure of photocatalyst, the number of electrons and protons transferred to the CO_2 molecule and reaction conditions (temperature and pH).

There are three possible reaction mechanisms proposed for the CO_2 reduction – formaldehyde, carbene and glyoxal pathways [2, 5], as shown in Fig. 4.1d. In the formaldehyde pathway, CO_2 reduction is carried out by binding a singly coordinated O atom in CO_2 to the active sites on the surface of the photocatalyst. The addition of a proton to the activated $CO_2^{\cdot-}$ results in the generation of carboxyl radical ($\cdot$COOH). Then, a proton and an electron are continuously added to $\cdot$COOH to produce formic acid. The formic acid is further reduced to formaldehyde through the intermediate dihydroxylmethyl radical by accepting two further electrons and protons. Next, the production of methanol and methane can be realized by promoting the subsequent reduction steps. This pathway cannot produce CO, which is one of the most common products during the CO_2 reduction reaction. According to the carbene pathway, CO_2 is bonded to the active center through the C atom. The $CO_2^{\cdot-}$ accepts one proton and one electron, which breaks a C = O bond and produces CO. CO is an intermediate or a side product, depending on the adsorption strength between CO and the photocatalyst surface. If the binding energy between the CO^* and photocatalysts is large, CO^* can accept further electrons and protons to form carbon ($:C:$). Otherwise, CO is a side product. By the subsequent hydrogenation process, the generated $\cdot CH_3$ radical reacts with hydroxyl anion to produce methanol or recombine with a proton and an electron to generate methane. In the glyoxal pathway, the two O atoms of CO_2 bind to the active site. The glyoxal pathway is mainly involved in the formation of multicarbon products in the CO_2 reduction process. In detail, $CO_2^{\cdot-}$ can obtain a proton and an electron to generate formate, which then reacts with another proton to produce formic acid. With the further addition of an electron and the elimination of hydroxyl anion, free formyl radicals ($HC\cdot O$) are generated. These formyl radicals can dimerize to generate glyoxal, which start the formation of C_2 and C_3 products.

It is thereby obvious that the adsorption mode of the CO_2 molecules largely determines the specific reaction pathway, thus resulting in different final reduction products. Moreover, the density of the transferred electrons and the number of protons as well as the binding energy of the intermediates affect the product selectivity.

4.2.3.3 Competition of H_2 evolution and O_2 reduction

H_2O is the ideal reactant for photocatalytic CO_2 reduction because it could serve as a source of electrons and protons. More importantly, the proton reduction reaction is a relatively facile reaction in terms of thermodynamics and kinetics. Thus, the side reaction of proton reduction usually occurs during the CO_2 reduction process, leading to a lowered efficiency and selectivity of the latter. Therefore, suppressing the H_2O reduction greatly improves the CO_2 conversion efficiency. However, considering that O_2 is the product of water oxidation, a possible consecutive O_2 reduction could also be competitive [19].

4.2.3.4 Water oxidation half-reaction

The water oxidation half-reaction is an important parallel reaction in photocatalytic CO_2 reduction because it could consume the photogenerated holes, which has a positive effect on promoting the electron-hole separation and prolonging the lifetime of the photoinduced electron. In general, electron donors such as triethanolamine or triethylamine are used to consume the photoinduced holes. However, H_2O is the ideal and mostly used electron donor, which serves as a proton source and reaction micro-environment for proton transport. The oxidation products of water are O_2, hydrogen peroxide (H_2O_2) or hydroxyl radical [5, 16]. Thermodynamically, the water oxidation reaction is feasible because the oxidation potential of this reaction is less positive than the valence band potential of most semiconductors. However, O_2 is hard to detect in the photocatalytic system, with H_2O as the electron donor. There are two possible reasons proposed for this phenomenon. One is that O_2 formation is difficult to accomplish because the evolution of O_2

a)

$\overset{\cdot}{C}$

$\overset{\delta-}{O}$ $\overset{\delta-}{O}$

$\overset{\cdot}{C}$

$\overset{\delta-}{O}$ $\overset{\delta-}{O}$

Oxygen coordination

b)

$\overset{\delta-}{O}$ $\overset{\delta-}{O}$

C

Carbon coordination

c)

O

$\overset{\delta-}{C=O}$

O

$\overset{\delta-}{C=O}$

Mixed coordination

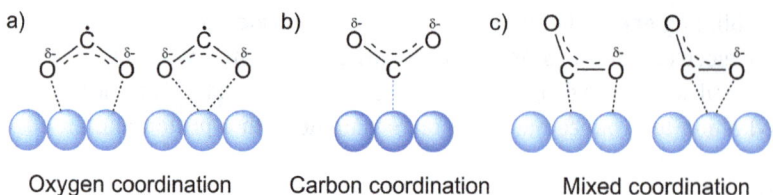

d) Formaldehyde pathway

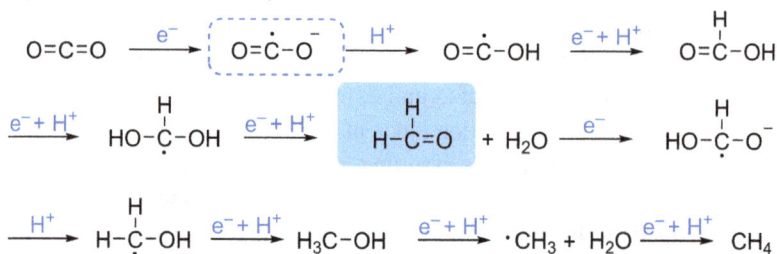

$O=C=O \xrightarrow{e^-} [O=\overset{\cdot}{C}-O^-] \xrightarrow{H^+} O=\overset{\cdot}{C}-OH \xrightarrow{e^-+H^+} O=\overset{H}{\underset{|}{C}}-OH$

$\xrightarrow{e^-+H^+} HO-\overset{H}{\underset{\cdot}{\underset{|}{C}}}-OH \xrightarrow{e^-+H^+} \overset{H}{\underset{|}{H-C=O}} + H_2O \xrightarrow{e^-} HO-\overset{H}{\underset{\cdot}{\underset{|}{C}}}-O^-$

$\xrightarrow{H^+} H-\overset{H}{\underset{\cdot}{\underset{|}{C}}}-OH \xrightarrow{e^-+H^+} H_3C-OH \xrightarrow{e^-+H^+} \cdot CH_3 + H_2O \xrightarrow{e^-+H^+} CH_4$

Carbene pathway

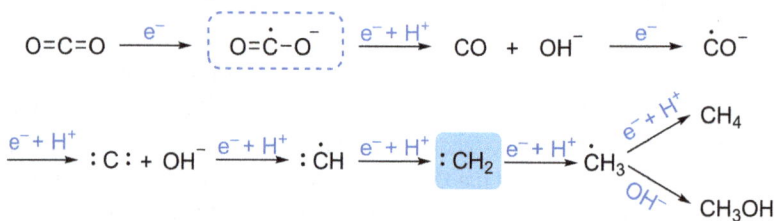

$O=C=O \xrightarrow{e^-} [O=\overset{\cdot}{C}-O^-] \xrightarrow{e^-+H^+} CO + OH^- \xrightarrow{e^-} \overset{\cdot}{C}O^-$

$\xrightarrow{e^-+H^+} :C: + OH^- \xrightarrow{e^-+H^+} :\overset{\cdot}{C}H \xrightarrow{e^-+H^+} :CH_2 \xrightarrow{e^-+H^+} \overset{\cdot}{C}H_3 \overset{\xrightarrow{e^-+H^+} CH_4}{\underset{\xrightarrow{OH^-} CH_3OH}{}}$

Glyoxal pathway

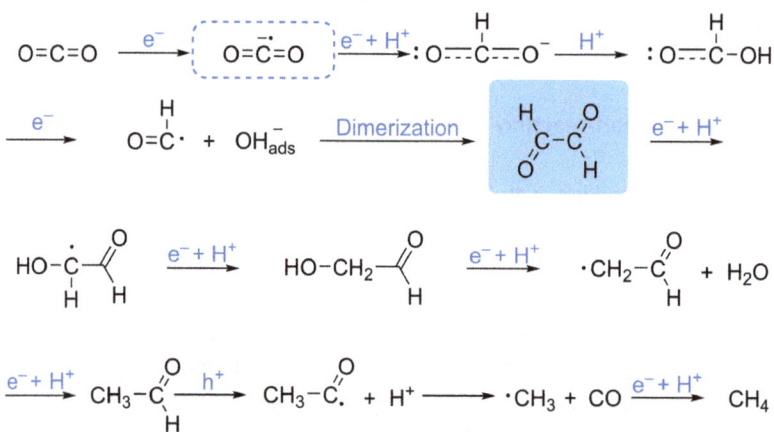

$O=C=O \xrightarrow{e^-} [O=\overset{-\cdot}{C}=O] \xrightarrow{e^-+H^+} :O\dot{=}\overset{H}{\underset{||}{C}}\dot{=}O^- \xrightarrow{H^+} :O\dot{=}\overset{H}{\underset{||}{C}}-OH$

$\xrightarrow{e^-} O=\overset{H}{\underset{|}{\overset{\cdot}{C}}} + OH^-_{ads} \xrightarrow{Dimerization} \overset{H}{\underset{O}{\overset{|}{C}-\overset{O}{\underset{||}{C}}\overset{}{\underset{H}{}}}} \xrightarrow{e^-+H^+}$

$HO-\overset{\cdot}{\underset{H}{\underset{|}{C}}}\overset{O}{\underset{H}{\diagdown}} \xrightarrow{e^-+H^+} HO-CH_2\overset{O}{\underset{H}{\diagdown}} \xrightarrow{e^-+H^+} \cdot CH_2-C\overset{O}{\underset{H}{\diagdown}} + H_2O$

$\xrightarrow{e^-+H^+} CH_3-C\overset{O}{\underset{H}{\diagdown}} \xrightarrow{h^+} CH_3-\overset{O}{\overset{||}{C}}\cdot + H^+ \longrightarrow \cdot CH_3 + CO \xrightarrow{e^-+H^+} CH_4$

Fig. 4.1: (a)–(c) Three different CO_2 adsorption modes on the surface of photocatalysts (reproduced with permission [18]. Copyright 2016, Royal Society of Chemistry). (d) Three proposed mechanisms of CO_2 reduction (reproduced with permission [2]. Copyright 2020, Royal Society of Chemistry).

molecule from H$_2$O requires four holes and the generated O$_2$ adsorbed on the surface of the photocatalyst is difficult to desorb [5, 16]. This process then only involves the formation and aggregation of peroxide intermediates [20]. However, the peroxide intermediates adsorbed on the surface of the photocatalyst could re-oxidize the CO$_2$ reduction intermediates, thus interrupting the reduction process. The other reason is that the generated O$_2$ consumes the photogenerated electrons and is back-reduced to H$_2$O$_2$ [19, 21]. Therefore, promoting the kinetics of proton-coupled electron transfer and eliminating the oxidation intermediates accumulated on the surface is an effective strategy to improve the H$_2$O oxidation and CO$_2$ reduction reaction.

4.2.3.5 Confirmation of CO$_2$ reduction products and selectivity

According to the above discussion, the photocatalytic CO$_2$ reduction reaction could produce various compounds. It is important to determine the composition of the reaction mixture and content of the products. In general, volatile products could be identified and quantified by gas chromatography (GC). The GC equipped with a thermal conductivity detector (TCD) detects H$_2$ and O$_2$, and the hydrogen flame ionization detector (FID) with a methanizer detects the hydrocarbons and CO. The nonvolatile products in the liquid phase of the reaction mixture can be analyzed by high-pressure liquid chromatography (HPLC). Apart from the GC and HPLC, mass spectrometry (MS) and GC–MS can also be used to confirm the formation of certain products. Moreover, the evaluation of selectivity of the photocatalytic CO$_2$ reduction reaction toward specific products is also an important indicator. The selectivity toward a specific product is calculated using the following equation:

$$\text{Selectivity of specific product }(\%) = \frac{X_i \cdot n(Y_i)}{\sum_i^n X_i \cdot n(Y_i)} \times 100\% \qquad (4.15)$$

where $n(Y_i)$ is the amounts (moles) of the reduction product formed within a certain time, while the X_i represents the number of electrons consumed by the multielectron reaction.

4.3 Strategies to boost the photocatalytic CO$_2$ reduction over carbon nitride-based photocatalysts

Due to its excellent properties, PCN has been a subject of extensive exploration as a candidate photocatalyst for this reaction. However, the catalytic efficiency of PCN in the first experiments was still far from satisfactory. To improve the CO$_2$ photoreduction performance of PCN, many strategies have been developed, such as nanostructure design, defect engineering, crystallinity modulation, heterostructure construction and cocatalysts engineering.

4.3.1 Nanostructure design

Nanostructure design of the expected architectures is an effective strategy to improve the photocatalytic performance by tuning the morphology to improve the light absorption and charge transport. To date, various nanoarchitectures, such as one-dimensional nanotubes/nanorods [22–25], two-dimensional nanosheets [26] and three-dimensional porous structures [27, 28], have been fabricated and applied for photocatalytic CO_2 reduction. For instance, hierarchical porous O-doped PCN nanotube (OCN-Tube), synthesized by thermal oxidation exfoliation and curling-condensation of bulk PCN, have been reported to enhance CO_2 reduction performance (Fig. 4.2a, 4.2b) [23]. A one-dimensional (1D) tubular nanostructure with a uniform diameter of 20–30 nm was confirmed by the field-emission scanning electron microscopic (FESEM) and transmission electron microscopic (TEM) images (Fig. 4.2c, 4.2d). Owing to high specific surface area, enhanced CO_2 adsorption capacity, improved visible light absorption as well as improved charge separation efficiency, the O-doped PCN nanotube showed improved photocatalytic performance for the target reaction.

Recently, many works have emphasized the advantages of delamination of bulk PCN into nanosheets for photocatalytic CO_2 reduction due to the higher specific surface area and shortened charge carrier migration distance, which could provide more active sites and suppress the electron-hole recombination, respectively. Up to now, PCN nanosheets were fabricated *via* many strategies such as thermal exfoliation, ultrasonic exfoliation, chemical exfoliation and other methods [29]. For example, Xia et al. prepared 2D nanosheets (NS-CN) *via* a thermal exfoliation method in an NH_3 atmosphere (Fig. 4.2e) [26]. Ultra-thin nanosheets were confirmed by taking FESEM images (Fig. 4.2f, 4.2g). The obtained PCN nanosheets possessed high specific surface area (Fig. 4.2h) and ultrathin thickness (ca. 2 nm), which provided more active sites and shortened carrier migration distance. Moreover, the PCN nanosheets showed a remarkably enlarged CO_2 uptake, attributed to improved physical adsorption as well as chemical interaction of CO_2 molecules with the amine groups (Fig. 4.2i). More importantly, the charge separation and transfer were promoted due to a reduced diffusion distance of charge carriers. As a result, the PCN nanosheets showed improved photocatalytic CO_2 reduction activity, affording CH_4 and CH_3OH formation rate of 1.39 and 1.87 $\mu mol\ g^{-1}\ h^{-1}$, respectively. In another work, hierarchical 3D porous PCN was found attractive for CO_2 reduction because it could harvest more incident light and promote rapid mass transport. Thus, Liu et al. synthesized hierarchical 3D porous PCN through a novel cold quenching strategy [27]. The hierarchical 3D porous PCN exhibited enhanced photocatalytic CO_2 reduction performance because of the boosted light absorption and improved electron–hole separation, as well as prolonged lifetime of the charge carriers.

a)

Heating zone

Condensation zone
A B C

Airflow

Exhaust gas

c)

100 nm

b)

Bulk g-C$_3$N$_4$

Airflow
Thermal exfoliation

g-C$_3$N$_4$ nanosheets

Curling
Self-assemble

g-C$_3$N$_4$ nanotubes

► N
► C

► N
► C
► O

d)

50 nm

e)

NH$_3$

Thermal shock

f)

500 nm

g)

100 nm

h)

i)

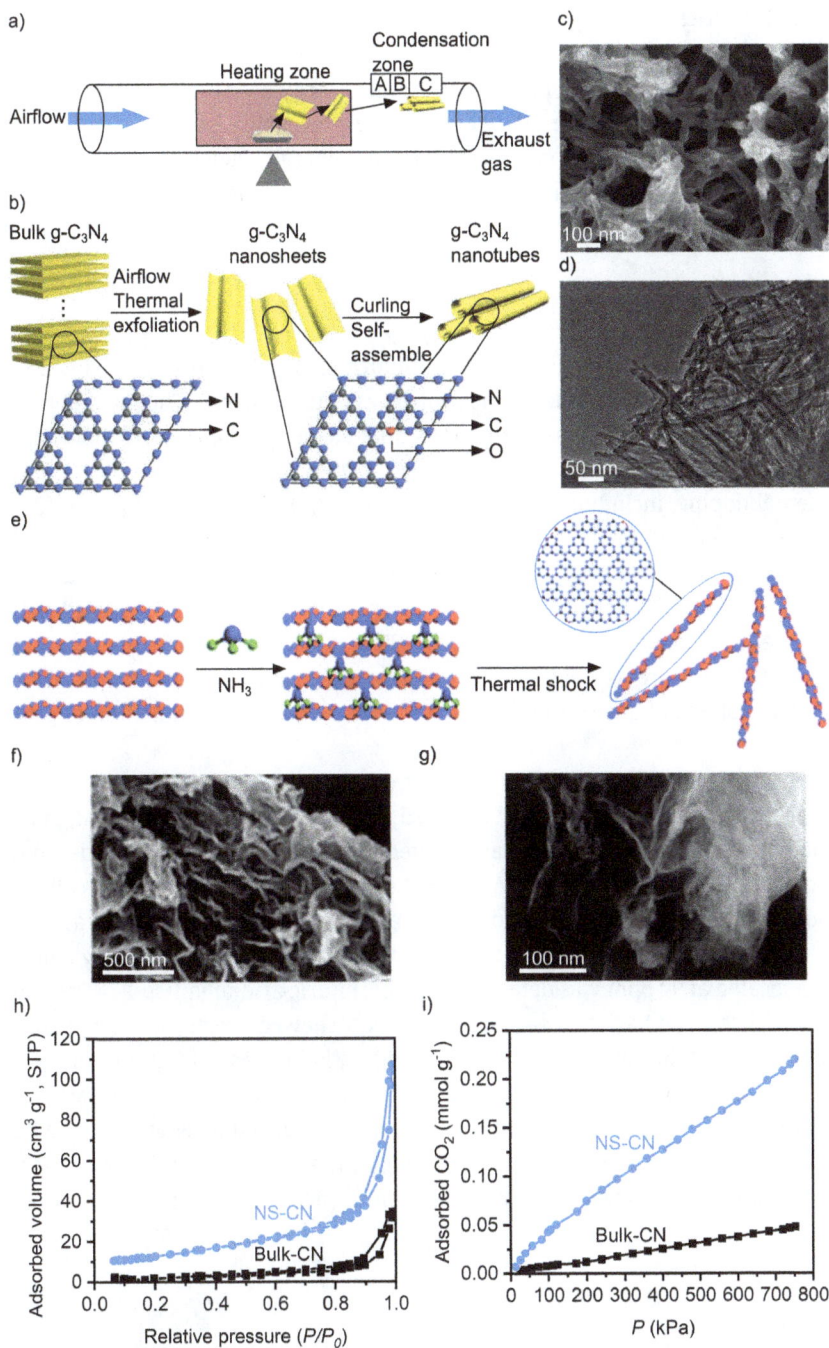

Fig. 4.2: (a) Schematic illustration of the synthesis process. (b) Illustration of OCN-Tube formation process. (c) SEM image of OCN-Tube. (d) TEM image of OCN-Tube (reproduced with permission [23]. Copyright

4.3.2 Defect engineering

Defect engineering has proven to be another useful strategy to improve the photocatalytic efficiency, which could tune the light harvesting capability, redox potential and number of active sites, as well as charge separation. Generally, defect engineering can be classified into two categories, namely doping and vacancy engineering.

4.3.2.1 Doping

Doping, a process of incorporating external impurities into the lattices of PCN, is a strategy to tune the optical and electronic properties of the material. The main doping strategies are rediscussed in this section, including elemental doping and molecular doping.

Elemental doping, including the B [30], C [31], O [23], S [32], P [33], Cl [34], K [35], have been implemented for PCN. For instance, Fu et al. reported that B was doped into a PCN scaffold between adjacent tri-s-triazine units by calcining a mixture of boric acid and urea [30]. Density functional theory (DFT) calculations revealed that the electron excitation from N to B in B-doped PCN is much easier than from N to C in pure PCN (Fig. 4.3a). Moreover, B doping could improve the charge transfer and localization. As a result, the CH_4 yield of B-doped PCN was 32 times higher than that of pristine PCN.

Apart from the elemental doping, molecular doping, anchoring another structure-matched organic molecule into the π-conjugated structure, has also been investigated to modulate the conventional π-conjugated system, intrinsic electronic properties and band structure [36–40]. For example, Wang and co-workers modified PCN *via* copolymerization of urea and barbituric acid [36]. The incorporation of barbituric acid in the PCN framework was confirmed by the solid-state ^{13}C NMR spectra, whereby a new weak peak located at 94 ppm was detected in the barbituric-acid-modified PCN sample (CNU-BA, Fig. 4.3b). The barbituric-acid-modified PCN showed a remarkable red shift of the optical absorption from 425 to 520 nm, suggesting improved light absorption (Fig. 4.3c). Moreover, the peak intensity of photoluminescence (PL) spectra significantly decreased with increasing amounts of barbituric acid, and together with the PL peaks, gradually shifted toward longer wavelength, indicating the suppressed electron-hole recombination because of the extension of the π-conjugation. In addition,

---

Fig. 4.2 (continued)
2017, Wiley-VCH. (e) Illustration of NS-CN photocatalysts formation process. (f) and (g) FESEM images of NS-CN. (h) N_2 adsorption–desorption isotherms measured for the bulk-CN and NS-CN. (i) CO_2 adsorption isotherms measured on the bulk-CN and NS-CN (reproduced with permission [26]. Copyright 2017, The Royal Society of Chemistry).

the dramatically enhanced photocurrent further confirmed the improved charge transfer efficiency. As a result, the barbituric-acid-modified PCN exhibited an improved photocatalytic CO$_2$ reduction performance, and the conversion rate of CO$_2$ to CO was 31.1 µmol h^{-1}, which is 15 times higher than that of pristine PCN. Other molecular doping strategies were also found effective to improve the photocatalytic efficiency. For example, Xue and co-workers demonstrated that the pyrene functional group could be covalently linked to the tri-s-triazine rings though a simple copolymerization strategy [39]. The pyrene-modified PCN exhibited higher hydrophobicity, thus promoting an unique biphasic photocatalytic activity. Recently, Yan and co-workers designed an intramolecular donor-acceptor system by introducing an organic donor into the PCN structure [40]. The formation of intramolecular donor-acceptor system could broaden the light absorption and lead to the spatial separation of electrons and holes, thereby resulting in a high CO$_2$ reduction performance.

4.3.2.2 Vacancy engineering

In addition to molecular doping, it is believed that the introduction of vacancies into the PCN matrix is another effective method to regulate the electronic band structure, improve charge separation, as well as to act as the active catalytic center. Therefore, the introduction of carbon and nitrogen vacancies into PCN has been extensively studied for CO$_2$ reduction [41–45]. Wang and co-workers created a carbon vacancy in the PCN matrix (MP-CVs) though a steam-etching approach to bind to the CO$_2$ [41]. Time-resolved PL spectra revealed that the lifetime of the charge carriers of MP-CVs was increased, suggesting that the carbon vacancies could capture the electrons and provide long-lived charges for the reduction reaction (Fig. 4.3d). CO$_2$ temperature-programmed desorption (CO$_2$-TPD) was performed to investigate the CO$_2$ adsorption behavior of CO$_2$ over the PCN and MP-CVs (Fig. 4.3e). The CO$_2$-TPD peak areas of MP-CVs were larger than that of PCN, indicating that the carbon vacancies could enhance CO$_2$ adsorption. Moreover, the CO$_2$-TPD peak of MP-CVs shifted to higher temperature, confirming that the activation energy barrier of CO$_2$ reduction was lowered due to the strong CO$_2$–C vacancy interaction. DFT results indicated that CO$_2$* was preferentially trapped in the vicinity of two NH$_2$ groups created by carbon loss, and the energy of the rate-determining step in MP-CVs was decreased by approximately 0.15 eV, thereby enhancing the reaction kinetics (Fig. 4.3f). Therefore, the optimized MP-CVs showed enhanced activity: the CO evolution rate was 54.6 µmol h^{-1}, which was 45 times higher than that of PCN (Fig. 4.3g). Apart from the C vacancy, the introduction of nitrogen vacancies into PCN also enhances CO$_2$ reduction activity because they could effectively dissociate excitons into free charges, and serve as active sites for the adsorption and activation of CO$_2$ [44].

a)

b)

c)

d)

e)

f)

g)

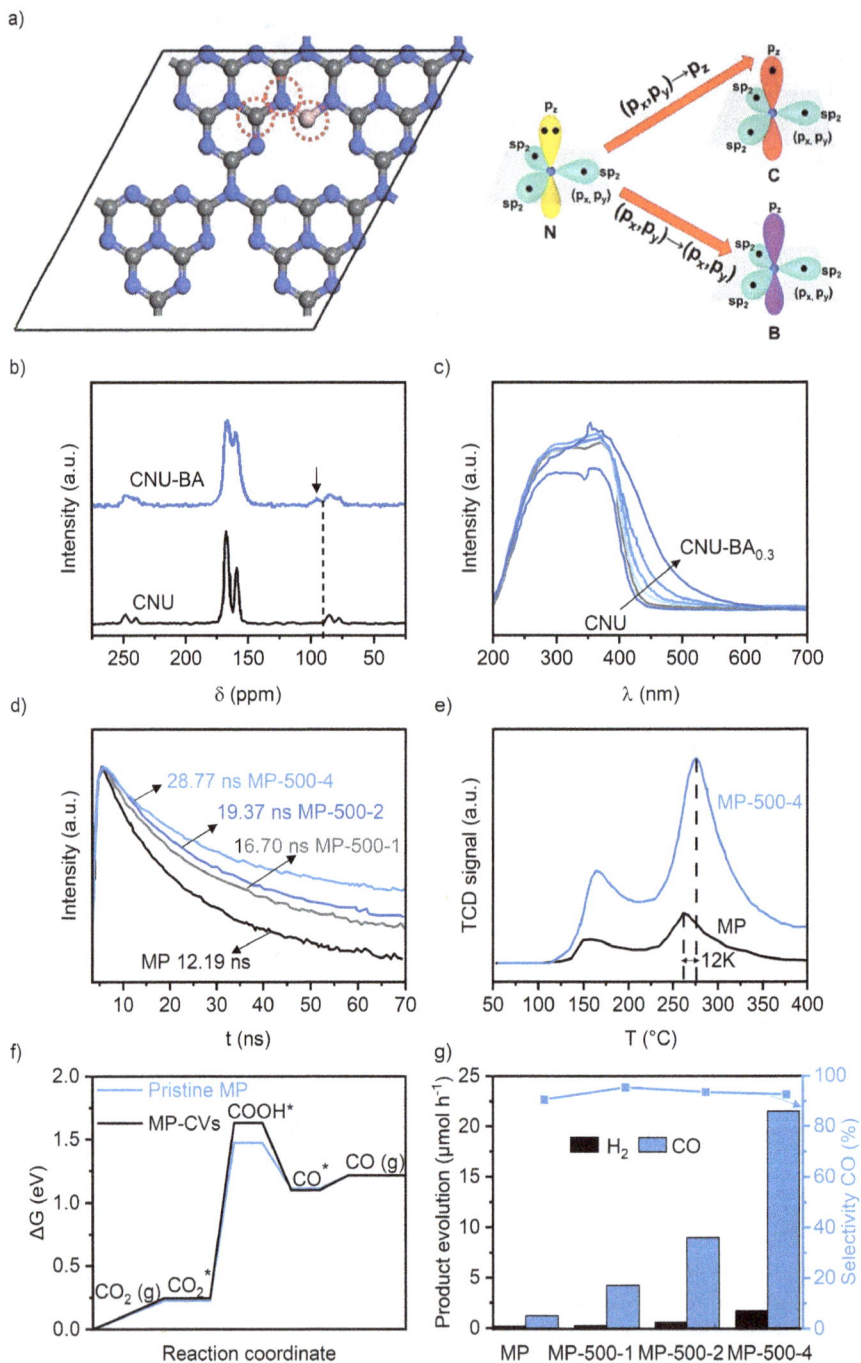

Fig. 4.3: (a) Atomic structure and schematic representation of electron excitation in B-doped PCN (reproduced with permission [30]. Copyright 2019, Wiley-VCH). (b) Solid-state ^{13}C NMR spectra of CNU and

4.3.3 Crystallinity modulation

The polymerization degree greatly affects the bulk charge transfer and the photocatalytic performance of PCN because high crystallinity reduces the bulk defects, narrows the band gap and promotes charge transfer [46]. Wang and co-workers prepared poly (heptazine imide)-based crystalline carbon nitride (CCN) from 5-aminotetrazole as the precursor *via* a facile ionothermal method [47]. The well-constructed poly(heptazine imide) (PHI) structure was confirmed by XRD data (Fig. 4.4a). The binding energy of C-K edge in CN-ATZ-NaK was positively shifted, which was attributed to the high crystallinity and extended conjugation of the heterocycles (Fig. 4.4b). Moreover, the clear lattice fringes in the TEM image further indicated the high crystallinity (Fig. 4.4c, 4.4d). Owing to the enhanced optical absorption and optimized electronic properties, CCN showed excellent photocatalytic CO$_2$ reduction performance.

Hydrothermal pretreatment is also a facile method to synthesize highly crystalline PCN. Wang and co-workers developed a locally crystallized PCN though hydrothermal pre-treatment with an amine-assisted nucleation strategy to selectively convert CO$_2$ to acetaldehyde [48]. The addition of amino-2-propanol could increase electron cloud density around the carbon atoms, resulting in high crystallinity. The locally crystallized PCN promoted the formation of the *OCCHO intermediate and suppressed the hydrogenation reaction of *CHO to *CH$_2$O. The formation of *OCCHO promoted the subsequent proton-coupled electron transfer process to form CH$_3$CHO rather than HCHO. As a result, the locally crystallized PCN exhibited excellent photocatalytic activity of 1814.7 μmol h^{-1} g^{-1} toward this rather unusual C$_2$-product and showed a quantum efficiency of 22.4% at 385 nm, which was higher than that of the state-of-art CO$_2$ photocatalysts.

4.4 Heterostructure construction

Construction of PCN-based heterojunction has become an effective approach to improve the photocatalytic activity by accelerating the spatial separation of charge carriers. Moreover, the formation of the heterojunction composites could provide several advantages: (i) improved light absorption, (ii) high photochemical stability, (iii) enhanced reactant adsorption and (iv) lowering the redox overpotential by the presence

---

Fig. 4.3 (continued)
CNU–BA samples. (c) UV–vis DRS spectra of the samples (reproduced with permission [36]. Copyright 2015, Elsevier). (d) Time-resolved PL (77 K) spectra of all samples. (e) CO$_2$-TPD of MP and MP-500-4.
(f) Gibbs free energy diagrams for the conversion of CO$_2$ into CO on pristine MP and MP-CVs.
(g) Photocatalytic activity of MP and MP-CVs (reproduced with permission [41]. Copyright, 2019, Wiley-VCH).

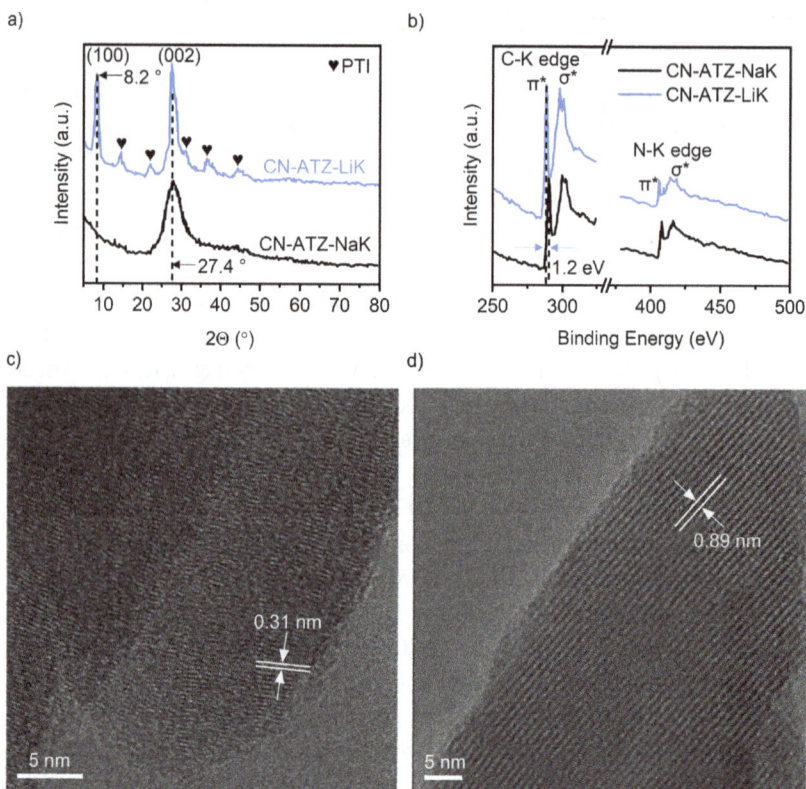

Fig. 4.4: (a) Powder XRD patterns of CN-ATZ-LiK and CN-ATZ-NaK. (b) Electron energy loss spectra (EELS) of CN-ATZ-LiK and CN-ATZ-NaK samples. (c) TEM and (d) HR-TEM images of CN-ATZ-NaK (reproduced with permission [47]. Copyright, 2019, Wiley-VCH).

of co-catalysts. According to reported literature, four main categories of PCN-based heterojunction have been extensively investigated: co-catalyst/PCN heterojunctions, type-II heterojunctions, Z-scheme heterojunctions and S-scheme heterojunctions. The details of the latest progress about heterojunctions are comprehensively discussed in this section.

4.4.1 Co-catalyst

The most common characteristic of the CO_2-reducing co-catalyst is high electron transfer capability and a more positive Fermi level than the excited conduction band of PCN, which boosts the separation of charge carriers. Within the reported literature, we can divide the co-catalyst/PCN heterojunctions into metals/PCN heterojunctions, carbon materials/PCN heterojunctions, molecular metal complex/PCN heterojunctions, metal carbide/sulfide/phosphide/PCN heterojunctions, MXene/PCN heterojunctions, black phosphorus/PCN heterojunctions and more diverse heterojunctions.

Loading of metal nanoparticles on the surface of PCN is the classical approach to improve the photocatalytic activity of PCN. Band bending occurs at the interface between the metal nanoparticles and PCN because of the different work function. This band bending zone, also known as the Schottky layer, creates the driving force for the directional charge migration. Various metals such as Pd [49, 50], Au [51] and Cu [52], have been widely studied to enhance the photocatalytic CO$_2$ reduction performance. Moreover, the exposed facets of the co-catalyst could also affect its performance in photocatalytic CO$_2$ conversion. For example, Cao et al. reported that the Pd nanoparticles featuring different exposed facets were deposited onto the surface of PCN though the electrostatic assembly method [50]. As shown in Fig. 4.5, Pd nanoparticles of different shapes, with different exposed facets, were uniformly dispersed on PCN. The transient photocurrent of T–CN (PCN with deposited tetrahedral Pd nanocrystals) was larger than that of C–CN (PCN with deposited cubic Pd nanocrystals), suggesting that the Pd {1 1 1} surface is more efficient as an electron sink than the Pd {1 0 0} surface due to the different work function (Fig. 4.5e). Importantly, DFT calculations confirmed that the Pd (111) facet showed improved CO$_2$ adsorption capability and CH$_3$OH desorption capability. Consequently, the tetrahedral Pd-deposited PCN showed enhanced photocatalytic CO$_2$ reduction activity (Fig. 4.5f).

In addition to the single-component metal nanoparticles, bimetallic alloy co-catalysts have also been explored for CO$_2$ photoreduction [53, 54]. For instance, Bai and co-workers demonstrated that the high-index PtCu loaded on PCN nanosheets exhibited higher photocatalytic activity and selectivity in CH$_4$ production than the PtCu with low-index facets [53]. This enhancement could be attributed to the fact that the high-index facets could provide greater number of low-coordinated metal active sites, promoting the adsorption and activation of CO$_2$.

Recently, modification of PCN with single-atom catalysts was found as an appealing approach to improve the photocatalytic CO$_2$ reduction efficiency due to the high atom utilization efficiency and distinct coordination environment [55]. Various single atoms, including Co [56–59], Cu [60, 61], Er [62], La [63], Ni [64, 65], Ru [66], Fe [67] and Au [68], have been successfully anchored on PCN to drive the CO$_2$ reduction reaction. For instance, Cao and co-workers successfully prepared single-atom copper-modified PCN though supramolecular preorganization, with subsequent condensation [61]. The single Cu atoms were uniformly dispersed on the PCN nanosheet, as confirmed by the HADDF-STEM image (Fig. 4.6a, 4.6c). The elemental mapping further indicated the homogeneous distribution of Cu atoms (Fig. 4.6b). Furthermore, EXAFS analysis was performed to reveal the coordination configuration of the isolated single Cu atoms. The main peak, around 1.5 Å, corresponding to the Cu–N bond was observed in the Cu/CN sample, suggesting that the Cu atoms were isolated and bonded to N or C atoms (Fig. 4.6d). The single Cu atom could serve not only as electron collector, but also act as a reactive binding center to activate CO$_2$ molecules. Therefore, the as-prepared

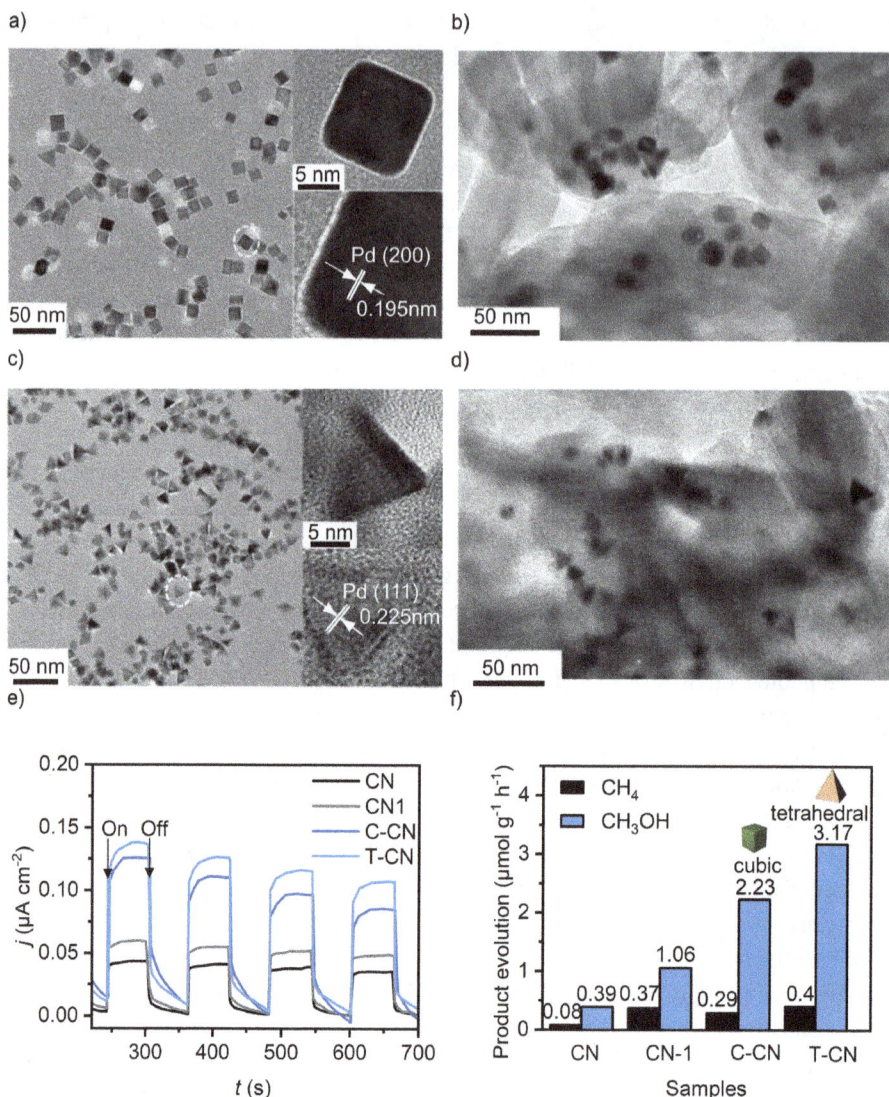

Fig. 4.5: (a) TEM and HRTEM images of Pd NCs. (b) TEM image of C–CN. (c) TEM and HRTEM images of Pd NTs. (d) TEM image of T–CN. (e) Transient photocurrent responses of all the samples. (f) CH_4 and CH_3OH production over reference CN and CN1 (prepared from CN by the same pH adjustment process, but without modification with Pd) samples, as well as C–CN and T–CN samples (reproduced with permission [50]. Copyright, 2017, Elsevier).

photocatalyst exhibited a much more efficient and selective CO_2 photoreduction. Furthermore, to integrate the representative merits of single atom catalysis, Xiang and co-workers designed atomically dispersed Co–Ru bimetal dopant into PCN for effective photocatalytic CO_2 reduction though an accessible self-seeded process [69]. Aberration-corrected high-angle annular dark-field scanning TEM (AC HAADF-STEM) images indi-cated the presence of atomically dispersed Ru and Co species, visualized as the bright dots (Fig. 4.6e). Moreover, the distance between two neighboring bright dots was 0.45 nm, which is close to the coordination environment of Co and Ru, further confirm-ing the coexistence of single-atom Ru–Co coupled over thr PCN support. (Fig. 4.6f). The near-edge shoulder of Co species was located between CoO and Co$_3$O$_4$, while the absorp-tion edge of Ru species is centered between Ru foil and RuO$_2$, suggesting the presence of positively charged Co and Ru atom (Fig. 4.6g, 4.6i). Meanwhile, the dominant Co–N peak centered at ~ 1.35 Å and Ru–N peak at ~ 1.45 and ~ 2.60 Å were found in the EXAFS spectra, further indicating the coexistence of the single-atom Ru and Co (Fig. 4.6h, j). The single-atom Co could promote the electron transfer, while the presence of the single-atom Ru on the surface is beneficial for selective CO$_2$ binding. Benefiting from the dual-functionalized Co and Ru reactive sites, the as-synthesized photocatalyst showed excellent performance in CO$_2$ reduction, with an AQE of 2.8% at the wavelength of 385 nm.

Apart from the metal co-catalyst, covalent materials with π-conjugated structures, including graphene [70–72], carbon nanosheets [73], carbon dots [74–76], carbon paper [77], graphdiyne [78] and similar, have been widely applied for improving the photoca-talytic CO$_2$ reduction due to their high electronic conductivity and large specific surface area. For instance, Ong et al. fabricated a sandwich-like graphene–PCN (GCN) nanocom-posite though a one-pot impregnation–thermal polymerization strategy [70]. A sheet-on-sheet structure in intimate contact was formed by the thermal polymerization of urea molecules pre-adsorbed on the graphene oxide (Fig. 4.7a). The GCN exhibited an en-hanced light absorption and an extended absorption edge, which was attributed to the role of graphene in the hybrid nanostructures. More importantly, graphene increases electron transfer *via* the heterojunction of graphene and PCN, thus suppressing the charge recombination. As a result, the as-prepared GCN showed improved CO$_2$ reduc-tion performance, which was 2.3 times higher than that of pure PCN. To achieve the intimate interfacial contact between the graphene and PCN at a relatively low tempera-ture, the same group reported a 2D/2D hybrid structure by incorporating reduced gra-phene oxide (rGO) and protonated PCN though an electrostatic self-assembly strategy, followed by a NaBH$_4$-reduction method [71]. Owing to the sufficient interfacial contact for effective charge transfer across the 2D/2D heterojunction, the as-prepared photoca-talyst again showed a remarkably enhanced photocatalytic CO$_2$ reduction activity.

Moreover, in view of the advantages of high crystallinity and graphitic structure, Cao and co-workers constructed a vertically aligned highly crystalline carbon nitride nano-rods on 2D graphene sheet (CNNA/rGO) through an ionothermal method (Fig. 4.7b). The crystalline PCN nanorods were tightly anchored at the surface of graphene to form a reg-ular composite structure (Fig. 4.7c). Moreover, another notable finding was that CNNA/rGO showed remarkably improved CO$_2$ adsorption capacity than that of CNNA, which was attributed to the heterojunction effect and the ordered 1D structure (Fig. 4.7d). The CNNA/rGO exhibited excellent activity, with a total CO$_2$ conversion of 12.63 μmol h^{-1} g^{-1},

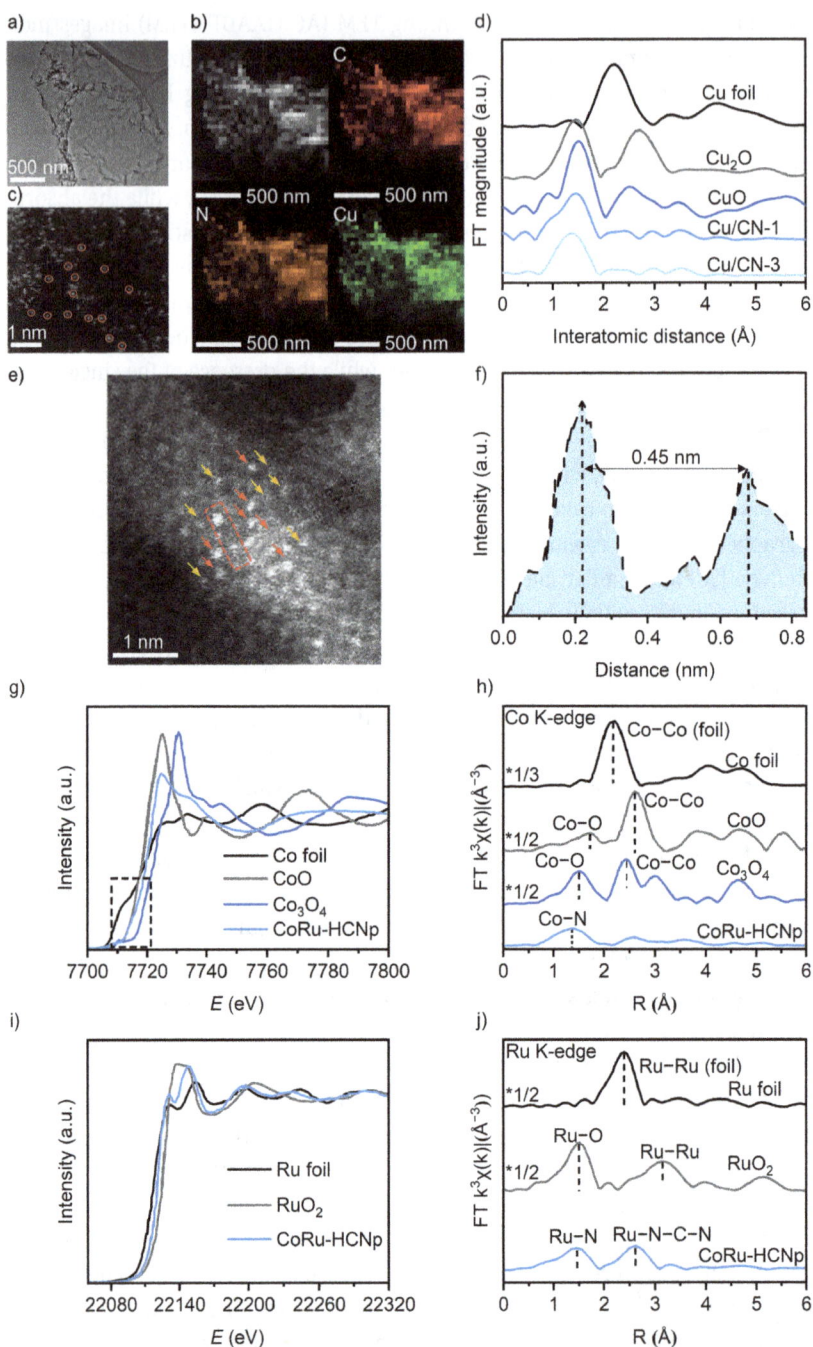

Fig. 4.6: (a) TEM image, (b) EDS mapping and (c) HADDF-STEM image of Cu/CN-3. (d) FT-EXAFS of Cu K-edge for the Cu/CN samples and Cu foil, with Cu$_2$O and CuO as the references (reproduced with

which was 4 times higher than that of crystalline PCN. This enhanced performance could be ascribed to the unique 2D/1D heterojunction, which improves the light absorption, CO$_2$ binding and interface charge transport.

Apart from graphene, carbon dots with a diameter of less than 10 nm are attractive for hybridization due to the excellent electron/hole transfer and broad optical absorption [76]. For example, Tang and co-workers prepared unique hole-accepting C dots and coupled them with CN for the selective CO$_2$ reduction to methanol [75]. These C dots (^{m}CD) were fabricated by a microwave method, while the other C dots (^{s}CD) synthesized by sonication-assisted oxidation served as electron acceptors. The as-prepared ^{m}CD acted as hole acceptors, which facilitated the electron accumulation on the surface of CN. This hole accepting property was confirmed by the diffuse reflectance transient absorption spectroscopy (Fig. 4.7e). Compared to CN, the electron signal of ^{m}CD/CN was higher, suggesting that the number of long-lived electrons was increased because the holes in CN were presumably transferred to ^{m}CD. After adding Ag$^+$ as an electron scavenger, the electron absorption signal of ^{m}CD/CN was decreased, further indicating that ^{m}CD indeed acts as the hole acceptor. In contrast, ^{s}CD/CN showed a signal of lower intensity, compared to pure CN, inferring the electron transfer from CN to ^{s}CD. The electron signal of ^{s}CD/CN did not change after the addition of AgNO$_3$, indicating that the electrons were already effectively extracted by the ^{s}CD. Moreover, the ^{m}CD could selectively absorb and oxidize H$_2$O instead of methanol, resulting in the high selectivity of CH$_3$OH production. As a result, the ^{m}CD/CN produced methanol at a rate of 13.9 µmol g^{-1} h^{-1} and a quantum efficiency of 2.1% at 420 nm.

Molecular metal complexes, based on homogeneous catalytic systems, potentially offer high efficiency and selectivity for photocatalytic CO$_2$ reduction [13]. However, the metal complexes suffer from poor stability and weak oxidation stability, which restricts their applicability. Therefore, various molecular metal complexes, including Ru complexes, [79–82] Co complexes [83, 84], Fe complexes [85] and Re complexes [86], have been coupled with PCN to overcome these disadvantages. For instance, Maeda and coworkers developed a heterogeneous photocatalytic system consisting of a ruthenium(Ru) complex and PCN for photocatalytic reduction of CO$_2$ into formic acid [80]. The Ru complex not only acted as the catalytic center, but also served as electron acceptor to promote charge separation. Benefiting from the improved electron transfer efficiency, the Ru complex/PCN hybrid material converted CO$_2$ into HCOOH with a selectivity > 80% and a turnover number (TON) greater than 1,000. The apparent quantum yield was 5.7%

Fig. 4.6 (continued)
permission [61]. Copyright, 2020, American Chemical Society). (e) Aberration-corrected HAADF-STEM image of CoRu-HCNp. (f) The intensity profile of the red-circled area in the panel (e). (g) Normalized XANES spectra and (h) Fourier-transform EXAFS spectra at the Co *K*-edge of CoRu-HCNp and references. (i) Normalized XANES spectra. (j) Fourier-transform EXAFS spectra at the Ru K-edge of CoRu-HCNp and references (reproduced with permission [69]. Copyright, 2021, Wiley-VCH).

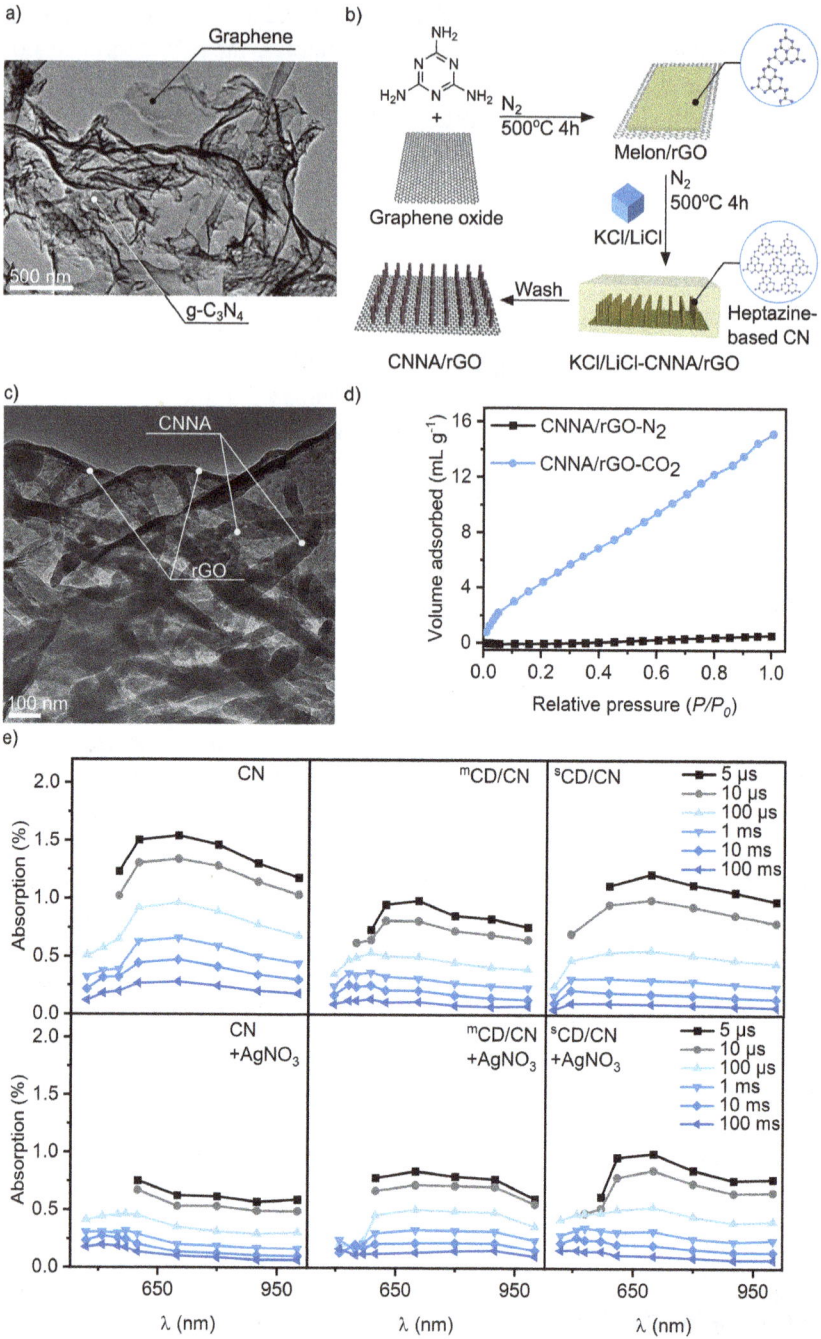

Fig. 4.7: (a) TEM images of GCN (reproduced with permission [70]. Copyright, 2014, The Royal Society of Chemistry). (b) The schematic illustration of CNNA/rGO composite formation in the molten salt medium.

at 400 nm, which was higher than that of the earlier reported heterogeneous photocatalyst. In a followed study, an artificial Z-scheme system based on a Ru binuclear complex and mesoporous PCN for CO$_2$ reduction was reported by the same group [82]. One Ru complex served as a catalytic center while the other acted as a photosensitizing unit. Mechanistic studies indicated that the electron is transferred according to the Z-scheme principle. In consequence, this photocatalytic system showed higher activity and high selectivity (87−99%) for HCOOH production, with turnover numbers >33,000. However, the noble metal-based complexes are expensive and rare. Therefore, Maeda and co-workers synthesized a photocatalytic system consisting of iron-complex and mesoporous PCN for the selective photocatalytic reduction of CO$_2$ to CO [85]. The [Fe(qpy)(H$_2$O)$_2$]$^{2+}$/mesoporous PCN exhibited high CO evolution rate, with a selectivity of 97% and an apparent quantum yield of about 4.2%. The enhanced performance was attributed to the high reactivity of Fe center, which activated CO$_2$ and suppressed the competing H$_2$ evolution.

Besides the previously discussed systems, metal carbides/sulfides/phosphides [87], MXenes [88] and black phosphorus [89] are also candidate materials for constructing composites with PCN. For instance, Ye and co-workers prepared black phosphorus quantum dots on PCN (BP@PCN) through a simple electrostatic attraction approach [89]. The as-prepared BP@PCN displayed significantly improved photocatalytic CO$_2$ reduction activity, which was ascribed to the efficient charge separation and facile exciton dissociation.

4.4.2 Type-II heterojunction

In PCN-based type-II heterojunctions, which are based on staggered band gaps, both the CB and VB positions of another semiconductor are either more positive or negative than PCN. The spatial separation of charge carriers is achieved here due to the built-in electric field but at the cost of lower reduction and oxidation potential. Therefore, various PCN-based type-II heterojunctions have been widely studied in photocatalytic CO$_2$ reduction, including UiO-66/PCN nanosheets [90], TiO$_2$/PCN [91], CsPbBr$_3$ QDs/PCN [92] and BiOBr nanosheets/PCN [93]. For example, Xu and co-workers anchored CsPbBr$_3$ QDs on the surface of NHx-rich porous PCN nanosheets to construct CPB-PCN heterojunction though N−Br chemical bonding for photocatalytic CO$_2$ reduction [92]. The amino groups on the edges of PCN interact strongly with CsPbBr$_3$ QDs to

Fig. 4.7 (continued)
(c) TEM images of CNNA/rGO. (d) Adsorption isotherms of CO$_2$ and N$_2$ on CNNA/rGO at 273 K and 1 atm (reproduced with permission [72]. Copyright 2019, Elsevier). (e) Diffuse reflectance TAS spectra for samples with and without 10 mM AgNO$_3$ (reproduced with permission [75]. Copyright 2020, Springer Nature).

form N–Br bonding. The unique N–Br interaction could promote charge separation at the interface and prolong the lifetime of charge carriers. Owing to the heterojunction and N–Br interaction, the CPB-PCN exhibited good stability and a significantly enhanced CO formation rate of 148.9 $\mu mol\ g^{-1}\ h^{-1}$, which was 15 and 3 times higher than that of pure QDs and PCN, respectively.

4.4.3 Z-Scheme heterojunction

The Z-scheme heterojunction, mimicking the natural photosynthesis of green plants, facilitates the spatial migration of charge carriers and maintains both the high reduction and oxidation abilities simultaneously, while relying on twice the photon number. The Z-scheme heterojunctions could be divided into three categories: traditional Z-scheme, indirect Z-scheme and the direct Z-scheme. In the traditional Z-scheme photocatalysts, shuttle redox ionic couples, including Fe^{3+}/Fe^{2+}, IO_3^-/I^- and $[Co\text{-}(bpy)_3]^{3+/2+}/[Co(phen)_3]^{3+/2+}$, are used in the liquid phase [94]. The Z-scheme structure with a solid electron mediator is called the indirect Z-scheme (Fig. 4.8a), and the third that does not rely on a mediator is known as the direct Z-scheme (Fig. 4.8b). Specifically, the electron transfer mechanism of direct Z-scheme is explained as follows. Under light irradiation, the photogenerated electrons from the CB of semiconductor I are transferred and recombine with the hole in the VB of the semiconductor II. Meanwhile, the photogenerated electrons and holes remain on semiconductor II and semiconductor I, respectively, and that maintain the high reduction and oxidation potential. For example, Wong and co-workers synthesized a hierarchical urchin-like $\alpha\text{-}Fe_2O_3/PCN$ direct Z-scheme heterojunction via an impregnation–hydrothermal method [95]. Figure 4.8c-e presented the TEM and HRTEM images of $\alpha\text{-}Fe_2O_3/PCN$ composite, confirming the intimate contact. Notably, the CO_2 absorption of the $\alpha\text{-}Fe_2O_3/PCN$ composite was improved due to the alkaline nature of iron and the synergistic effect between the two components. Moreover, the direct Z-scheme heterojunction could facilitate not only the separation of charge carriers, but also improve the reduction ability of photoinduced electrons (Fig. 4.8f). This Z-scheme charge transfer mechanism was confirmed by EPR results. As a result, the $\alpha\text{-}Fe_2O_3/PCN$ composite showed a CO evolution rate of 27.2 $\mu mol\ g^{-1}\ h^{-1}$, which is 2.2 times higher than that of pristine PCN (Fig. 4.8g). Moreover, various PCN-based direct Z-scheme heterojunctions have been developed for the enhanced photocatalytic CO_2 reduction performance, such as ZnO/PCN [96], Sn_2S_3/PCN [97], Cu_2O/PCN [98], Bi_2WO_6/PCN [99], $Cs_2AgBiBr/PCN$ [100] or $BiVO_4/PCN$ [101, 102].

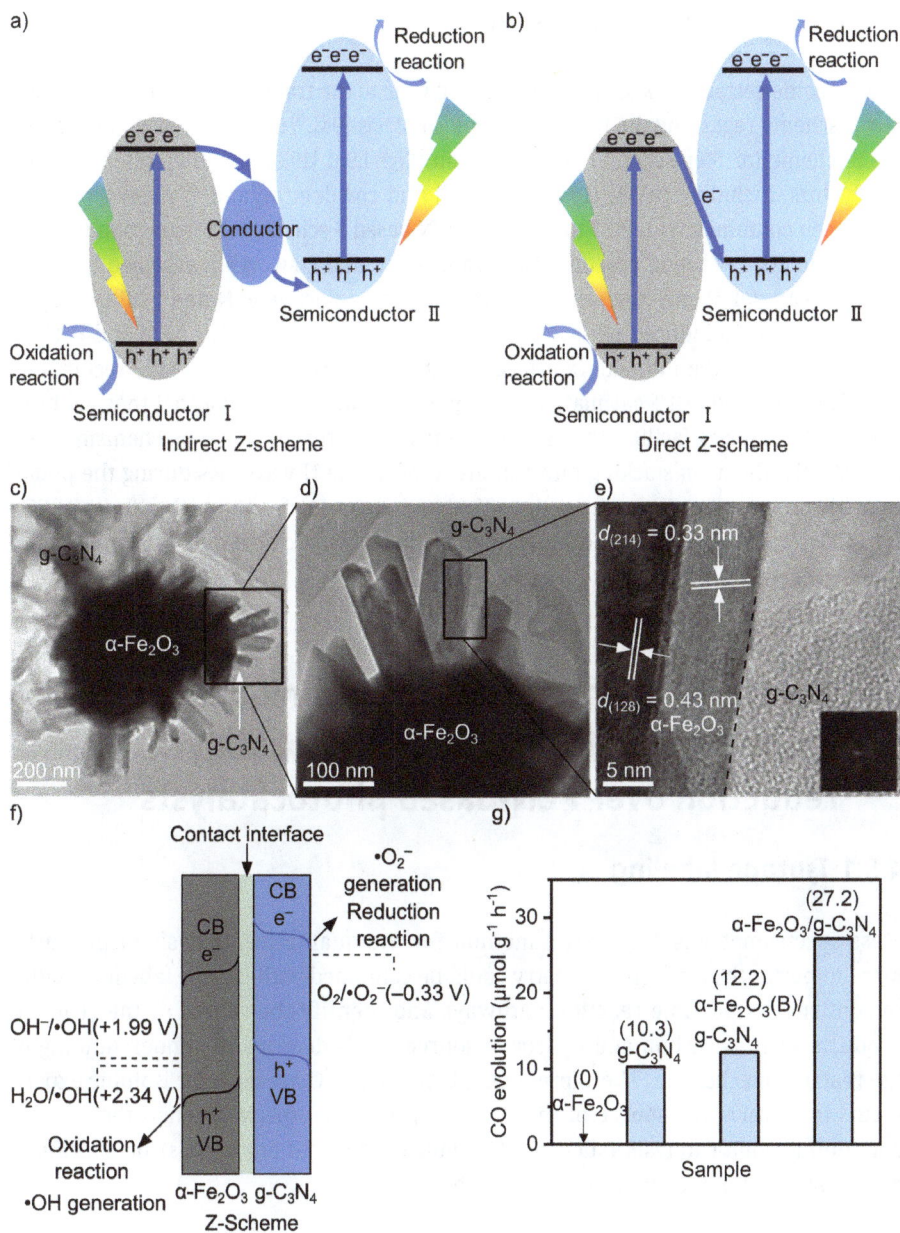

Fig. 4.8: (a) Schematic representation of indirect Z-scheme photocatalytic system. (b) Schematic representation of direct Z-scheme photocatalytic system. (c) and (d) TEM image of α-Fe$_2$O$_3$/PCN. (e) HRTEM image of α-Fe$_2$O$_3$/PCN. (f) Z-scheme photocatalytic system. (g) Average CO production rates of PCN, α-Fe$_2$O$_3$ and α-Fe$_2$O$_3$/PCN hybrid (reproduced with permission [95]. Copyright, 2018, Wiley-VCH).

4.4.4 S-scheme heterojunction

To overcome certain critical shortcomings intrinsic to the traditional Z-scheme and indirect Z-scheme regarding the thermodynamic analysis [94], Yu and co-workers proposed the S-scheme (or Step-scheme) heterojunction (Fig. 4.9a) [103]. Recently, different semiconductors, including $InVO_4$ [104], TiO_2 [105] and covalent organic frameworks [106], have been combined with PCN to fabricate PCN-based S-scheme heterojunction for photocatalytic CO_2 reduction reaction. For instance, Yu and co-workers designed and constructed van der Waals heterojunction by coupling defective PCN and COF through a facile evaporation-induced self-assembly method (Fig. 4.9b) [106]. The TEM and HRTEM image confirmed the intimate interface, which promotes the electron-hole recombination (Fig. 4.9c, 4.9d). DFT calculations and experimental results indicated that the presence of N vacancies facilitate the electron transfer *via* the S-scheme mechanism. More importantly, the $\pi - \pi$ stacking interactions stabilize the N vacancies during the photocatalytic process. Benefitting from the efficient charge transfer and high affinity of COF for CO_2, the PCN/COF exhibited much higher photocatalytic activity and stability than the single components.

4.5 Advanced characterization techniques for the mechanistic insight of photocatalytic CO_2 reduction over PCN-based photocatalysts

4.5.1 Isotope labeling

Mass spectrometry is a powerful method for the qualitative analysis of products. More importantly, mass spectrometry could be combined with isotope labeling studies to confirm the possible reaction pathways and identify the source of the reaction products such as carbon and hydrogen sources. This deepens the understanding of the reaction mechanism. Therefore, $^{13}CO_2$, D_2O and $H_2^{18}O$ are used for isotopic experiments to confirm the source of the reaction products, especially when the carbon-containing photocatalysts (such as PCN and carbon nanomaterials) or sacrificial agents are used in the photocatalytic system.

4.5.2 In situ characterization

The in situ/operando characterization technique is an effective approach to investigate the reaction mechanism and the pathway, monitor the intermediates, identify the active centers, unravel the structural changes and electronic environment [107].

a)

b)

c)

d)

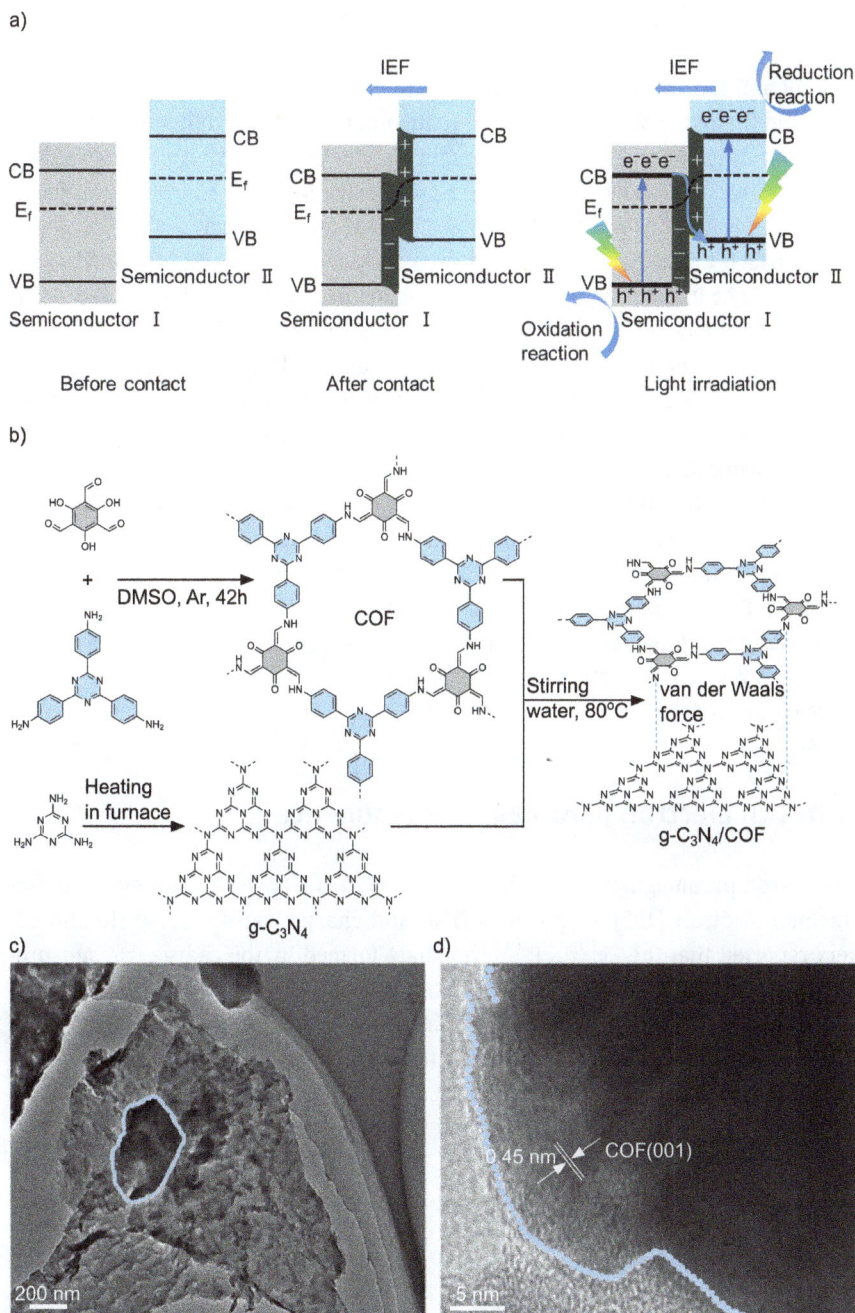

Fig. 4.9: (a) Schematic of charge-transfer processes in an S-scheme heterojunction. (b) Schematic illustration for the preparation of PCN/COF van der Waals heterojunction. (c) TEM image of PCN/COF and (d) HRTEM image of PCN/COF (reproduced with permission [106]. Copyright, 2022, Elsevier).

4.5.3 In situ FTIR and Raman spectroscopy

In situ FTIR and Raman spectroscopy are effective tools to identify the reaction intermediates and the pathway during the CO_2 reduction process, which helps to study the reaction mechanism. For instance, the photocatalytic CO_2 reduction process on PCN nanosheet is investigated by using in situ FTIR spectra [26]. The characteristic peaks of carbonate species appear at 1,512, 1,438, 1,631, 1,265 and 1,048 cm^{-1}, which originate from the surface binding interaction between the adsorbed CO_2 and the PCN nanosheet. Under light irradiation, the formate species (peaks at 1,286, 1,326, 1,386, 1,532, 1,558, 1,641 and 1,695 cm^{-1}), formaldehyde species (peaks at 1,419, 1,458, 1,607 and 1,657 cm^{-1}) and methanol species (peaks at 1,103 and 1,732 cm^{-1}) are detected, and their content increase, while the carbonate species decrease. These results indicated that the CO_2 reduction reaction proceeded though multiple steps involving the intermediates of formic acid and formaldehyde.

Moreover, in situ Raman spectroscopy has also been adopted in plasmonic photocatalyst to monitor the reaction intermediates and the catalyst surface state [108]. The representative Raman peaks of hydrocarbons, alcohols, acids and esters were observed in the Raman spectra, suggesting the formation of transient adsorbates and intermediates during the CO_2 reduction process. Although the noble metals are naturally suited for Raman spectroscopy characterization, this technique could be extended to other photocatalysts.

4.5.4 In situ electron paramagnetic resonance

In situ electron paramagnetic resonance (in situ EPR) is a powerful strategy to detect intermediate radicals [109], active sites [110] and charge transfer [111]. He and co-workers reported that the Fe^I TCPP intermediate formed by the charge transfer process on the PCN/FeTCPP was confirmed by the *in situ* EPR spectra [109]. Apart from the detection of intermediate radicals, the in situ EPR could also detect active sites [110]. The in situ EPR results suggested that the metal center (Co) is the active site in the photocatalytic CO_2 reduction process because the intensity of Co^{2+} signal did not show an observable decline as the adsorbed CO_2 molecules could accept the electron from Co^+ and be activated.

4.5.5 DFT calculations

DFT calculations is a critical and important tool to reveal the possible reaction pathway and the intermediates of the photocatalytic CO_2 reduction reaction. Moreover, DFT calculations could give the adsorption and activation energy of CO_2 and the reaction intermediates, providing theoretical guidance. For example, to understand the product distribution on the single-atom Er, DFT calculations were performed [63]. The favorable reaction pathway for the formation of CO and CH_4 was proposed by calculating the free energy. The energy required for the formation of gaseous CO was lower than that of CH_4 formation, suggesting the favorable production of CO rather than CH_4. Therefore, the systematic theoretical reaction mechanism calculations provided rational atomic understanding of the product distribution on the single-atom Er catalysts.

4.5.6 Transient absorption spectroscopy

Transient absorption spectroscopy is a widely applicable tool to study the charge carrier dynamics in solar energy conversion field [112]. The photon absorption, charge transfer, separation and recombination could be detected by the transient absorption spectroscopy. More importantly, the direction of the charge carrier transfer was also confirmed by the transient absorption spectroscopy [76, 113]. For instance, in the C dots/PCN heterojunction, the flow of charges could completely reverse, as exemplified by the two types of C dots prepared by different methods [75]. Transient absorption spectroscopy confirmed that the C dots prepared by the different methods could have either hole-accepting or electron-accepting function, as described in Section 4.4.1.

4.6 Summary and outlook

In summary, recent developments of PCN-based photocatalysts for photocatalytic CO_2 reduction were presented and systematized. The fundamentals and mechanism of photocatalytic CO_2 reduction were discussed. Furthermore, various strategies to improve the photocatalytic CO_2 reduction activity were comprehensively delineated, including nanostructure design, defect engineering, crystallinity modulation and heterostructure construction. At the end, advanced characterization techniques for mechanistic insight into photocatalytic CO_2 reduction reaction were recapitulated.

Although the development of PCN-based photocatalysts for photocatalytic CO_2 reduction progressed substantially, the solar energy conversion efficiency is still comparably low. Herein, the challenges and insights are summarized as follows.

(1) It is necessary to establish a comprehensive and reliable product testing system. Photocatalytic CO_2 reduction is a complex multistep process. At present, the possible products, such as HCOOH and multicarbon molecules, are usually not fully identified. Moreover, the oxidation product of H_2O is O_2 or hydrogen peroxide. It should be confirmed whether the H_2O_2 is the oxidation product or a reduction product in which the generated O_2 is back-reduced [21].

(2) The CO_2 photoconversion reaction mainly focuses on the reduction half-reaction, whereas the oxidation half-reaction is neglected. It is however of great significance to accelerate the oxidation half-reaction because, otherwise, this is the rate-determining step, and the realized progress in the reduction stays invisible. A higher oxidation rate avoids charge storage effects and thereby minimizes the recombination of electrons and holes so that more electrons could participate in the reduction half-reaction. The photogenerated holes are used to oxidize pure H_2O or sacrificial agents in the conventional CO_2 photoreduction reaction system, which makes the comparison between the performance of the catalysts of different groups very difficult. It is, in addition, rather desirable to use photogenerated holes to produce high value-added chemicals. Recently, the photocatalytic CO_2 reduction has been coupled with selective organic synthesis to sufficiently utilize the photogenerated electrons and holes to achieve a double value and realize the economic and sustainability goals more quickly [114].

(3) Even though significant efforts have been made in the conversion of CO_2 into C_1 chemicals, the production of multicarbon molecules is more desirable and less explored. It is meaningful to design photocatalysts with active sites for C–C coupling reaction. Moreover, it is necessary to develop photocatalysts with different functions to construct a cascade catalytic system for realizing multicarbon hydrocarbons, also by engineering means and optimized reactor constructions.

(4) The development of more in-depth and advanced in situ/*operando* characterization techniques is wanted. Although great achievements have been made in exploring the reaction pathways, the exact reaction mechanism and active centers at the nano-to-atomic level remain to be further investigated and clarified. Moreover, the interface dynamics and coordination and electronic structures of the photocatalysts during the reduction deserve more detailed studies. A deep understanding of these interface dynamics is needed to achieve superior CO_2 photocatalytic performance.

The design and construction of technically optimized electrode structures and CO_2 reduction device is crucial for potential scalable applications. The photocatalytic device should be easy to deconstruct, refurbish or recycled and re-used. Furthermore, external fields, including electric fields, thermal fields or photovoltaic support devices, could be integrated into the CO_2 reduction setups.

References

[1]   Li X, Yu JG, Jaroniec M, Chen XB. Cocatalysts for selective photoreduction of CO_2 into solar fuels. Chem Rev 2019, 119(6), 3962–4179.

[2]   Kong TT, Jiang YW, Xiong YJ. Photocatalytic CO_2 conversion: What can we learn from conventional COx hydrogenation? Chem Soc Rev 2020, 49(18), 6579–91.

[3]   Halmann M. Photoelectrochemical reduction of aqueous carbon dioxide on *p*-type gallium phosphide in liquid junction solar cells. Nature 1978, 275(5676), 115–16.

[4]   Inoue T, Fujishima A, Konishi S, Honda K. Photoelectrocatalytic reduction of carbon dioxide in aqueous suspensions of semiconductor powders. Nature 1979, 277(5698), 637–38.

[5]   Habisreutinger SN, Schmidt-Mende L, Stolarczyk JK. Photocatalytic reduction of CO$_2$ on TiO$_2$ and Other Semiconductors. Angew Chem Int Ed 2013, 52(29), 7372–408.

[6]   Li R, Zhang W, Zhou K. Metal-organic-framework-based catalysts for photoreduction of CO$_2$. Adv Mater 2018, 30(35), 1705512.

[7]   Chen J, Dong CW, Idriss H, Mohammed OF, Bakr OM. Metal halide perovskites for solar-to-chemical fuel conversion. Adv Energy Mater 2020, 10(13), 1902433.

[8]   Wang JJ, Lin S, Tian N, Ma TY, Zhang YH, Huang HW. Nanostructured metal sulfides: Classification, modification strategy, and solar-driven CO$_2$ reduction application. Adv Funct Mater 2021, 31(9), 2008008.

[9]   Wang X, Maeda K, Thomas A, Takanabe K, Xin G, Carlsson JM, et al. A metal-free polymeric photocatalyst for hydrogen production from water under visible light. Nat Mater 2009, 8(1), 76–80.

[10]   Cao SW, Low JX, Yu JG, Jaroniec M. Polymeric photocatalysts based on graphitic carbon nitride. Adv Mater 2015, 27(13), 2150–76.

[11]   Ong WJ, Tan LL, Ng YH, Yong ST, Chai SP. Graphitic carbon nitride (g-C$_3$N$_4$)-based photocatalysts for artificial photosynthesis and environmental remediation: Are we a step closer to achieving sustainability? Chem Rev 2016, 116(12), 7159–329.

[12]   Dong GH, Zhang LZ. Porous structure dependent photoreactivity of graphitic carbon nitride under visible light. J Mater Chem 2012, 22(3), 1160–66.

[13]   Lu QQ, Eid K, Li WP, Abdullah AM, Xu GB, Varma RS. Engineering graphitic carbon nitride (g-C$_3$N$_4$) for catalytic reduction of CO$_2$ to fuels and chemicals: Strategy and mechanism. Green Chem 2021, 23(15), 5394–428.

[14]   Tu WG, Zhou Y, Zou ZG. Photocatalytic conversion of CO$_2$ into renewable hydrocarbon fuels: State-of-the-art accomplishment, challenges, and prospects. Adv Mater 2014, 26(27), 4607–26.

[15]   Morris AJ, Meyer GJ, Fujita E. Molecular Approaches to the photocatalytic reduction of carbon dioxide for solar fuels. Acc Chem Res 2009, 42(12), 1983–94.

[16]   Sun ZY, Talreja N, Tao HC, Texter J, Muhler M, Strunk J, et al. Catalysis of carbon dioxide photoreduction on nanosheets: Fundamentals and challenges. Angew Chem Int Ed 2018, 57(26), 7610–27.

[17]   D'Alessandro DM, Smit B, Long JR. Carbon dioxide capture: Prospects for new materials. Angew Chem Int Ed 2010, 49(35), 6058–82.

[18]   Chang XX, Wang T, Gong JL. CO$_2$ photo-reduction: Insights into CO$_2$ activation and reaction on surfaces of photocatalysts. Energy Environ Sci 2016, 9(7), 2177–96.

[19]   Teng ZY, Zhang QT, Yang HB, Kato K, Yang WJ, Lu YR, et al. Atomically dispersed antimony on carbon nitride for the artificial photosynthesis of hydrogen peroxide. Nat Catal 2021, 4(5), 374–84.

[20]   Fu JW, Jiang KX, Qiu XQ, Yu JG, Liu M. Product selectivity of photocatalytic CO$_2$ reduction reactions. Mater Today 2020, 32, 222–43.

[21]   Yu XX, Yang ZZ, Qiu B, Guo SE, Yang P, Yu B, et al. Eosin Y-functionalized conjugated organic polymers for visible-light-driven CO$_2$ reduction with H$_2$O to CO with high efficiency. Angew Chem Int Ed 2019, 58(2), 632–36.

[22]   Zheng Y, Lin LH, Ye XJ, Guo FS, Wang XC. Helical graphitic carbon nitrides with photocatalytic and optical activities. Angew Chem Int Ed 2014, 53(44), 11926–30.

[23]   Fu J, Zhu B, Jiang C, Cheng B, You W, Yu J. Hierarchical porous O-doped g-C$_3$N$_4$ with enhanced photocatalytic CO$_2$ reduction activity. Small 2017, 13(15), 1603938.

[24]   Mo Z, Zhu X, Jiang Z, Song Y, Liu D, Li H, et al. Porous nitrogen-rich g-C$_3$N$_4$ nanotubes for efficient photocatalytic CO$_2$ reduction. Appl Catal: B 2019, 256, 117854.

[25]   Liu Q, Chen CC, Yuan KJ, Sewell CD, Zhang ZG, Fang XM, et al. Robust route to highly porous graphitic carbon nitride microtubes with preferred adsorption ability *via* rational design of one-

dimension supramolecular precursors for efficient photocatalytic CO_2 conversion. Nano Energy 2020, 77, 105104.

[26] Xia PF, Zhu BC, Yu JG, Cao SW, Jaroniec M. Ultra-thin nanosheet assemblies of graphitic carbon nitride for enhanced photocatalytic CO_2 reduction. J Mater Chem A 2017, 5(7), 3230–38.

[27] Liu MJ, Wageh S, Al-Ghamdi AA, Xia PF, Cheng B, Zhang LY, et al. Quenching induced hierarchical 3D porous g-C_3N_4 with enhanced photocatalytic CO_2 reduction activity. Chem Commun 2019, 55(93), 14023–26.

[28] Wang XW, Li QC, Gan L, Ji XF, Chen FY, Peng XK, et al. 3D macropore carbon-vacancy g-C_3N_4 constructed using polymethylmethacrylate spheres for enhanced photocatalytic H_2 evolution and CO_2 reduction. J Energy Chem 2021, 53, 139–46.

[29] Zhang JS, Chen Y, Wang XC. Two-dimensional covalent carbon nitride nanosheets: Synthesis, functionalization, and applications. Energy Environ Sci 2015, 8(11), 3092–108.

[30] Fu JW, Liu K, Jiang KX, Li HJW, An PD, Li WZ, et al. Graphitic carbon nitride with dopant induced charge localization for enhanced photoreduction of CO_2 to CH_4. Adv Sci (Weinheim, Ger) 2019, 6(18), 1900796.

[31] Samanta S, Yadav R, Kumar A, Sinha AK, Srivastava R. Surface modified C, O co-doped polymeric g-C_3N_4 as an efficient photocatalyst for visible light assisted CO_2 reduction and H_2O_2 production. Appl Catal: B 2019, 259, 118054.

[32] Wang K, Li Q, Liu BS, Cheng B, Ho WK, Yu JG. Sulfur-doped g-C_3N_4 with enhanced photocatalytic CO_2-reduction performance. Appl Catal: B 2015, 176, 44–52.

[33] Liu XL, Wang P, Zhai HS, Zhang QQ, Huang BB, Wang ZY, et al. Synthesis of synergetic phosphorus and cyano groups (–C≡N) modified g-C_3N_4 for enhanced photocatalytic H_2 production and CO_2 reduction under visible light irradiation. Appl Catal: B 2018, 232, 521–30.

[34] Liu CY, Zhang YH, Dong F, Reshak AH, Ye LQ, Pinna N, et al. Chlorine intercalation in graphitic carbon nitride for efficient photocatalysis. Appl Catal: B 2017, 203, 465–74.

[35] Wang K, Fu JL, Zheng Y. Insights into photocatalytic CO_2 reduction on C_3N_4: Strategy of simultaneous B, K co-doping and enhancement by N vacancies. Appl Catal: B 2019, 254, 270–82.

[36] Qin J, Wang S, Ren H, Hou Y, Wang X. Photocatalytic reduction of CO_2 by graphitic carbon nitride polymers derived from urea and barbituric acid. Appl Catal: B 2015, 179, 1–8.

[37] Xia J, Karjule N, Mondal B, Qin J, Volokh M, Xing L, et al. Design of Melem-based supramolecular assemblies for the synthesis of polymeric carbon nitrides with enhanced photocatalytic activity. J Mater Chem A 2021, 9(33), 17855–64.

[38] Hayat A, Khan J, Rahman MU, Mane SB, Khan WU, Sohai M, et al. Synthesis and optimization of the trimesic acid modified polymeric carbon nitride for enhanced photocatalytic reduction of CO_2. J Colloid Interface Sci 2019, 548, 197–205.

[39] Gong X, Yu S, Guan M, Zhu X, Xue C. Pyrene-functionalized polymeric carbon nitride with promoted aqueous-organic biphasic photocatalytic CO_2 reduction. J Mater Chem A 2019, 7(13), 7373–79.

[40] Song XH, Zhang XY, Wang M, Li X, Zhu Z, Huo PW, et al. Fabricating intramolecular donor-acceptor system via covalent bonding of carbazole to carbon nitride for excellent photocatalytic performance towards CO2 conversion. J Colloid Interface Sci 2021, 594, 550–60.

[41] Yang PJ, Zhuzhang HY, Wang RR, Lin W, Wang XC. Carbon vacancies in a Melon polymeric matrix promote photocatalytic carbon dioxide conversion. Angew Chem Int Ed 2019, 58(4), 1134–37.

[42] Shen M, Zhang LX, Wang M, Tian JJ, Jin XX, Guo LM, et al. Carbon-vacancy modified graphitic carbon nitride: Enhanced CO_2 photocatalytic reduction performance and mechanism probing. J Mater Chem A 2019, 7(4), 1556–63.

[43] Li H, Zhu BC, Cao SW, Yu JG. Controlling defects in crystalline carbon nitride to optimize photocatalytic CO_2 reduction. Chem Commun 2020, 56(42), 5641–44.

[44] Li F, Yue XY, Zhang DN, Fan JJ, Xiang QJ. Targeted regulation of exciton dissociation in graphitic carbon nitride by vacancy modification for efficient photocatalytic CO$_2$ reduction. Appl Catal: B 2021, 292, 120179.

[45] Yang PJ, Shang L, Zhao JH, Zhang M, Shi H, Zhang HX, et al. Selectively constructing nitrogen vacancy in carbon nitrides for efficient syngas production with visible light. Appl Catal: B 2021, 297, 120496.

[46] Lin LH, Yu ZY, Wang XC. Crystalline Carbon Nitride Semiconductors for photocatalytic water splitting. Angew Chem Int Ed 2019, 58(19), 6164–75.

[47] Zhang GG, Li GS, Heil T, Zafeiratos S, Lai FL, Savateev A, et al. Tailoring the grain boundary chemistry of polymeric carbon nitride for enhanced solar hydrogen production and CO$_2$ reduction. Angew Chem Int Ed 2019, 58(11), 3433–37.

[48] Liu Q, Cheng H, Chen TX, Lo TWB, Xiang ZM, Wang FX. Regulating the *OCCHO intermediate pathway towards highly selective photocatalytic CO$_2$ reduction to CH$_3$CHO over locally crystallized carbon nitride. Energy Environ Sci 2022, 15(1), 225–33.

[49] Bai S, Wang XJ, Hu CY, Xie ML, Jiang J, Xiong YJ. Two-dimensional g-C$_3$N$_4$: An ideal platform for examining facet selectivity of metal co-catalysts in photocatalysis. Chem Commun 2014, 50(46), 6094–97.

[50] Cao SW, Li Y, Zhu BC, Jaroniec M, Yu JG. Facet effect of Pd cocatalyst on photocatalytic CO$_2$ reduction over g-C$_3$N$_4$. J Catal 2017, 349, 208–17.

[51] Li HL, Gao Y, Xiong Z, Liao C, Shih K. Enhanced selective photocatalytic reduction of CO$_2$ to CH$_4$ over plasmonic Au modified g-C$_3$N$_4$ photocatalyst under UV-vis light irradiation. Appl Surf Sci 2018, 439, 552–59.

[52] Sun ZM, Fang W, Zhao L, Wang HL. 3D porous Cu-NPs/g-C$_3$N$_4$ foam with excellent CO$_2$ adsorption and Schottky junction effect for photocatalytic CO$_2$ reduction. Appl Surf Sci 2020, 504, 144347.

[53] Lang QQ, Yang YJ, Zhu YZ, Hu WL, Jiang WY, Zhong SX, et al. High-index facet engineering of PtCu cocatalysts for superior photocatalytic reduction of CO$_2$ to CH$_4$. J Mater Chem A 2017, 5(14), 6686–94.

[54] Wang ZY, Lee H, Chen JN, Wu MY, Leung DYC, Grimes CA, et al. Synergistic effects of Pd-Ag bimetals and g-C$_3$N$_4$ photocatalysts for selective and efficient conversion of gaseous CO$_2$. J Power Sources 2020, 466, 228306.

[55] Gao C, Low JX, Long R, Kong TT, Zhu JF, Xiong YJ. Heterogeneous Single-Atom Photocatalysts: Fundamentals and Applications. Chem Rev 2020, 120(21), 12175–216.

[56] Huang PP, Huang JH, Pantovich SA, Carl AD, Fenton TG, Caputo CA, et al. Selective CO$_2$ reduction catalyzed by single cobalt sites on carbon nitride under visible-light irradiation. J Am Chem Soc 2018, 140(47), 16042–47.

[57] Zhang PP, Zhan XN, Xu LB, Fu XZ, Zheng TY, Yang XY, et al. Mass production of a single-atom cobalt photocatalyst for high-performance visible-light photocatalytic CO$_2$ reduction. J Mater Chem A 2021, 9(46), 26286–97.

[58] Ma MZ, Huang ZA, Doronkin DE, Fa WJ, Rao ZQ, Zou YZ, et al. Ultrahigh surface density of Co-N$_2$C single-atom-sites for boosting photocatalytic CO$_2$ reduction to methanol. Appl Catal: B 2022, 300, 120695.

[59] Gong YN, Shao BZ, Mei JH, Yang W, Zhong DC, Lu TB. Facile synthesis of C$_3$N$_4$-supported metal catalysts for efficient CO$_2$ photoreduction. Nano Res 2022, 15(1), 551–56.

[60] Li Y, Li BH, Zhang DN, Cheng L, Xiang QJ. Crystalline carbon nitride supported copper single atoms for photocatalytic CO$_2$ reduction with nearly 100% CO Selectivity. ACS Nano 2020, 14(8), 10552–61.

[61] Wang J, Heil T, Zhu BC, Tung CW, Yu JG, Chen HM, et al. A single Cu-center containing enzyme-mimic enabling full photosynthesis under CO$_2$ Reduction. ACS Nano 2020, 14(7), 8584–93.

[62] Ji SF, Qu Y, Wang T, Chen YJ, Wang GF, Li X, et al. Rare-earth single erbium atoms for enhanced photocatalytic CO$_2$ Reduction. Angew Chem Int Ed 2020, 59(26), 10651–57.

[63] Chen P, Lei B, Dong XA, Wang H, Sheng JP, Cui W, et al. Rare-earth single-atom La-N charge-transfer bridge on carbon nitride for highly efficient and selective photocatalytic CO$_2$ reduction. ACS Nano 2020, 14(11), 15841–52.

[64] Cheng L, Yin H, Cai C, Fan JJ, Xiang QJ. Single Ni atoms anchored on porous few-layer g-C$_3$N$_4$ for photocatalytic CO$_2$ reduction: The role of edge confinement. Small 2020, 16(28), 2002411.

[65] Wang YY, Qu Y, Qu BH, Bai LL, Liu Y, Yang ZD, et al. Construction of six-oxygen-coordinated single Ni sites on g-C$_3$N$_4$ with boron-oxo species for photocatalytic water-activation-induced CO$_2$ reduction. Adv Mater 2021, 33(48), 2105482.

[66] Sharma P, Kumar S, Tomanec O, Petr M, Chen JZ, Miller JT, et al. Carbon nitride-based ruthenium single atom photocatalyst for CO$_2$ reduction to methanol. Small 2021, 17(16), 2006478.

[67] Zhao ZY, Liu W, Shi YT, Zhang HM, Song XD, Shang WZ, et al. An insight into the reaction mechanism of CO$_2$ photoreduction catalyzed by atomically dispersed Fe atoms supported on graphitic carbon nitride. Phys Chem Chem Phys 2021, 23(8), 4690–99.

[68] Yang YL, Li F, Chen J, Fan JJ, Xiang QJ. Single Au atoms anchored on amino-group-enriched graphitic carbon nitride for photocatalytic CO$_2$ reduction. Chemsuschem 2020, 13(8), 1979–85.

[69] Cheng L, Yue XY, Wang LX, Zhang DN, Zhang P, Fan JJ, et al. Dual-single-atom tailoring with bifunctional integration for high-performance CO$_2$ photoreduction. Adv Mater 2021, 33, 2105135.

[70] Ong WJ, Tan LL, Chai SP, Yong ST. Graphene oxide as a structure-directing agent for the two-dimensional interface engineering of sandwich-like graphene-g-C$_3$N$_4$ hybrid nanostructures with enhanced visible-light photoreduction of CO$_2$ to methane. Chem Commun 2015, 51(5), 858–61.

[71] Ong WJ, Tan LL, Chai SP, Yong ST, Mohamed AR. Surface charge modification *via* protonation of graphitic carbon nitride (g-C$_3$N$_4$) for electrostatic self-assembly construction of 2D/2D reduced graphene oxide (rGO)/g-C$_3$N$_4$ nanostructures toward enhanced photocatalytic reduction of carbon dioxide to methane. Nano Energy 2015, 13, 757–70.

[72] Xia Y, Tian ZH, Heil T, Meng AY, Cheng B, Cao SW, et al. Highly selective CO$_2$ capture and Its direct photochemical conversion on ordered 2D/1D heterojunctions. Joule 2019, 3(11), 2792–805.

[73] Wang YG, Xia QN, Bai X, Ge ZG, Yang Q, Yin CC, et al. Carbothermal activation synthesis of 3D porous g-C$_3$N$_4$/carbon nanosheets composite with superior performance for CO$_2$ photoreduction. Appl Catal: B 2018, 239, 196–203.

[74] Feng HJ, Guo QQ, Xu YF, Chen T, Zhou YY, Wang YG, et al. Surface nonpolarization of g-C$_3$N$_4$ by decoration with sensitized quantum dots for improved CO$_2$ photoreduction. Chemsuschem 2018, 11(24), 4256–61.

[75] Wang Y, Liu X, Han XY, Godin R, Chen JL, Zhou WZ, et al. Unique hole-accepting carbon-dots promoting selective carbon dioxide reduction nearly 100% to methanol by pure water. Nat Commun 2020, 11(1), 2531.

[76] Wang Y, Godin R, Durrant JR, Tang JW. Efficient hole trapping in carbon dot/oxygen-modified carbon nitride heterojunction photocatalysts for enhanced methanol production from CO$_2$ under neutral conditions. Angew Chem Int Ed 2021, 60(38), 20811–16.

[77] Zhang QH, Xia Y, Cao SW. "Environmental phosphorylation" boosting photocatalytic CO$_2$ reduction over polymeric carbon nitride grown on carbon paper at air-liquid-solid joint interfaces. Chin J Catal 2021, 42(10), 1667–76.

[78] Wang Y, Zhang Y, Wang YM, Zeng CX, Sun M, Yang DY, et al. Constructing van der Waals heterogeneous photocatalysts based on atomically thin carbon nitride sheets and graphdiyne for highly efficient photocatalytic conversion of CO$_2$ into CO. ACS Appl Mater Interfaces 2021, 13(34), 40629–37.

[79] Maeda K, Sekizawa K, Ishitani O. A polymeric-semiconductor-metal-complex hybrid photocatalyst for visible-light CO$_2$ reduction. Chem Commun 2013, 49(86), 10127–29.

[80] Kuriki R, Sekizawa K, Ishitani O, Maeda K. Visible-light-driven CO$_2$ reduction with carbon nitride: Enhancing the activity of ruthenium catalysts. Angew Chem Int Ed 2015, 54(8), 2406–09.

[81]   Kuriki R, Yamamoto M, Higuchi K, Yamamoto Y, Akatsuka M, Lu D, et al. Robust binding between carbon nitride nanosheets and a binuclear ruthenium(II) complex enabling durable, selective CO$_2$ reduction under visible light in aqueous solution. Angew Chem Int Ed 2017, 56(17), 4867–71.

[82]   Kuriki R, Matsunaga H, Nakashima T, Wada K, Yamakata A, Ishitani O, et al. Nature-inspired, highly durable CO$_2$ reduction system consisting of a binuclear ruthenium(II) complex and an organic semiconductor uing visible light. J Am Chem Soc 2016, 138(15), 5159–70.

[83]   Roy S, Reisner E. Visible-light-driven CO$_2$ reduction by mesoporous carbon nitride modified with polymeric cobalt phthalocyanine. Angew Chem Int Ed 2019, 58(35), 12180–84.

[84]   Ma B, Chen G, Fave C, Chen LJ, Kuriki R, Maeda K, et al. Efficient visible-light-driven CO$_2$ reduction by a cobalt molecular catalyst covalently linked to mesoporous carbon nitride. J Am Chem Soc 2020, 142(13), 6188–95.

[85]   Cometto C, Kuriki R, Chen LJ, Maeda K, Lau TC, Ishitani O, et al. A carbon nitride/Fe quaterpyridine catalytic system for photostimulated CO$_2$-to-CO Conversion with visible light. J Am Chem Soc 2018, 140(24), 7437–40.

[86]   Yu H, Haviv E, Neumann R. Visible-light photochemical reduction of CO$_2$ to CO coupled to hydrocarbon dehydrogenation. Angew Chem Int Ed 2020, 59(15), 6219–23.

[87]   Tang JY, Yang D, Zhou WG, Guo RT, Pan WG, Huang CY. Noble-metal-free molybdenum phosphide co-catalyst loaded graphitic carbon nitride for efficient photocatalysis under simulated irradiation. J Catal 2019, 370, 79–87.

[88]   Yang C, Tan Q, Li Q, Zhou J, Fan J, Li B, et al. 2D/2D Ti$_3$C$_2$ MXene/g-C$_3$N$_4$ nanosheets heterojunction for high efficient CO$_2$ reduction photocatalyst: Dual effects of urea. Appl Catal: B 2020, 268, 118738.

[89]   Han C, Li J, Ma Z, Xie H, Waterhouse GIN, Ye L, et al. Black phosphorus quantum dot/g-C$_3$N$_4$ composites for enhanced CO$_2$ photoreduction to CO. Sci China Mater 2018, 61(9), 1159–66.

[90]   Shi L, Wang T, Zhang HB, Chang K, Ye JH. Electrostatic self-assembly of nanosized carbon nitride nanosheet onto a zirconium metal-organic framework for enhanced photocatalytic CO$_2$ reduction. Adv Funct Mater 2015, 25(33), 5360–67.

[91]   Raziq F, Sun LQ, Wang YY, Zhang XL, Humayun M, Ali S, et al. Synthesis of large surface-area g-C$_3$N$_4$ comodified with MnO$_x$ and Au-TiO$_2$ as efficient visible-light photocatalysts for fuel production. Adv Energy Mater 2018, 8(3), 1701580.

[92]   Ou M, Tu W, Yin S, Xing W, Wu S, Wang H, et al. Amino-assisted anchoring of CsPbBr$_3$ perovskite quantum dots on porous g-C$_3$N$_4$ for enhanced photocatalytic CO$_2$ reduction. Angew Chem Int Ed 2018, 57(41), 13570–74.

[93]   Liu D, Chen D, Li N, Xu Q, Li H, He J, et al. Surface engineering of g-C$_3$N$_4$ by stacked BiOBr sheets rich in oxygen vacancies for boosting photocatalytic performance. Angew Chem Int Ed 2020, 59(11), 4519–24.

[94]   Xu QL, Zhang LY, Cheng B, Fan JJ, Yu JG. S-scheme heterojunction photocatalyst. Chem 2020, 6(7), 1543–59.

[95]   Jiang Z, Wan W, Li H, Yuan S, Zhao H, Wong PK. A hierarchical Z-scheme alpha-Fe$_2$O$_3$/g-C$_3$N$_4$ hybrid for enhanced photocatalytic CO$_2$ reduction. Adv Mater 2018, 30(10), 1706108.

[96]   Yu W, Xu D, Peng T. Enhanced photocatalytic activity of g-C$_3$N$_4$ for selective CO$_2$ reduction to CH$_3$OH *via* facile coupling of ZnO: a direct Z-scheme mechanism. J Mater Chem A 2015, 3(39), 19936–47.

[97]   Huo Y, Zhang J, Dai K, Li Q, Lv J, Zhu G, et al. All-solid-state artificial Z-scheme porous g-C$_3$N$_4$/Sn$_2$S$_3$-DETA heterostructure photocatalyst with enhanced performance in photocatalytic CO$_2$ reduction. Appl Catal: B 2019, 241, 528–38.

[98]   Zhao X, Fan Y, Zhang W, Zhang X, Han D, Niu L, et al. Nanoengineering construction of Cu$_2$O nanowire arrays encapsulated with g-C$_3$N$_4$ as 3D spatial reticulation all-solid-state direct Z-scheme photocatalysts for photocatalytic reduction of carbon dioxide. ACS Catal 2020, 10(11), 6367–76.

[99]   Li M, Zhang L, Fan X, Zhou Y, Wu M, Shi J. Highly selective CO$_2$ photoreduction to CO over g-C$_3$N$_4$/Bi$_2$WO$_6$ composites under visible light. J Mater Chem A 2015, 3(9), 5189–96.

[100] Wang Y, Huang H, Zhang Z, Wang C, Yang Y, Li Q, et al. Lead-free perovskite $Cs_2AgBiBr_6$@g-C_3N_4 Z-scheme system for improving CH_4 production in photocatalytic CO_2 reduction. Appl Catal: B 2021, 282, 119570.

[101] Wu J, Xiong L, Hu Y, Yang Y, Zhang X, Wang T, et al. Organic half-metal derived erythroid-like $BiVO_4$/ hm-C_4N_3 Z-scheme photocatalyst: Reduction sites upgrading and rate-determining step modulation for overall CO_2 and H_2O conversion. Appl Catal: B 2021, 295, 120277.

[102] Bian J, Zhang ZQ, Feng JN, Thangamuthu M, Yang F, Sun L, et al. Energy platform for directed charge transfer in the cascade Z-scheme heterojunction: CO_2 photoreduction without a cocatalyst. Angew Chem Int Ed 2021, 60(38), 20906–14.

[103] Fu JW, Xu QL, Low JX, Jiang CJ, Yu JG. Ultrathin 2D/2D WO_3/g-C_3N_4 step-scheme H_2 production photocatalyst. Appl Catal: B 2019, 243, 556–65.

[104] Gong S, Teng X, Niu Y, Liu X, Xu M, Xu C, et al. Construction of S-scheme 0D/2D heterostructures for enhanced visible-light-driven CO_2 reduction. Appl Catal: B 2021, 298, 120521.

[105] He F, Zhu B, Cheng B, Yu J, Ho W, Macyk W. 2D/2D/0D TiO_2/C_3N_4/Ti_3C_2 MXene composite S-scheme photocatalyst with enhanced CO_2 reduction activity. Appl Catal: B 2020, 272, 119006.

[106] Wang J, Yu Y, Cui J, Li X, Zhang Y, Wang C, et al. Defective g-C_3N_4/covalent organic framework van der Waals heterojunction toward highly efficient S-scheme CO_2 photoreduction. Appl Catal: B 2022, 301, 120814.

[107] Cao XY, Tan DX, Wulan B, Hui KS, Hui KN, Zhang JT. In situ characterization for boosting electrocatalytic carbon dioxide reduction. Small Methods 2021, 5(10), 2100700.

[108] Devasia D, Wilson AJ, Heo J, Mohan V, Jain PK. A rich catalog of C-C bonded species formed in CO_2 reduction on a plasmonic photocatalyst. Nat Commun 2021, 12(1), 2612.

[109] Lin L, Hou CC, Zhang XH, Wang YJ, Chen Y, He T. Highly efficient visible-light driven photocatalytic reduction of CO_2 over g-C_3N_4 nanosheets/tetra(4-carboxyphenyl)porphyrin iron(III) chloride heterogeneous catalysts. Appl Catal: B 2018, 221, 312–19.

[110] Liu HH, Zhang F, Wang HF, Xue JR, Guo YM, Qian QZ, et al. Oxygen vacancy engineered unsaturated coordination in cobalt carbonate hydroxide nanowires enables highly selective photocatalytic CO_2 reduction. Energy Environ Sci 2021, 14(10), 5339–46.

[111] Wang ZW, Wan Q, Shi YZ, Wang H, Kang YY, Zhu SY, et al. Selective photocatalytic reduction CO_2 to CH_4 on ultrathin TiO_2 nanosheet via coordination activation. Appl Catal: B 2021 288, 120000.

[112] Mitchell E, Law A, Godin R. Experimental determination of charge carrier dynamics in carbon nitride heterojunctions. Chem Commun 2021, 57(13), 1550–67.

[113] Chen XJ, Wang J, Chai YQ, Zhang ZJ, Zhu YF. Efficient photocatalytic overall water splitting induced by the giant internal electric field of a g-C_3N_4/rGO/PDIP Z-scheme heterojunction. Adv Mater 2021, 33(7), 2007479.

[114] Yuan L, Qi MY, Tang ZR, Xu YJ. Coupling strategy for CO_2 valorization integrated with organic synthesis by heterogeneous photocatalysis. Angew Chem Int Ed 2021, 60(39), 21150–72.

Qi Xiao and Jingsan Xu*

Chapter 5
Carbon nitride as noninnocent catalyst support

5.1 Introduction

Graphitic carbon nitride (g-C_3N_4, CN) materials are very attractive for many applications, especially for catalysis. The chemistry enabled by CN in popular applications, such as photocatalytic water splitting, CO_2 transformation, and single atom catalysis, etc., have been described well and summarized in the other chapters. In all kinds of chemical transformations, CN acts as either catalyst itself or as the support material to hold catalytic metal active sites. In full water splitting and other reactions mediated by catalysts immobilized at the CN, the latter serves: 1) as the light absorber to provide upon excitation with light charge carriers to the catalyst for the subsequent redox process and 2) as the heterogeneous support. However, the CN's role in other catalytic systems, particularly those applied in dark, have been largely overlooked. In particular, it is the case in heterogeneous catalysis for organic transformations in which CN is usually considered simply as an "innocent" platform to hold the metal particles – the catalytic sites. In fact, the catalytic activity is defined by the nature of both the metal particles and the support material. It has already been recognized that it is a unique feature of CN to act not only as the platform for the dispersion of catalytic sites, but also as the assistant to impact the catalytic performance. In this chapter, we highlight the CN's noninnocent role in heterogeneous catalysis – it is not just a support.

CN has rich surface properties that are important for catalysis, such as basic surface functionalities, electron-deficient nature, H-bonding motifs, etc., as shown in Fig. 5.1 [1, 2]. It is well known that CNs are a class of polymeric materials consisting mainly of carbon and nitrogen. But, in fact, real CN materials, like those prepared by polycondensation of cyanamide, usually contain a small amount of hydrogen (amine groups on the terminating edges). The presence of hydrogen indicates that the real CN is incompletely condensed, and that a number of surface defects exist, which can be used in catalysis. Moreover, its high thermal stability (it is stable up to 600 °C in air) and hydrothermal stability (it is not soluble in acidic, neutral or basic media) enable

*Corresponding author: Jingsan Xu, School of Chemistry and Physics, Science and Engineering Faculty, Queensland University of Technology, Brisbane, QLD 4000, Australia,
e-mail: jingsan.xu@qut.edu.au
Qi Xiao, State Key Laboratory for Modification of Chemical Fibers and Polymer Materials, College of Materials Science and Engineering, Donghua University, Shanghai 201620, China

https://doi.org/10.1515/9783110746976-005

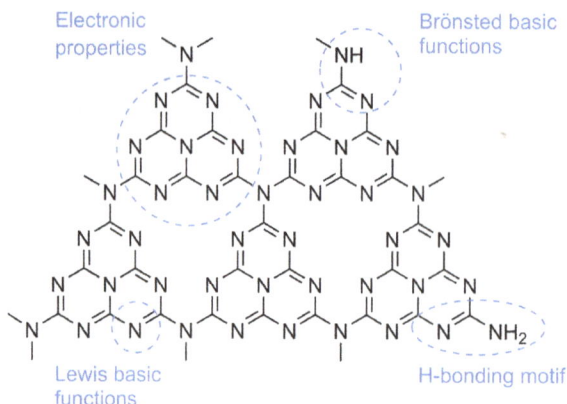

Fig. 5.1: Multiple surface functionalities found on CN (reprinted with permission from ref. [1, 2]. Adapted with permission from Copyright © 2008, the Royal Society of Chemistry, and Copyright © 2014, the American Chemical Society).

the material to be applied in various reaction environments (in different solvents and atmosphere), potentiating CN as excellent support material in heterogeneous catalysis.

In the supported catalysis, it is generally accepted that material activity and durability depend on the metal nanoparticle size and dispersion, as well as the interactions between the metal particles and the support. Therefore, the support plays a crucial role in the catalytic performance. The unique electronic and surfaces properties of CN, as mentioned above, make it very appealing as the support. From the interaction point of view, charges flow between the CN support and the metal particles upon their contact, which results in electron-rich or electron-poor CN and hence affects the reaction efficiency and selectivity. Analogical to the concept of noninnocent ligands used in metal-organic complexes, we will term CN as the noninnocent support in the abovementioned conditions. To avoid repetition with the other chapters on the extensively studied photocatalytic water splitting, CO_2 reduction, single atom catalysis and heterojunction catalysis, in this chapter, we will focus exclusively on the role of CN as the noninnocent catalyst support. We will highlight its actual role in the unique interactions with the supported metal particles as well as the reaction medium.

5.2 The adsorption property of CN (the interaction of reactants with CN)

Wang and coworkers developed supported Pd@mpg-C_3N_4 (Pd@CN) catalyst for the hydrogenation of phenol to cyclohexanone with both high conversion (99%) and selectivity (>99%), under atmospheric pressure of H_2 at room temperature, using water as a clean solvent [3]. The Pd@CN catalyst shows unique basic properties and electronic behavior. The mechanistic studies suggest that in the initial stage, phenol is easily adsorbed on the surface of the catalyst, and H_2 is activated by the electronically supported Pd. The abundant N atoms provide stable anchoring sites for Pd particles, while electron transfer from CN to Pd particles promotes hydrogen dissociation (Fig. 5.2a). Phenol was suggested to be adsorbed in a nonplanar fashion through the hydroxyl group at the basic sites of the CN catalyst surface due to stronger O–H . . . N bonding compared to the weaker O–H . . . π interaction (Fig. 5.2a). The benzene ring of phenol is partially hydrogenated to the enol (Fig. 5.2b), which isomerizes rapidly to give cyclohexanone. As the cyclohexanone molecule is a weaker H-donor, it leaves the surface of the catalyst quickly and is replaced by a more strongly binding new phenol molecule (Fig. 5.2c). As a result, further hydrogenation of cyclohexanone to cyclohexanol is avoided, while high selectivity is achieved. This study clearly indicates that the interaction of the reactant phenol with the CN surface through the hydroxyl group affects the reaction selectivity. In other words, CN plays a critical role in the catalytic process.

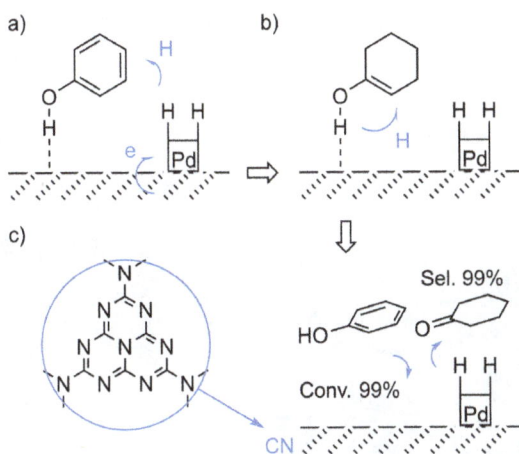

Fig. 5.2: Possible reaction mechanism of phenol hydrogenation to cyclohexanone over Pd@CN (adapted with permission from the American Chemical Society, Copyright © 2011 [3]).

Further study has shown that the adsorption capacity of CN toward the phenol molecules is ~ 20 times higher than that towards cyclohexanone. The produced cyclohexanone leaves the catalyst surface after the hydrogenation of phenol to cyclohexanone, which contributes to high selectivity and avoids formation of the deep hydrogenation side products [4].

A specially designed CN catalyst for hydrogenation of nitro-compounds has been developed by Ding and coworkers [5]. The catalyst possesses a unique structure composed of CN with underlying Ni supported on Al_2O_3 (CN/Ni/Al_2O_3). The composite catalyst shows good catalytic performance for hydrogenation of nitro-compounds to anilines under strongly acidic conditions (Fig. 5.3). It is worth noting that the catalyst efficiently drives the one-step hydrogenation of nitrobenzene to p-aminophenol in 1.5 M H_2SO_4. Compared to previous studies, in which strongly acidic conditions restricted the scope of catalysts to noble metals only, in this example, Ni is protected from corrosion or poisoning by the encapsulating robust CN layer. The mechanistic study has revealed that in the composite catalyst, Ni donates electrons to the CN, which, in turn, adsorbs and activates hydrogen to drive the hydrogenation reaction with high selectivity, even under strongly acidic conditions. This work highlights the unique electronic properties of CN with the ability to adsorb and activate hydrogen for hydrogenation reaction. The encapsulation ability of CN to maintain a robust stable catalyst will be useful particularly for the catalytic reactions under harsh conditions.

Unlike the conventional metal oxide supports, such as ZrO_2, TiO_2 and others, scientists find that CN tends to give rise more uniform catalyst nanoparticles upon chemical deposition. For example, as reported by Yi and coworkers [6], the CN support supplies anchor centers for the more uniform in situ deposition of Cu_2O nanoparticles from aqueous solution, which enables exposure of more catalytically active sites in metal oxide (Fig. 5.4). In addition, the CN support demonstrates good adsorption capacity for CO_2

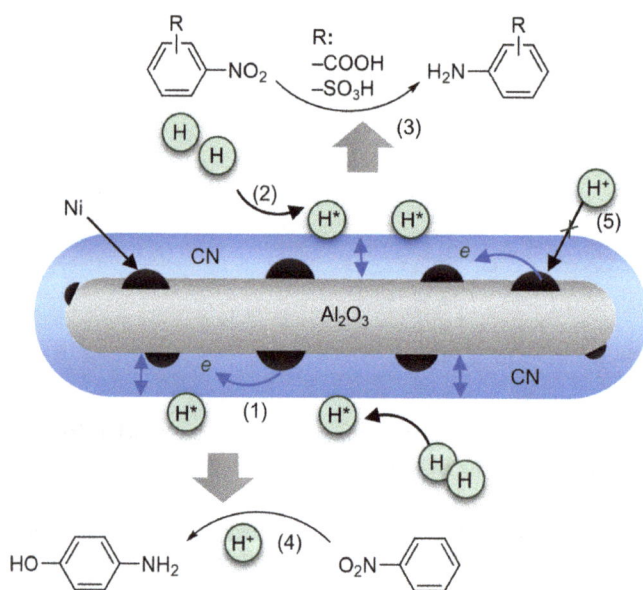

Fig. 5.3: Schematic mechanism of nitroarenes hydrogenation to anilines by the CN composite catalyst with underlying nickel. The arrows correspond to the following processes: (1) electron donation from nickel to CN; (2) dissociative adsorption of the hydrogen molecule followed by binding to the CN; (3) reduction of the nitro compound by the activated hydrogen; (4) reduction of nitrobenzene by the adsorbed hydrogen under strongly acidic conditions, followed by hydroxylation and rearrangement of the intermediate; and (5) protection of nickel from corrosion or poisoning by the CN layer (reproduced with permission from ref. [5], Copyright © 2014, American Chemical Society [5]).

owing to the rich N sites; the strong adsorption enhances the CO_2 concentration surrounding the catalyst. Besides, the authors have revealed that the use of CN is beneficial for the creation of abundant C1 intermediates (e.g., *CO, *CHO, CH_2O) for the downstream C–C coupling. In particular, *CO has been identified as a key intermediate in the C–C coupling during the formation of C_2H_4. Therefore, it is reasonable to conclude that the *CO intermediate formed at the surface of CN participates in the C–C coupling over the Cu_2O nanoparticles, and the synergistic effect between the CN and Cu_2O enhances the intrinsic activity of the catalyst for the accelerated kinetic rate of C_2H_4 production.

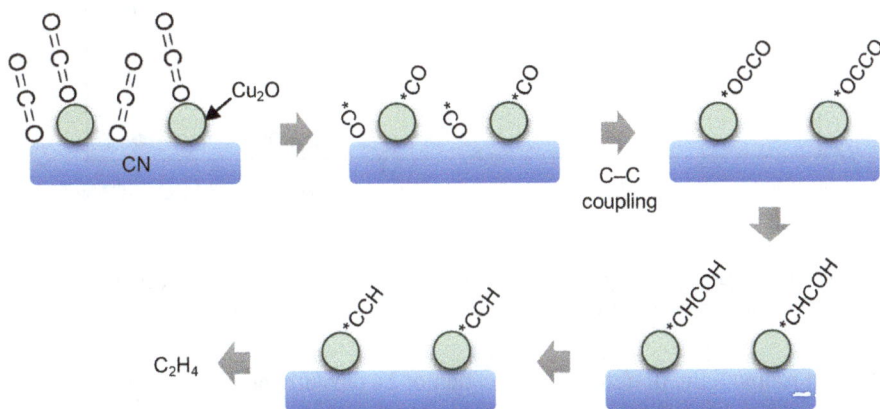

Fig. 5.4: Key reaction steps of CO_2 reduction to ethylene (reproduced with permission from ref. [6], Copyright © 2020 Wiley-VCH GmbH).

In another work, researchers have introduced defects into CN support to modify the Pd-CN interaction and, as a result, to improve the electrocatalytic conversion of 2,4-dichlorophenol to phenol [7]. The CN featuring terminal $N \equiv C$ groups and N vacancies as the defects was prepared by alkali-assisted (KOH) thermal condensation of urea. Compared to the pristine CN support, which has strong affinity toward the phenol molecules, thus hindering adsorption and activation of 2,4-dichlorophenol, the defective CN prohibits effective binding of the phenol molecules at the Pd sites. As a result, 2,4-dichlorophenol molecules are allowed to approach the catalyst surface. Density functional theory (DFT) calculations confirmed the above analysis (Fig. 5.5). The phenol (P) binds much stronger to Pd/CN than the 2,4-dichlorophenol (2,4-DCP) for the selected adsorption configurations. The preferential phenol adsorption at the Pd/CN surface hampers the effective adsorption of 2,4-dichlorophenol, which is essential for the subsequent hydrodechlorination reaction. When the catalyst surface contains single - $C \equiv N$ defect or single nitrogen vacancy ($-N_V$) defect, the relative adsorption strength of 2,4-dichlorophenol increases. Thus, the Pd/CN material containing $-C \equiv N$ defects can effectively mediate the spatial orientation and electronic structure of Pd nanoparticles

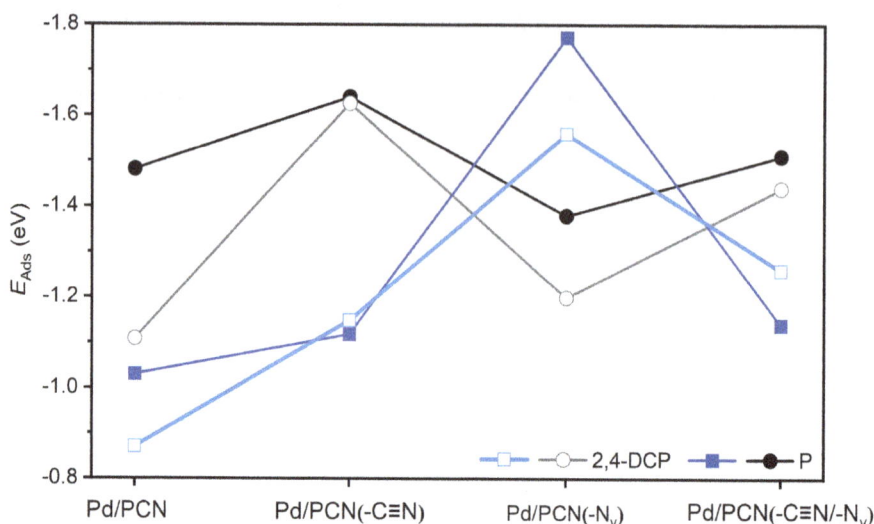

Fig. 5.5: The adsorption energies of 2,4-DCP and P on different surfaces (blue indicates that the adsorption site is adjacent to the cyano group, and gray/black indicates that the adsorption site is adjacent to the nitrogen vacancy) (adapted with permission from ref [7] Copyright © Elsevier B.V.).

(NPs) and promote the preferential adsorption of 2,4-dichlorophenol rather than phenol, resulting in an enhanced catalytic efficiency.

Shuai et al. developed CN-supported ultrafine Pd and Pd–Cu NPs for catalytic hydrogenation of waterborne contaminants, nitrite and nitrate ions [8]. The optimized catalysts show high activity in contaminant degradation, excellent selectivity for nitrogen gas compared to ammonium production and stability over multiple reaction cycles. The CN supports produce very small metal NPs (1–5 nm in diameter) for catalytic reactions, by taking advantage of CN's hydrophilic nature, abundant N-containing groups and porous structure. The performance of the catalysts is attributed to the porous structure of the CN support, which facilitates mass transfer of reactants. For example, fast diffusion of N-containing reactants and intermediates in nitrite and nitrate reduction promotes N–N coupling and preferential formation of nitrogen gas over ammonia/ammonium.

5.3 The electronic property of CN (the interaction of metal NPs with CN)

The nitrogen surface chemistry of carbon-based materials strongly affects interaction of supported metal NPs and thereby improves the catalytic activity. Prati et al. prepared the supported Pd catalysts on N-free activated carbon (AC*), covalent triazine framework (CTF) and CN as the support, respectively, with the defined nitrogen content

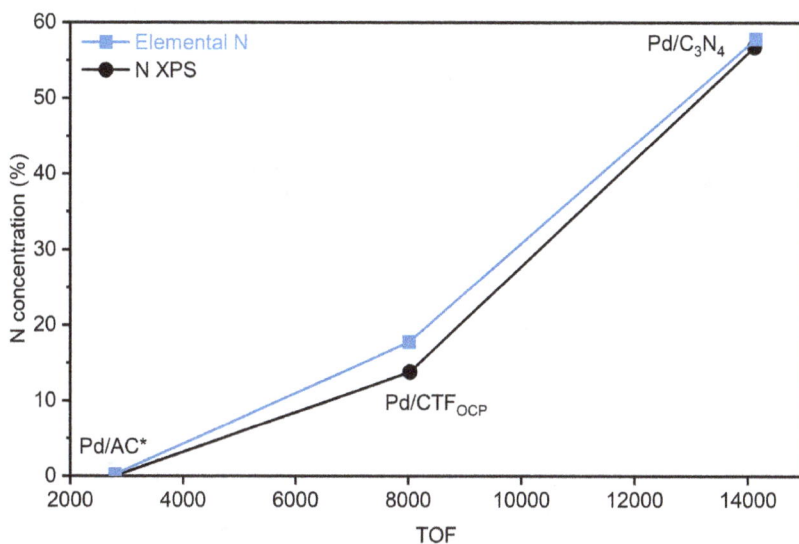

Fig. 5.6: Activity of the supported Pd catalyst as a function of nitrogen concentration [9] (adapted with permission from ref. [9], Copyright © 2012 Wiley-VCH Verlag GmbH&Co. KGaA, Weinheim).

(Fig. 5.6) [9]. The catalysts were used for oxidation of benzyl alcohol. It was shown that the activity does not correlate with the Pd particle diameter, oxidation state or the surface area of the catalysts. Instead, the activity correlates almost linearly with the concentration of nitrogen species on the support. A distinct decrease in nitrogen content from 57% in CN to 14% in CTF determined by XPS strongly support the fact that the coordination of Pd by greater number of nitrogen atoms contributes to higher catalytic activity.

In another study, the Pd@CN catalyst shows excellent catalytic performance in the selective hydrogenation of quinoline to 1,2,3,4-Tetrahydroquinoline (THQ). The texture of the CN influences the catalytic activity of the material in the reaction. The ordered mesoporous CN (ompg-CN) remarkably enhances both the catalytic conversion (100%) and the selectivity (100%) to THQ in the hydrogenation of quinoline with molecular hydrogen as a reductant (Fig. 5.7a), which is much higher than the traditional Pd/C catalyst (48% conversion and 88% selectivity in 3 h) and mpg-CN catalyst (54% conversion and 100% selectivity in 3 h) in the same reaction conditions [10]. The enhanced catalytic activity is attributed to the transfer of electron from the CN to Pd, which enriches the electron density of the metallic Pd and, thereby, accelerates the hydrogenation reaction. As shown in Fig. 5.7b, mechanistically, the adsorption of quinoline onto the CN is promoted by the formation of N . . . H–N hydrogen bonds between the quinoline molecule and the NH and NH_2 groups on the surface of CN. Subsequently, the quinoline molecule interacted with the active H produced over Pd nanoparticles, resulting in the formation of THQ. The introduction of ordered mesoporous CN was suggested to facilitate the mass transfer, achieving good accessibility of

a)

b)

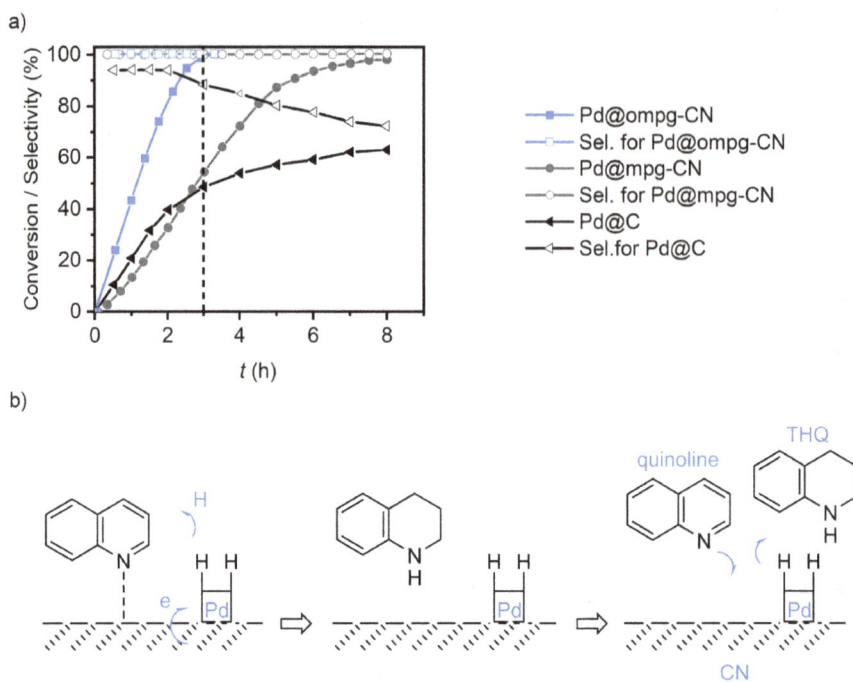

Fig. 5.7: (a) The catalytic performance of various supported Pd catalysts in hydrogenation of quinoline to THQ. (b) Possible reaction mechanism of quinoline hydrogenation over Pd@CN catalyst [10] (adapted with permission from ref [10], Copyright © 2012 Elsevier Inc.).

reactant molecule to the formed high concentration of metallic Pd active sites with high electron density, which therefore leads to better catalytic performance.

Pd@CN has also been demonstrated as a catalyst completing the CO_2-based hydrogen storage cycle using formic acid (FA) by Yoon and coworkers [11]. FA has recently been recognized as a safe and reversible hydrogen storage material. Hydrogen gas can be liberated from FA even at room temperature ($HCOOH \rightarrow CO_2 + H_2$). For hydrogen storage, FA is regenerated upon hydrogenation of CO_2 ($CO_2 + H_2 \rightarrow HCOOH$). The Pd nanoparticles with an average diameter of 1.7 nm were dispersed uniformly onto CN, without any agglomeration, and exhibit superior activity in the dehydrogenation of FA, with a turnover frequency of 144 h^{-1} at room temperature. In addition, the developed Pd@CN catalyst is able to synthesize FA selectively upon CO_2 hydrogenation with the aid of triethylamine. The basic sites of the CN support bind CO_2 for the synthesis of FA. Therefore, the hydrogen storage cycle could be completed with the developed catalyst (Fig. 5.8a). DFT calculations were conducted using three Melem units. The optimized structure contains three different types of nitrogen sites (N, N' and N") inside the CN (Fig. 5.8b). The analysis of calculated molecular orbitals indicates that highest occupied molecular orbital (HOMO) is localized at the sp^2-hybridized nitrogen atoms of the triangular macrocycle composed of three Melem units (Fig. 5.8b). Notably, the HOMO is

a)

b)

c)

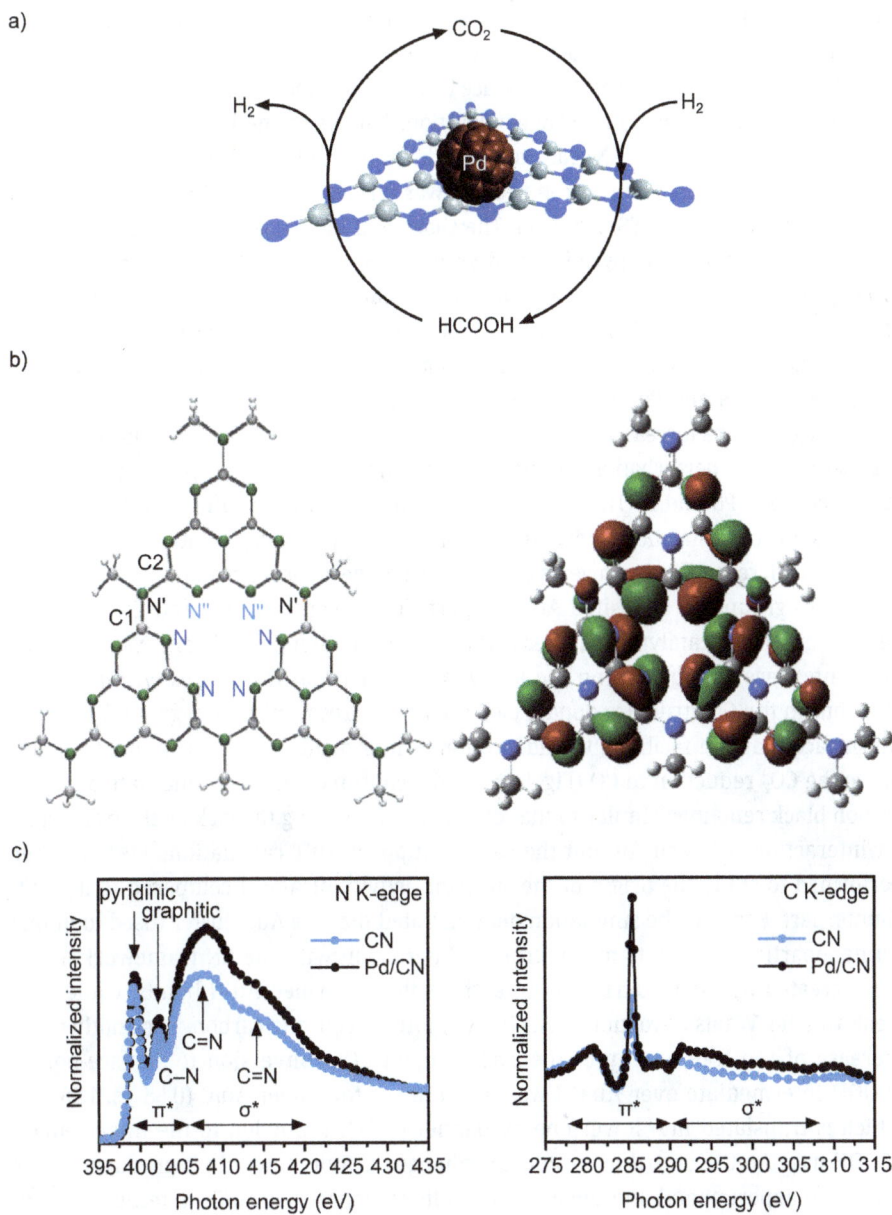

Fig. 5.8: (a) Schematic of Pd/CN-catalyzed hydrogen production and storage. (b) DFT-optimized structure using a truncated CN unit, and the calculated HOMO of the CN optimized structure. (c) The XANES spectra of CN and Pd/CN, N K-edge and C K-edge (adapted with permission from ref. [11], Copyright © 2014 The Royal Society of Chemistry [11]).

more diffused at sp^2-N (or N″) than the N′ sites, implying that during Pd NPs deposition, Pd^{2+} ions from the precursor or acidic proton of FA in H_2 generation are likely to interact with N or N″ sites. These N sites may induce formation of small-sized Pd NPs, stabilize the obtained Pd NPs and avoid their agglomeration. Moreover, the interaction of Pd^0 NPs with the nitrogen atoms at CN leads to the redistribution of electron density from the N atoms of CN to Pd NPs. The electron transfer was further confirmed by X-ray absorption near edge structure (XANES, Fig. 5.8c). The sharp and intense π^* peaks in the N K-edge spectra correspond to the pyridinic and graphitic N species of CN. In the σ^* region, a broad peak at >405 eV was observed, which is attributed to C–N conjugated double bonds (407–410 eV) and/or C–N double bonds (411–415 eV). The increased peak intensities after Pd deposition indicate that charge transfer from N sites to Pd occurred via orbital hybridization between N and Pd. In addition, the peak intensities associated with C species in the C K-edge spectra increased following Pd deposition in both π^* and σ^* regions, suggesting that C orbital perturbation resulted from electronic interactions between Pd and C and/or between Pd and N. The electronic properties of CN with Pd nanoparticles to improve the overall catalytic activity suggest that CN is a promising catalyst support for a CO_2-mediated, reversible hydrogen storage system to meet energy requirements.

Yang's group has deposited Au nanoparticles on carbon black and CN and compared their electrocatalytic activities toward CO_2 reduction [12]. XPS and XANES measurements were conducted and the analysis revealed that the N sites with lone-pair electrons in the CN structure donate electrons to Au, resulting in a negatively charged Au surface that can facilitate the adsorption of intermediates (*COOH) and, thus, improve the CO_2 reduction to CO (Fig. 5.9). On the contrary, the electronic state of Au on carbon black remained similar to that of bulk Au, indicating the lack of electron transfer/interaction between Au and the carbon support. DFT calculations were further performed to study the origin of the superior activity of Au/CN compared to its Au/C counterpart. For one, the simulation demonstrated that the Au_8 cluster (used to mimic Au nanoparticles) exhibits much stronger interaction with the CN framework at the voids created by the tri-s-triazine units through the formed Au-N bonds, while only weak van der Waals attraction exists between Au cluster and carbon. For another, the increase of the free energy of the system upon CO_2 conversion to surface-bound *COOH intermediate over Au/CN was much lower than over Au/C (0.58 vs. 1.26 eV), which is translated into lower energy barrier of CO_2 reduction to the intermediate *COOH over Au/CN, and, thus, results in enhanced activity toward CO production. The combined experimental and theoretical results evidence that Au/CN interaction induces the formation of negatively charged Au surface, which could stabilize the key intermediate *COOH. Similar results were also observed for the CN-supported Ag NPs for the CO_2 reduction.

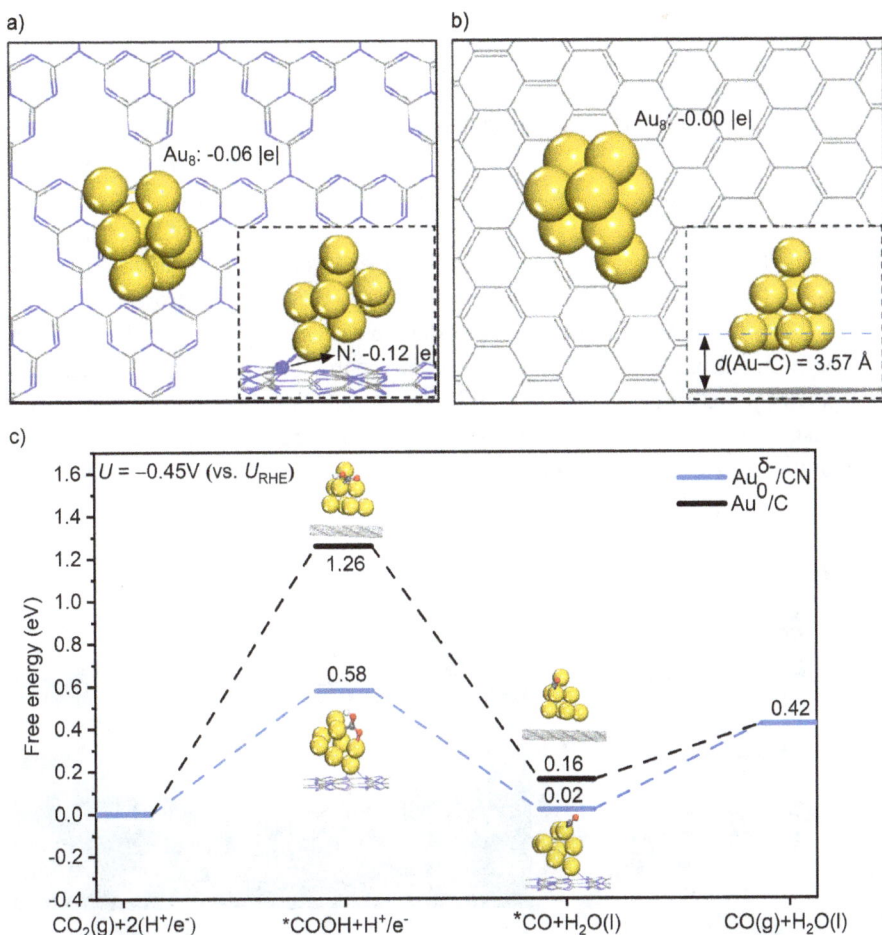

Fig. 5.9: Top and side (inserted) views of the optimized Au_8 cluster adsorbed on (a) CN at the hollow site and (b) at the graphene-carbon support. The key charge transfers are labeled. (c) The calculated Gibbs free-energy profile of CO_2 reduction into CO catalyzed by Au/CN and Au/C, respectively. Yellow, white, red, gray and blue balls represent Au, H, O, C and N atoms, respectively (adapted with permission from ref [12], Copyright © 2018 American Chemical Society [12]).

5.4 Hydrogen spillover in CN

In heterogeneous catalysis that involves hydrogen, the molecules of the latter are adsorbed and dissociated by the metal catalyst, normally platinum, palladium or other transition metals. One very interesting and useful phenomenon is that the hydrogen atoms dissociated at the metal catalysts migrate to the nonmetal support materials, documented as the concept of hydrogen spillover. Hydrogen spillover has been observed for many carbon-based support materials, such as graphite, carbon nanotubes and graphene. Recently, researchers have studied the hydrogen spillover on noble metal-loaded CNs, both experimentally and computationally. Pan et al. predicted by first-principles calculation that CN nanotube is a potential high-performing material in hydrogen storage due to the porous structure and the rich active sites for the adsorption – double-bonded nitrogen atoms and the pore edges, with hydrogen uptake up to 5.45 wt% of the CN nanotube bundles under ambient conditions [13]. Moreover, their calculations indicate that light metal doping with, for example, Ti, may induce hydrogen spillover and, thus, increase the storage capacity (Fig. 5.10a). Then, it was experimentally confirmed that decoration with Pd nanoparticles significantly improves the hydrogen uptake capacity, owing to the hydrogen spillover mechanism. The Pd/CN composite with mesopores demonstrates a hydrogen storage capacity up to 3.4 wt% at 25 °C and 4 MPa [14]. By alloying cobalt with Pd, the hydrogen uptake of the Pd_3Co/CN composite was further increased, reaching 5.3 wt% at room temperature and 3 MPa, which is attributed to the facile spillover process (Fig. 5.10b) [15].

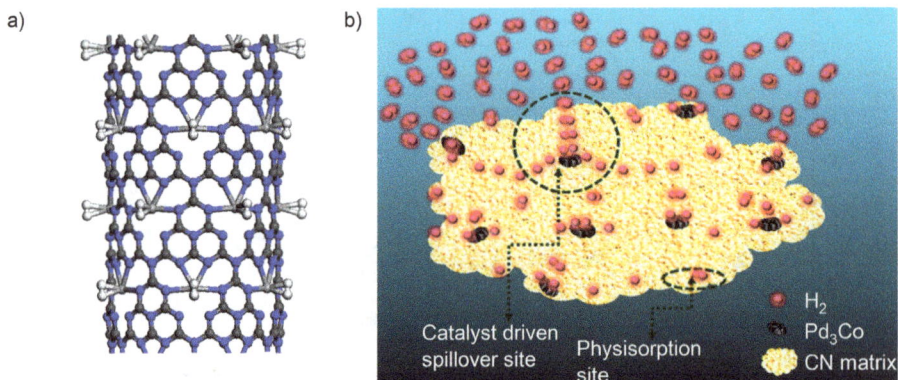

Fig. 5.10: (a) Representative structure of Ti-functionalized graphitic carbon nitride nanotube, with one H_2 molecule adsorbed at each triangular pore. C, N, H and Ti atoms are represented in gray, blue, white and silver, respectively [13] (Copyright © 2011 Elsevier Ltd). (b) Schematic representation of the hydrogen adsorption mechanism in Pd_3Co/CN structure [15] (adapted with permission from ref. [15], Copyright © 2016 American Chemical Society).

Xu and coworkers designed a Pd/CN stabilized Pickering emulsion microreactor to couple release of hydrogen from ammonia borane with hydrogenation of alkenes at the oil–water interface of the emulsion droplets, taking advantage of the high hydrogen content in ammonia borane (AB, 19.4 wt%) [16]. One remarkable result is that the utilization efficiency of the hydrogen in ammonia borane reaches 100%, i.e., the hydrogen liberated from AB is fully used for hydrogenation reactions. The hydrogen utilization efficiency can reach unity in such an emulsion system, largely due to the localized production, storage and consumption of hydrogen on the Pd nanoparticles: the Pickering emulsion can store hydrogen via adsorption in the Pd nanoparticles and microbubble formation inside the liquid phases (Fig. 5.11). The stored hydrogen then reacts with the alkene molecules until completely consumed. The hydrogen spillover process was suggested to occur between the Pd and CN in the Pickering emulsion microreactor system.

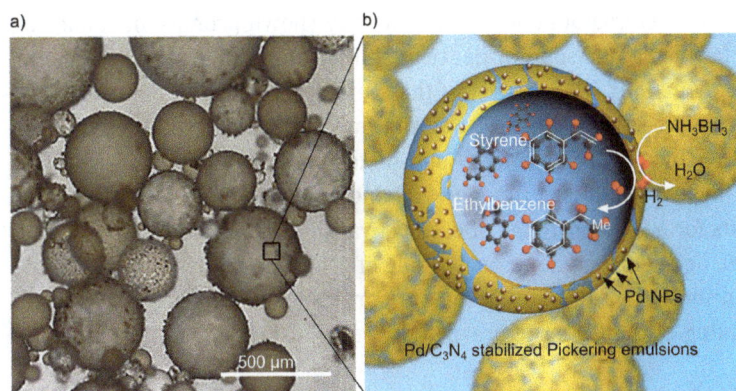

Fig. 5.11: (a) Light microscopy image of Pickering emulsions made from 1 mL of styrene and 1 mL of 2 mg mL^{-1} 2 wt% Pd/g-CN dispersion. (b) Schematic illustration of AB hydrolysis combined with styrene hydrogenation in the Pickering emulsion system, catalyzed by the Pd NPs at the droplet interfaces [16] (Copyright © 2018 Wiley-VCH Verlag GmbH&Co. KGaA, Weinheim).

5.5 Summary and outlook

In this chapter, we have briefly summarized the role of CN as the noninnocent catalyst support in various catalytic reactions. Among those reported CN-supported catalysts, significantly enhanced activity and selectivity were achieved in almost all cases. Apart from the fact that a CN framework provides perfect sites to hold metal NPs and ensure the uniform dispersion of the NPs, a significant role is ascribed to the electronic properties and adsorption of reactants. The lone pairs of nitrogen atoms in CN facilitate electron transfer to the supported metal NPs, which, similar to the Schottky barrier, prevents the reverse electron flow from the metal NPs to CN. As a result, reduction reactions proceed more effectively at the catalytic active sites. Attractively, the unique structural features of CN enable specific adsorption modes of many reactants, especially organic compounds, thus offering advantages for high selectivity. Thus, the unique surface and electronic properties of CN should not be overlooked – the rich nitrogen content, inherent chemical and mechanical stability provide sufficient foundation for the employment of CN as excellent catalyst support. Nevertheless, the origin of the improvement of activity and selectivity remains a topic of debate. More detailed advanced characterization of the interaction between CN and the supported metal NPs and the interaction between CN and the reactant would give a better understanding of the CN-based catalysts and provide guidance for catalysts design. Therefore, much effort should be put in this direction. In situ characterizations to determine the structure of real active sites and further investigation of the abovementioned interactions with the CN-based catalysts would be a meaningful research topic, not only for the heterogeneous catalysis but also for materials science. Although the CNs with defects and heteroatom doping have already shown promise in enhancing the catalytic performance, the solid evidence is still lacking. Elaborate experimental and theoretical calculations with respect to those effects are desired. Moreover, the unique adsorption properties of the reactant and the intermediates on CN provide a promising way to achieve good selectivity and better understanding of the reaction mechanism. We believe that it is of fundamental and practical importance to develop CN-based catalysts for a broader range of organic transformations with desired product selectivity. We hope that the noninnocent role of CN as the catalyst support would draw more attention when developing the supported heterogeneous catalysts. In-depth understanding of the CN structure–activity relationships could be revealed in the near future.

References

[1] Thomas A, Fischer A, Goettmann F, Antonietti M, Müller J-O, Schlögl R, et al. Graphitic carbon nitride materials: Variation of structure and morphology and their use as metal-free catalysts. J Mater Chem 2008, 18(41), 4893–908.

[2] Zhu J, Xiao P, Li H, Carabineiro SA. Graphitic carbon nitride: Synthesis, properties, and applications in catalysis. ACS Appl Mater Interfaces 2014, 6(19), 16449–65.

[3] Wang Y, Yao J, Li H, Su D, Antonietti M. Highly selective hydrogenation of phenol and derivatives over a Pd@carbon nitride catalyst in aqueous media. J Am Chem Soc 2011, 133(8), 2362–5.

[4] Li Y, Xu X, Zhang P, Gong Y, Li H, Wang Y. Highly selective Pd@mpg-C_3N_4 catalyst for phenol hydrogenation in aqueous phase. RSC Adv 2013, 3(27), 10973–82.

[5] Fu T, Wang M, Cai W, Cui Y, Gao F, Peng L, et al. Acid-resistant catalysis without use of noble metals: Carbon nitride with underlying nickel. ACS Catal 2014, 4(8), 2536–43.

[6] Zhang J, Guo Y, Shang B, Fan T, Lian X, Huang P, et al. Unveiling the synergistic effect between graphitic carbon nitride and Cu_2O toward CO_2 electroreduction to C_2H_4. ChemSusChem 2021, 14(3), 929–37.

[7] Shu S, Wang P, Zhang W, Wang W, Li J, Chu Y, et al. Pd nanoparticles on defective polymer carbon nitride: Enhanced activity and origin for electrocatalytic hydrodechlorination reaction. Chin Chem Lett 2020, 31(10), 2762–8.

[8] Ye T, Durkin DP, Banek NA, Wagner MJ, Shuai D. Graphitic carbon nitride supported ultrafine Pd and Pd-Cu catalysts: Enhanced reactivity, selectivity, and longevity for nitrite and nitrate hydrogenation. ACS Appl Mater Interfaces 2017, 9(33), 27421–6.

[9] Chan-Thaw CE, Villa A, Veith GM, Kailasam K, Adamczyk LA, Unocic RR, et al. Influence of periodic nitrogen functionality on the selective oxidation of alcohols. Chem – Asian J 2012, 7(2), 387–93.

[10] Gong Y, Zhang P, Xu X, Li Y, Li H, Wang Y. A novel catalyst Pd@ompg-C_3N_4 for highly chemoselective hydrogenation of quinoline under mild conditions. J Catal 2013, 297, 272–80.

[11]   Lee JH, Ryu J, Kim JY, Nam S-W, Han JH, Lim T-H, et al. Carbon dioxide mediated, reversible chemical hydrogen storage using a Pd nanocatalyst supported on mesoporous graphitic carbon nitride. J Mater Chem A 2014, 2(25), 9490–5.

[12]   Zhang L, Mao F, Zheng LR, Wang HF, Yang XH, Yang HG. Tuning metal catalyst with metal–C_3N_4 interaction for efficient CO_2 electroreduction. ACS Catal 2018, 8(12), 11035–41.

[13]   Koh G, Zhang Y-W, Pan H. First-principles study on hydrogen storage by graphitic carbon nitride nanotubes. Int J Hydrogen Energy 2012, 37(5), 4170–8.

[14]   Nair AAS, Sundara R, Anitha N. Hydrogen storage performance of palladium nanoparticles decorated graphitic carbon nitride. Int J Hydrogen Energy 2015, 40(8), 3259–67.

[15]   Nair AAS, Sundara R. Palladium cobalt alloy catalyst nanoparticles facilitated enhanced hydrogen storage performance of graphitic carbon nitride. J Phys Chem C 2016, 120(18), 9612–8.

[16]   Han C, Meng P, Waclawik ER, Zhang C, Li XH, Yang H, et al. Palladium/graphitic carbon nitride (g-C_3N_4) stabilized emulsion microreactor as a store for hydrogen from ammonia borane for use in alkene hydrogenation. Angew Chem Int Ed Engl 2018, 57(45), 14857–61.

Ivo Freitas Teixeira* and Gabriel Ali Atta Diab

Chapter 6
Carbon nitride-based materials: the ultimate support for single-atom catalysis

6.1 Introduction

Nowadays, catalysts are employed in 90% of all chemical processes carried out in chemical industries [1]. Without a doubt, sustainable economic development will require new and more efficient catalytic technologies. Berzelius, in 1835, defined catalysts as substances that increase the rate of chemical reactions without being consumed [2]. Typically, catalysts are divided into two categories: homogeneous catalysts, those that work in the same phase as the reactants (e.g., enzymes and soluble organometallic and inorganic complexes), and heterogeneous catalysts, which are solids and work in a different phase of the reactant (e.g., MOFs, COFs, metal oxides, supported metals and carbon nitrides) [1].

Recently, a potential missing link between homogeneous and heterogeneous catalysis seemed to be found, and it was named single-atom catalysis (SAC). The catalytic performance of metal-based catalysts is dictated by the active sites on the metal structure. Considerable efforts have been put in to increase the number of active sites to enhance the intrinsic reactivity of heterogeneous, metal-based catalysts [3]. It is generally recognized that the metal particle size in metal-based catalysts determines the catalytic performance and, in general, the catalytic activity increases with decrease in particle size. These observations inspire the catalysis research community to tune the structures of catalysts as well as the chemical environment of active sites down to the single-atom level [4]. A large number of publications in the last decade have shown the possibility of uniting the robustness of heterogeneous systems with the well-defined metal environment of homogeneous catalysts in SAC. However, the field of SAC has also been limited by the availability of appropriate analytical methods. Only with recent gains, it became possible to investigate single-atoms sites in a more systematic way [1, 5, 6].

Herein, we will present the most recent advances in metal single-atom catalysts (SACs) that are supported by carbon nitride-based materials (C_3N_4). Throughout this chapter, the mostly employed synthetic methods will be introduced and discussed for

*Corresponding author: Ivo Freitas Teixeira, Department of Chemistry, Federal University of São Carlo, 13565-905 São Carlos, São Paulo, Brazil, e-mail: ivo@ufscar.br
Gabriel Ali Atta Diab, Department of Chemistry, Federal University of São Carlo, 13565-905 São Carlos, São Paulo, Brazil

https://doi.org/10.1515/9783110746976-006

different carbon nitride-based materials. Furthermore, characterization techniques and the challenges that scientists encounter in them will be discussed. Lastly, the catalytic applications of SACs based on carbon nitrides will be reviewed, with a special focus on photocatalytic applications.

6.2 Single-atom catalysis (SAC)

Single-atom catalysts display distinctive performance in different chemical reactions due to their unique coordination environments, high atom utilization, high activity and selectivity [4]. Emerging as a new and innovative research topic, SAC can play a pivotal role in improved chemical synthesis as it provides a catalytic state that resembles advanced metal complexes, however, in immobile, stable and robust hosts [1, 7–11]. The reduction of the particle size, down to isolated atoms, enhances the catalytic turnover per gram of metal, which is important to reduce the amount of transition metals required by a reaction, without a loss of efficiency [1, 8–10].

Although the concept has been existing for a long time, only recently have SAC gained prominence, mostly due to the availability of appropriate host structures. Heterogeneous host materials in SAC fulfill similar functions as ligands in homogeneous catalysts – tuning the local environment, and influencing the stability and electronic properties of the metal centers. Therefore, host materials provide a means for tailoring the catalytic activity of these single-atom heterogeneous catalysts for targeted applications [1]. In the last few years, we have witnessed many examples of SACs; their performance disrupting the diverse fields of heterogeneous catalysis with their unique reactivity; but we also substantially improved our understanding of these processes at the molecular level. The term SAC, to date, has been employed mostly referring to late transition metal-based systems; however, plentiful examples exist in which isolated atoms of other elements play key catalytic roles [1]. When Zhang et al. reported the first example of heterogeneous single-atom catalyst and introduced the concept of SAC, they referred to cases in which the active sites consisted of metal atoms [7, 8, 12]. This terminology was essentially kept by the catalysis community, which employs the term SAC and refers particularly to the late transition metals (LTM), mainly Pt, Pd, Fe, Co, Ni and Cu [3, 13]. However, there is a recent trend to use the term SAC in a broader context as, for example, done by Pérez-Ramírez and collaborators in their review. They defined SAC as "An atom of any element (metal, metalloid, non-metal, halogen) that is spatially isolated from atoms of the same chemical identity (e.g., adsorbed, embedded, immobilized, grafted or confined) and directly bound to a solid carrier." [1] As this broader definition would include doped carbon nitrides; in this chapter, we will define SAC only as supported metal single-atoms.

6.3 Carbon nitride-based materials (C_3N_4): the (possibly) ultimate support for single-atom catalysts

Single-atom heterogeneous catalysts, commonly defined as individual atoms confined in a solid framework serving as active centers without metal-metal interaction, have been acknowledged by the catalysis community as a most promising class of heterogeneous catalysts [4]. However, reducing the metal particle size from the nanometer to the sub-nanometer scale (atomic dispersion) is accompanied by an abruptly increasing surface energy; thus, ultra-small particles are more likely to aggregate to larger clusters resulting in a challenge to maintain the atomic level dispersion [8]. The key factor to protect and make this atomic dispersion energetically feasible is controlling the interaction/bonding between the metal single atoms and the support [4].

C_3N_4 has emerged as one of the most promising two-dimensional supports to prepare SACs, due to the presence of nitrogen-rich heteropores in their lattices. These, indeed, can anchor metal atoms firmly [14]. In comparison with another most commonly used support for SACs preparation, nitrogen-doped carbons, which, in general, expose significant structural heterogeneity – the more uniform and controlled structure, united with the higher content of nitrogen pores within C_3N_4 materials, offers abundant and more precisely defined coordination sites [14–18].

C_3N_4 materials are nitrogen-rich polymeric organic semiconductors, with a unique layered structure and remarkable intrinsic properties, such as high chemical and thermal stability. The C_3N_4 materials can be synthesized in different dimensions (0D, 1D, 2D, 3D), with different molecular structures (i.e., heptazine- or triazine-based), different crystallinities and different charge balances in their structure (e.g., Li-PTI and K- or Na-PHI and g-C_3N_4) [4, 14]. These properties promote the C_3N_4 materials to compete for the position of the ultimate candidate to support metal single-atoms.

6.4 Characterization techniques

Broad work in the last decade has shown the possibility to stabilize metal single-atoms in supports and exploit the high activity of the single-atom metal sites. However, the field of SAC has been limited by the resolution of the analytical methods. Only with recent gains it has become possible to investigate the single-atoms sites in a more systematic way [1, 5]. The characterization of single-atom catalysts demands a series of complementary techniques, including high-resolution electron microscopy (e.g., scanning electron transmission microscopy (STEM) and scanning tunneling microscopy (STM)) and spectroscopies (e.g., X-ray absorption spectroscopy (XAS), extended X-ray absorption fine structure spectroscopy (EXAFS), energy-dispersive spectroscopy and electron energy loss spectroscopy). In this section, we will summarize the most widely used characterization techniques applied in SAC, and discuss their limitations.

6.4.1 Electron microscopy

Confirming the presence of well-dispersed isolated single atoms is the first step in the investigation of SAC. The most common way to demonstrate the successful obtention of metal single-atoms sites is their visualization by STM and STEM, usually combined with the spherical aberration-correction and detected in high-angle annular dark-field mode to further increase the contrast for lighter elements [1].

Electron microscopy techniques are the most widely used to investigate SACs, and are appealing, as they allow visualizing metal single-atoms sites. Nevertheless, electron microscopy images must be interpreted carefully since it is well known that single-atoms can be generated and stabilized during the electron microscopy analysis due to the generation of vacancies and scattering of metal atoms from contaminants, nanoparticles or clusters [19]. Although electron microscopy techniques are a very important tool to investigate SACs, they should be combined with other techniques to confirm unambiguously the presence of metal single-atoms.

6.4.2 X-ray absorption spectroscopy

X-ray absorption near-edge structure (XANES) spectroscopy and EXAFS permit to determine directly the coordination environment of metal atoms in a framework and their chemical state [20]. These techniques have been crucial to investigate the relationship between the local environment of active metal sites and their catalytic activity in heterogeneous catalysis [4]. EXAFS and XANES can successfully provide information about the atomic and electronic structures of metal single-atoms, including interatomic distance, the coordination number/geometry, oxidation states, among others. Thus, they have been widely used in the characterization of SACs based on C_3N_4. EXAFS is especially useful to indicate the absence of metal-metal bonds, which is crucial to confirm the atomic dispersion of the metal. However, EXAFS results must also be interpreted cautiously – the technique, for instance, cannot differentiate M–O and M–N bonds. Therefore, clusters and nanoparticles of metal oxide/metal hydroxides can be easily misinterpreted as SAC M–N bonds. Nevertheless, the importance and value of XANES and EXAFS to investigate SACs are undeniable, especially with the further development of in situ and *operando* experiments that allow following the metal single-atoms stabilization in real-time [1]. Another great advantage of these techniques is that they provide average characteristics of the sample, which is complementary to the electron microscopy techniques. On the other hand, their great drawback is their accessibility as synchrotron-based techniques. Therefore, to accelerate the development of the SAC topic, it is important to develop effective lab-based techniques that allow confirmation of the presence of metal single-atoms. One promising technique is diffuse reflectance Fourier transform infrared spectroscopy (DRIFT-FTIR) of probe molecules. This technique will be introduced and discussed in the next section.

6.4.3 Infrared spectroscopy

DRIFT-FTIR, with suitable probe molecules, allows evaluating the existence of metal single-atoms [20]. The CO molecule is particularly useful to assess the nature of supported metals, since its vibration frequency in the IR is highly sensitive to the coordination mode. Over the past years, in situ FTIR was applied to atomically dispersed metal atoms in supported SACs, which has been especially successful for Pt metal [5, 7, 21, 22]. However, very few examples have been reported for SACs based on C_3N_4. In 2021, Teixeira's group reported the use of CO probe molecule to confirm the presence of Ni as single atoms, stabilized in PHI using DRIFT-FTIR (Fig. 6.1) [23]. The authors claimed that only peaks related to linearly adsorbed CO (2040 and 2,179 cm^{-1}) were observed and no indication of bridging CO was noticed [23].

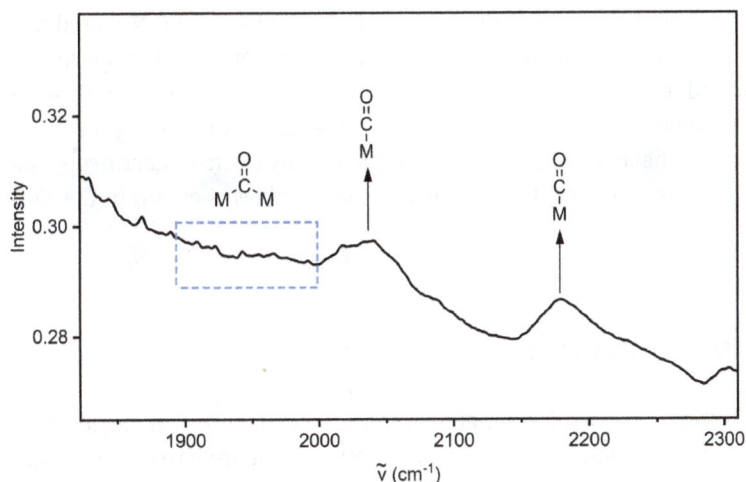

Fig. 6.1: CO DRIFTS FT-IR spectrum of the Ni–PHI sample (adapted from ref. [23] with permission from the Royal Society of Chemistry, copyright 2021).

Theoretically, probe molecule DRIFT-FTIR permits not only to confirm the atomic dispersion of the metal in the support but also to quantify it by comparing the signals of atomically distributed metals against those of clusters and nanoparticles. Despite very promising first experiments, to apply this technique properly, it is crucial to differentiate the bands related to probe molecules adsorbed at the support from the bands of probe molecules adsorbed at the metal sites. Although thermal treatment before analysis and the experiment setup allow minimizing the contribution of the absorption bands from the support, it is still essential to systematically investigate the support to properly understand its interaction with the probe molecules. The investigation of polymeric and defective carbon nitrides as supports, with their variety of structures, make the application of this technique even more challenging. To sum up a too detailed discussion, one can state that further development is necessary to allow the systematic investigation of SACs based on C_3N_4 by DRIFT-FTIR of the probe molecules. We believe that the highly crystalline supports, such as PHI and PTI, are particularly strong candidates to be investigated, and to create a database needed in the future to reference and quantify the single-atoms sites in C_3N_4-based SACs.

6.4.4 X-ray diffraction (XRD)

X-ray diffraction (XRD), in general, is not very useful to investigate single-atoms sites, unless they are incorporated regularly into a crystalline lattice. In this case, single-crystal XRD analysis can give important information about the local environment of single-atoms and the general structure of the material. It is therefore particularly useful to characterize MOFs [24] and zeolitic materials [1, 25]. In terms of C_3N_4-based materials, powder XRD is more important to confirm the absence of nanoparticles. Atomically dispersed metals must show the absence of peaks in the XRD, which seem to be an obvious statement. However, it is common that studies reporting metals in high loadings claim to have metal single-atoms supported by electron microscopy images, while at the same time peaks of metals and metal oxides show up in the XRD patterns.

6.5 Synthesis methods

A range of metals (Ni, Co, Fe, Cu, Pd, Ag, Pt or Ir) have been stabilized as single atoms on C_3N_4-based materials by employing a series of methods that are based on different approaches, such as direct (e.g., copolymerization) [26] or post-synthetic (e.g., wet deposition and cation exchange) [23]. In the preparation of SACs, it is of paramount importance to guarantee the effective dispersion of isolated single atoms on supports. In general, the stabilization and effective dispersion of metal single-atoms depend on the formation of strong bonding interaction between the metal centers and the support scaffolds. However, it is still a great challenge to prevent the single atoms from spontaneous aggregation into larger clusters, due to the high surface free energy [4].

The advances in the synthesis of SACs supported by C_3N_4-based materials will be presented in this section, organized according to the method (i.e., wet chemical route, cation exchange, atomic layer deposition, pyrolysis and microwave irradiation-assisted deposition).

6.5.1 Wet chemical route

Wet-chemical route impregnation starts with immersing the C_3N_4 substrate in the solution of a metal precursor [4]. In some cases, the method relies only on the adsorption of the metal from the solution media at the substrate's surface, which can be separated and dried and, if necessary, calcined afterwards [27, 28]. For example, Kim et al. have successfully synthesized cobalt single-atoms by stirring the mixture of C_3N_4 in the solution of $Co(NO_3)_2$ for 18 h. Then, the sample was separated by centrifugation and annealed at 400 °C [29]. In other cases, the metal deposition/precipitation can be pushed to the support, chemically (e.g., pH modification) or by the evaporation of the solvent [30–32]. For example, Lee et al. prepared well-dispersed Pd and Pt single-atoms with 0.5% loading by incipient wetness impregnation in ethanol, followed by the complete evaporation of the solvent at 60 °C. Similarly, Peng and collaborator obtained copper single-atoms by pushing the precipitation of copper ions by adding sulfuric acid, followed by calcination at 550 °C [33]. Despite the different strategies, strong interactions between the substrate scaffold and the metal species are essential to avoid aggregation [34].

Chai et al. reported the preparation of Pt single atoms by sonicating a dispersion of hollow porous carbon nitride nanospheres in a solution of $H_2PtCl_6 \cdot 6H_2O$ for 30 min. Then, the solid was separated by centrifugation, and the adsorbed Pt species were further reduced by the addition of a solution of $NaBH_4$. As shown in Fig. 6.2a, the authors claimed to improve the metal dispersion by using a regulator (i.e., 1-butyl -3-methylimidazole hexafluorophosphate) [35]. They also applied a similar strategy to Ni and Fe; however, advanced characterization data that could confirm the presence of Pt as single-atom sites were not provided. Similarly, Vilé *et. al.* reported in 2015 the stabilization of Pd single atom in mpg-CN by the in situ reduction of palladium salt solution with $NaBH_4$.

Shi et al. reported a simple impregnation of C_3N_4 with $NiCl_2 \cdot 6H_2O$ solution that led to the formation of Ni single atoms after an activation step at 200 °C for 2 h in 5% O_2/Ar atmosphere. The Pérez-Ramírez group compared the in situ reduction and the direct synthesis (i.e., thermal co-polymerization) of Pd, Ag, Au and Ir single atoms in graphitic carbon nitride (Fig. 6.2b). In summary, they concluded that both methods may lead to metal single-atom stabilization [36]. However, based on the characterizations presented, both methods seem to fail to drive the formation of only single atoms; the presence of metal clusters is evident in the STEM HAADF images.

a)

b)

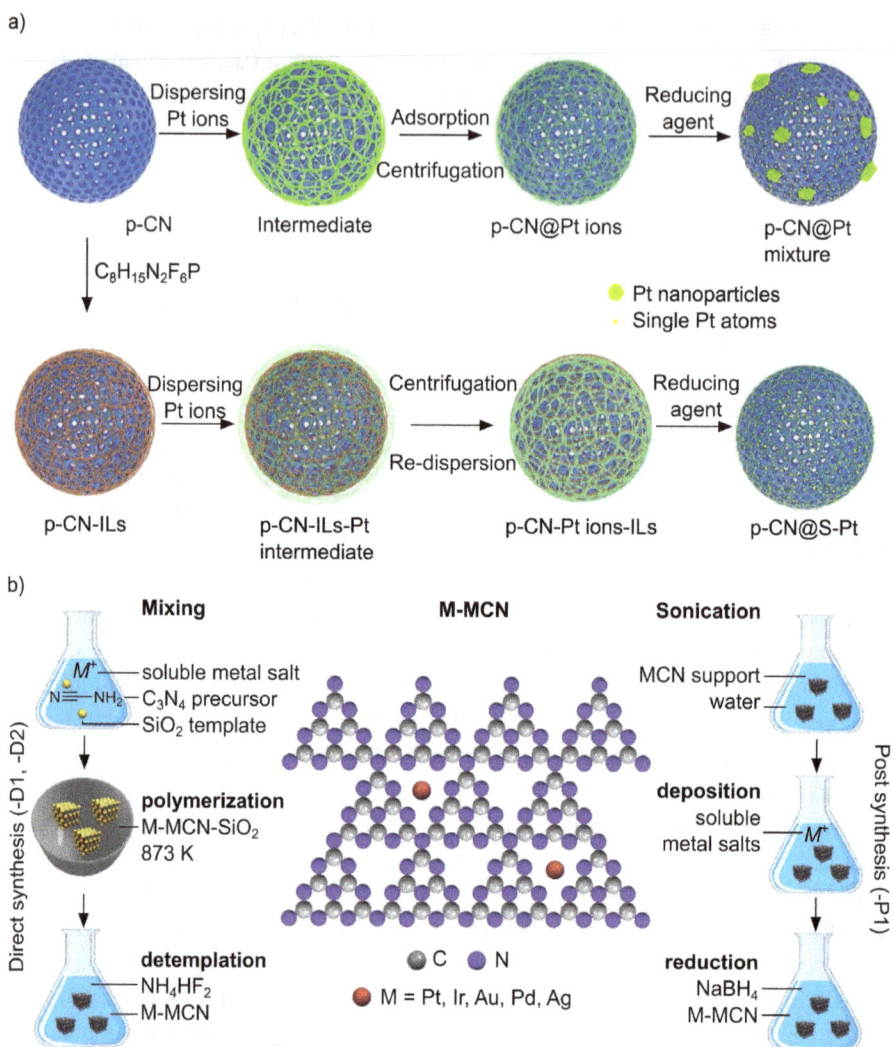

Fig. 6.2: (a) Schematic representation of the synthetic process applied to the formation of a mixture of Pt nanoparticles and single atoms on C_3N_4 and the p-CN@S–Pt prepared in the presence of the regulator $C_8H_{15}N_2F_6P$ (adapted from ref. [35] with permission from American Chemical Society, copyright 2021). (b) Direct synthesis and post-synthetic approaches to stabilize metal atoms within g-C_3N_4 (adapted from ref. [36] with permission from John Wiley and Sons, copyright 2017).

Although the wet chemical impregnation method gives a fair control of the dispersion of the single atoms, especially when aggregation inhibitors are used, it is feasible and may be scaled up simply. Generally, it tends to work better with lower metal concentrations. Analysis of data indicates that the wet-chemistry route is usually associated with low metal loading to secure the high dispersed character (<0.5%) [4].

6.5.2 Cation exchange

The preparation of single atoms by cation exchange is very recent. Firstly, this method necessarily requires the use of carbon nitride-based materials, with charges in their structure. Typical examples are sodium or potassium poly(heptazine imides) (Na-PHI or K-PHI), but also lithium poly(triazine imide)s (Li-PTI). PHI and PTI-based materials are of particular interest as they are crystalline with a well-defined structures in which the negative charge of poly(heptazine imide) or poly(triazine imide) 2D-layers is compensated by Na^+, K^+ or Li^+ cations [37, 38]. The possibility to exchange alkali ions in the matrix by some other metals such as the transition metals (i.e., Mg, Co, Ni, Ru, Pt and Ag) was first investigated by Savateev *et al* [39]. However, the stabilization of single atoms was not investigated. Chen et al. have used the cation exchange method with magnesium as an intermediate step to enable the subsequent stabilization of palladium in PHI and PTI by microwave-assisted deposition. In 2021, Teixeira et al. [23] were the first to propose the stabilization of transition metals (i.e., Ru, Ni and Pt) using the direct exchange of sodium in Na-PHI by transition metal cations. Despite some indications of single atoms being found by the authors using, for example FTIR-DRIFT of adsorbed CO, this work does not pull in advanced characterizations. Later, in 2021, Teixeira et al. reported the same cation exchange method to stabilize iron cations in the PHI structure (Fig. 6.3a), and this time the presence of single-atom transition metals was confirmed by EXAFS and aberration corrected annular dark-field scanning transmission electron microscopy (Fig. 6.3b and c).

Although Teixeira et al. confirmed the stabilization of iron single atoms in PHI at low loading (i.e., 0.1%) and reported that the cation exchange method can yield material with metal loading up to 5%, [40] the authors also observed the presence of iron hydroxide clusters when metal loadings exceeded 2%. It is worth highlighting that the authors propose the stabilization of the iron single-atoms by the formation of tetra-coordinated metal sites between the PHI layers, while their catalytic results in the photooxidation of C-H bonds corroborate their hypothesis. Consequently, the cation exchange method in PHI leads to the formation of active catalytic sites, mainly in the bulk of the material, instead of at the surface, which is different from most of the reports on single atoms supported by C_3N_4.

6.5.3 Atomic layer deposition

Atomic layer deposition (ALD) is a bottom-up synthesis strategy commonly applied for the fabrication of thin films using gas-solid surface reaction. It is considered one of the most effective methods to accurately deposit metal single-atom. Typically, an ALD cycle consists of four steps: (1) a thermally stable precursor is absorbed on the exposed surface from the gas phase until it reaches adsorption saturation; (2) the precursor pulse is stopped and an inert gas is used to remove the excess of the unreacted

a)

b) c)

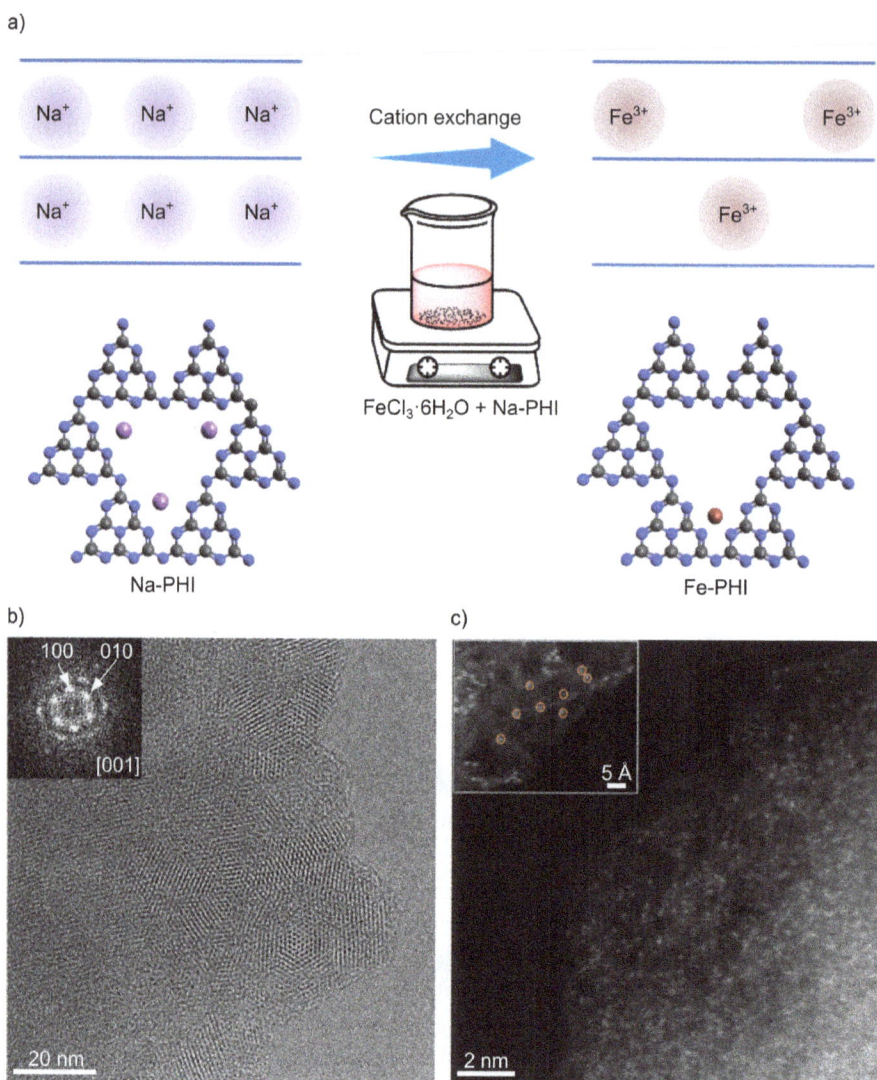

Fig. 6.3: (a) Schematic representation of the controlled nonthermal cation exchange method used to replace the Na^+ cations in the structure of PHI with Fe^{3+} cations (adapted from ref. [40] with permission from Elsevier, copyright 2022). (b) HR-TEM image of Fe-PHI (0.1%) and the corresponding FFT (left) indexed in a hexagonal lattice. (c) Annular dark-field scanning transmission electron microscopy image of Fe-PHI showing metal single atoms (adapted from ref. [40] with permission from Elsevier, copyright 2022).

precursor and volatile by-products; (3) the second precursor is introduced in the system and reacts with the saturated adsorbed layer; (4) then, the residual unreacted precursor and by-products are removed again from the reaction chamber to complete the ALD cycle [4, 41, 42]. Theoretically, the ALD method is a self-limiting layer-by-layer method that differs from CVD and allows to precisely control the metal deposition on the surface. However, the SACs' synthesis depends not only on the deposition of the metal as an atomic dispersion, but also on the stabilization of these metals to prevent aggregation.

There are only few examples in the literature that report the successful deposition and stabilization of single-atom metals on carbon nitride by the ALD method [27, 43]. One of them was reported by Cao et al. where the ALD method was applied to deposit Co on the surface of P-doped PCN using cobalt bis(cyclopentadienyl) (Co(Cp)$_2$) as a gaseous precursor [43]. After the deposition of the cobalt complex, the sample was submitted to a treatment with O$_3$ to remove the Cp ligand (Fig. 6.4a). The authors confirmed the stabilization of the Co single-atom species by HAADF-STEM (Fig. 6.4b,c and d) and XANES/EXAFS spectroscopy [43].

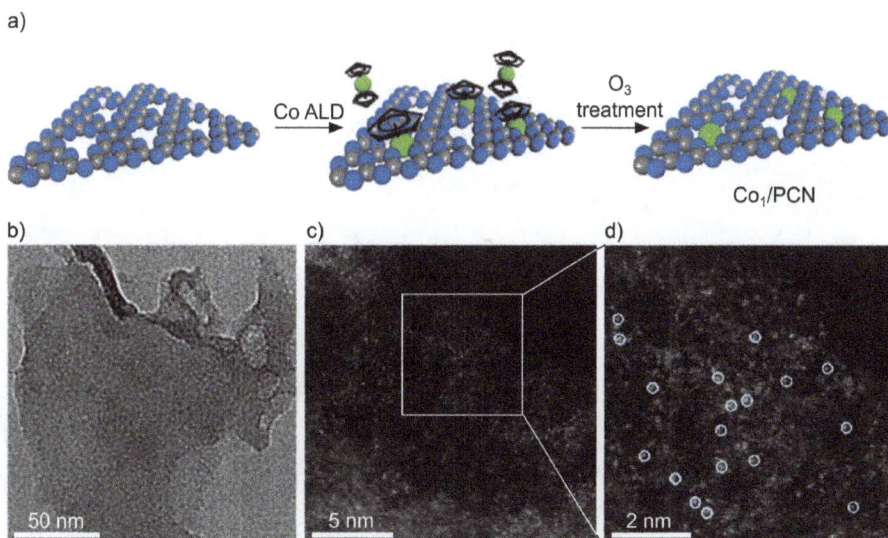

Fig. 6.4: (a) Schematic representation of the synthesis of a single-site Co/PCN catalyst by the ALD method. (b) TEM image of Co/PCN. (c) and (d) HAADF-STEM images of Co/PCN at low (in panel (c)) and high magnifications, respectively (reproduced from ref. [43] with permission from John Wiley and Sons, copyright 2017).

6.5.4 Microwave irradiation-assisted deposition

Microwave irradiation-assisted deposition consists of using electromagnetic waves with the wavelength in the microwave range (i.e., 0.1–1,000 mm) as a source of energy to promote the formation of SACs. This method presents some advantages, compared with the conventional heating approach, which is the precise control of the reaction temperature as well as pressure in the microwave reactor. The dielectric heating process, in general, permits a uniform heating of the whole reaction system, consequently, improving the product purity and minimizing side reactions [44]. However, it is important to point out that despite microwave-assisted methods being very useful for laboratory scale, they are not ideal for large-scale applications since the microwave reactors are limited to small sizes and their operation on large scale involves high costs.

This technique was applied mainly by the Pérez-Ramírez group in the synthesis of SACs (Fig. 6.5a) [14, 36, 45]. For example, they stabilized Pd single atoms in C_3N_4 materials and reached metal loading of 0.5 wt. % (Fig. 6.5b and c).

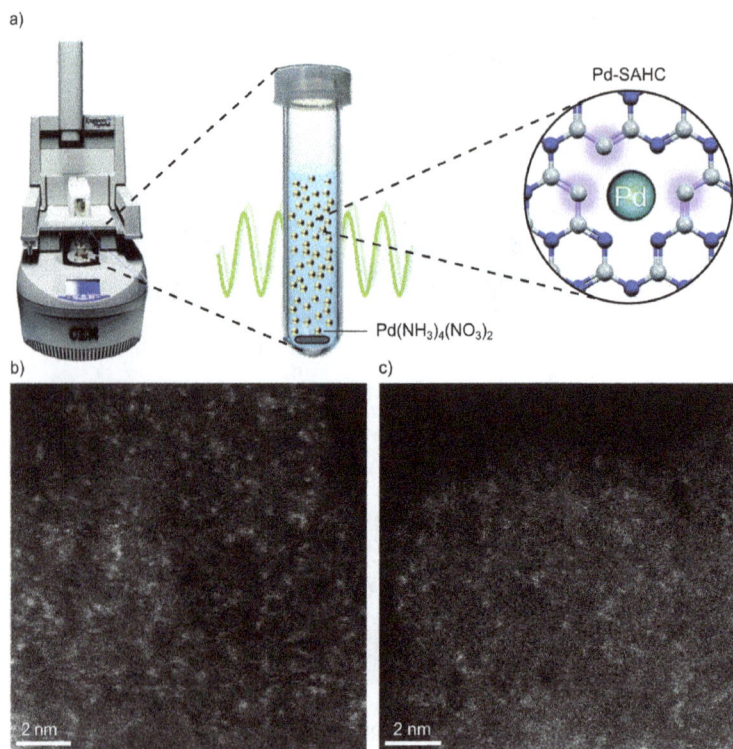

Fig. 6.5: (a) Schematic representation of the introduction of palladium single atoms by microwave-irradiation-assisted metal deposition (adapted from ref. [46] with permission from the Royal Society of Chemistry, copyright 2017). AC-STEM images of the fresh (b) and used (c) Pd-C_3N_4 catalyst (adapted from ref. [45] with permission from the Springer Nature, copyright 2018).

6.5.5 Thermal co-polymerization

Very likely inspired by the many examples of nitrogen doped carbon-based heterogeneous SACs [1], thermal co-polymerization became the most employed method to synthesize C_3N_4-based SACs. Typically, C_3N_4-based materials can be easily synthesized by thermal condensation of carbon- and nitrogen-rich precursors (e.g., melamine, urea and cyanimide). A common approach in the synthesis of C_3N_4-based SACs is the addition of a metal precursor to the reaction mixture before thermal treatment. There are two typical strategies to introduce the metal precursor: (1) molecular assembly and (2) physical mixing. In the first strategy, the metal and the nitrogen-rich precursors are previously mixed in order to form an assembled structure, which is further calcined, usually at temperature from 500 to 600 °C under an inert gas atmosphere (i.e., N_2 or Ar) [47–49]. The second approach consists of physical mixing of the metal and the nitrogen-rich precursors before the calcination step [50]. Commonly, the thermal co-polymerization is conducted under the same conditions used in the synthesis of C_3N_4-based materials. However, it is important to highlight that the presence of a metal might catalyze the thermal decomposition of C_3N_4-based materials. Therefore, it is important to consider this aspect before selecting the temperature that will be used in the synthesis of SACs by the thermal co-polymerization method.

Kumru et al. reported the formation of tungsten single atoms by assembling silicontungstic acid with melamine, followed by a thermal treatment at 550 °C under a nitrogen atmosphere. Similarly, Wang et al. successfully obtained Cu single atoms by the molecular assembly approach, introducing copper nitrate to the prepared melamine-cyanuric acid supramolecular aggregates. The Cu-containing melamine-cyanuric acid supramolecular aggregates were then filtrated, washed, dried and further calcined at 550 °C [47]. On the other hand, Chen and collaborators reported La single atoms, stabilized in C_3N_4 by a simple direct physical mixing of urea and $La_2(CO_3)_3$, followed by calcination (Fig. 6.6) [50].

In summary, several methods are available to synthesize SACs that are based on C_3N_4 materials. The topic of single-atom synthesis is still a very recent and more systematic studies are needed to properly guarantee that all metals incorporated in the C_3N_4 support are atomically dispersed. Furthermore, control of the metal loading remains a challenge – very few studies were able to synthesize SACs with different metal loadings without losing the atomic dispersion control. It is of paramount importance for the next steps in the SAC topic, which will certainly focus more on controlling the distance between metal single sites, the coordination sphere, and the synthesis of bimetallic SACs. All these challenges will only be controllably addressed when further development in characterization methods is achieved, not only in terms of resolution, but especially in terms of multispecies characterization and accessibility.

6.6 Applications

SACs have already proven to be a most promising material class in heterogeneous catalysis due to their maximum atomic utilization, tunable electronic structure and unique control of the coordination environment of the transition metal single atoms

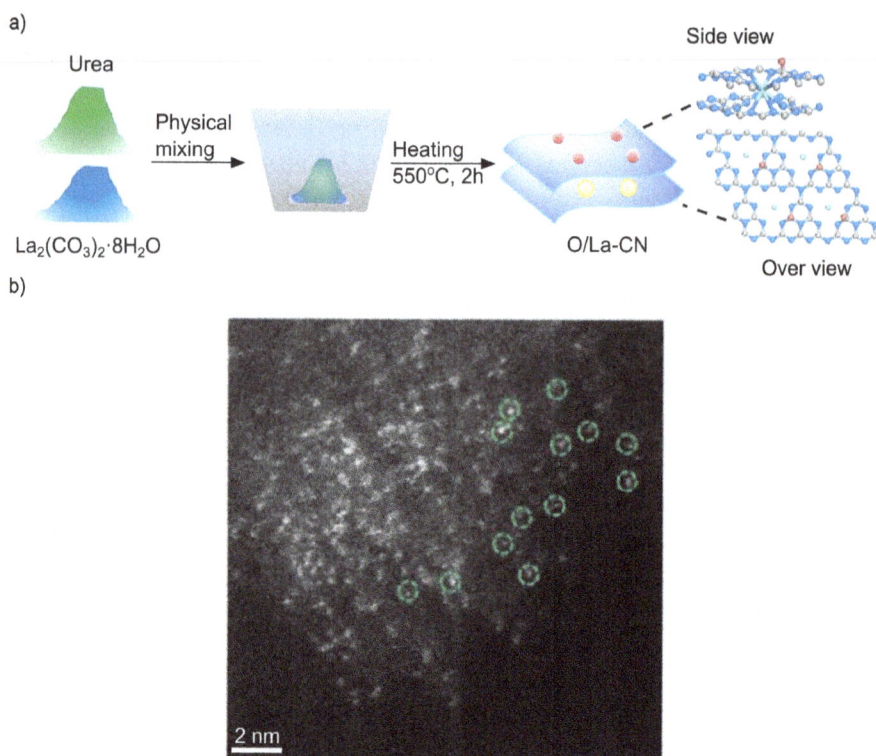

Fig. 6.6: (a) Schematic representation of the catalyst synthesis by direct physical mix approach. (b) Spherical aberration-corrected HAADF-STEM images of O/La-CN catalyst (adapted from ref. [50] with permission from the American Chemical Society, copyright 2020).

[1, 5]. SACs can play a pivotal role in improved chemical synthesis as they provide a unique environment that resembles metal complexes, but in immobile, stable, and robust hosts [1, 7–11]. The presence of isolated atoms enhances the catalytic turnover per g metal, which is important to reduce the amount of transition metals required by a reaction, without loss of efficiency [1, 8–10]. In the past few years, SACs have been applied in different topics, ranging from catalysis, such as synthesis of fine chemicals, to energy storage and energy conversion, among others [4, 51].

Due to suitable potentials of valence and conduction bands, the optical band gap of ca.2.7 eV and other features that are typical for polymeric semiconductors, C_3N_4-based SACs are applied mainly in photocatalysis, which will be highlighted in this chapter. However, few other examples will be presented, such as hydrogenations and C-C coupling reactions. The application of C_3N_4-based SACs in electrocatalysis will not be covered in this part.

6.6.1 Photocatalysis

A high-performing photocatalyst should be able to i) absorb light efficiently; ii) separate charges and transfer them to the reactant species; iii) act as catalyst by lowering the redox overpotential at the active sites and helping the activation of the reactant molecules [52]. C_3N_4 materials have proven to be very promising candidates for photocatalytic applications due to their ability to absorb light efficiently in the visible and near-infrared ranges, chemical stability, non-toxicity, straightforward synthesis using only earth-abundant elements and versatility as a platform for constructing hybrid materials, such as SACs. In terms of electronic structure, C_3N_4-based materials possess appropriate band gaps, ranging from 2.5 (e.g., K-PHI) [53] up to 3.2 eV (e.g., PTI), [54] which can be tuned by different strategies [52]. The stabilization of metal single atoms in C_3N_4 can modify the electronic structure of these materials, consequently influencing their light absorption and charge-separation efficiency. Furthermore, the metal single atoms can also act as efficient active sites for catalytic reactions, which can greatly improve the reaction kinetics. In the following section, we will cover the latest progress in the application of metal single atoms, anchored at C_3N_4 materials for water splitting, hydrogen and oxygen evolution.

6.6.1.1 Water splitting, hydrogen and oxygen evolution

Hydrogen is considered a clean fuel, with an extreme energy density of 140 MJ kg^{-1}. Besides that, hydrogen presents no harm to the environment, as its combustion product is only water [55–57]. However, to consider H_2 as an ultimately clean fuel, it has to be necessarily generated by renewable energy sources. The generation of H_2 from water by employing a photocatalyst and solar light as the energy input is an especially promising approach [58, 59]. A range of semiconductor photocatalysts have been reported to catalyze the evolution of H_2 from water. However, the application is still limited by the partial recombination of photogenerated electron-hole pairs within the photocatalysts, limitation in charge transfer to the medium and poor interaction between the semiconductors and the co-catalysts [60]. Therefore, the use of metal single atoms to tailor the semiconductors' electronic properties while simultaneously acting as the active site or co-catalyst is a strategy that can contribute enormously to the improvement of activity and quantum efficiency (QE) of this important reaction.

Water splitting involves two half-reactions, the hydrogen evolution reaction (HER) and the oxygen evolution reaction (OER), which are rarely investigated simultaneously. Mostly, sacrificial reagents are used, and only one half-reaction is studied [61]. Concerning the reduction side of the reaction (HER), platinum is one of the most efficient co-catalysts for H_2 generation and has been widely used for enhancing the performance of C_3N_4-based materials [52, 62, 63]. Li et al. reported Pt single atom as a co-catalyst for g-C_3N_4 in HER. They showed that isolated Pt single atoms induce an

a)

b)

c)

d)

e)

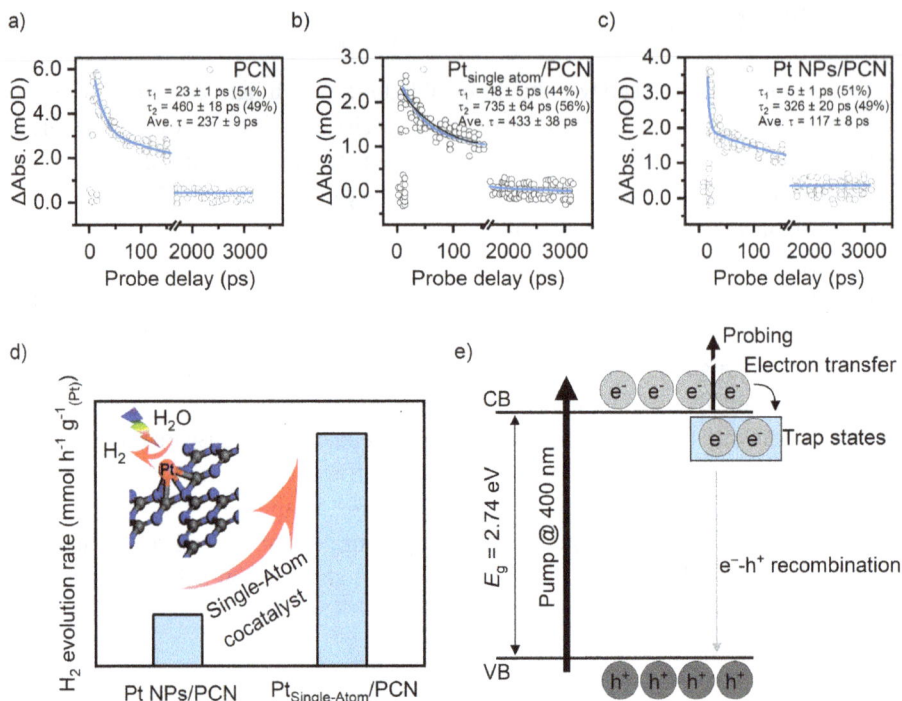

Fig. 6.7: Representative ultrafast transient absorption (TA) kinetics probed at 750 nm (pump at 400 nm) for (a) g-C_3N_4, (b) $Pt_{single\ atom}$/g-C_3N_4 and (c) Pt NPs/g-C_3N_4. (d) Schematic model of $Pt_{single\ atom}$/g-C_3N_4 and photocatalytic comparison with Pt NPs/g-C_3N_4. (e) Schematic illustration of the mechanisms which led to a longer lifetime of photogenerated electrons (adapted with permission from ref. [64]. Copyright 2016 John Wiley and Sons).

intrinsic change of the surface trap states, leading to a longer lifetime of photogenerated electrons and thereby improving the photocatalytic performance (Fig. 6.7). Single-atom Pt co-catalyst led to remarkable enhancements in the photogeneration of H_2; Pt single atom/g-C_3N_4 presents 8.6 times higher activity compared to Pt nanoparticles (NPs), and nearly 50-fold compared to pristine g-C_3N_4.[64]

Another example of Pt single atoms improving the activity in H_2 evolution was reported by Su and collaborators. The authors claimed that the improved activity is due to the modification of the band structure through "high-valence metal single-atom confinement" [65]. The authors reported an almost 10-fold increment for the Pt single-atom catalyst compared with the one based on Pt NPs [65]. These results were obtained in the absence of a sacrificial reagent. Consequently, it is an example of water splitting leading to the generation of H_2 and O_2 simultaneously. However, very little information was provided about the reaction conditions. Similar results were reported by Cao et al., who showed that Pt single atoms supported on g-C_3N_4 are 13 times more active than their nanocrystal counterpart [66]. However, both works could not apply expensive and not available characterization techniques to prove that the Pt single atoms remained atomically dispersed when illuminated under the reduction conditions.

Cao et al. surpassed the activity of Pt single atoms by using Pd single atoms instead. The authors reported the preparation of atomically dispersed Pd on g-C_3N_4, which lead to the formation of two distinct sites, bridging the Pd atoms located between the layers and the surface Pd atoms – the active sites [67]. They claimed that 1.8 times higher activity observed for the Pd/g-C_3N_4 compared to Pt/g-C_3N_4 is due to the better vertical charge migration mediated by the interlayer Pd sites and the improved charge separation due to the capture of separated electrons by the surface Pd atoms [67]. Nickel single atoms were also successfully employed to promote the hydrogen evolution reaction. Jin et al. reported that the anchoring of Ni single atoms at PCN improves the photoactivity, similar to Pd, elevating the catalyst activity by over 30-fold [68].

A beautiful example of using cobalt single atoms anchored at phosphine-doped PCN in full water splitting was provided by Wei et al. [43] In this work, no sacrificial reagent or noble-metal co-catalyst was employed, and still excellent results were obtained. The authors reported a H_2 evolution rate of 410 $\mu mol \cdot h^{-1} \cdot g^{-1}$. They attributed the improved activity to the effective suppression of the photoexcited charge recombination and facile charge separation and transfer, prolonged by about 20 times the charge carriers lifetime compared to pristine PCN, which consequently boosted both half-reactions of the water splitting [43].

The oxygen evolution reaction using C_3N_4-based materials is mainly performed via electrochemistry [4]. However, DFT simulations indicate that single atoms anchored at C_3N_4 are promising photocatalysts for this reaction [69, 70]. Chu et al. showed that Co single atoms can promote the evolution of oxygen using $AgNO_3$ as the electron acceptor [29]. A similar example was reported by Teng et al. using atomically dispersed antimony in PCN [71]. In both cases, authors aimed to use the generated O_2 to back- reduce it into H_2O_2.

6.6.1.2 CO_2 reduction

Society relies on fossil resources, for example, coal, gas and oil to provide energy and chemicals [72]. The CO_2 produced by these non-renewable carbon is altering the climate on the Earth [73]. Recently, countermeasures have become a part of sciences, and great effort toward reducing CO_2 emissions is observed. Inspired by natural photosynthesis, the photocatalytic conversion of CO_2 into energy-bearing products has been considered as an alternative option to address greenhouse gas emissions while generating valuable chemicals [74, 75]. In this artificial photosynthesis approach, a photocatalyst makes use of the solar input energy to activate the CO_2 molecules and convert them into hydrocarbon fuels and chemicals.

Pure PCN (without any co-catalyst) already promotes reduction of CO_2 into other chemicals (e.g., CO, [76] CH_3CHO [77] and CH_4 [77]); however, with very low efficiency. This process can be highly improved by the stabilization of metal single atoms at C_3N_4-based materials. For example, Gao et al. investigated photocatalytic CO_2 reduction over Pd or Pt single atoms anchored at g-C_3N_4 by DFT simulations. Comparing the rate-determining barriers of different possible reaction pathways, the authors reported that HCOOH was the lowest energy product of CO_2 reduction over the Pd/g-C_3N_4 catalyst, while CH_4 was favored over Pt/g-C_3N_4. Cao's group reported the synthesis of single-atom copper-modified PCN via supramolecular preorganization with subsequent condensation. The authors were able to produce methanol, methane and CO using only water as a source of hydrogen, with an apparent overall photoefficiency of 1.32% [78]. Li et al. showed that Cu single atoms in g-C_3N_4 under a different set of conditions exhibit high selectivity in CO_2 reduction toward CO (nearly 100%), under the absence of any other co-catalyst or sacrificial agent [79]. Similar results were reported by Chen et al. using La single atoms on C_3N_4. The author showed 80.3% selectivity to CO from CO_2, and reported a high productivity of 92 $\mu mol \cdot g^{-1} \cdot h^{-1}$ [50].

Yang et al. successfully demonstrated that Au single atoms can improve the yield of CO by 1.97 and CH_4 by 4.15 times, compared with pure g-C_3N_4. The authors attributed the enhanced activity to the lowered energy barrier in CO_2 reduction to CH_4, narrowed band gap and suppressed recombination of charge carriers [80]. Sharma et al. used microwave-assisted deposition to stabilize Ru single atoms in mesoporous C_3N_4 (mC_3N_4) and tested the obtained material in the photoreduction of CO_2 to methanol. The authors claimed that the Ru single atoms/mC_3N_4 can separate charges, where the holes are used to oxidize water into O_2 and H^+, and the electrons can reduce the H^+/CO_2 into methanol. They reported an impressive yield of methanol (1,500 $\mu mol \cdot g^{-1}$) under illumination and optimized conditions [81].

6.6.2 C–C bond formation and hydrogenation

Deposition of metal single atoms at C_3N_4-based materials can not only tune and improve their properties in photocatalysis; the materials can also mediate reactions in the dark, such as hydrogenations and Suzuki cross-coupling [1]. In these reactions, the main role of the support is to coordinate firmly the metal single atoms, providing stability and regulating its activity. For example, Pérez-Ramírez and collaborators successfully stabilized palladium Pd single atoms at g-C_3N_4, which then efficiently catalyze the Suzuki coupling. The catalyst tolerates a broad scope of substrates. The authors reported that heterogeneous Pd single-atom/g-C_3N_4 catalyst surpasses the performance of the state-of-the-art homogeneous catalysts and conventional heterogeneous catalysts, based on nanoparticles or grafted molecular complexes, without apparent metal leaching or aggregation after the reaction. They attributed the improved activity and stability to the versatile coordination environment within the macroheterocycles of the g-C$_3$N$_4$, which assists each catalytic step [82].

Another interesting class of reactions that can be assisted by adaptive coordination environment provided by different C$_3$N$_4$ materials, are hydrogenations [83]. Noble metals nanoparticles, in general, display excellent catalytic activity in hydrogenation reactions. However, they tend to have poor selectivity to partial hydrogenations, such as acetylene to ethylene. Furthermore, the replacement of noble metal particles (e.g., Pt and Pd) by single-atom sites allows lower loading and higher number of active sites – all properties desired for catalysts that rely on very expensive metals. Huang et al. investigated Pd and Pt single atoms, supported at g-C$_3$N$_4$, Al$_2$O$_3$, SiO$_2$ and graphene. They reported that Pd single atoms/g-C$_3$N$_4$ displayed a higher selectivity in the hydrogenation of acetylene into ethylene, and largely suppressed the formation of coke, compared with Pd NPs. It is worth mentioning that these properties were not observed for the Pd single atoms/graphene, which showed a selectivity of only 78% to ethylene, compared with the 98% displayed by the Pd single atoms at g-C$_3$N$_4$. This indicates that the g-C$_3$N$_4$ support might induce a charge transfer to the metal single atoms, changing its electron density and, consequently, alters the selectivity of the catalyst [27].

Pérez-Ramírez's group efficiently stabilized Pd single sites confined in mpg-C$_3$N$_4$, and applied it in hydrogenations of alkynols. The authors reported a greater selectivity to alkenol (>90%) compared to Pd NPs. Furthermore, the prepared catalyst displayed good stability, which in general, is not observed when metal particle is minimized on other supports (e.g., SiO$_2$ and Al$_2$O$_3$). The authors also investigated how the changes in the support composition (i.e., C/N ratio) would affect the reaction. They observed that an increase in the C/N ratio in the support disfavors the activity of the Pd single sites [46]. The same research group, in a similar study, investigated the influence of the g-C$_3$N$_4$ support structure doping on the activity of the Pd single sites in the same hydrogenation reaction. They reported an optimized catalytic performance in the semihydrogenation of 2-methyl-3-butyn-2-ol into the alkenol when the g-C$_3$N$_4$ support was doped with 0.25 wt. % P, leading to a reaction rate that is over 5 times higher compared to Pd single atoms on pristine g-C$_3$N$_4$ [84].

6.7 Summary

SAC is a very promising topic, but is still in its early stage of development. Certainly, carbon nitride-based materials can play a key role in the further advances of this emerging field. We believe that it is fair to compare the single atoms field nowadays with the nanomaterials a few decades ago. With the progress in characterization techniques and improving their resolution, scientists became able to not only understand and control the properties of the nanomaterials, but we realized that nanomaterials were already present in the human-made objects and used by the society for centuries. A notable example is carbon nanotubes as well as cementite nanowires found in Damascus sabre steel from the seventeenth century, which was observed by high-resolution TEM only in 2006 [85]. Single atoms is another frontier that depends on the development of characterization techniques and improving their resolution. The progress in this adjacent field will enable us to understand and control synthesis of materials, manipulating single atoms. However, we believe that the results will drive us to the same conclusion – despite not properly understood; single atoms were present already in many catalytic studies reported in the past. Today, with more advanced characterization techniques and improved synthetic methods, a completely new horizon has opened in the fine control of single atoms and their investigation in catalytic reactions.

Recent findings in the anchoring of metal single atoms in C_3N_4 materials have clearly indicated that macroheterocycles present in the structures of these supports and aligned with their unique electronic properties promote the metal monodispersion, prevent the aggregation and improve stability. Thus, the layered 2D structure displayed by these supports provides an ideal platform to maximize the stability and accessibility of metal single atoms, leading to a metal single site that is strongly bonded to the C_3N_4 framework. However, at the same time, reactants are able to easily reach the metal single site during the catalytic process. It is also important to highlight that very often the C_3N_4 acts as a non-innocent support – its semiconducting properties in several examples promote the activity of the metal single atoms. The back-influence is also true – the coordination of metal single atoms in C_3N_4 materials can tune their electronic properties, leading to better charge separation, more effective charge transfer and finally improved photocatalytic activities.

In summary, the field of SAC based on C_3N_4 materials holds a bright future with many challenges to be faced, but also with exciting opportunities. The deep understanding and fine design of catalysts suitable for different chemical targets that was an exclusivity of the homogenous catalysis in the past has now become possible in the context of heterogeneous catalysis with the possibility of tailoring active sites at atomic-scale precision. However, further improvements in the precision of materials synthesis and advanced characterizations are still needed for the further exploitation of this promising class of catalysts.

References

[1]   Kaiser SK, Chen Z, Faust Akl D, Mitchell S, Pérez-Ramírez J. Single-atom catalysts across the periodic table. Chem Rev 2020, 120(21), 11703–11809.
[2]   Berzelius JJ. Sur un force jusqu'ici peu remarquée qui est probablement active dans la formation des composes organiques, Section on Vegetable Chemistry. Jahresber 1835, 14, 237.

[3] Liu L, Corma A. Metal catalysts for heterogeneous catalysis: From single atoms to nanoclusters and nanoparticles. Chem Rev 2018, 118(10), 4981–5079.

[4] Zhao M, Feng J, Yang W, Song S, Zhang H. Recent advances in graphitic carbon nitride supported single-atom catalysts for energy conversion. Chem Cat Chem 2021, 13(5), 1250–70.

[5] Li X, Yang X, Zhang J, Huang Y, Liu B. In situ/operando techniques for characterization of single-atom catalysts. ACS Catal 2019, 9(3), 2521–31.

[6] Mitchell S, Pérez-Ramírez J. Atomically precise control in the design of low-nuclearity supported metal catalysts. Nat Rev Mater 2021, 1–17.

[7] Qiao B, Wang A, Yang X, Allard LF, Jiang Z, Cui Y et al. Single-atom catalysis of CO oxidation using Pt_1/FeO_x. Nat Chem 2011, 3(8), 634–41.

[8] Wang A, Li J, Zhang T. Heterogeneous single-atom catalysis. Nat Rev Chem 2018, 2(6), 65–81.

[9] Zhang L, Ren Y, Liu W, Wang A, Zhang T. Single-atom catalyst: A rising star for green synthesis of fine chemicals. Natl Sci Rev 2018, 5(5), 653–72.

[10] Mitchell S, Vorobyeva E, Pérez-Ramírez J. The multifaceted reactivity of single-atom heterogeneous catalysts. Angew Chem Int Ed 2018, 57(47), 15316–29.

[11] Mondelli C, Gözaydın G, Yan N, Pérez-Ramírez J. Biomass valorisation over metal-based solid catalysts from nanoparticles to single atoms. Chem Soc Rev 2020, 49, 3764–3782.

[12] Su X, Yang X-F, Huang Y, Liu B, Zhang T. Single-atom catalysis toward efficient CO_2 conversion to CO and formate products. Acc Chem Res 2018, 52(3), 656–64.

[13] Li X, Yang X, Huang Y, Zhang T, Liu B. Supported noble-metal single atoms for heterogeneous catalysis. Adv Mater 2019, 31(50), 1902031.

[14] Chen Z, Vorobyeva E, Mitchell S, Fako E, López N, Collins SM et al. Single-atom heterogeneous catalysts based on distinct carbon nitride scaffolds. Natl Sci Rev 2018, 5(5), 642–52.

[15] Zhu Y, Zhang B, Liu X, Wang DW, Su DS. Unravelling the structure of electrocatalytically active Fe–N complexes in carbon for the oxygen reduction reaction. Angew Chem 2014, 126(40), 10849–53.

[16] Sa YJ, Seo D-J, Woo J, Lim JT, Cheon JY, Yang SY et al. A general approach to preferential formation of active Fe–N_x sites in Fe–N/C electrocatalysts for efficient oxygen reduction reaction. J Am Chem Soc 2016, 138(45), 15046–56.

[17] Xie J, Jin R, Li A, Bi Y, Ruan Q, Deng Y et al. Highly selective oxidation of methane to methanol at ambient conditions by titanium dioxide-supported iron species. Nat Catal 2018, 1(11), 889.

[18] Wu ZY, Xu XX, Hu BC, Liang HW, Lin Y, Chen LF, et al. Iron carbide nanoparticles encapsulated in mesoporous Fe-N-doped carbon nanofibers for efficient electrocatalysis. Angew Chem 2015, 127(28), 8297–301.

[19] Dyck O, Kim S, Kalinin SV, Jesse S. Placing single atoms in graphene with a scanning transmission electron microscope. Appl Phys Lett 2017, 111(11), 113104.

[20] Fu J, Wang S, Wang Z, Liu K, Li H, Liu H et al. Graphitic carbon nitride based single-atom photocatalysts. Front Phys 2020, 15(3), 1–14.

[21] Yang S, Tak YJ, Kim J, Soon A, Lee H. Support effects in single-atom platinum catalysts for electrochemical oxygen reduction. ACS Catal 2017, 7(2), 1301–7.

[22] Asokan C, DeRita L, Christopher P. Using probe molecule FTIR spectroscopy to identify and characterize Pt-group metal based single atom catalysts. Chin J Catal 2017, 38(9), 1473–80.

[23] Colombari FM, da Silva MAR, Homsi MS, de Souza BRL, Araujo M, Francisco JL, et al. Graphitic carbon nitrides as platforms for single-atom photocatalysis. Faraday Discuss 2021, 227, 306–320.

[24] Otake K-I, Cui Y, Buru CT, Li Z, Hupp JT, Farha OK. Single-atom-based vanadium oxide catalysts supported on metal–organic frameworks: Selective alcohol oxidation and structure–activity relationship. J Am Chem Soc 2018, 140(28), 8652–6.

[25] Jiao X, Wang Y, Mu Y, Sun Y, Li J. A new magnesium-containing aluminophosphate with a zeolite-like structure. RSC Adv 2016, 6(2), 1098–102.

[26] Liu J, Zou Y, Cruz D, Savateev A, Antonietti M, Vilé G. Ligand–metal charge transfer induced *via* adjustment of textural properties controls the performance of single-atom catalysts during photocatalytic degradation. ACS Appl Mater Interfaces 2021, 13(22), 25858–25867.

[27] Huang X, Yan H, Huang L, Zhang X, Lin Y, Li J et al. Toward understanding of the support effect on Pd$_1$ single-atom-catalyzed hydrogenation reactions. J Phys Chem C 2018, 123(13), 7922–30.

[28] Kim HE, Lee IH, Cho J, Shin S, Ham HC, Kim JY et al. Palladium single-atom catalysts supported on C@C$_3$N$_4$ for Electrochemical Reactions. Chem Electro Chem 2019, 6(18), 4757–64.

[29] Chu C, Zhu Q, Pan Z, Gupta S, Huang D, Du Y et al. Spatially separating redox centers on 2D carbon nitride with cobalt single atom for photocatalytic H$_2$O$_2$ production. Proc Natl Acad Sci U S A 2020, 117(12), 6376–82.

[30] Rivera-Cárcamo C, Serp P. Single atom catalysts on carbon-based materials. Chem Cat Chem 2018, 10(22), 5058–91.

[31] Yu W, Chen J, Shang T, Chen L, Gu L, Peng T. Direct Z-scheme g-C$_3$N$_4$/WO$_3$ photocatalyst with atomically defined junction for H$_2$ production. Appl Catal: B 2017, 219, 693–704.

[32] Tian S, Wang Z, Gong W, Chen W, Feng Q, Xu Q et al. Temperature-controlled selectivity of hydrogenation and hydrodeoxygenation in the conversion of biomass molecule by the Ru$_1$/mpg-C$_3$N$_4$ catalyst. J Am Chem Soc 2018, 140(36), 11161–4.

[33] Büker J, Huang X, Bitzer J, Kleist W, Muhler M, Peng B. Synthesis of Cu Single Atoms Supported on Mesoporous Graphitic Carbon Nitride and Their Application in Liquid-Phase Aerobic Oxidation of Cyclohexene. ACS Catal 2021, 11, 7863–75.

[34] Bayatsarmadi B, Zheng Y, Vasileff A, Qiao SZ. Recent advances in atomic metal doping of carbon-based nanomaterials for energy conversion. Small 2017, 13(21), 1700191.

[35] Zuo Y, Li T, Zhang N, Jing T, Rao D, Schmuki P et al. Spatially confined formation of single atoms in highly porous carbon nitride nanoreactors. ACS Nano 2021, 15(4), 7790–8.

[36] Chen Z, Mitchell S, Vorobyeva E, Leary RK, Hauert R, Furnival T et al. Stabilization of single metal atoms on graphitic carbon nitride. Adv Funct Mater 2017, 27(8), 1605785.

[37] Savateev A, Pronkin S, Epping JD, Willinger MG, Wolff C, Neher D et al. Potassium poly(heptazine imides) from aminotetrazoles: Shifting band gaps of carbon nitride-like materials for more efficient solar hydrogen and oxygen evolution. Chem Cat Chem 2017, 9(1), 167–74.

[38] Chen Z, Savateev A, Pronkin S, Papaefthimiou V, Wolff C, Willinger MG et al. "The easier the better" preparation of efficient photocatalysts – Metastable poly(heptazine imide) salts. Adv Mater 2017, 29(32), 1700555.

[39] Savateev A, Pronkin S, Willinger MG, Antonietti M, Dontsova D. Towards organic zeolites and inclusion catalysts: Heptazine imide salts can exchange metal cations in the solid state. Chem – Asian J 2017, 12(13), 1517–22.

[40] da Silva MAR, Silva IF, Xue Q, Lo BTW, Tarakina NV, Nunes BN et al. Sustainable oxidation catalysis supported by light: Fe-poly(heptazine imide) as a heterogeneous single-atom photocatalyst. Appl Catal: B 2022, 304, 120965.

[41] Khan R, Shong B, Ko BG, Lee JK, Lee H, Park JY et al. Area-selective atomic layer deposition using Si precursors as inhibitors. Chem, Mater 2018, 30(21), 7603–10.

[42] O'Neill BJ, Jackson DHK, Lee J, Canlas C, Stair PC, Marshall CL et al. Catalyst design with atomic layer deposition. ACS Catal 2015, 5, 1804–1825.

[43] Cao Y, Chen S, Luo Q, Yan H, Lin Y, Liu W et al. Atomic-level insight into optimizing the hydrogen evolution pathway over a Co$_1$-N$_4$ single-site photocatalyst. Angew Chem Int Ed 2017, 56(40), 12191–6.

[44] Baghbanzadeh M, Carbone L, Cozzoli PD, Kappe CO. Microwave-assisted synthesis of colloidal inorganic nanocrystals. Angew Chem Int Ed 2011, 50(48), 11312–59.

[45] Chen Z, Vorobyeva E, Mitchell S, Fako E, Ortuño MA, López N et al. A heterogeneous single-atom palladium catalyst surpassing homogeneous systems for Suzuki coupling. Nat Nanotechnol 2018, 13(8), 702–7.

[46] Vorobyeva E, Chen Z, Mitchell S, Leary RK, Midgley P, Thomas JM et al. Tailoring the framework composition of carbon nitride to improve the catalytic efficiency of the stabilised palladium atoms. J Mater Chem A 2017, 5(31), 16393–403.

[47] Wang G, Zhang T, Yu W, Si R, Liu Y, Zhao Z. Modulating location of single copper atoms in polymeric carbon nitride for enhanced photoredox catalysis. ACS Catal 2020, 10(10), 5715–22.

[48] Li Y, Gong F, Zhou Q, Feng X, Fan J, Xiang Q. Crystalline isotype heptazine-/triazine-based carbon nitride heterojunctions for an improved hydrogen evolution. Appl Catal: B 2020, 268, 118381.

[49] Kumru B, Cruz D, Heil T, Antonietti M. *In situ* formation of arrays of tungsten single atoms within carbon nitride frameworks fabricated by one-step synthesis through monomer complexation. Chem, Mater 2020, 32(21), 9435–43.

[50] Chen P, Lei B, Dong X, Wang H, Sheng J, Cui W et al. Rare-earth single-atom La–N charge-transfer bridge on carbon nitride for highly efficient and selective photocatalytic CO_2 reduction. ACS Nano 2020, 14(11), 15841–52.

[51] Sun T, Xu L, Wang D, Li Y. Metal organic frameworks derived single atom catalysts for electrocatalytic energy conversion. Nano Res 2019, 12(9), 2067–80.

[52] Teixeira IF, Barbosa ECM, Tsang SCE, Camargo PHC. Carbon nitrides and metal nanoparticles: From controlled synthesis to design principles for improved photocatalysis. Chem Soc Rev 2018, 47(20), 7783–817.

[53] Sahoo SK, Teixeira IF, Naik A, Heske J, Cruz D, Antonietti M et al. Photocatalytic water splitting reaction catalyzed by ion-exchanged salts of potassium poly(heptazine imide) 2D materials. J Phys Chem C 2021, 125(25), 13749–58.

[54] Lin L, Lin Z, Zhang J, Cai X, Lin W, Yu Z et al. Molecular-level insights on the reactive facet of carbon nitride single crystals photocatalysing overall water splitting. Nat Catal 2020, 3(8), 649–55.

[55] Lewis NS, Nocera DG. Powering the planet: Chemical challenges in solar energy utilization. Proc Natl Acad Sci U S A 2006, 103(43), 15729–35.

[56] Yu J, Wang S, Cheng B, Lin Z, Huang F. Noble metal-free $Ni(OH)_2$–g-C_3N_4 composite photocatalyst with enhanced visible-light photocatalytic H_2-production activity. Catal Sci Technol 2013, 3(7), 1782–9

[57] Zhao Z, Sun Y, Dong F. Graphitic carbon nitride based nanocomposites: A review. Nanoscale 2015, 7(1), 15–37.

[58] Li Y, Cheng X, Ruan X, Song H, Lou Z, Ye Z et al. Enhancing photocatalytic activity for visible-light-driven H_2 generation with the surface reconstructed $LaTiO_2N$ nanostructures. Nano Energy 2015, 12, 775–84.

[59] Mettenbörger A, Gönüllü Y, Fischer T, Heisig T, Sasinska A, Maccato C et al. Interfacial insight in multi-junction metal oxide photoanodes for water-splitting applications. Nano Energy 2016, 19, 415–27.

[60] Teixeira IF, Tarakina NV, Silva IF, López-Salas N, Savateev A, Antonietti M. Overcoming electron transfer efficiency bottlenecks for hydrogen production in highly crystalline carbon nitride-based materials. Adv Sustainable Syst 2022, 6(3), 2100429.

[61] Fina F, Menard H, Irvine JTS. The effect of Pt NPs crystallinity and distribution on the photocatalytic activity of Pt–g-C_3N_4. Phys Chem Chem Phys 2015, 17(21), 13929–36.

[62] Zhou P, Lv F, Li N, Zhang Y, Mu Z, Tang Y et al. Strengthening reactive metal-support interaction to stabilize high-density Pt single atoms on electron-deficient g-C_3N_4 for boosting photocatalytic H_2 production. Nano Energy 2019, 56, 127–37.

[63] Zeng Z, Su Y, Quan X, Choi W, Zhang G, Liu N et al. Single-atom platinum confined by the interlayer nanospace of carbon nitride for efficient photocatalytic hydrogen evolution. Nano Energy 2020, 69, 104409.

[64] Li X, Bi W, Zhang L, Tao S, Chu W, Zhang Q et al. Single-atom Pt as co-catalyst for enhanced photocatalytic H_2 evolution. Adv Mater 2016, 28(12), 2427–31.

[65] Su H, Che W, Tang F, Cheng W, Zhao X, Zhang H et al. Valence band engineering *via* Pt[II] single-atom confinement realizing photocatalytic water splitting. J Phys Chem C 2018, 122(37), 21108–14.

[66] Cao Y, Wang D, Lin Y, Liu W, Cao L, Liu X et al. Single Pt atom with highly vacant *d*-orbital for accelerating photocatalytic H_2 evolution. ACS Appl Energy Mater 2018, 1(11), 6082–8.

[67] Cao S, Li H, Tong T, Chen H-C, Yu A, Yu J et al. Single-atom engineering of directional charge transfer channels and active sites for photocatalytic hydrogen evolution. Adv Funct Mater 2018, 28(32), 1802169.

[68] Jin X, Wang R, Zhang L, Si R, Shen M, Wang M et al. Electron configuration modulation of nickel single atoms for elevated photocatalytic hydrogen evolution. Angew Chem Int Ed 2020, 59(17), 6827–31.

[69] Roy P, Pramanik A, Sarkar P. Graphitic carbon nitride sheet supported single-atom metal-free photocatalyst for oxygen reduction reaction: A first-principles analysis. J Phys Chem Lett 2021, 12(11), 2788–95.

[70] Li X, Cui P, Zhong W, Li J, Wang X, Wang Z et al. Graphitic carbon nitride supported single-atom catalysts for efficient oxygen evolution reaction. Chem Commun 2016, 52(90), 13233–6.

[71] Teng Z, Zhang Q, Yang H, Kato K, Yang W, Y-r L et al. Atomically dispersed antimony on carbon nitride for the artificial photosynthesis of hydrogen peroxide. Nat Catal 2021, 4(5), 374–84.

[72] Brandt A, Graesvik J, Hallett JP, Welton T. Deconstruction of lignocellulosic biomass with ionic liquids. Green Chem 2013, 15, 550–83.

[73] Ragauskas AJ, Williams CK, Davison BH, Britovsek G, Cairney J, Eckert CA et al. The path forward for biofuels and biomaterials. Science 2006, 311, 484–9.

[74] Tan -L-L, Ong W-J, Chai S-P, Mohamed AR. Noble metal modified reduced graphene oxide/TiO_2 ternary nanostructures for efficient visible-light-driven photoreduction of carbon dioxide into methane. Appl Catal: B 2015, 166, 251–9.

[75] Chang X, Wang T, Gong J. CO_2 photoreduction: Insights into CO_2 activation and reaction on surfaces of photocatalysts. Energy Environ Sci 2016, 9(7), 2177–96.

[76] Dong G, Zhang L. Porous structure dependent photoreactivity of graphitic carbon nitride under visible light. J Mater Chem 2012, 22(3), 1160–6.

[77] Niu P, Yang Y, Yu JC, Liu G, Cheng H-M. Switching the selectivity of the photoreduction reaction of carbon dioxide by controlling the band structure of a g-C_3N_4 photocatalyst. Chem Commun 2014, 50(74), 10837–40.

[78] Wang J, Heil T, Zhu B, Tung C-W, Yu J, Chen HM et al. A single Cu-center containing enzyme-mimic enabling full photosynthesis under CO_2 reduction. ACS Nano 2020, 14(7), 8584–93.

[79] Li Y, Li B, Zhang D, Cheng L, Xiang Q. Crystalline carbon nitride supported copper single atoms for photocatalytic CO_2 reduction with nearly 100% CO Selectivity. ACS Nano 2020, 14(8), 10552–61.

[80] Yang Y, Li F, Chen J, Fan J, Xiang Q. Single Au atoms anchored on amino-group-enriched graphitic carbon nitride for photocatalytic CO_2 reduction. Chem Sus Chem 2020, 13(8), 1979–85.

[81] Sharma P, Kumar S, Tomanec O, Petr M, Zhu Chen J, Miller JT et al. Carbon nitride-based ruthenium single atom photocatalyst for CO_2 reduction to methanol. Small 2021, 17(16), 2006478.

[82] Chen Z, Vorobyeva E, Mitchell S, Fako E, Ortuño MA, López N et al. A heterogeneous single-atom palladium catalyst surpassing homogeneous systems for Suzuki coupling. Nat Nanotechnol 2018, 13, 702–7.

[83] Vilé G, Albani D, Nachtegaal M, Chen Z, Dontsova D, Antonietti M et al. A stable single-site palladium catalyst for hydrogenations. Angew Chem Int Ed 2015, 54(38), 11265–9.

[84] Chen Z, Mitchell S, Krumeich F, Hauert R, Yakunin S, Kovalenko MV et al. Tunability and scalability of single-atom catalysts based on carbon nitride. ACS Sustainable Chem Eng 2019, 7(5), 5223–30.

[85] Reibold M, Paufler P, Levin A, Kochmann W, Pätzke N, Meyer D C. Carbon nanotubes in an ancient Damascus sabre. Nature 2006, 444, 286.

Xiu Lin, Shi-Nan Zhang, Dong Xu, Lu-Han Sun, Guang-Yao Zhai,
Peng Gao and Xin-Hao Li*

Chapter 7
Carbon nitride for heterojunction catalysis

7.1 Introduction

Rational design of g-C_3N_4-based heterostructure catalysts with metal or semiconductor materials could integrate the superiority of multiple components to simultaneously alter electron structure, optical response and redox ability of the material to be applied for the efficient production of high value-added compounds in many catalytic fields [1–3]. Reasonable utilization of heterojunction effect [4] can improve the activity of g-C_3N_4, reduce the cost of catalyst and, most importantly, meet the requirements for a sustainable catalyst [5–9]. With the development of various functional catalyst components (e.g., a metal or a semiconductor), some classical concepts [4, 10–12] have been proposed to describe the synergy between them and g-C_3N_4 and guide the design of more effective heterogeneous catalysts [4]. Lately, the possible candidates [13–19] for functional catalyst components for g-C_3N_4 heterojunction have been extended from metals to semiconductors, further expanding the synthesis paths of g-C_3N_4-based heterostructure catalysts. The interface structure between these novel catalyst components and g-C_3N_4 components has not changed or reconstructed significantly. The unique synergistic ways inspired us to reconsider the key importance of g-C_3N_4-based heterostructure interface charge exchange to its final catalytic activity. The rectifying contact (the contact between metal and semiconductor) allows us to adjust the electron density at g-C_3N_4 and the functional catalyst components through unidirectional interfacial charge transfer, which is driven by the difference of their work functions [20–22] so as to reasonably design catalytically effective heterojunctions for specific reactions. Among all rectifying contacts, the g-C_3N_4-based heterostructures, with predictable and designable interfacial synergistic effects, are becoming the rising stars in the next-generation catalysis. These heterostructures are compatible with the supported metal catalysts widely used both in laboratories and in industry.

The formation of g-C_3N_4-based heterostructure catalysts depends on the difference of the Fermi level (E_F, at 0 K electrons cannot occupy states above this energy level)/

---

*Corresponding author: Xin-Hao Li, School of Chemistry and Chemical Engineering, Shanghai Jiao Tong University, Shanghai 200240, P.R. China, e-mail: xinhaoli@sjtu.edu.cn
Xiu Lin, Shi-Nan Zhang, Dong Xu, Lu-Han Sun, Guang-Yao Zhai, Peng Gao, School of Chemistry and Chemical Engineering, Shanghai Jiao Tong University, Shanghai 200240, P.R. China

https://doi.org/10.1515/9783110746976-007

work function (Φ, the minimum amount of energy required to remove an electron from a metal or semiconductor) or energy of the conduction band (E_C) and the valence band (E_V) [23, 24] between g-C_3N_4 and the functional catalyst components. Even under zero bias, the electrons will automatically flow from the component with relatively low work function to the other side until the work functions on both sides of the interface reach equilibrium, thus forming electron depletion region and electron enrichment region, which can greatly change the adsorption/desorption behavior and/or activation energy of specific reactants in some catalytic reactions [25]. The electronic band structure of typical g-C_3N_4-based heterostructure catalysts is shown in Fig. 7.1, including type I heterojunction with a straddling gap [26–30], type II heterojunction with a staggered gap [31–33] and Schottky heterojunction [34–38]. In type I heterojunction, the conduction band (CB) and the valence band (VB) levels of a semiconductor are in between those of g-C_3N_4 (Fig. 7.1a). In the case of type I heterojunction, electrons in a semiconductor flow to g-C_3N_4 until the Fermi levels are aligned, which leads to the formation of electron enrichment region on the g-C_3N_4 side and the electron depletion region on the semiconductor side. In type II heterojunction, both the CB and the VB levels of g-C_3N_4 are more negative than the corresponding levels of a semiconductor (Fig. 7.1b). The electrons from the semiconductor are transferred to the g-C_3N_4 side, which is driven by the Fermi levels difference, and further construct the electron enrichment region on the g-C_3N_4 side and the electron depletion region on the semiconductor side. In a special case, the CB level of g-C_3N_4 is more negative than the VB levels of the semiconductor (Fig. 7.1c). No electrons and holes are able to pass the interface and reach the respective bands of another semiconductor – no space charge zone is formed at the interface. Figure 7.1d describes the type II heterojunction formed between the *p*-type semiconductor and the *n*-type g-C_3N_4 (Fig. 7.1d). [39–41] Electrons from g-C_3N_4 flow to the semiconductor side due to the Fermi levels difference, further resulting in the electron enrichment region on the semiconductor side and the electron depletion region on the g-C_3N_4 side. Moreover, *p*-type and *n*-type g-C_3N_4 can also be used to construct rectifying contacts with metals or semimetals, based on the difference in their Fermi levels. In Fig. 7.1e, electrons flow toward the metal because of the more negative Fermi level/work function of g-C_3N_4 (Fig. 7.1e). Conversely, due to the more negative Fermi level/work function of the metal, electrons flow to the g-C_3N_4 side (Fig. 7.1f). Accordingly, the formed electron enrichment/depletion region on the metal and semiconductor sides is the main factor that ultimately changes the catalytic activity.

As a result, in g-C_3N_4-based heterostructure catalysts, the electronic and optical properties are effectively adjusted by constructing electron depletion region/electron enrichment region at the interface, which greatly enhance the redox ability of the g-C_3N_4 and/or develops a new catalytic function for a specific reaction. Such features facilitate the development of the next-generation sustainable heterojunction-based catalysts without environmental detriment.

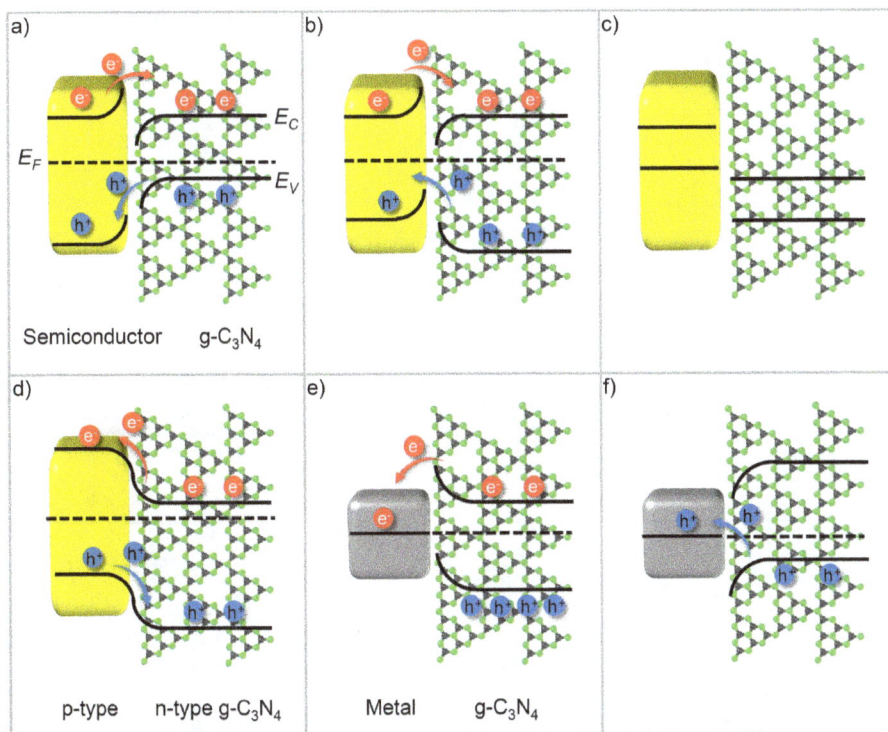

Fig. 7.1: Schematic illustration of electronic structures of the typical g-C$_3$N$_4$-based heterojunction. (a) Type I heterojunction. (b) Type II heterojunction. (c) A special case of heterojunction in which electrons are not able to pass the interface. (d) Type II heterojunction between p-type semiconductor and g-C$_3$N$_4$ (n-type semiconductor). (e) Interface between metal and g-C$_3$N$_4$. The Fermi level in g-C$_3$N$_4$ is more negative than that in metal. (f) Interface between metal and g-C$_3$N$_4$. The Fermi level in metal is more negative than that in g-C$_3$N$_4$.

7.2 g-C$_3$N$_4$ heterojunctions for photocatalysis

g-C$_3$N$_4$, as a polymeric semiconductor, has received considerable attention in the search for visible-light-active semiconductor photocatalysts. It shows high thermal and chemical stability, suitable band structure and it may be obtained from nitrogen-containing precursors. A moderate band gap (2.4–2.8 eV) leads to the onset of visible light absorption at ~ 450 nm [42, 43]. Since the first report on photocatalytic H$_2$ production by Antonietti group in 2009 [44, 45], the design of g-C$_3$N$_4$-based photocatalysts has become a hotspot in water splitting. Further functionalization of g-C$_3$N$_4$ with other components, which are already active in specific organic reactions, may also boost the light-driven organic synthesis on the as-formed nanocomposites with enhanced activities or even provide novel functions.

7.2.1 g-C₃N₄ heterojunctions for photocatalytic C–C coupling reactions

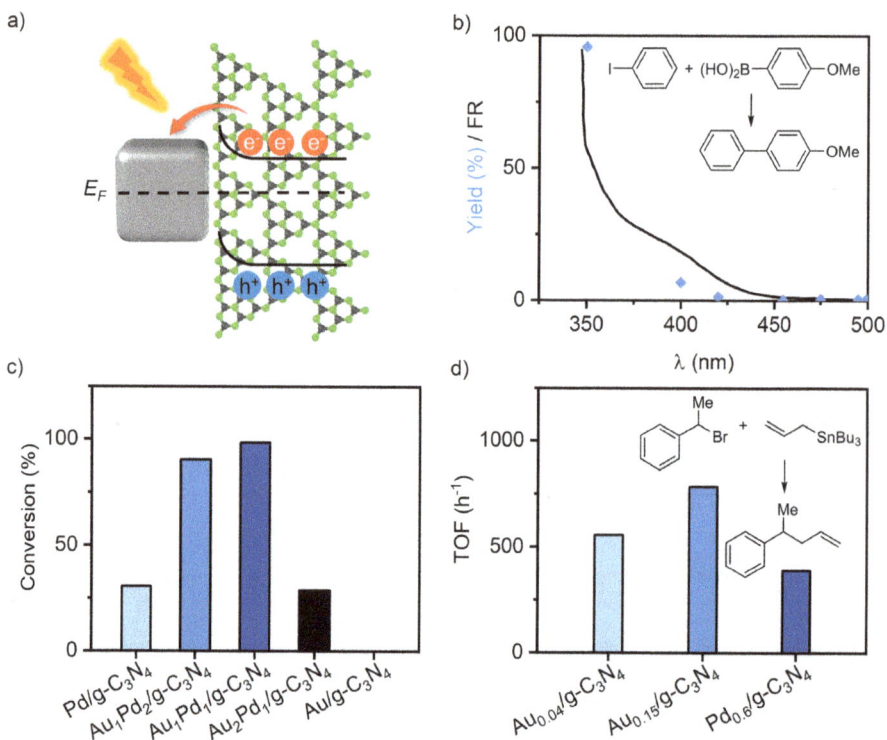

Fig. 7.2: (a) Schematic representation of electron transfer between metal and g-C₃N₄ support. (b) UV–vis absorbance spectrum of g-C₃N₄ and corresponding activity of Pd NPs@g-C₃N₄ (reproduced with permission [46]. Copyright 2013, Springer Nature). (c) Conversion in Suzuki cross-coupling over different Au_xPd_y/g-C₃N₄ catalysts (x and y represent the mass ratio between Au and Pd) (reproduced with permission [47]. Copyright 2020, Elsevier). (d) The turnover frequency (TOF) of one Au atom in Stille cross-coupling over Au_x/g-C₃N₄ (x is wt% of Au) (reproduced with permission [48]. Copyright 2020, Springer Nature).

Heterojunction catalysts composed of g-C₃N₄ and metal particles have been widely used in photocatalytic C–C cross-coupling reactions owing to excellent reusability and activity (Fig. 7.2a). Li and coworkers [46] synthesized Pd NPs@g-C₃N₄ hybrids where nano-sized palladium nanoparticles (Pd NPs) were evenly distributed on the g-C₃N₄ support (Fig. 7.2b). Pd NPs@g-C₃N₄ hybrids, as Schottky photocatalysts (a class of photocatalysts that could form Schottky barrier with a potential energy barrier for electrons at a metal-semiconductor junction), can effectively catalyze Suzuki cross-coupling reaction. Under mild conditions, at room temperature, palladium NPs are activated by the electron transfer from g-C₃N₄ to noble Pd. To enhance the catalytic performance, Dibiri and coworkers [47] loaded AuPd array nanoparticles on g-C_3N_4 sheets (AuPd/g-C_3N_4) to synthesize Schottky catalysts for this transformation. Local surface plasmon resonance (LSPR, an optical phenomenon for the electromagnetic radiation trapped in the metal NPs smaller than the wavelength of the electromagnetic radiation used to excite the plasmon) helps Au to absorb light strongly and transfer the activated electrons with higher energy to Pd. The electron-rich Pd showed a significantly enhanced catalytic activity in Suzuki cross-coupling, compared to the Pd_1/g-C_3N_4 sample (Fig. 7.2c). Besides this transformation, Stille cross-coupling is also an important targeted reaction for g-C_3N_4 heterojunction photocatalysts. Li and coworkers [48] synthesized the Au/g-C_3N_4 Schottky catalysts for application in Stille cross-coupling reaction under illumination with visible light. The catalytic performance can be enhanced by adjusting the loading of Au (Fig. 7.2d). The rectifying contact results in electron-rich Au nanoparticles, further improving the catalytic performance.

7.2.2 g-C_3N_4 heterojunctions for photocatalytic hydrogenation reactions

Catalytic hydrogenation is an important process in laboratories and industry. However, most of the used transition metals or organic catalysts usually require high temperature and other harsh conditions. As a consequence, photocatalytic hydrogenation has drawn much attention for its milder conditions. Metal/g-C_3N_4 heterojunction catalysts have been proved to be effective in the hydrogenation of unsaturated hydrocarbons, which is a vital reaction in organic synthesis [49, 50]. Nadagouda [49] used g-C_3N_4 to immobilize magnetic Fe for application in photocatalytic hydrogenation of alkynes and alkenes (Fig. 7.3a). With the help of noncovalent interactions between g-C_3N_4 and iron ferrite, Fe@g-C_3N_4 catalyst could complete hydrogenation of alkynes and alkenes without the help of high-pressure hydrogen at room temperature. Photocatalytic conversion is a potential approach to capture CO_2 from atmosphere [37, 51, 52].

Yu and coworkers [37] synthesized Pd/g-C_3N_4 hybrid catalysts and demonstrated the importance of the facet effect (the physical and chemical properties of the crystal facets are different because of the different arrangement or exposure of atoms) of Pd nanoparticles for CO_2 photoreduction. Both theoretical and experimental results proved that Pd(100) surface in T-CN performs better as the electron sink. It also has higher capability for CO_2 adsorption and CH_3OH desorption, compared with Pd(111) surface in C–CN, which strongly affects its performance in CO_2 photoreduction (Fig. 7.3b). It is still a challenge to efficiently catalyze the hydrogenation of nitroaromatics to aniline under wide conditions. Using the mixed solvothermal method, Yang [53] firstly designed and then prepared the P-C_3N_4/$ZnIn_2S_4$ catalysts with a heterojunction interface (Fig. 7.3c). The as-formed heterojunction interface resulted in a higher separation rate of photo-induced electron–hole pairs, boosting the photocatalytic performance in nitrobenzene reduction (Fig. 7.3d).

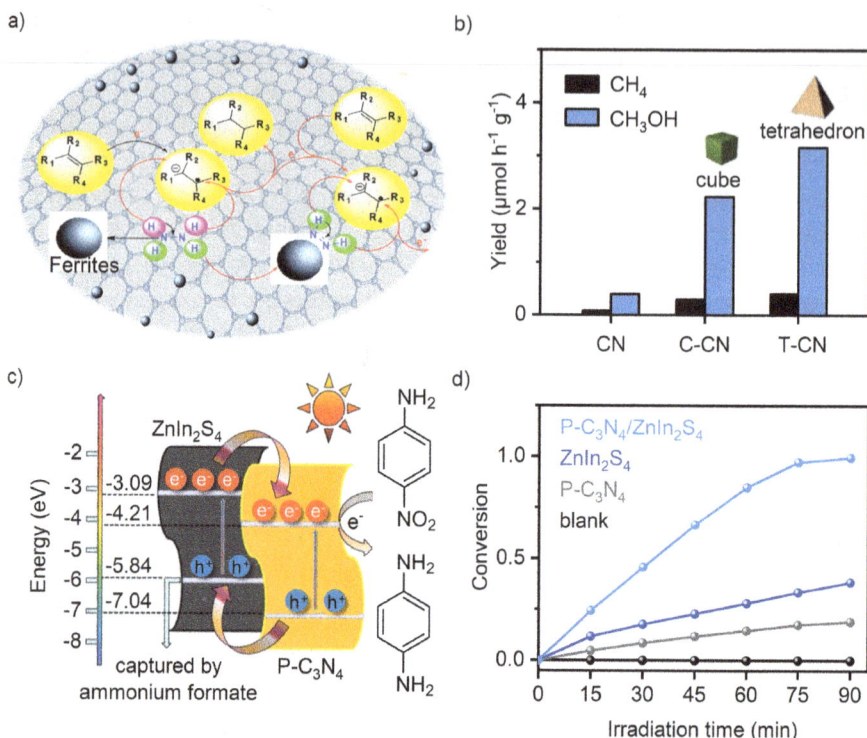

Fig. 7.3: (a) Reaction mechanism of alkene hydrogenation over Fe@g-C_3N_4 (reproduced with permission [49]. Copyright 2016, American Chemical Society). (b) Production yield of CO_2 hydrogenation over g-C_3N_4 (CN), Pd nanocube/g-C_3N_4 (C–CN), Pd nanotetrahedron/g-C_3N_4 (T-CN) (reproduced with permission [37]. Copyright 2017, Elsevier). (c) Schematic view of electron transfer between $ZnIn_2S_4$ and P-doped graphitic carbon nitrogen (P-C_3N_4) (The vacuum level is defined as 0 eV). (d) Conversion of 4-nitroaniline to 4-phenylenediamine over P-C_3N_4, $ZnIn_2S_4$, and P-C_3N_4/$ZnIn_2S_4$ photocatalysts under visible light ($\lambda > 400$ nm). Panels (c) and (d) are reproduced with permission [53] (Copyright 2016, Royal Society of Chemistry).

7.2.3 g-C_3N_4 heterojunctions for formic acid dehydrogenation

As formic acid (FA) is non-toxic, stable, and due to its high hydrogen content, research has been conducted to use it as a reagent in H_2 production and storage [54, 55]. The decomposition mechanism has been studied, and two possible reaction pathways, including the dehydrogenation and dehydration reactions, were outlined:

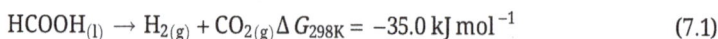

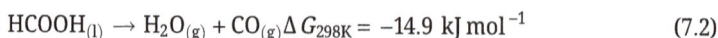

$$HCOOH_{(l)} \rightarrow H_{2(g)} + CO_{2(g)} \Delta G_{298K} = -35.0 \text{ kJ mol}^{-1} \tag{7.1}$$

$$HCOOH_{(l)} \rightarrow H_2O_{(g)} + CO_{(g)} \Delta G_{298K} = -14.9 \text{ kJ mol}^{-1} \tag{7.2}$$

It is generally known that carbon monoxide (CO) can poison and inactivate catalysts during reactions [56, 57]. However, this undesirable process can be avoided by modifying the elemental compositions of catalysts and changing the reaction conditions.

g-C_3N_4 was usually used as a support for mono/bi/multi metal-based materials and semiconductors, and doped with heteroatoms to boost catalytic performance in photo-degradation of HCOOH (Fig. 7.4a). Li and coworkers [56] reported a new series of Schottky g-C_3N_4-based photocatalysts to enhance greatly catalytic performance in the decomposition of HCOOH. The heterojunction catalysts were constructed via the rectifying contact between Pd NPs and g-C_3N_4 (Pd@CN), resulting in a markedly boosted rate of hydrogen evolution over the catalyst under visible light (Fig. 7.4b). More importantly, the work function of Pd is located between the CB and VB of g-C_3N_4, straddling the energy difference of 2.7 eV (Fig. 7.4c) [58], which leads to the formation of electronically enriched Pd NPs upon electron transfer from g-C_3N_4. The electron-rich Pd exhibited the highest catalytic activity in the generation of H_2, with a TOF of 49.8 h^{-1} at 288 K [59], which is more than six times higher than that (2.3 and 6.7) of the carbon-supported Pd catalysts under the same conditions. In addition, the observed excellent catalytic activity was increased further, upon visible light irradiation (TOF = 71.0 h^{-1}).

Following Schottky photocatalysts, Liu and coworkers [60] reported bimetallic g-C_3N_4-based heterojunction catalysts of plasmonic AuPd alloy NPs, supported on super small g-C_3N_4 nanospheres (Au$_x$Pd$_y$/CNS) for photocatalytic degradation of HCOOH to H_2 generation. It is well known that Au can capture electromagnetic radiation in the visible range to induce the production of high-energy surface electrons by LSPR effects [63–65]. Consequently, the high-energy electrons on the Au surface migrate to Pd-active sites. The AuPd/CNS exhibited remarkable catalytic activity under visible light irradiation, compared to dark conditions, which possesses the highest TOF value up to 1,017.8 h^{-1}, based on the largest Au/Pd ratio of 1.52 (Fig. 7.4d). In addition, the AuPd with CNS support catalyst shows higher activity than that of bulk CN, indicating that the ultra-small CNS could effectively facilitate the electron transport between the metal and the support. It was also found that the catalytic activity of bimetallic AuPd was higher than that of monometallic Pd under the same conditions because of the presence of charge redistribution between Au and Pd in the alloy, which accelerates the decomposition of HCOOH.

Metin and coworkers [61] fabricated a multi-metallic Z-scheme heterojunction photocatalyst comprising g-C_3N_4, Ag$_3$PO$_4$, Ag and AgPd alloy nanoparticles (g-CN/Ag/Ag$_3$PO$_4$-AgPd) to enhance the dehydrogenation of HCOOH. Ag$_3$PO$_4$, as a second semiconductor with narrow band gap, facile preparation and strongly oxidative nature, can form Ag NPs spontaneously, which acts as a recombination center in the mechanism, based on the Z-scheme [66]. As shown in Fig. 7.4e, g-CN/Ag/Ag$_3$PO$_4$/AgPd exhibited the highest TOF value of 2,107 h^{-1} in the control group, which is far higher than monometallic and bimetallic catalysts by generating more oxidative holes and improving the charge separation efficiency in two distinct semiconductors. This study clearly revealed the importance of a Z-scheme heterojunction photocatalyst to enhance the decomposition of

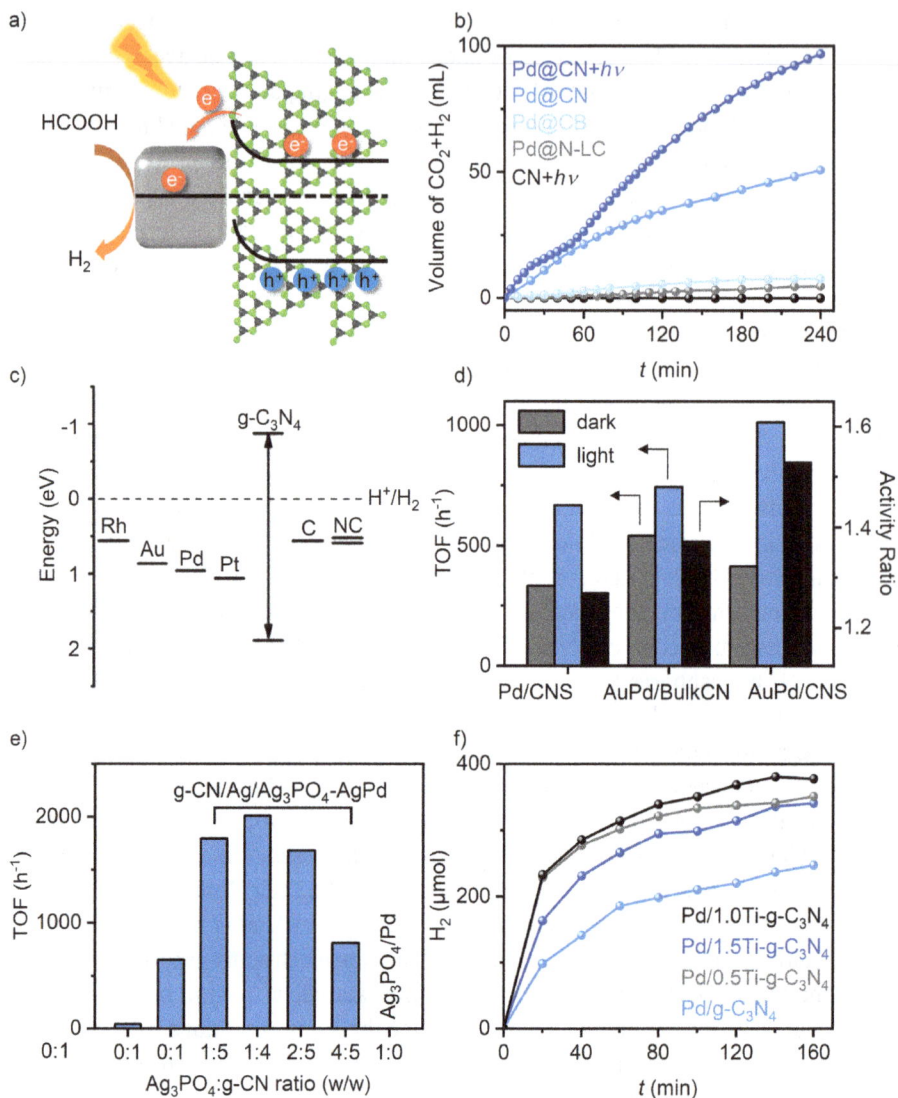

Fig. 7.4: (a) Schematic diagrams of Schottky heterojunction for HCOOH dehydrogenation. (b) Decomposition of HCOOH over different catalysts (CN, carbon nitride; CB, carbon black; N-LC, layered carbon; hv, irradiation with visible light ($\lambda \geq 400$ nm)). Typical conditions: 1 M aqueous HCOOH solution (10 mL), Pd@CN-8% (20 mg), 288 K. (c) Work functions of typical metals, and band structure of carbon nitride and N-doped carbon (NC). Panels (b) and (c) are reproduced with permission [56] (Copyright 2014, Wiley-VCH). (d) TOFs and activity ratio of different catalysts in the dark and under visible light irradiation (reproduced with permission [60]. Copyright 2019, Elsevier). (e) TOFs of catalysts with different nanocomposites under visible light irradiation at 50 °C (reproduced with permission [61]. Copyright 2021, Elsevier). (f) Comparison of catalytic performance of Pd/xTi-g-C_3N_4 catalysts for HCOOH dehydrogenation (reproduced with permission [62]. Copyright 2017, Wiley-VCH).

HCOOH under visible light irradiation. Z-scheme heterojunction is a heterojunction formed between two semiconductors (A and B), of which both the CB and VB of semiconductor A are higher than those of semiconductor B. In such an architecture, the photogenerated electrons remain on semiconductor A and provide higher reduction potential while the photogenerated holes remain on semiconductor B with a higher oxidation potential.

Yamashita and coworkers [62] supported Pd nanoparticles by Ti-doped g-C_3N_4 by deposition–precipitation and subsequent reduction route. Doping of Ti into g-C_3N_4 nanotubes increases the H_2 storage capacity due to the possible hydrogen spillover effect (the migration of hydrogen atoms from the metal catalyst onto the nonmetallic support or adsorbate), as predicted by first-principles calculations [67]. The Pd/Ti-g-C_3N_4 Pd/g-C_3N_4 exhibited the best activity (77.0 h^{-1}) in HCOOH dehydrogenation compared to that of Pd/g-C_3N_4 (Fig. 7.4f). These results suggested that the synergistic effect between $[TiO_4]^{4-}$ moieties and Pd NPs played a pivotal role in enhancing the catalytic activity in HCOOH dehydrogenation. Pd/Ti-g-C_3N_4 with optimum Ti content (1.47%) and with small Pd particles (2.9 nm) showed a superior catalytic activity in the production of H_2 from HCOOH, originating from the interaction of Pd NPs and $[TiO_4]^{4-}$ moieties in g-C_3N_4 structure.

7.2.4 g-C_3N_4 heterojunctions for ammonia borane hydrolysis

Ammonia borane (NH_3BH_3) has drawn great attention because of its various characteristics, such as the protic and hydridic hydrogens, the nitrogen and boron atoms, and the heteropolar dihydrogen bonding [68]. Recently, NH_3BH_3 was used as a hydrogen carrier for the production of H_2 to solve the energy crisis as a result of its low molecular weight (30.87 g mol^{-1}), high gravimetric hydrogen capacity (19.6 wt%), high stability in solid form and non-toxicity under ambient conditions [69], making it a highly attractive and environmental-friendly candidate as a hydrogen carrier. The hydrolysis process of NH_3BH_3 ($NH_3BH_3 + 2H_2O \longrightarrow NH_4BO_2 + 3H_2$) can proceed rapidly in the highly efficient catalytic system at room temperature. In order to improve the diffusion of solid NH_3BH_3 for enhancing the dehydrogenation reaction rate, solvents are usually necessary to ensure the high output of H_2 for practical use. Furthermore, a large number of relevant studies reflect the importance of using NH_3BH_3 as hydrogen storage compound, designing and optimizing catalytic systems to accelerate the hydrolysis reaction.

Schottky junction photocatalysts have been reported to hydrolyze NH_3BH_3 with an enhanced dehydrogenation rate. The metal-loaded g-C_3N_4-based catalysts also accelerated the charge separation and transfer of photogenerated carriers to semiconductors (Fig. 7.5a). Yu and coworkers [70] prepared monodisperse Ni nanoparticles anchored on g-C_3N_4 nanosheets via a self-assembly route and studied their photocatalytic activity in NH_3BH_3 hydrolysis under visible light irradiation. The H_2 production rate improved

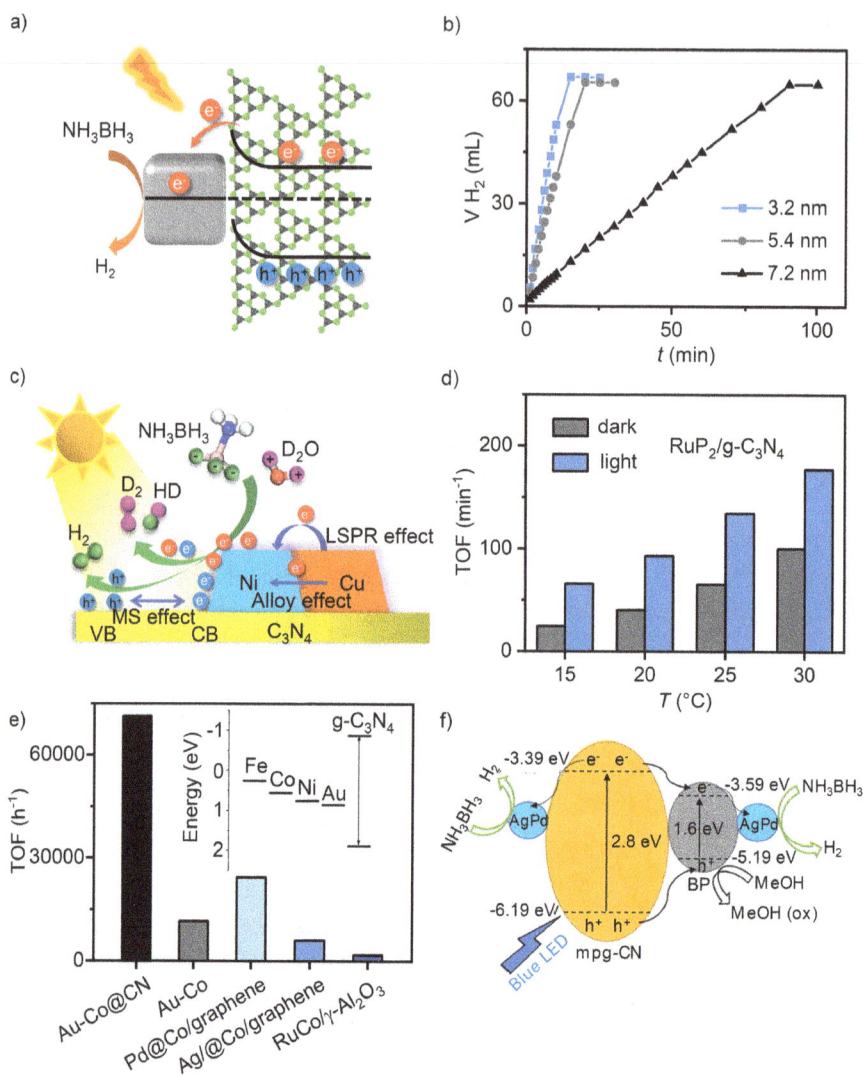

Fig. 7.5: (a) Schematic diagrams of Schottky heterojunction for NH_3BH_3 hydrolysis. (b) Plot volume of hydrogen generated from NH_3BH_3 hydrolysis catalyzed by $Ni/g-C_3N_4$ composite catalysts with Ni NP of different diameter under visible light irradiation ([Ni] = 1.5 mM, [NH_3BH_3] = 110 mM, T = 25 °C) (reproduced with permission [70]. Copyright 2019, Royal Society of Chemistry). (c) Schematic diagrams of Schottky heterojunction between NiCu alloy and CNS applied in NH_3BH_3 hydrolysis (reproduced with permission [71]. Copyright 2020, American Chemical Society). (d) TOFs of $RuP_2/g-C_3N_4$ for NH_3BH_3 hydrolysis under different temperature (reproduced with permission [74]. Copyright 2021, Elsevier). (e) TOFs for hydrogen evolution via decomposition of NH_3BH_3 over Au-Co@CN described in this work and various benchmark catalysts. Inset is work functions of typical metals and band structure of carbon nitride illustrating rectifying contact based on Au–Co hybrid NPs (reproduced with permission [72]. Copyright 2014, American Chemical Society). (f) The overall electron migration mechanism in mpg-CN/BP-AgPd NPs under blue light (BP represents black phosphorus) (reproduced with permission [73]. Copyright 2020, American Chemical Society).

significantly with the decrease of Ni NPs diameter (Fig. 7.5b). Among the series of Ni/g-C_3N_4 catalysts, Ni/g-C_3N_4 with 3.2 nm Ni NPs had the highest activity, with TOF of 18.7 min^{-1} in the decomposition of NH_3BH_3. The possible mechanism was proposed that the electron-hole pairs in g-C_3N_4 were generated in the CB and VB simultaneously under visible light irradiation. The electron could rapidly transfer to the Ni NPs from g-C_3N_4 through the heterojunction interface. Therefore, the electron-rich surface of Ni NPs active sites could facilitate the dehydrogenation of NH_3BH_3 efficiently. Additionally, Wei and coworkers [74] reported application of ruthenium phosphide (RuP_2) quantum dots (QDs), supported on g-C_3N_4, via an in situ phosphorization method toward highly efficient photocatalytic NH_3BH_3 hydrolysis. QDs-based g-C_3N_4 catalyst has shown excellent catalytic activity in NH_3BH_3 hydrolysis due to its smaller diameters of several nanometers. Under visible light irradiation, the initial TOF value of NH_3BH_3 hydrolysis was improved by more than 100% (134 min^{-1}) at ambient temperature (Fig. 7.5d). The activation energy was calculated and it showed decrease from 67.7 ± 0.9 to 47.6 ± 1.0 kJ mol^{-1}.

Regulating electron density at the surface of metal active sites is an effective strategy to facilitate reaction rates. Wang and coworkers [71] systematically studied the NiCu alloy-loaded g-C_3N_4 nanosheets in NH_3BH_3 hydrolysis. In the comparison experiments, NiCu NPs, without the supporting CNS, showed a much lower production volume of hydrogen than that of NiCu/CNS. Volume of hydrogen accumulated in 10 min by employing only NiCu was less than a half of that of NiCu/CNS, which demonstrates the importance of g-C_3N_4 support to increase the decomposition activity of NH_3BH_3. The Schottky heterojunction between the CuNi alloy and g-C_3N_4, by integrating the alloying effect and the LSPR effect under visible light irritation, could act as electron trapping sites (Fig. 7.5c). Research suggested that the redistribution of electrons between Ni and Cu significantly facilitates NH_3BH_3 adsorption, H_2O molecules activation and H atoms desorption, effectively promoting the cleavage of O–H and B–H bonds and the release of H_2. Furthermore, mass spectrometry, infrared spectra, and isotope labeling experiments were conducted to track the hydrogen source and reveal the reaction mechanism.

Rational design of Schottky heterojunction between the semiconductor support and the metal nanoparticles, combined with the high activity of bimetallic nanoparticles, may principally increase the catalytic performance to some extent. Li and coworkers [72] reported the preparation of Au-Co bimetallic nanoparticles supported on multifunctional g-C_3N_4 (Au-Co@CN) for highly efficient dehydrogenation of NH_3BH_3 under ambient atmosphere. After optimizing a series of catalytic parameters in NH_3BH_3 hydrolysis, the Au-Co@CN with 30 wt% of metal loading and Au/Co of 4:96 atomic ratio achieved the best TOF value – up to 1,704 h^{-1}, which is more than 3 times higher than that of Au@Co nanoparticles (Fig. 7.5e). The photocatalytic response was owing to the synergetic effect between Au and Co nanoparticles in Au-Co@CN and rectifying contacts, resulting in highly efficient NH_3BH_3 hydrolysis. Metin and coworkers [73] reported a ternary nanocomposite of mpg-CN/BP-AgPd by assembling AgPd alloy nanoparticles (NPs) on mesoporous g-C_3N_4/black phosphorus (mpg-CN/BP). This nanocomposite consisted of heterojunction supports formed by two diverse nonmetallic semiconductors (black phosphorus (BP) and mpg-CN) with adjustable band gaps and edge potentials, resulting in enhanced catalytic activity of AgPd alloy NPs for hydrogen production, upon NH_3BH_3 hydrolysis. AgPd alloy NPs exhibited the highest activity with mpg-CN/BP, compared to the pristine mpg-CN, pristine BP and mpg-CN/BP binary composite supports, which revealed the formation of a heterojunction interface, with the Schottky barrier between BP and mpg-CN to enhance the photocatalytic performance in NH_3BH_3 dehydrogenation.

7.3 g-C$_3$N$_4$ heterojunctions for organic synthesis

Carbon nitride (C_3N_4) has unique properties such as facile synthetic strategy, abundance of raw materials, unique layered structure and chemical stability [75–77], which has become a research hotspot. The electronic state of C_3N_4 is quite sensitive to the materials it contacts, and it could either accept electrons or donate electrons because of its moderate band gap. There is a strong interaction between the metal and the C_3N_4 [78, 79], when physical contact is built in the metal–C_3N_4 heterojunction. Due to its unique electronic and structural properties, application of C_3N_4 in organic synthesis [80] has spurred worldwide attention in the past decade.

7.3.1 g-C$_3$N$_4$ heterojunctions for hydrogenation reactions

Hydrogenation of unsaturated compounds is important for the synthesis of fine chemicals, pharmaceuticals, fragrances and flavorants [81]. Fu and coworkers [82] introduced the CN/Ni/Al$_2$O$_3$, consisting of carbon nitride (CN) and the underlying nickel, which exhibits excellent performance in the hydrogenation of nitro compounds with H_2 under strongly acidic conditions due to the active site being isolated from the reaction environment (Fig. 5.3). The CN is inactive against hydrogenation without the help of nickel. As nickel donates its electrons to the CN, forming the electron-deficient Ni, hydrogen is directly adsorbed and activated by CN. This structure protects the active nickel from acid corrosion by CN, and the inert CN is activated by nickel for catalytic hydrogenation.

Gaseous hydrogen is difficult to handle due to its hazardous properties. It is environmentally friendly approach to use ammonia borane complex for the hydrogenation of unsaturated bonds as the reagent is stable and composed of abundant elements. Li and coworkers [83] studied cobalt nanoparticles featuring Co^{2+} species on the surface, which in turn are supported by carbon nitride. This architecture serves as a heterogeneous catalyst for effective hydrogenation reactions in aqueous ammonia borane at room temperature (Fig. 7.6b). The metal core of Co nanoparticle is covered by a surface amorphous layer, which can further prevent the oxidation of the embedded Co nanoparticles. The catalytic activity of Co/CN was facilitated by the interaction between Co NP and CN

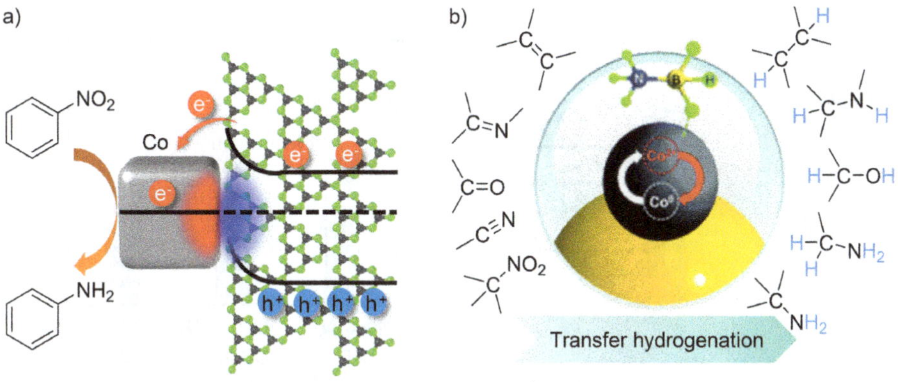

Fig. 7.6: (a) Electron exchange at the interface of Co and C_3N_4. (b) The schematic mechanism of transfer hydrogenation reactions over Co/CN using ammonia-borane as hydrogen source (reproduced with permission [83]. Copyright 2015, Royal Society of Chemistry).

(Fig. 7.6a). The enhanced catalytic activity is attributed to the rectifying contact between the CN and the metal.

7.3.2 g-C_3N_4 heterojunctions for dehydrogenation reactions

As a version of the eco-friendly energy carrier, hydrogen is becoming a research hotspot in the energy field. There is a great challenge to develop an approach to effectively generate and store hydrogen. There are two possible reaction pathways for the decomposition of HCOOH, including dehydrogenation and dehydration. It is very important to design catalysts for the efficient and selective hydrogen generation from HCOOH. Li and coworkers [84] studied Pd@g-C_3N_4/SiO$_2$ with an adjustable band structure, which improved the rate of hydrogen production from HCOOH, and allowed to reach a TOF of 306 h^{-1} (Fig. 7.7b). After Pd nanoparticles were embedded, Pd@g-C_3N_4/SiO$_2$ exhibited high catalytic performance in the dehydrogenation of HCOOH. There is a linear relationship between the band gap and the catalytic activity. As the synthesis temperature decreases, the wide band gap of the g-C_3N_4/SiO$_2$ support leads to higher performance. The semiconductor support increased the electron density of Pd nanoparticles due to the rectifying effect and the electrons flow from g-C_3N_4/SiO$_2$ to metal Pd (Fig. 7.7a), leading to the electron redistribution at the metal-semiconductor interface. Adjusting rationally the band structure and regulating the work function of the semiconductor support is a powerful way to improve the catalytic activity. Yamashita and coworkers [85] reported that the electron-rich Pd obtained upon electron transfer from Co to Pd, in combination with the basic features of the supporting g-C_3N_4 enhance catalytic activity of the material in hydrogen production upon HCOOH dehydrogenation (Fig. 7.7c).

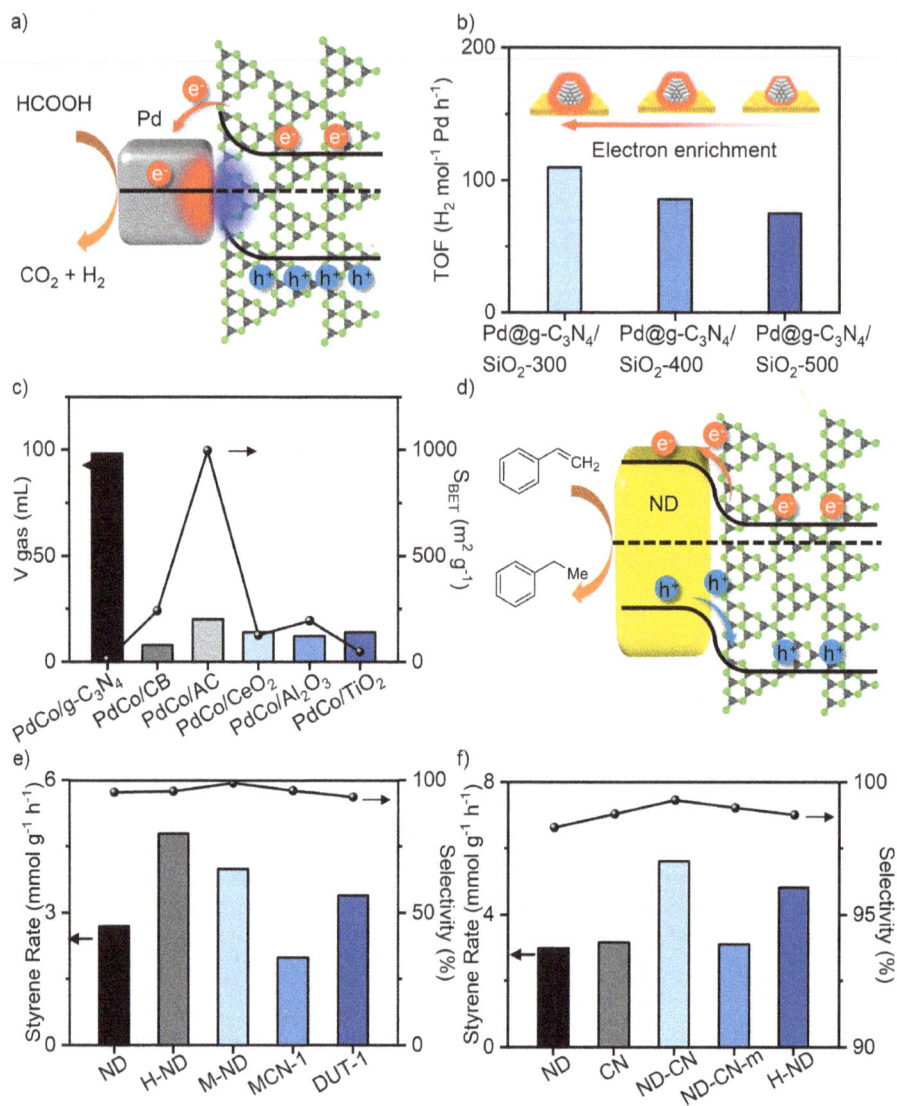

Fig. 7.7: (a) Electron exchange at the interface of Pd nanoparticles and C_3N_4. (b) Band structure-dependent activities for the dehydrogenation of HCOOH over Pd@CN/SiO_2 samples and corresponding TOF values (reproduced with permission [84]. Copyright 2016, Royal Society of Chemistry). (c) Gas volume of H_2 and CO_2 generated by various PdCo catalysts and apparent surface area of different supports (reproduced with permission [85]. Copyright 2019, Elsevier). (d) Electron exchange at the interface of carbon nitride and nanodiamond. (e, f) Steady-state rate of styrene production and selectivity for direct dehydrogenation of ethylbenzene to styrene over different catalysts. Panel (e) is reproduced with permission [86] (Copyright 2015, American Chemical Society). Panel (f) is reproduced with permission [87] (Copyright 2019, Elsevier).

Direct dehydrogenation (DDH) of ethylbenzene has been regarded as the green-way of producing styrene. Zhao and coworkers [86] synthesized carbon nitride-wrapped nanodiamond hybrids (H-ND), which showed high catalytic activity in DDH of ethylbenzene due to the enhanced synergistic effect between carbon nitride and nanodiamonds (ND). For comparison, the classical ND-carbon-nitride (M-ND), mesoporous carbon nitride (MCN-1) and the developed mesoporous carbon nitride (DUT-1) were also prepared. The carbon nitride layer deagglomerates the ND in H-ND, which results in a larger surface area compared to the original ND. The synergistic effect can be attributed to the improvement of C = O nucleophilicity in the presence of electron-rich N atoms in carbon nitride [88] and the enhancement of the catalytic activity of the C = O active sites (Fig. 7.7d). The increase in the number of defects resulting from the unique microstructure and high surface area of H-ND, and the synergistic effect between carbon nitride and ND enhances the H-ND catalytic performance in the DDH reaction, compared to the parent ND (Fig. 7.7e).

However, the active sites of ND were closely covered by the carbon nitride layer, which reduced the catalytic performance. Research shows that the ketone carbonyl groups and defects in the surface structure can activate C–H bonds. Nanodiamond@-carbon nitride hybrid (ND-CN) [87] was introduced by ammonium chloride to generate more defects and, as a result, to achieve higher catalytic performance, superior to the mixture of ND and CN(ND-CN-m). The loose porous carbon nitride layer released more active sites belonging to the ND and sufficiently dispersed ND densely, without weakening the synergy between the nanodiamond and the carbon nitride. It resulted in 1.9- and 1.2-folds higher styrene rates than the original ND and H-ND in the vapor-free DDH of ethylbenzene (Fig. 7.7f).

7.3.3 g-C$_3$N$_4$ heterojunctions for oxidation reactions

It is important to catalyze oxidation of the saturated C–H bonds in inexpensive feedstock to high value-added chemicals in fine chemical industry. The abundant and stable solid g-C$_3$N$_4$ under ambient conditions can meet the requirements as an industrial catalyst. Due to the poor thermal stability of organic materials, they are rarely used in gas phase reactions. This problem can be solved by supporting vanadium oxide on g-C$_3$N$_4$ because of its stability against oxidation and thermal stability, which can then efficiently catalyze toluene vapor phase oxidation [89]. Huang and coworkers suggested that nitrogen-rich properties of g-C$_3$N$_4$ provide abundant anchoring sites for the dispersion of vanadium oxide. The binding energy of V $2p_{3/2}$ in vanadium oxide is 516.9 eV, which is slightly lower than the binding energy of V $2p_{3/2}$ in V–g-C$_3$N$_4$, as inferred from the high-resolution vanadium 2p X-ray photoelectron spectrum. This observation is attributed to the interaction of vanadium with nitrogen, which increases the electron cloud density of vanadium (Fig. 7.8a). When used as a catalyst for the oxidation of toluene, V-g-C$_3$N$_4$ showed significantly improved selectivity and maintained high activity.

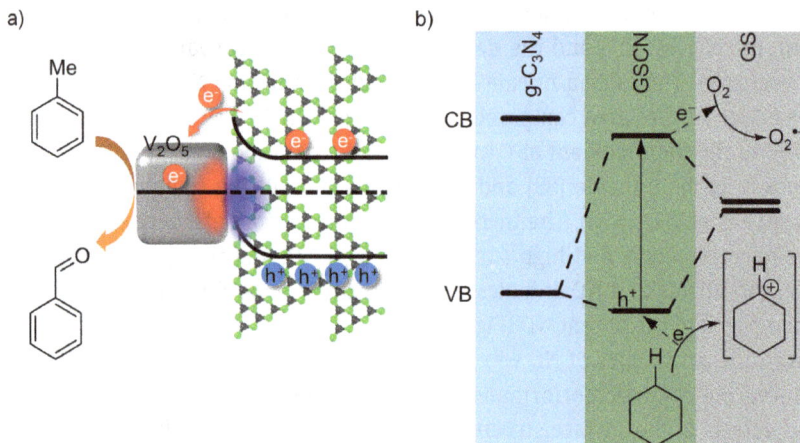

Fig. 7.8: (a) The schematic representation of electron exchange at the interface of V$_2$O$_5$ and C$_3$N$_4$. (b) The schematic mechanism of cyclohexane oxidation (reproduced with permission [90]. Copyright 2011, American Chemical Society).

During the catalytic oxidation process by O$_2$, one of the biggest challenges is to control the selectivity of the reaction without forming the overoxidized products. Li and coworkers [90] reported graphene sheet (GS)/polymeric carbon nitride (g-C$_3$N$_4$) complexes for the activation of molecular oxygen to catalyze selective oxidation of C-H bonds under heterogeneous conditions. The graphene sheet/carbon nitride (GSCN) complex as an economic metal-free catalyst exhibited high selectivity towards ketones due to the synergistic effect between g-C$_3$N$_4$ and GS to improve its ability towards O$_2$ activation for the subsequent C-H bonds' selective oxidation in cyclohexane. Single-layer GS was selected as the electron donor for g-C$_3$N$_4$ because of its similar layered structure and suitable electronic and chemical properties. O$_2$ is reduced to O$_2$$^{\bullet-}$ by electrons excited from the VB to the CB of C$_3$N$_4$, and remains bonded to the surface of C$_3$N$_4$, compensating the positive charge of the hole. Meanwhile, the cyclohexane is oxidized by the positive hole in the VB of C$_3$N$_4$, and subsequently reacts with O$_2$$^{\bullet-}$ (Fig. 7.8b). The introduction of GS can significantly lower the CB energy of C$_3$N$_4$ via charge transfer or π–π^* interactions. GSCN offers excellent conversion and high selectivity toward ketone by adjusting the content of graphene with respect to the C$_3$N$_4$ component.

7.4 g-C₃N₄ heterojunctions for electrocatalytic and photoelectrocatalytic reactions

Recently, as a promising polymeric semiconductor, $g\text{-}C_3N_4$ has attracted great interest in novel electrode and photoelectrode materials, owing to its outstanding properties such as high physicochemical stability and adjustable electronic band structure [91]. The $g\text{-}C_3N_4$-based catalysts have been applied in various electrocatalytic processes, including electroreduction reaction, electro-oxidation reaction and photoelectrocatalysis.

7.4.1 g-C₃N₄ heterojunctions for electrocatalytic reduction reactions

For cathodic hydrogen evolution reaction (HER), an electron-rich metallic surface could enhance the adsorption of proton and promote catalytic activity. In this regard, a molybdenum boride/carbon nitride ($MoB/g\text{-}C_3N_4$) Schottky catalyst [92] (Fig. 7.9a) was reported by Mai and coworkers for efficient HER by constructing electron-rich metallic MoB across a rectifying contact with an n-type $g\text{-}C_3N_4$. X-ray photoelectron spectroscopy (XPS) was used to characterize the electronic structures of MoB before and after the rectifying contact. As shown in Fig. 7.9b, compared with pristine of MoB, a negative shift in the XPS of typical Mo $3d_{5/2}$ peaks was observed in $MoB/g\text{-}C_3N_4$, confirming that the electrons were transferred from $g\text{-}C_3N_4$ to MoB [93]. As a result, the electron-rich $MoB/g\text{-}C_3N_4$ Schottky catalyst provided excellent performance with a higher exchange current density of 17 mA cm^{-2} (Fig. 7.9c), even comparable to state-of-the-art noble metal-free catalysts.

Besides HER, the $g\text{-}C_3N_4$-based catalysts were also used in the nitrogen reduction reaction (NRR) by the interface engineering strategy. Chu and coworkers [94] designed a two-dimensional material with face-to-face contact by the strong interfacial electronic interactions as an effective NRR catalyst. The highly coupled MoS_2/C_3N_4 heterostructure resulted in more pronounced electron-rich regions in MoS_2, owing to the interfacial charge transport from C_3N_4 to MoS_2 (Fig. 7.9d). The chemical state of MoS_2/C_3N_4 was studied by XPS. The Mo 3d peak of MoS_2/C_3N_4 was shifted negatively in comparison with the MoS_2, clearly demonstrating the existence of electronic coupling and the electron transfer from C_3N_4 to MoS_2 (Fig. 7.9e). As a result, the electron-rich MoS_2/C_3N_4 catalyst boosted the NRR performances with Faradaic efficiency (FE) [95] of 17.8% and an NH_3 yield of 18.5 μg h^{-1} mg^{-1} for NH_3 production, which are much better than that of MoS_2 or C_3N_4 component (Fig. 7.9f). Density functional theory calculations revealed that the electronic coupling between C_3N_4 and MoS_2 could promote the stabilization of the key intermediate and enhance the NRR activity.

Electrochemical CO_2 reduction reaction (CO_2RR) to high value-added fuels has attracted great interest. It has been reported that electron-rich metal nanoparticles could accelerate the electron transfer to CO_2 and enhance CO_2RR [98]. Motivated by these

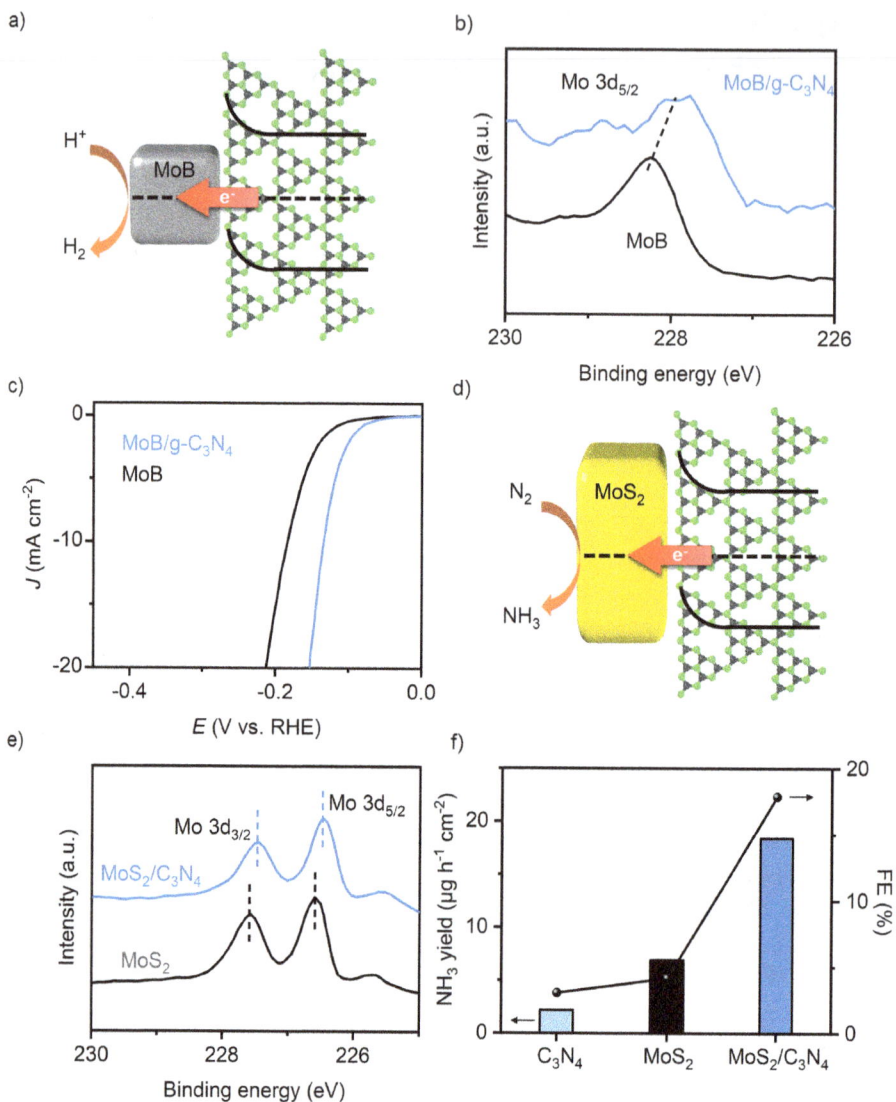

Fig. 7.9: (a) The schematic representation of MoB/g-C$_3$N$_4$ electronic structure. (b) Mo 3d$_{5/2}$ XPS spectra for pure MoB and MoB/g-C$_3$N$_4$ catalyst. (c) HER performance of pure MoB and MoB/g-C$_3$N$_4$ catalysts. Panels (a)–(c) are reproduced with permission [92] (Copyright 2018, Wiley-VCH). (d) The schematic representation of MoS$_2$/g-C$_3$N$_4$ electronic structure. (e) Mo 3d$_{5/2}$ XPS spectra for pure MoB and MoS$_2$/g-C$_3$N$_4$ catalysts. (f) NRR performances of MoS$_2$, C$_3$N$_4$, and MoS$_2$/C$_3$N$_4$ under identical conditions. Panels (d)–(f) are reproduced with permission [94] (Copyright 2020, American Chemical Society).

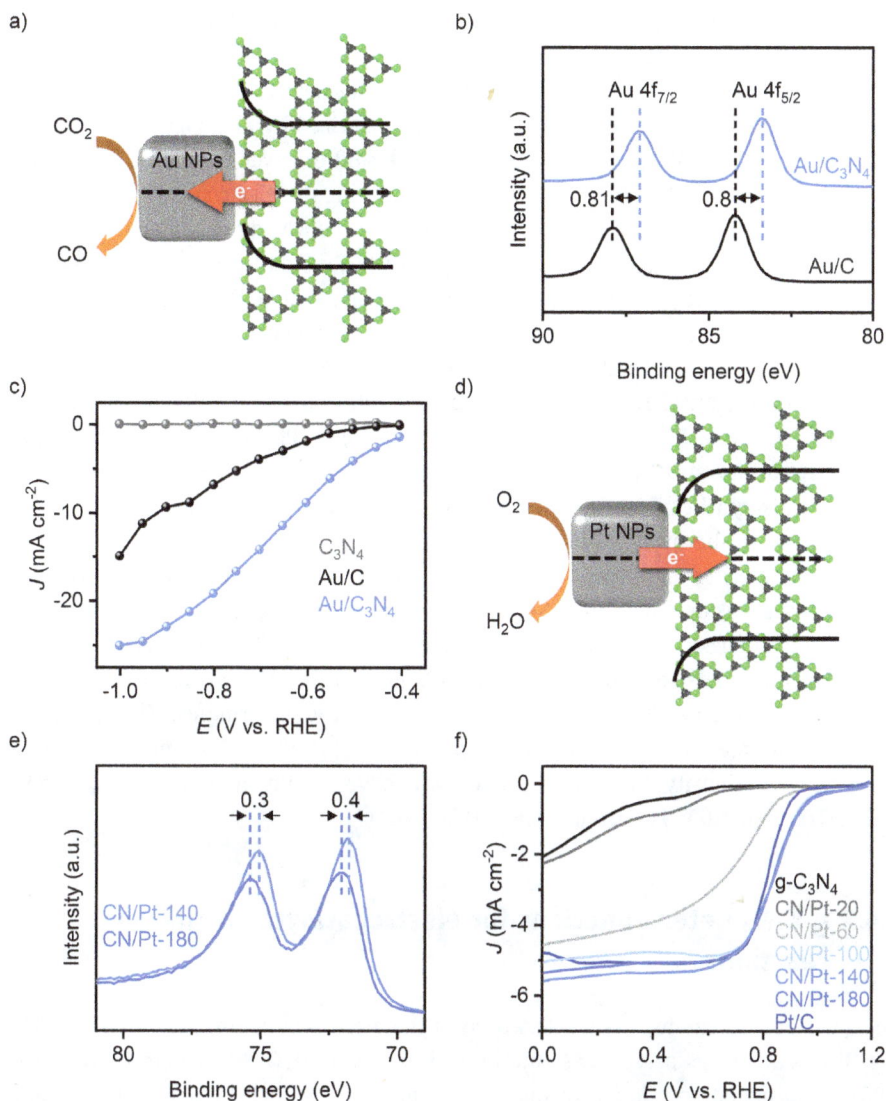

Fig. 7.10: (a) The schematic representation of Au/C_3N_4 electronic structure. (b) Au X-ray photoelectron spectra for Au/C and Au/C_3N_4 catalyst. (c) CO partial current densities for Au/C_3N_4, Au/C and C_3N_4 at different potentials. Panels (a)–(c) are reproduced with permission [96] (Copyright 2018, Wiley-VCH). (d) The schematic representation of CN/Pt electronic structure. (e) X-ray photoelectron spectra of Pt 4f for CN/Pt-140 and CN/Pt-180. (f) LSV profiles of g-C_3N_4, CN/Pt-20, CN/Pt-60, CN/Pt-100, CN/Pt-140, CN/Pt-180 and Pt/C catalysts at a sweep speed of 1,600 rpm. Panels (d)–(f) are reproduced with permission [97] (Copyright 2018, American Chemical Society).

studies, Yang and coworkers [96] designed an electron-rich metal Ru-based catalyst by dispersing Au nanoparticles on C_3N_4 (Au/C_3N_4) featuring electron transfer from C_3N_4 to Au (Fig. 7.10a), which showed excellent electrocatalytic activity in CO_2RR. The electronic structure of the formed Au/C_3N_4 catalyst was characterized by XPS. Figure 7.10b shows that the Au $4f_{5/2}$ and Au $4f_{7/2}$ peaks of Au/C_3N_4 both shift to lower binding energies compared with that of carbon-black support (Au/C), proving the electron-enriched Au. The electron-enriched Au surface boosted the CO_2RR performance and exhibited a higher CO partial current density (j_{CO}) (Fig. 7.10c), compared to Ru/C and pristine C_3N_4 catalysts. Based on these results, we conclude that in the Au/C_3N_4 system, the C_3N_4 support could transfer electrons to the Au surface and enhance the CO_2RR performance.

Besides electron-rich catalysts, the electron deficiency of metal surface induced by electron transferred from metal to carbon nitride could also enhance the electroreduction reaction. Xu and coworkers [97] constructed Pt/graphitic carbon nitride (GCN) electrocatalysts (CN/Pt-x, x refers to the absorbed dose of gamma radiation during materials synthesis), with electrons flowing from metal to carbon nitride that were applied in the oxygen reduction reaction (ORR) (Fig. 7.10d). Higher content of metal could transfer more electrons to the support. The electron-deficient nature of Pt nanoparticles was demonstrated by the slightly lower binding energy of Pt 4 f peak of CN/Pt-140 compared to CN/Pt-180 with higher Pt loading (Fig. 7.10e), suggesting that the electronic interface exists between Pt NPs and g-C_3N_4. The strong metal-support interaction between Pt NPs and g-C_3N_4 enhances the ORR performance. The optimized CN/Pt-140 samples provide a half-wave potential ($E_{1/2}$) of 0.827 V and a diffusion-limited current density (J) of 5.56 mA $\cdot$ cm^{-2}, which were equivalent to that of commercial Pt/C ($E_{1/2} = 0.816$ V, $J = 5.01$ mA $\cdot$ cm^{-2}) (Fig. 7.10 f).

7.4.2 g-C_3N_4 heterojunctions for electrocatalytic oxidation reactions

The g-C_3N_4-based catalysts were also used in the electro-oxidation reaction [99]. Taking OER reaction as a typical example, iridium oxide (IrO_2)/GCN heterostructures [100] are designed with low-coordinate IrO_2 nanoparticles (NPs) confined on graphic CN nanosheets for an efficient acidic OER (Fig. 7.11a). The X-ray photoelectron spectroscopy (XPS) survey spectra showed that the binding energy of Ir 4f in the IrO_2/GCN containing 40 wt% IrO_2 (40-IG) shifts negatively by 0.2 eV as compared with bare IrO_2 NPs, confirming the strong electronic interaction between IrO_2 and GCN (Fig. 7.11b). Meanwhile, the lower binding energies indicated electron transfers from GCN to IrO_2, which results in a higher electron density around Ir atoms. The electrochemical test was carried out to evaluate the OER performance. As shown in Fig. 7.11c, the 40-IG sample provided the lowest overpotential of 276 mV at 10 mA cm^{-2}, compared with the other samples. Meanwhile, the 40-IG possesses improved OER kinetics, with the lowest Tafel slope of 57 mV dec^{-1} (Fig. 7.11d). Notably, compared with pristine IrO_2

NPs, all the IrO_2/GCN heterostructures have higher activity normalized by the mass of the catalyst (Fig. 7.11e), indicating that the key importance of GCN nanosheets in accelerating the oxygen evolution kinetics at IrO_2. In particular, the optimized sample 40-IG exhibits the highest mass activity of 1,280 mA mg^{-1} at 1.6 V. More significantly, as

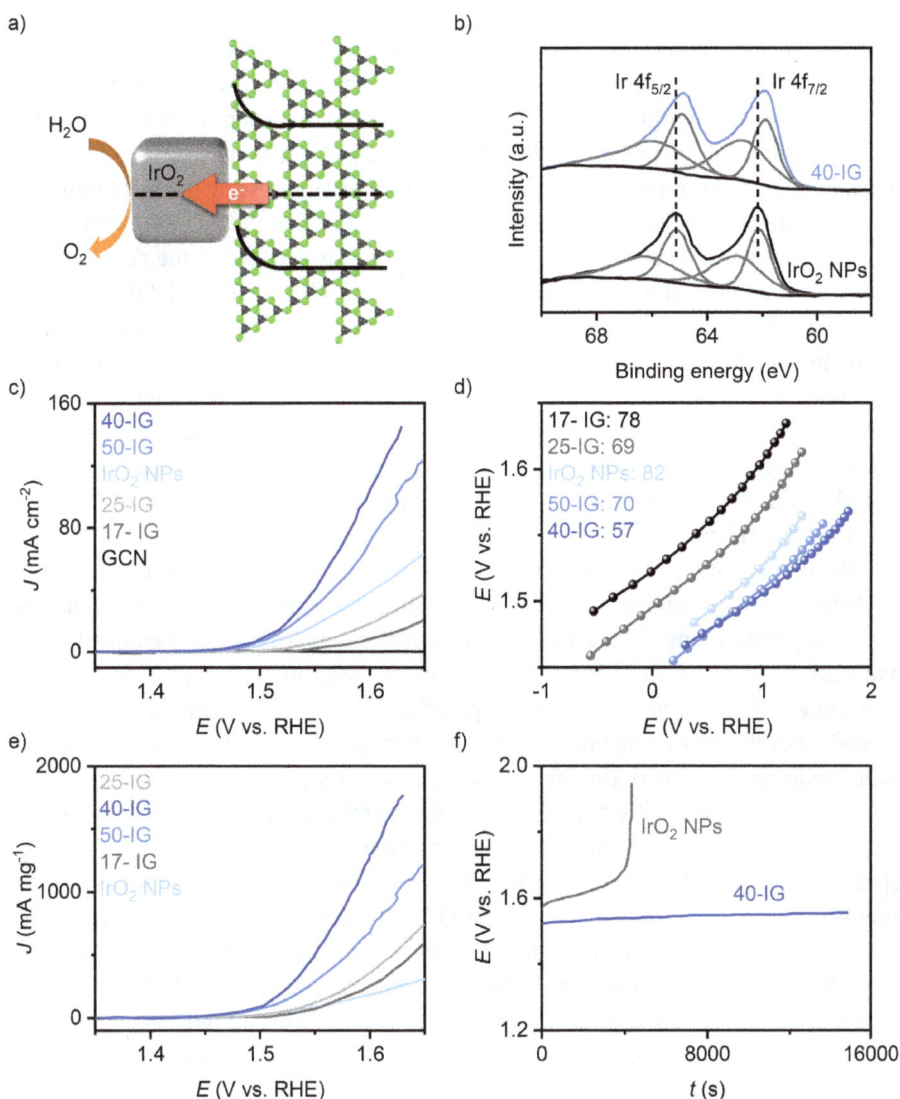

Fig. 7.11: (a) The schematic representation of IrO_2/g-C_3N_4 electronic structure. (b) Ir 4f X-ray photoelectron spectra of 40-IG and IrO_2. (c) LSV polarization curves measured at a scan rate of 5 mV s^{-1}. (d) Tafel plots derived from LSV curves. (e) LSV curves plotted based on IrO_2 mass-normalized current density. (f) Chronopotentiometry of IrO_2 NPs and 40-IG at a current density of 20 mA cm^{-2}. Panels (a)–(f) are reproduced with permission [100] (Copyright 2020, American Chemical Society).

shown in Fig. 7.11f, the 40-IG sample has excellent stability, with only 35 mV change after 4 h chronopotentiometry test, while the IrO_2 quickly loses its activity after 1.2 h.

7.4.3 g-C_3N_4 heterojunctions for photoelectrocatalytic reactions

Besides the application of g-C_3N_4-based catalysts in electrochemistry, photoelectrochemical CO_2RR and water splitting are also important fields [101]. For example, the fabrication of zinc phthalocyanine/carbon nitride (ZnPc/carbon nitride) composites was presented and utilized in PEC-CO_2RR [102]. The as-formed ZnPc/carbon nitride efficiently promotes the electrocatalytic performance by suppressing the recombination of the photogenerated electrons and holes (Fig. 7.12a). The PEC-CO_2RR experiments were carried out in 0.1 M $KHCO_3$ solution. Methanol was detected as the main product with a small amount of formic acid. Compared with carbon nitride and ZnPC, the composites of ZnPc/carbon nitride provided a much higher yield of 13 mmol cm^{-2} in 8 h at the optimal voltage of –1.0 V (Fig. 7.12b), demonstrating the key importance of the energy band matching between the ZnPc and carbon nitride for boosting PEC-CO_2RR. Moreover, the ZnPc/carbon nitride exhibited a highly synergic effect between PC-CO_2RR and EC-CO_2RR as shown in Fig. 7.12c. The methanol yield rate at ZnPc/carbon nitride nanosheets in PEC-CO_2RR is higher than PC, EC and PC + EC (the yield rate of PC + EC is equal to the yield rate of PC plus the yield rate of EC).

Metal oxide array (TiO_2 and ZnO) coupled with g-C_3N_4 are considered as promising heterojunction composites in PEC water splitting, owing to their strong ability for charge separation and visible light adsorption [104]. As a result, the coupled metal oxide array/g-C_3N_4 exhibit better PEC performance than individual photoanode composed of g-C_3N_4 or metal oxide array only. For example, Xiao and coworkers [103] fabricated a novel direct Z-scheme TiO_2/g-C_3N_4 nanotube arrays for photoelectrochemical water splitting (Fig. 7.12d). Notably, the oxygen vacancy (OV) layer formed by $NaBH_4$ reduction is critical for Z-scheme TiO_2/g-C_3N_4 heterojunctions. The introduced OV in g-C_3N_4/OV-TiO_2 results in a lower charge transfer resistance as shown by the Nyquist circle (Fig. 7.12e). The g-C_3N_4/OV-TiO_2 sample showed very high photocurrent responses compared to that of TiO_2, OV-TiO_2 and g-C_3N_4/TiO_2 (Fig. 7.12f). Taken together, the g-C_3N_4 possessed higher PEC performance in water splitting compared to the other samples. The improved PEC performance derives mainly from the improved visible light absorption and carrier separation, promoted by constructing highly coupled Z-scheme heterojunction between OV-TiO_2 and g-C_3N_4.

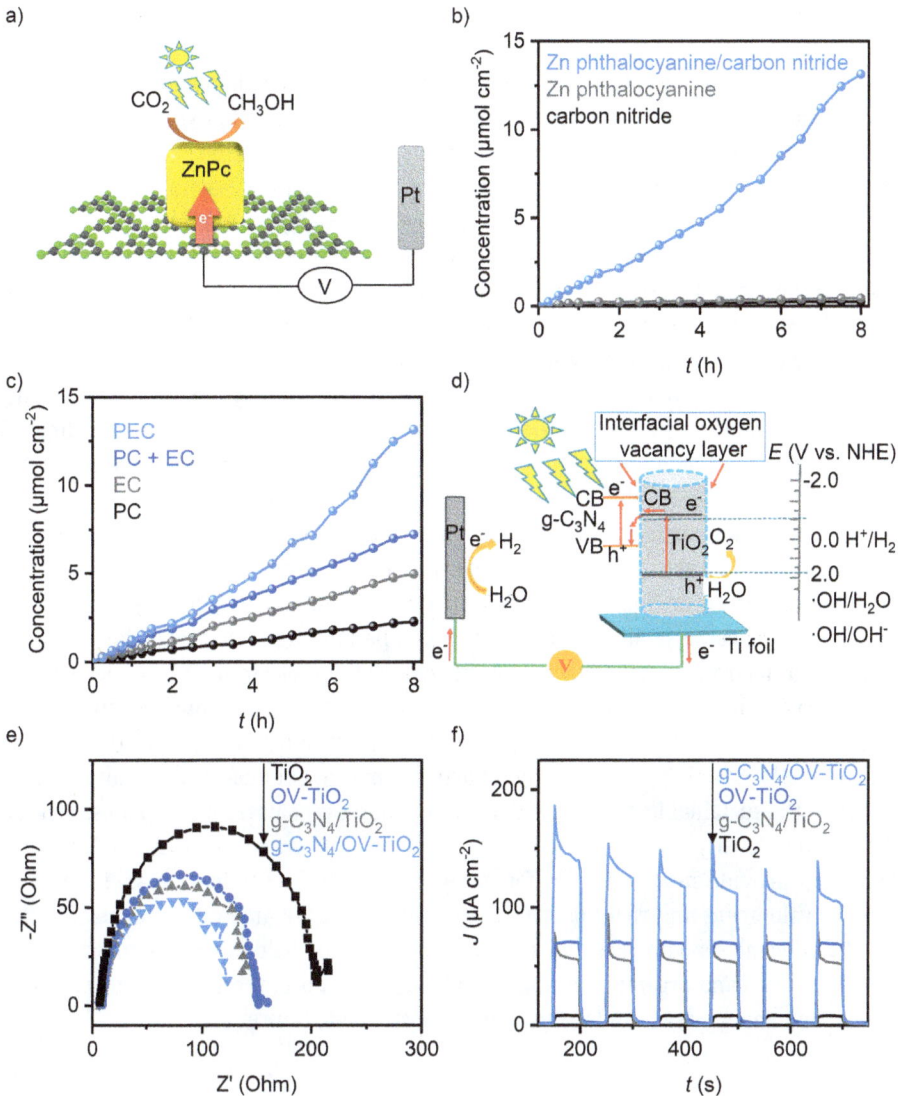

Fig. 7.12: (a) The schematic representation of PEC CO_2RR. (b) Time-dependent curves of methanol yield at -1.0 V on carbon nitride nanosheets, ZnPc, ZnPc/carbon nitride nanosheets for PEC-CO_2RR. (c) Time-dependent yield of methanol obtained under different conditions – PC, EC, PC + EC and PEC-CO_2RR on ZnPc/carbon nitride nanosheets. Panels (a)–(c) are reproduced with permission [102] (Copyright 2019, Elsevier). (d) Schematic illustration of carrier transfer in g-C_3N_4/OV-TiO_2 heterostructure. (e) EIS plots and (f) transient photocurrent responses of g-C_3N_4/OV-TiO_2, g-C_3N_4/TiO_2, OV-TiO_2, TiO_2. Panels (d)–(f) are reproduced with permission [103] (Copyright 2018, American Chemical Society).

7.5 Summary

The rational design of g-C$_3$N$_4$-based heterostructure catalysts with metal or semiconductor materials for green photocatalysis (e.g., C–C coupling reaction, hydrogenation reaction, dehydrogenation reaction and hydrolysis reaction), electrocatalysis (e.g., HER, OER, NRR and CO$_2$RR) and organic synthesis (e.g., hydrogenation reaction, dehydrogenation reaction and oxidation reaction) have developed rapidly as an emerging frontier field. This chapter has provided a brief overview of the recent studies that have centered on the rational design, synthesis and catalytic applications of g-C$_3$N$_4$-based heterojunctions, which is regarded as an excellent candidate material for basic research and potential catalytic applications due to their unique electronic structures and interfacial effects. The support effect bottomed on the g-C$_3$N$_4$-based Schottky heterojunctions leads to the adjustment of the electron density of the active metal center or the bending of the semiconductor side energy band to a great extent, which directly triggers the improvement of the activity of the catalytic systems. The strong electron interaction between metal nanoparticles or semiconductors and g-C$_3$N$_4$ carrier has been found as charge transfer in the catalysis system to ensure excellent activity for various catalytic reactions. Based on the mutual promotion principle of multi-component g-C$_3$N$_4$ heterojunction-based catalysts discussed above, it can be envisaged that well-elaborated g-C$_3$N$_4$-based heterojunctions with enhanced chemical and electronic properties will be widely used to improve significantly the performance of systems in photocatalysis, electrocatalysis and organic synthesis. Combining the multifunctionality of g-C$_3$N$_4$ and excellent catalytic activity of metal nanoparticles or semiconductor materials, further design of the sustainable g-C$_3$N$_4$-based heterojunction dyads based on the adjustment of the electron density of active metal center or the bending of semiconductor side energy band will provide new strategies for green chemistry. To summarize, owing to the rapid development and rich knowledge accumulated in the area of g-C$_3$N$_4$ heterojunction-based catalysts with excellent catalytic activity in the field of green photocatalysis, electrocatalysis and organic synthesis, it will play an important role in solving environmental and energy issues and will lead to the sustainable future.

References

[1]   Ye S, Wang R, Wu MZ, Yuan YP. A review on g-C$_3$N$_4$ for photocatalytic water splitting and CO$_2$ reduction. Appl Surf Sci 2015, 358, 15–27.

[2]   Zhao TJ, Feng WJ, Zhang JJ, Zhang B, Liu YX, Lin YX, et al. Electrostatically mediated selectivity of Pd nanocatalyst *via* rectifying contact with semiconductor: Replace ligands with light. Appl Catal, B 2018, 238, 404–09.

[3]   Cai YY, Li XH, Zhang YN, Wei X, Wang KX, Chen JS. Highly efficient dehydrogenation of formic acid over a palladium-nanoparticle-based Mott-Schottky photocatalyst. Angew Chem Int Ed Engl 2013, 52(45), 11822–25.

[4]   Li H, Chen C, Yan DF, Wang YY, Chen R, Zou YQ, et al. Interfacial effects in supported catalysts for electrocatalysis. J Mater Chem A 2019, 7(41), 23432–50.

[5]   He F, Wang ZX, Li YX, Peng SQ, Liu B. The nonmetal modulation of composition and morphology of g-C$_3$N$_4$-based photocatalysts. Appl Catal, B 2020, 269, 118828.

[6]   Mamba G, Mishra AK. Graphitic carbon nitride (g-C$_3$N$_4$) nanocomposites: A new and exciting generation of visible light driven photocatalysts for environmental pollution remediation. Appl Catal, B 2016, 198, 347–77.

[7]   Wang HH, Lin QY, Yin LT, Yang YL, Qiu YA, Lu CH, et al. Biomimetic design of hollow flower-like g-C$_3$N$_4$@PDA organic framework nanospheres for realizing an efficient photoreactivity. Small 2019, 15(16), e1900011.

[8]   Che W, Cheng WR, Yao T, Tang FM, Liu W, Su H, et al. Fast photoelectron transfer in (C$_{ring}$)-C$_3$N$_4$ plane heterostructural nanosheets for overall water splitting. J Am Chem Soc 2017, 139(8), 3021–26.

[9]   Liu DN, Chen DY, Li NJ, Xu QF, Li H, He JH, et al. Surface engineering of g-C$_3$N$_4$ by stacked BiOBr sheets rich in oxygen vacancies for boosting photocatalytic performance. Angew Chem Int Ed 2020, 59(11), 4519–24.

[10]   Tauster SJ, Fung SC, Garten RL. Strong Metal-Support Interactions. Group 8 Noble Metals Supported on TiO$_2$. J Am Chem Soc 1978, 4, 170–75.

[11]   Tang L, Meng XG, Deng DH, Bao XH. Confinement catalysis with 2D materials for energy conversion. Adv Mater 2019, 31(50), e1901996.

[12]   Kwon T, Jun M, Joo J, Lee K. Nanoscale hetero-interfaces between metals and metal compounds for electrocatalytic applications. J Mater Chem A 2019, 7(10), 5090–110.

[13]   Shiraishi Y, Kofuji Y, Kanazawa S, Sakamoto H, Ichikawa S, Tanaka S, et al. Platinum nanoparticles strongly associated with graphitic carbon nitride as efficient co-catalysts for photocatalytic hydrogen evolution under visible light. Chem Commun 2014, 50(96), 15255–58.

[14]   Gu K, Pan XT, Wang WW, Ma JJ, Sun Y, Yang HL, et al. In situ growth of Pd nanosheets on g-C$_3$N$_4$ nanosheets with well-contacted interface and enhanced catalytic performance for 4-nitrophenol reduction. Small 2018, 4, e1801812.

[15]   Lei CS, Zhou W, Shen LJ, Zheng XH, Feng QG, Liu Y, et al. Enhanced selective H$_2$S oxidation performance on Mo$_2$C-Modified g-C$_3$N$_4$. ACS Sustainable Chem Eng 2019, 7(19), 16257–63.

[16]   Xu M, Han L, Dong SJ. Facile fabrication of highly efficient g-C$_3$N$_4$/Ag$_2$O heterostructured photocatalysts with enhanced visible-light photocatalytic activity. ACS Appl Mater Interfaces 2013, 5(23), 12533–40.

[17]   Yang LQ, Huang JF, Shi L, Cao LY, Liu HM, Liu YY, et al. Sb doped SnO$_2$-decorated porous g-C$_3$N$_4$ nanosheet heterostructures with enhanced photocatalytic activities under visible light irradiation. Appl Catal, B 2018, 221, 670–80.

[18]   Zhang KX, Su H, Wang HH, Zhang JJ, Zhao SY, Lei W, et al. Atomic-scale Mott-Schottky heterojunctions of boron nitride monolayer and graphene as metal-free photocatalysts for artificial photosynthesis. Adv Sci 2018, 5(7), 1800062.

[19]   Shen RC, Xie J, Lu XT, Chen XB, Bifunctional LX. Cu$_3$P decorated g-C$_3$N$_4$ nanosheets as a highly active and robust visible-light photocatalyst for H$_2$ production. ACS Sustainable Chem Eng 2018, 6(3), 4026–36.

[20]   Zhang Z, Yates JT Jr. Band bending in semiconductors: Chemical and physical consequences at surfaces and interfaces. Chem Rev 2012, 112(10), 5520–51.

[21]   Theophilos I, VX E. Charge transfer in metal catalysts supported on doped TiO$_2$: A theoretical approach based on metal–semiconductor contact theory. J Catal 1996, 161, 560–69.

[22]   Li YF, Zhou MH, Cheng B, Shao Y. Recent advances in g-C$_3$N$_4$-based heterojunction photocatalysts. J Mater Sci Technol 2020, 56, 1–17.

[23]   Zhang LP, Jaroniec M. Toward designing semiconductor-semiconductor heterojunctions for photocatalytic applications. Appl Surf Sci 2018, 430, 2–17.

[24] Wang HL, Zhang LS, Chen ZG, Hu JQ, Li SJ, Wang ZH, et al. Semiconductor heterojunction photocatalysts: Design, construction, and photocatalytic performances. Chem Soc Rev 2014, 43(15), 5234–44.

[25] Moniz SJA, Shevlin SA, Martin DJ, Guo ZX, Tang JW. Visible-light driven heterojunction photocatalysts for water splitting – A critical review. Energy Environ Sci 2015, 8(3), 731–59.

[26] Bafaqeer A, Tahir M, Amin NAS. Well-designed ZnV_2O_6/g-C_3N_4 2D/2D nanosheets heterojunction with faster charges separation *via* pCN as mediator towards enhanced photocatalytic reduction of CO_2 to fuels. Appl Catal, B 2019, 242, 312–26.

[27] Wang WJ, Li GY, An TC, Chan DKL, Yu JC, Wong PK. Photocatalytic hydrogen evolution and bacterial inactivation utilizing sonochemical-synthesized g-C_3N_4/red phosphorus hybrid nanosheets as a wide-spectral-responsive photocatalyst: The role of type-I band alignment. Appl Catal, B 2018, 238, 126–35.

[28] Kokane SB, Sasikala R, Phase DM, Sartale SD. In_2S_3 nanoparticles dispersed on g-C_3N_4 nanosheets: Role of heterojunctions in photoinduced charge transfer and photoelectrochemical and photocatalytic performance. J Mater Sci 2017, 52(12), 7077–90.

[29] Zhu MS, Kim S, Mao L, Fujitsuka M, Zhang JY, Wang XC, et al. Metal-free photocatalyst for H_2 Evolution in visible to near-infrared region: Black phosphorus/graphitic carbon nitride. J Am Chem Soc 2017, 139(37), 13234–42.

[30] Lin B, Li H, An H, Hao WB, Wei JJ, Dai YZ, et al. Preparation of 2D/2D g-C_3N_4 nanosheet@$ZnIn_2S_4$ nanoleaf heterojunctions with well-designed high-speed charge transfer nanochannels towards high-efficiency photocatalytic hydrogen evolution. Appl Catal, B 2018, 220, 542–52.

[31] Cai HR, Wang B, Xiong LF, Bi JL, Yuan LY, Yang GD, et al. Orienting the charge transfer path of type-II heterojunction for photocatalytic hydrogen evolution. Appl Catal, B 2019, 256, 117853.

[32] Ma R, Zhang S, Li L, Gu PC, Wen T, Khan A, et al. Enhanced visible-light-induced photoactivity of type-II CeO_2/g-C_3N_4 nanosheet toward organic pollutants degradation. ACS Sustainable Chem Eng 2019, 7(10), 9699–708.

[33] He C, Zhang JH, Zhang WX, Li TT. Type-II InSe/g-C_3N_4 heterostructure as a high-efficiency oxygen evolution reaction catalyst for photoelectrochemical water splitting. J Phys Chem Lett 2019, 10(11), 3122–28.

[34] Liu EZ, Jin CY, Xu CH, Fan J, Hu XY. Facile strategy to fabricate Ni_2P/g-C_3N_4 heterojunction with excellent photocatalytic hydrogen evolution activity. Int J Hydrogen Energy 2018, 43(46), 21355–64.

[35] Sun CZ, Zhang H, Liu H, Zheng XX, Zou WX, Dong L, et al. Enhanced activity of visible-light photocatalytic H_2 evolution of sulfur-doped g-C_3N_4 photocatalyst *via* nanoparticle metal Ni as cocatalyst. Appl Catal, B 2018, 235, 66–74.

[36] Sun N, Zhu YX, Li MW, Zhang J, Qin JN, Li YX, et al. Thermal coupled photocatalysis over Pt/g-C_3N_4 for selectively reducing CO_2 to CH_4 *via* cooperation of the electronic metal–support interaction effect and the oxidation state of Pt. Appl Catal, B 2021, 298, 120565.

[37] Cao SW, Li Y, Zhu BC, Jaroniec M, Yu JG. Facet effect of Pd cocatalyst on photocatalytic CO_2 reduction over g-C_3N_4. J Catal 2017, 349, 208–17.

[38] Cao SW, Jiang J, Zhu BC, Yu JG. Shape-dependent photocatalytic hydrogen evolution activity over a Pt nanoparticle coupled g-C_3N_4 photocatalyst. Phys Chem Chem Phys 2016, 18(28), 19457–63.

[39] Li Y, Shen JF, Quan WX, Diao Y, Wu MJ, Zhang B, et al. 2D/2D p-n heterojunctions of $CaSb_2O_6$/g-C_3N_4 for visible light-driven photocatalytic degradation of tetracycline. Eur J Inorg Chem 2020, 2020(40), 3852–58.

[40] Chen FF, Chen JB, Li LY, Peng F, Yu Y. g-C_3N_4 microtubes@$CoNiO_2$ nanosheets p-n heterojunction with a hierarchical hollow structure for efficient photocatalytic CO_2 reduction. Appl Surf Sci 2021, 151997.

[41]   Fan GD, Du BH, Zhou JJ, Yu WW, Chen ZY, Yang SW. Stable Ag_2O/g-C_3N_4 p-n heterojunction photocatalysts for efficient inactivation of harmful algae under visible light. Appl Catal, B 2020, 265, 118610.

[42]   Dong F, Zhao ZW, Xiong T, Ni ZL, Zhang WD, Sun YJ, et al. *In situ* construction of g-C_3N_4/g-C_3N_4 metal-free heterojunction for enhanced visible-light photocatalysis. ACS Appl Mater Interfaces 2013, 5(21), 11392–401.

[43]   Huang H, Yang SB, Vajtai R, Wang X, Ajayan PM. Pt-decorated 3D architectures built from graphene and graphitic carbon nitride nanosheets as efficient methanol oxidation catalysts. Adv Mater 2014, 26(30), 5160–65.

[44]   Wang XC, Maeda K, Thomas A, Takanabe K, Xin G, Carlsson JM, et al. A metal-free polymeric photocatalyst for hydrogen production from water under visible light. Nat Mater 2009, 8(1), 76–80.

[45]   Ong WJ, Tan LL, Ng YH, Yong ST, Chai SP. Graphitic carbon nitride (g-C_3N_4)-based photocatalysts for artificial photosynthesis and environmental remediation: Are we a step closer to achieving sustainability?. Chem Rev 2016, 116(12), 7159–329.

[46]   Li XH, Baar M, Blechert S, Antonietti M. Facilitating room-temperature Suzuki coupling reaction with light: Mott-Schottky photocatalyst for C-C-coupling. Sci Rep 2013, 3(1), 10.1038/srep01743.

[47]   Movahed SK, Miraghaee S, Dabiri M. AuPd alloy nanoparticles decorated graphitic carbon nitride as an excellent photocatalyst for the visible-light-enhanced Suzuki–Miyaura cross-coupling reaction. J Alloys Compd 2020, 819, 152994.

[48]   Yu QY, Lin X, Li XH, Chen JS. Photocatalytic stille cross-coupling on gold/g-C_3N_4 nano-heterojunction. Chem Res Chin Univ 2020, 36(6), 1013–16.

[49]   Baig RBN, Verma S, Varma RS, Nadagouda MN. Magnetic Fe@g-C3N4: A photoactive catalyst for the hydrogenation of alkenes and alkynes. ACS Sustainable Chem Eng 2016, 4(3), 1661–64.

[50]   Sharma P, Sasson Y. Sustainable visible light assisted in situ hydrogenation *via* a magnesium–water system catalyzed by a Pd-g-C_3N_4 photocatalyst. Green Chem 2019, 21(2), 261–68.

[51]   He YM, Wang Y, Zhang LH, Teng BT, Fan MH. High-efficiency conversion of CO_2 to fuel over ZnO/g-C_3N_4 photocatalyst. Appl Catal, B 2015, 168-169, 1–8.

[52]   Zhang RY, Huang ZA, Li Cj, Zuo YS, Zhou Y. Monolithic g-C_3N_4/reduced graphene oxide aerogel with *in situ* embedding of Pd nanoparticles for hydrogenation of CO_2 to CH_4. Appl Surf Sci 2019, 475, 953–60.

[53]   Chen W, Liu TY, Huang T, Liu XH, Yang XJ. Novel mesoporous P-doped graphitic carbon nitride nanosheets coupled with $ZnIn_2S_4$ nanosheets as efficient visible light driven heterostructures with remarkably enhanced photo-reduction activity. Nanoscale 2016, 8(6), 3711–19.

[54]   Johnson TC, Morris DJ, Wills M. Hydrogen generation from formic acid and alcohols using homogeneous catalysts. Chem Soc Rev 2010, 39(1), 81–88.

[55]   Enthaler S, von Langermann J, Schmidt T. Carbon dioxide and formic acid – The couple for environmental-friendly hydrogen storage?. Energy Environ Sci 2010, 3(9), 1207–17.

[56]   Ye TN, Lv LB, Li XH, Xu M, Chen JS. Strongly veined carbon nanoleaves as a highly efficient metal-free electrocatalyst. Angew Chem Int Ed 2014, 53(27), 6905–09.

[57]   Ji XL, Lee KT, Holden R, Zhang L, Zhang J, Botton GA, et al. Nanocrystalline intermetallics on mesoporous carbon for direct formic acid fuel cell anodes. Nat Chem 2010, 2(4), 286–93.

[58]   Li XH, Antonietti M. Metal nanoparticles at mesoporous N-doped carbons and carbon nitrides: Functional Mott-Schottky heterojunctions for catalysis. Chem Soc Rev 2013, 42(16), 6593–604.

[59]   TOF is a measure of the instantaneous efficiency of a catalyst. TOF = $n_{product}/(n_{catalyst} \times t)$, where $n_{product}$ is the number of moles of product, $n_{catalyst}$ is the number of moles of catalyst, t is the reaction time.

[60]   Zhang SB, Li M, Zhao JK, Wang H, Zhu XL, Han JY, et al. Plasmonic AuPd-based Mott-Schottky photocatalyst for synergistically enhanced hydrogen evolution from formic acid and aldehyde. Appl Catal, B 2019, 252, 24–32.

[61]   Altan O, Metin Ö. Boosting formic acid dehydrogenation *via* the design of a Z-scheme heterojunction photocatalyst: The case of graphitic carbon nitride/Ag/Ag$_3$PO$_4$-AgPd quaternary nanocomposites. Appl Surf Sci 2021, 535, 147740.

[62]   Wu YM, Wen MC, Navlani-Garcia M, Kuwahara Y, Mori K, Yamashita H. Palladium nanoparticles supported on titanium-doped graphitic carbon nitride for formic acid dehydrogenation. Chem Asian J 2017, 12(8), 860–67.

[63]   Wang F, Li CH, Chen HJ, Jiang RB, Sun LD, Li Q, et al. Plasmonic harvesting of light energy for Suzuki coupling reactions. J Am Chem Soc 2013, 135(15), 5588–601.

[64]   Liu HM, Li M, Dao TD, Liu YY, Zhou W, Liu LQ, et al. Design of PdAu alloy plasmonic nanoparticles for improved catalytic performance in CO$_2$ reduction with visible light irradiation. Nano Energy 2016, 26, 398–404.

[65]   Aslam U, Rao VG, Chavez S, Linic S. Catalytic conversion of solar to chemical energy on plasmonic metal nanostructures. Nat Catal 2018, 1(9), 656–65.

[66]   Li XP, Xu P, Chen M, Zeng GM, Wang DB, Chen F, et al. Application of silver phosphate-based photocatalysts: Barriers and solutions. Chem Eng J 2019, 366, 339–57.

[67]   Koh GY, Zhang YW, Pan H. First-principles study on hydrogen storage by graphitic carbon nitride nanotubes. Int J Hydrogen Energy 2012, 37(5), 4170–78.

[68]   Demirci UB. Ammonia borane, a material with exceptional properties for chemical hydrogen storage. Int J Hydrogen Energy 2017, 42(15), 9978–10013.

[69]   Eberle U, Felderhoff M, Schuth F. Chemical and physical solutions for hydrogen storage. Angew Chem Int Ed 2009, 48(36), 6608–30.

[70]   Gao MY, Yu YS, Yang WW, Li J, Xu SC, Feng M, et al. Ni nanoparticles supported on graphitic carbon nitride as visible light catalysts for hydrolytic dehydrogenation of ammonia borane. Nanoscale 2019, 11(8), 3506–13.

[71]   Zhang SB, Li M, Li LS, Dushimimana F, Zhao JK, Wang S, et al. Visible-light-driven multichannel regulation of local electron density to accelerate activation of O–H and B–H bonds for ammonia borane hydrolysis. ACS Catal 2020, 10(24), 14903–15.

[72]   Guo LT, Cai YY, Ge JM, Zhang YN, Gong LH, Li XH, et al. Multifunctional Au–Co@CN nanocatalyst for highly efficient hydrolysis of ammonia borane. ACS Catal 2014, 5(1), 388–92.

[73]   Eken Korkut S, Kucukkececi H, Metin O. Mesoporous graphitic carbon nitride/black phosphorus/ AgPd alloy nanoparticles ternary nanocomposite: A highly efficient catalyst for the methanolysis of ammonia borane. ACS Appl Mater Interfaces 2020, 12(7), 8130–39.

[74]   Wei L, Yang YM, Yu YN, Wang XM, Liu HY, Lu YH, et al. Visible-light-enhanced catalytic hydrolysis of ammonia borane using RuP$_2$ quantum dots supported by graphitic carbon nitride. Int J Hydrogen Energy 2021, 46(5), 3811–20.

[75]   Wang Y, Wang XC, Antonietti M. Polymeric graphitic carbon nitride as a heterogeneous organocatalyst: From photochemistry to multipurpose catalysis to sustainable chemistry. Angew Chem Int Ed 2012, 51(1), 68–89.

[76]   Kaner RB, Gilman JJ, Tolbert SH. Designing superhard materials. Science 2005, 308, 1268–69.

[77]   Liang JN, Yang XH, Wang Y, He P, Fu HT, Zhao Y, et al. A review on g-C$_3$N$_4$ incorporated with organics for enhanced photocatalytic water splitting. J Mater Chem A 2021, 9(22), 12898–922.

[78]   Wang QH, Jin Z, Kim KK, Hilmer AJ, Paulus GL, Shih CJ, et al. Understanding and controlling the substrate effect on graphene electron-transfer chemistry *via* reactivity imprint lithography. Nat Chem 2012, 4(9), 724–32.

[79]   Zhang JS, Zhang MW, Sun RQ, Wang XC. A facile band alignment of polymeric carbon nitride semiconductors to construct isotype heterojunctions. Angew Chem Int Ed 2012, 51(40), 10145–49.

[80]   Jin X, Balasubramanian VV, Selvan ST, Sawant DP, Chari MA, Lu GQ, et al. Highly ordered mesoporous carbon nitride nanoparticles with high nitrogen content: A metal-free basic catalyst. Angew Chem Int Ed 2009, 48(42), 7884–87.

[81] Wang Y, Yao J, Li H, Su DS, Antonietti M. Highly selective hydrogenation of phenol and derivatives over a Pd@carbon nitride catalyst in aqueous media. J Am Chem Soc 2011, 133(8), 2362–65.

[82] Fu T, Wang M, Cai WM, Cui YM, Gao F, Peng LM, et al. Acid-resistant catalysis without use of noble metals: Carbon nitride with underlying nickel. ACS Catal 2014, 4(8), 2536–43.

[83] Zhao TJ, Zhang YN, Wang KX, Su J, Wei X, Li XH. General transfer hydrogenation by activating ammonia-borane over cobalt nanoparticles. RSC Adv 2015, 5(124), 102736–40.

[84] Wang HH, Zhang B, Li XH, Antonietti M, Chen JS. Activating Pd nanoparticles on sol–gel prepared porous g-C_3N_4/SiO_2 *via* enlarging the Schottky barrier for efficient dehydrogenation of formic acid. Inorg Chem Front 2016, 3(9), 1124–29.

[85] Navlani-García M, Salinas-Torres D, Mori K, Kuwahara Y, Yamashita H. Enhanced formic acid dehydrogenation by the synergistic alloying effect of PdCo catalysts supported on graphitic carbon nitride. Int J Hydrogen Energy 2019, 44(53), 28483–93.

[86] Zhao ZK, Li WZ, Dai YT, Ge GF, Guo XW, Wang GR. Carbon nitride encapsulated nanodiamond hybrid with improved catalytic performance for clean and energy-saving styrene production *via* direct dehydrogenation of ethylbenzene. ACS Sustainable Chem Eng 2015, 3(12), 3355–64.

[87] Ge G, Zhao Z. Nanodiamond@carbon nitride hybrid with loose porous carbon nitride layers as an efficient metal-free catalyst for direct dehydrogenation of ethylbenzene. Appl Catal, A 2019, 571, 82–88.

[88] Zhang J, Su DS, Blume R, Schlogl R, Wang R, Yang XG, et al.. Surface chemistry and catalytic reactivity of a nanodiamond in the steam-free dehydrogenation of ethylbenzene. Angew Chem Int Ed 2010, 49(46), 8640–44.

[89] Huang ZJ, Yan FW, Yuan GQ. Vapor-phase selective oxidation of toluene catalyzed by graphitic carbon nitride supported vanadium oxide. Catal Lett 2016, 147(2), 509–16.

[90] Li XH, Chen JS, Wang X, Sun J, Antonietti M. Metal-free activation of dioxygen by graphene/g-C_3N_4 nanocomposites: Functional dyads for selective oxidation of saturated hydrocarbons. J Am Chem Soc 2011, 133(21), 8074–77.

[91] Cao SW, Low JX, Yu JG, Jaroniec M. Polymeric photocatalysts based on graphitic carbon nitride. Adv Mater 2015, 27(13), 2150–76.

[92] Zhuang ZC, Li Y, Li ZL, Lv F, Lang ZQ, Zhao KN, et al. MoB/g-C_3N_4 interface materials as a Schottky catalyst to boost hydrogen evolution. Angew Chem Int Ed 2018, 57(2), 496–500.

[93] In XPS, low-energy X-rays are irradiated to the sample to induce the photoelectric effect. According to the Einstein's equation of the photoelectric effect, the kinetic energy of a photoelectron is related to the binding energy of the atom through the X-ray energy: $h\upsilon + E_i = E_k(e^-) + E_f$ and $E_b = E_f - E_i = h\upsilon - E_k(e^-)$. In photoemission process, the total energy of the system before photoelectrons are emitted is the energy of the X-ray photon ($h\upsilon$) plus the energy of the target atom in the initial state (E_i), while the total energy of the system after the emission of photoelectrons is the kinetic energy of the photoelectrons (E_k) plus the energy of ionized atoms in the final state (E_f). Comparing the total energy before and after photoemission with Einstein's equation, it can be seen that the "binding energy" E_b of an electron is the difference between the final state energy and the initial state energy of the target atom $E_f - E_i$. The increase in binding energy means a decrease in the kinetic energy of the collected photoelectrons. In other words, the electron cloud around the atom is less dense, so it has a binding effect on the photoelectrons.

[94] Chu K, Liu YP, Li YB, Guo YL, Tian Y. Two-dimensional (2D)/2D interface engineering of a MoS_2/C_3N_4 heterostructure for promoted electrocatalytic nitrogen fixation. ACS Appl Mater Interfaces 2020, 12(6), 7081–90.

[95] Faradaic efficiency (FE): $FE = (m \times n \times F)/(i \times t)$, where m is the number of moles of product, n is the number of reacting electrons, and F is Faraday constant (96485 C mol^{-1}), i is the reaction current, t is the reaction time.

[96] Zhang L, Mao FX, Zheng LR, Wang HF, Yang XH, Yang HG. Tuning metal catalyst with metal–C_3N_4 interaction for efficient CO_2 electroreduction. ACS Catal 2018, 8(12), 11035–41.

[97] Shi X, Wang W, Miao XR, Tian F, Xu ZW, Li N, et al. Constructing conductive channels between platinum nanoparticles and graphitic carbon nitride by gamma irradiation for an enhanced oxygen reduction reaction. ACS Appl Mater Interfaces 2020, 12(41), 46095–106.

[98] Geng ZG, Kong XD, Chen WW, Su HY, Liu Y, Cai F, et al. Oxygen vacancies in ZnO nanosheets enhance CO_2 electrochemical reduction to CO. Angew Chem Int Ed 2018, 57(21), 6054–59.

[99] Ma TY, Cao JL, Jaroniec M, Qiao SZ. Interacting carbon nitride and titanium carbide nanosheets for high-performance oxygen evolution. Angew Chem Int Ed 2016, 55(3), 1138–42.

[100] Chen JY, Cui PX, Zhao GQ, Rui KR, Lao MM, Chen YP, et al. Low-coordinate iridium oxide confined on graphitic carbon nitride for highly efficient oxygen evolution. Angew Chem Int Ed 2019, 58(36), 12540–44.

[101] Wang LQ, Si WP, Tong YY, Hou F, Pergolesi D, Hou JG, et al. Graphitic carbon nitride (g-C_3N_4)-based nanosized heteroarrays: Promising materials for photoelectrochemical water splitting. Carbon Energy 2020, 2(2), 223–50.

[102] Zheng JG, Li XJ, Qin YH, Zhang SQ, Sun MS, Duan XG, et al. Zn phthalocyanine/carbon nitride heterojunction for visible light photoelectrocatalytic conversion of CO_2 to methanol. J Catal 2019, 371, 214–23.

[103] Xiao LM, Liu TF, Zhang M, Li QY, Yang JJ. Interfacial construction of zero-dimensional/one-Dimensional g-C_3N_4 nanoparticles/TiO_2 nanotube arrays with Z-scheme heterostructure for improved photoelectrochemical water splitting. ACS Sustainable Chem Eng 2019, 7(2), 2483–91.

[104] Volokh M, Peng GM, Barrio J, Shalom M. Carbon nitride materials for water splitting photoelectrochemical cells. Angew Chem Int Ed 2019, 58(19), 6138–51.

Oleksandr Savateev* and Stefano Mazzanti

Chapter 8
Carbon nitride organic photocatalysis

8.1 Introduction

Throughout history, humanity has been relying on natural substances. Small and large organic molecules were extracted from natural sources and used either as-obtained or processed to synthesize derivatives. With the progress in understanding of how the structure of organic molecules affects their properties, it became clear that molecules, which would be optimal for certain applications simply do not exist in nature. One example is the CF_3 group. It strongly affects the properties of pharmaceuticals and drugs, but it does not occur in natural compounds. High demand in "artificial" molecules from different branches of science and technology stimulates development of new and more efficient methodologies in synthetic organic chemistry.

Photons in the visible range of the electromagnetic spectrum (1.7–3.1 eV) deliver energy to the reaction mixture in the most concentrated fashion. Therefore, photochemistry and photocatalysis, either as standalone techniques or in combination with other methods, have been contributing to the toolbox of synthetic organic chemistry. Today, they are primary approaches to enable thermodynamically challenging processes via photoinduced electron transfer (PET).

Photocatalytic production of hydrogen from the aqueous solution of triethanolamine by graphitic carbon nitride, reported in 2009, served as a solid proof of concept that this class of materials can also mediate photocatalytic reactions with organic molecules in place of water. In 2010, oxidation of benzylic alcohol to benzaldehyde over mesoporous graphitic carbon nitride (mpg-CN) was reported. Today, *carbon nitride organic photocatalysis* is a very diverse area of research. It includes selective oxidation of organic compounds, dual transition metal photocatalysis, chromoselective catalysis and many more approaches. Due to low cost, graphitic carbon nitrides are also promising semiconductors for photocatalysis on larger scale. The aim of this chapter is to provide an up-to-date summary of carbon nitrides features that make this class of materials, in the context of photocatalytic synthesis of organic molecules, complementary

Acknowledgments: The authors gratefully acknowledge the European Commission (DECADE, grant agreement ID: 862030).

*Corresponding author: Oleksandr Savateev,** Colloid Chemistry Department, Max Planck Institute of Colloids and Interfaces, Am Muehlenberg 1, 14476 Potsdam, Germany,
e-mail: oleksandr.savatieiev@mpikg.mpg.de
Stefano Mazzanti, Colloid Chemistry Department, Max Planck Institute of Colloids and Interfaces, Am Muehlenberg 1, 14476 Potsdam, Germany

https://doi.org/10.1515/9783110746976-008

to the earlier developed method and, in many cases, also unique. Chemical transformations are arranged into groups to illustrate these features of carbon nitrides.

8.2 Photoinduced electron transfer in photochemistry and photocatalysis

This section introduces a central concept in photochemistry and photocatalysis –PET. To understand why photochemistry and photocatalysis are so versatile and powerful in enabling different and often very challenging processes via PET, we need to analyze the challenges that chemists encounter when performing chemical reactions.

8.2.1 Thermochemistry

The Gibbs free energy of a reaction is composed of two terms – standard enthalpy of a reaction (ΔH^0), and the product of standard entropy of a reaction (ΔS^0) and absolute temperature (T):

$$\Delta G^0 = \Delta H^0 - T\Delta S^0 \tag{8.1}$$

The standard enthalpy of a reaction may be calculated as a difference between the standard enthalpies of formation of products and reagents:

$$\Delta H^0 = \sum \Delta H_f^0(\text{product}_i) - \sum \Delta H_f^0(\text{reagent}_i) \tag{8.2}$$

Similarly, the standard entropy of a reaction is the difference between the standard entropies of products and reagents:

$$\Delta S^0 = \sum \Delta S^0(\text{product}_i) - \sum \Delta S^0(\text{reagent}_i) \tag{8.3}$$

Exergonic (or downhill) reactions are spontaneous reactions characterized by negative values of the Gibbs free energy, $\Delta G^0 < 0$. Endergonic (or uphill) processes are processes that do not occur spontaneously in Nature. In these reactions, $\Delta G^0 > 0$. Photosynthesis, which is intrinsically an endergonic process, proceeds in nature only because the positive ΔG value is thoroughly compensated by the highly exergonic thermonuclear reaction in the Sun. Therefore, the overall process that is composed of the binding of CO_2 and H_2O into carbohydrates (endergonic reaction) in the garden and the fusion of protons into α-particle in the Sun (highly exergonic reaction) millions kilometers away is still highly exergonic. In other words, for every intrinsically endergonic reaction that nevertheless proceeds in Nature or in a chemistry laboratory, there is another quite often a hidden more exergonic process. Overall, the Universe gradually consumes its 'fuel' and reaches a thermodynamically more stable state. On this path, some endergonic processes might become possible.

Synthetic organic chemistry relies strongly on highly reactive reagents – compounds with weakly negative, zero, and especially on those with positive enthalpy of formation, for example, $CH_{4(g)}$ has $\Delta H_f^0 = -74.2$ kJ mol^{-1}, which is significantly less negative compared to, for example, $CO_{2(g)}$, $\Delta H_f^0 = -393.5$ kJ mol^{-1}. All elements in their standard states, for example, Li, Na, K metals, $Cl_{2(g)}$, etc. have $\Delta H_f^0 = 0$ kJ mol^{-1}. Examples of chemical reagents with positive enthalpy of formation, which means that energy is actually consumed upon formation of such compound from elements, are ozone ($\Delta H_f^0 = 143$ kJ mol^{-1}) and NO_2 ($\Delta H_f^0 = 33.2$ kJ mol^{-1}). Addition of such reagents into the reaction mixture decreases the enthalpy term (ΔH_f^0), as seen from eq. (8.2). Provided that the entropy term, $-T\Delta S^0$, is sufficiently negative, i.e., ΔS^0 is positive, employing highly reactive reagents in organic synthesis lowers ΔH^0 to such an extent that ΔG^0 becomes negative. In this case, a chemical reaction becomes downhill or spontaneous. This principle is illustrated in Fig. 8.1 – Path I is uphill while Path II – downhill.

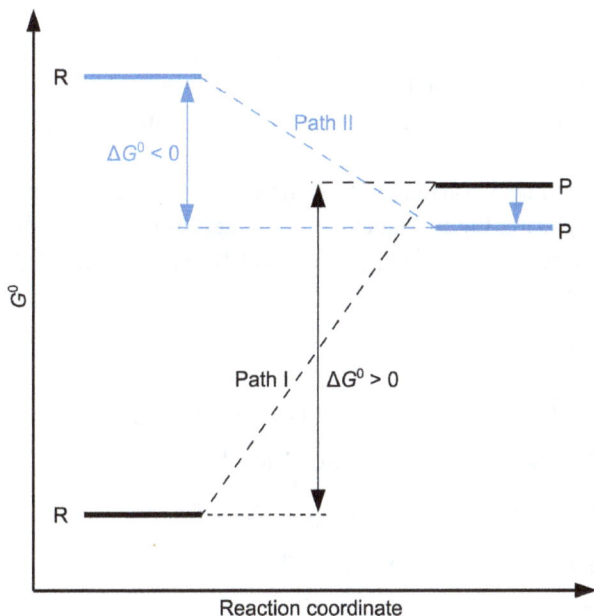

Fig. 8.1: The role of reactive reagents in organic synthesis to enable thermodynamically challenging reactions. Activation barriers of both paths are not shown.

To explain Fig. 8.1 and the role of highly reactive reagents in organic synthesis, let us consider the synthesis of chloromethane from methane. The reaction of methane with sodium chloride is hypothetical, but nevertheless highly appealing, as it would give two valuable products – chloromethane and sodium hydride (eq. (8.4)). Even without rigorous calculations, it is clear that the reaction is endergonic – under standard conditions, natural gas does not react with kitchen salt, but to be consistent, the calculation of ΔG^0 for this reaction is given below [1]:

$$CH_{4(g)} + NaCl_{(s)} = CH_3Cl_{(g)} + NaH_{(s)}$$

$$\Delta H^0 = \Delta H_f^0\left(CH_3Cl_{(g)}\right) + \Delta H_f^0\left(NaH_{(s)}\right) - \Delta H_f^0\left(NaCl_{(s)}\right) - \Delta H_f^0\left(CH_{4(g)}\right) =$$

$$= -82.0 - 56.4 - (-411.1) - (-74.2) = 346.9 \text{ kJ mol}^{-1}$$

$$\Delta S^0 = \Delta S^0\left(CH_3Cl_{(g)}\right) + \Delta S^0\left(NaH_{(s)}\right) - \Delta S^0\left(NaCl_{(s)}\right) - \Delta S^0\left(CH_{4(g)}\right) =$$

$$= 234.4 + 40.0 - 72.1 - 188.7 = 13.6 \text{ J mol}^{-1}$$

$$\Delta G^0 = \Delta H^0 - T\Delta S^0 = 346.9 - 298.15 \cdot 0.0136 = 342.8 \text{ kJ mol}^{-1} \tag{8.4}$$

While the entropy factor facilitates this reaction ($\Delta S^0 > 0$), the highly negative enthalpy of sodium chloride formation, $\Delta H_f^0(NaCl_{(s)}) = -411.1$ kJ mol^{-1}, is the parameter that makes this reaction impossible under the standard conditions. In Fig. 8.1, a hypothetical reaction between methane and sodium chloride corresponds to the endergonic path I.

We can obtain chloromethane by a reaction of methane with elemental chlorine. Calculations suggest that by replacing the thermodynamically very stable NaCl with elemental chlorine in its standard state ($\Delta H_f^0(Cl_{2(g)}) = 0$ kJ mol^{-1}), synthesis of CH_3Cl becomes a spontaneous reaction:

$$CH_{4(g)} + Cl_{2(g)} = CH_3Cl_{(g)} + HCl_{(g)}$$

$$\Delta H^0 = \Delta H_f^0\left(CH_3Cl_{(g)}\right) + \Delta H_f^0\left(HCl_{(g)}\right) - \Delta H_f^0\left(Cl_{2(g)}\right) - \Delta H_f^0\left(CH_{4(g)}\right) =$$

$$= -82.0 - 92.3 - (0) - (-74.2) = -100.1 \text{ kJ mol}^{-1}$$

$$\Delta S^0 = \Delta S^0\left(CH_3Cl_{(g)}\right) + \Delta S^0\left(HCl_{(g)}\right) - \Delta S^0\left(Cl_{2(g)}\right) - \Delta S^0\left(CH_{4(g)}\right) =$$

$$= 234.4 + 186.9 - 223.0 - 188.7 = 9.6 \text{ J mol}^{-1}$$

$$\Delta G^0 = \Delta H^0 - T\Delta S^0 = -100.1 - 298.15 \cdot 0.0096 = -103.0 \text{ kJ mol}^{-1} \tag{8.5}$$

Path II in Fig. 8.1 reflects the thermochemistry of methane interaction with elemental chlorine. Another important consequnce of substituing NaH by HCl on the right side of the eq. (8.5), is the stabilization of the system by 35.9 kJ mol^{-1}, which also contributes to the overall negative ΔG^0. Therefore, in Fig. 8.1, energetically, products of Path II are below the products of Path I. In some cases, stabilization of the system upon the formation of a thermodynamically very stable product is what drives a chemical reaction. Readers are invited to verify this statement by using eqs. (8.1)–(8.3) to determine the corresponding ΔG^0 values of two reactions that lead to evolution of hydrogen gas: (1) water dissociation into H_2 and O_2 (uphill) and (2) the reaction of water with sodium metal (downhill). Formation of thermodynamically stable NaOH is the driving force of the second reaction. Another example is the reduction of carboxylic acids and esters to the corresponding alcohols by $LiAlH_4$, which is driven by the formation of very strong Al–O bond(s) – energy released at this step can thoroughly compensate cleavage of carboxylic C = O bond.

Of course, a hypothetical reaction of CH_4 with NaCl (eq. (8.4)) and CH_4 with Cl_2 (eq. (8.5)) are different reactions, but both yield CH_3Cl, which is the target molecule in these cases. Organic synthesis is full of examples where such formalism – destabilization of a chemical system by adding thermodynamically less stable reagents, stabilization of the system by forming thermodynamically very stable (by-)products or a combination of both – is exceptionally useful to design chemical reactions and access compounds by coupling their synthesis with other highly exergonic reactions. In fact, a vast majority of chemical reactions performed on the lab scale are downhill, while many industrial large-scale reactions are intrinsically uphill – the energy supplied to drive such reactions is then stored in highly reactive compounds, which are then used on a smaller scale.

It must be noted that the primary role of heating, commonly employed in organic synthesis, is to overcome the activation energy barrier of exergonic reactions rather than enabling endergonic ones. In the context of the reaction described by eq. (8.4), assuming that ΔH and ΔS are not functions of temperature, readers can verify that the temperature of the reaction mixture must be several tens of thousands K(!) in order for the entropy factor become dominant and ΔG – negative, which would imply a spontaneous reaction.

All examples of the formalism discussed above used to enable intrinsically exergonic reactions (except photochemistry and photocatalysis) are based on changing the *chemical composition* of the reaction mixture. This fact makes photochemistry, photocatalysis and especially photosynthesis remarkable as Nature uses only photons generated millions of kilometers away from the actual "photoreactor" to lift the thermodynamic potential of the, system followed by its spontaneous relaxation into a more stable state.

Before we move to the key concept of photocatalysis and photochemistry, readers are invited once again to examine eq. (8.4). ΔG^0 value of the chemical reaction described by this equation, when converted into electron-Volt scale and normalized per a couple of reacting molecules is ~ 3.6 eV, which corresponds to the energy of a 344 nm photon (see Planck–Einstein relation, eq. (8.9). It leads to a logical question: "Would illumination of CH_4 and NaCl mixture with UV photons be sufficient to lift the thermodynamic potential of the system and enable this intrinsically uphill process?"

8.2.2 Photoinduced electron transfer

Overall, water splitting into H_2 and O_2 (Chapter 3) and "artificial photo*synthesis*" – conversion of CO_2 and H_2O mixture into thermodynamically less stable products, such as CO, CH_3OH, HCOOH and O_2, are endergonic or uphill reactions (Chapter 4). All photo*catalytic* processes are intrinsically exergonic or downhill reactions [2]. Oxidation of benzyl alcohol to aldehyde, which is the most studied reaction in semiconductor photocatalysis and one of the first organic reactions mediated by carbon nitrides, is an example of the exergonic reaction, as supported by thermochemical calculations:

$$PhCH_2OH_{(l)} + O_{2(g)} = PhC(O)H_{(l)} + H_2O_{2(l)}$$

$$\Delta H^0 = \Delta H_f^0\left(Ph(O)H_{(l)}\right) + \Delta H_f^0\left(H_2O_{2(l)}\right) - \Delta H_f^0\left(PhCH_2OH_{(l)}\right) - \Delta H_f^0\left(O_{2(g)}\right) =$$

$$= -87.1 - 190.8 - (-155.0 + 0) = -122.9 \, kJ \, mol^{-1}$$

$$\Delta S^0 = \Delta S^0\left(Ph(O)H_{(l)}\right) + \Delta S^0\left(H_2O_{2(l)}\right) - \Delta S^0\left(PhCH_2OH_{(l)}\right) - \Delta S^0\left(O_{2(g)}\right) =$$

$$= 221.2 + 110.0 - 216.7 - 205.2 = -90.7 \, J \, mol^{-1}$$

$$\Delta G^0 = \Delta H^0 - T\Delta S^0 = -122.9 - 298.15 \cdot (-0.0907) = -95.9 \ kJ \, mol^{-1}$$

$$(8.6)$$

Although the oxidation of benzyl alcohol by molecular oxygen is exergonic, it does not proceed at any significant rate under the standard conditions due to the very first step – single electron transfer (SET) or proton coupled electron transfer (PCET, see Section 8.3.3) between benzyl alcohol and O_2 being uphill. This fact renders the whole process under the standard conditions being kinetically sluggish.

When considering the SET between a general electron donor and an acceptor, we can estimate the driving force of this process as follows:

$$\Delta G^0 = e(E(D^{\cdot+}/D) - E(A/A^{\cdot-}))$$ (8.7)

where e is the elemenary charge; $E(D^{\cdot+}/D)$ is oxidation potential of an electron donor, V, versus reference electrode (RE); $E(A/A^{\cdot-})$ is the reduction potential of an electron acceptor, V, versus RE.

The left part of Fig. 8.2 serves as a graphical representation of eq. (8.7) [3] where the oxidation potential of an electron donor, defined by its HOMO level, is more positive than the reduction potential of an electron acceptor, which in turn is defined by its LUMO level.

Excitation of an electron donor promotes an electron from HOMO to LUMO, and generates a specie *[Donor], excited state. Electron transfer from SOMO of *[Donor] (former LUMO of the donor) to the LUMO of the electron acceptor is thermodynamically feasible. In this case, chemists use a term "photoinduced electron transfer" – electron transfer that is induced by photon absorption. The magnitude of the PET driving force is calculated as follows [4]:

$$\Delta G = e[E(D^{\cdot+}/D) - E(A/A^{\cdot-})] - E(^*D) - \frac{e^2}{4\pi\varepsilon_0\varepsilon_s r} \tag{8.8}$$

where $E(^*D)$ is the energy of the donor excited state, eV; r is the distance between $A^{\cdot-}$ and $D^{\cdot+}$ in the charge transfer complex, m; ε_0 is vacuum permittivity, 8.85×10^{-12} F m^{-1}; ε_s is permittivity of solvent.

The last term in eq. (8.8) corresponds to the energy that the system gains upon formation of a charge transfer complex $[D^{\cdot+}][A^{\cdot-}]$. In acetonitrile, for example, it is only ~ 0.06 eV [5]. Therefore, the energy required to overcome the barrier of SET between an electron donor and acceptor in dark is primarily provided by their electronically excited states.

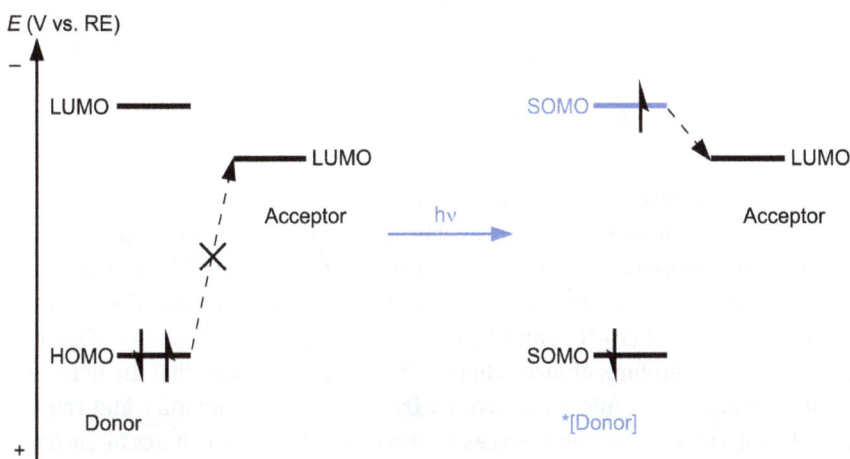

Fig. 8.2: Schematic alignment of frontier orbitals in the electron donor and acceptor ground and excited states. Transfer of electron between the molecular orbitals (energy levels) is denoted by dashed arrows.

8.2.3 Photochemistry and photocatalysis

Benzyl alcohol, O_2 and the vast majority of small organic molecules and inorganic ions are transparent to visible light – they absorb photons with wavelength <400 nm. In order to accomplish the PET, such molecules must be excited with UV photons. For example, photochemical trifluoromethylation of caffeine is accomplished by employing RSO_2–CF_3 compounds that absorb photons with $\lambda \leq 380$ nm – a photochemical approach in Fig. 8.3 [6]. Trifluoromethylation of caffeine is also enabled by mesoporous graphitic carbon nitride (mpg-CN, a type of heptazine-based carbon nitride featuring mesoporous structure) using either CF_3SO_2Cl (under redox neutral conditions) or CF_3SO_2Na (under net-oxidative conditions) under visible light – a photocatalytic approach in Fig. 8.3 [7, 8].

Photochemical approach

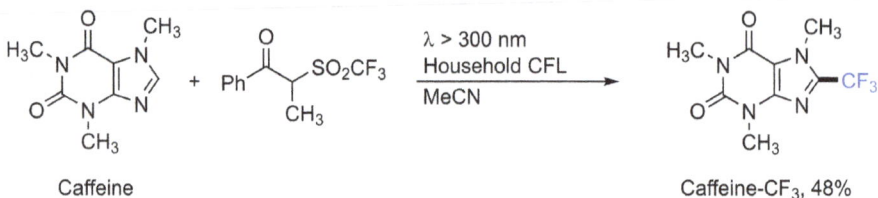

Caffeine   Caffeine-CF$_3$, 48%

Photocatalytic approach

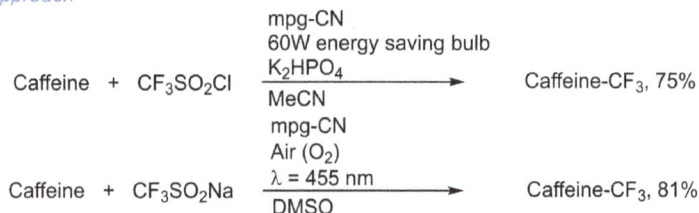

Fig. 8.3: Photochemical and photocatalytic trifluoromethylation of caffeine.

While photochemistry can exclude the third component, the photocatalyst, according to the principle of photochemical activation, the presence of a chromophore in a molecule is required to harvest the electromagnetic radiation. Photocatalysis allows using small molecules lacking any chromophores as reagents. Note that in Fig. 8.3, compared to CF$_3$SO$_2$Na and CF$_3$SO$_2$Cl, RSO$_2$–CF$_3$ possesses phenyl ring as a chromophore and, therefore, absorbs photons in near UV. Provided that the heterogeneous photocatalyst is completely recovered from the reaction mixture and reused, using small molecules or inorganic ions as a source of radicals contributes to the overall atom economy of photocatalytic processes, compared to photochemical ones.

In the following sections we will see that the role of carbon nitrides is not limited to PET, which may also be enabled by a large scope of organic dyes and transition metal complexes. Due to the nitrogen-rich structure, the carbon nitride surface functions as a catalyst, which activates reagents and stabilizes intermediates generated upon PET, and thus lowers the activation barrier.

8.3 Organic photocatalysis with carbon nitrides

Carbon nitride organic photocatalysis started developing in the first decade of the twenty-first century. It is not surprising that many concepts have been borrowed by the community from the semiconductor metal oxide photocatalysis, originally employed in water splitting as well as in molecular photocatalysis with organic dyes, and eventually translated into carbon nitride organic photocatalysis. However, in many cases, the function of carbon nitrides is oversimplified. For example, the reaction mechanism of many reactions implies pure electron transfer – photogenerated holes in the valence band oxidize the substrate via SET, while photogenerated electrons in the conduction band synchronously reduce the substrate. However, it is a rare case, when the rate of electron transfer from one substrate to the carbon nitride excited state valence band and from carbon nitride excited state conduction band to another substrate match each other. Typically, the carbon nitride excited state undergoes either reductive or oxidative quenching that produces short-lived (still detectable by fast spectroscopic techniques) either reduced or oxidized state, respectively. On the other hand, several examples indicate that carbon nitrides mediate not only the flow of electrons, but also charge-compensating protons. Therefore, the explanation of photocatalytic mechanism using either excited-state proton-coupled electron transfer (ES-PCET) or hydrogen atom transfer (HAT), in many cases, is more realistic, compared to PET. Overall, depending on the structure of a reagent and due to the complex structure of carbon nitrides, these materials serve, for instance, as a sensitizer and a photobase as well as a platform to design chromoselective transformations.

8.3.1 Optical gap and band edges

Historically, water splitting has been one of the earliest reactions mediated by semiconductors. Assuming zero kinetic overpotential, a semiconductor with the optical gap of only 1.23 eV, which is defined by the conduction band minimum at 0 V versus SHE and the valence band maximum at +1.23 V versus SHE, is, in principle, capable of splitting water. Compared to water, organic molecules are characterized by more positive oxidation potentials and more negative reduction potentials. Analysis of redox potentials obtained for a set of 175 organic compounds (Fig. 8.4) [9] indicates that 83% of organic compounds in the sample have oxidation potential more positive than the oxidation potential of water, and 98% have reduction potential more negative than that of water. The first important conclusion is that in organic photocatalysis, semiconductors with wider optical gap are more applicable and universal as they can enable oxidation and reduction of a wider scope of organic compounds via PET.

The intrinsic optical gap in carbon nitrides, defined by $\pi-\pi^*$ transitions, is ~ 2.7 eV. In covalent carbon nitrides, such as mpg-CN, the conduction band is located at –1.5 V versus SCE and the valence band at +1.2 V versus SCE [8]. In ionic carbon nitrides, such as potassium poly(heptazine imide) (K-PHI), both potentials are shifted to more positive values by ~ 0.7 V [11]. By looking again at the distribution of the redox potentials, mpg-CN is able to oxidize 28% of the organic compounds and reduce 31% of the organic compounds in the sample via PET. K-PHI is able to oxidize 72% and reduce 8% of organic compounds via PET (Fig. 8.4). This example illustrates that despite these two classes of carbon nitrides being built of heptazine units, small structural changes do not affect their optical gap, but afford materials particularly useful to mediate either a demanding reduction or oxidation of substrates via PET.

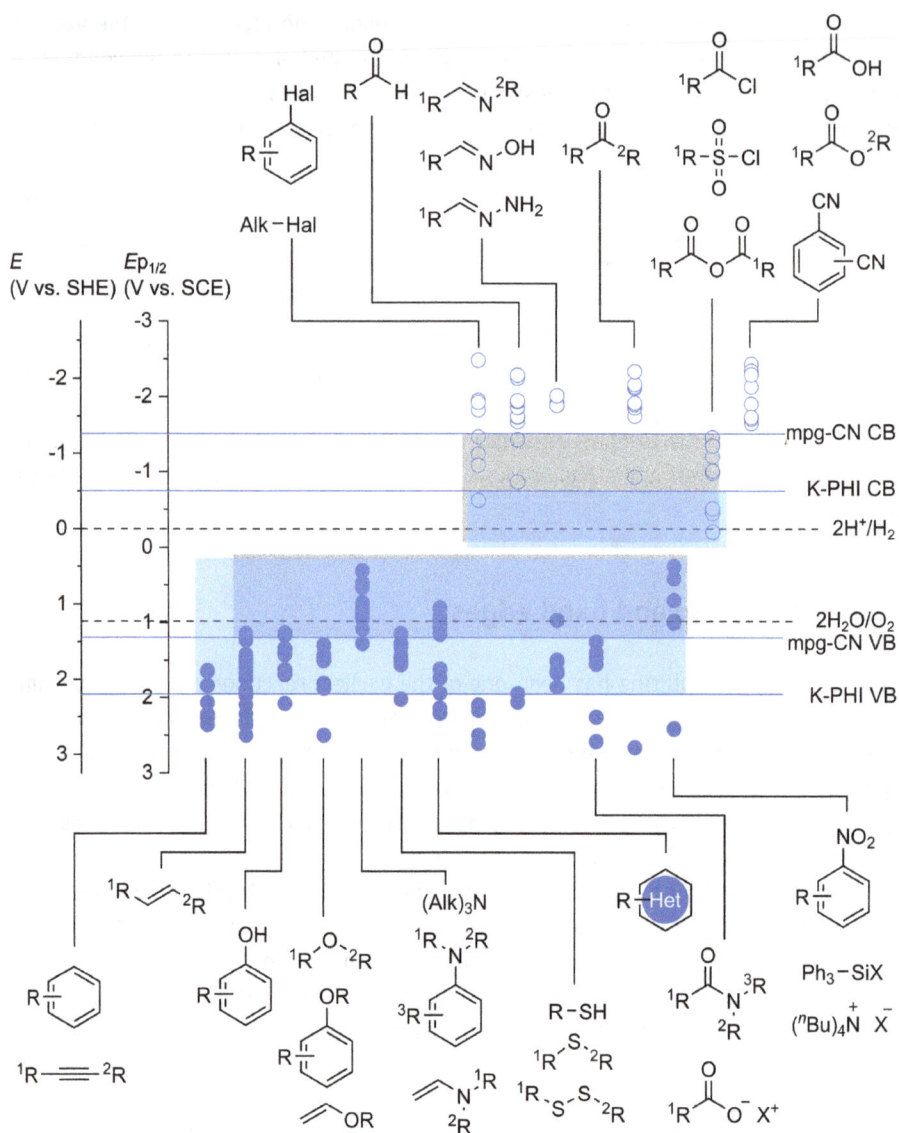

Fig. 8.4: Half-peak reduction (opened circles) and oxidation (filled circles) potentials of organic compounds, determined by cyclic voltammetry [9]. Standard potential of proton reduction and water oxidation are shown with horizontal dashed lines. Potentials of mpg-CN and K-PHI conduction band and valence band are denoted with solid horizontal lines. Blue and gray rectangles highlight the range of half-peak potentials accessible for PET with K-PHI and mpg-CN excited states respectively. SCE scale and SHE scale are corelated via equation $E(SHE) = E(SCE) + 0.244$ [10].

A list of reactions that are based on SET from a substrate to the photoexcited carbon nitrides is given below. Thus, covalent carbon nitride, g-C₃N₄, prepared by calcining guanidine hydrochloride, enables a redox-neutral reaction between activated alkenes and α-aminoalkyl radicals, which are derived from trialkylsilanes and α-aminoacids, upon one-electron oxidation (Fig. 8.5) [12].

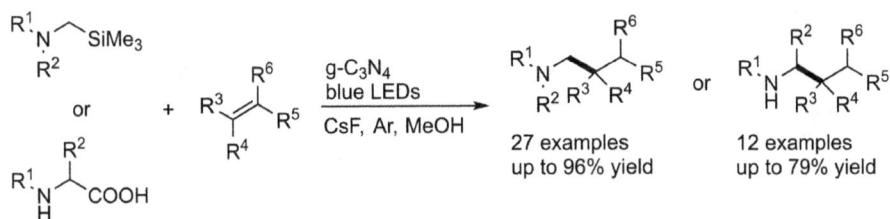

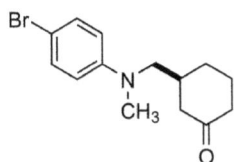

75%

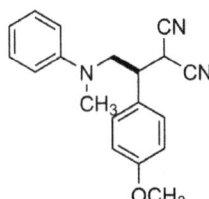

83%

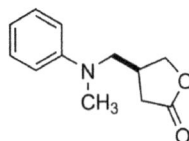

61%

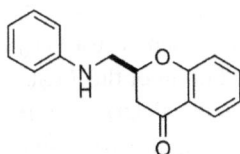

78%

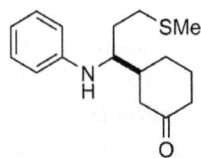

71%, 1:1 d.r.

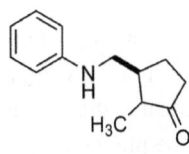

73%, 2.6:1 d.r.

Fig. 8.5: Redox-neutral desilylative and decarboxylative coupling of α-silylamines and α-aminoacids with alkenes mediated by g-C₃N₄. Selected structures of products are shown.

Ionic carbon nitride, CN-K, enables perfluoroalkylation-distal functionalization of alkenes (Fig. 8.6) [13]. Air O_2 is used as electron acceptor to generate $R_FC^•$ radical upon one-electron oxidation of R_FSO_2Na via PET. CN-K photocatalyst gave a considerably higher yield of the product compared to several homogeneous sensitizers and semiconductors. Substrates with $n = 0, 2, 3$ gave products with moderate and high yields, while those with $n = 1$ and 4 failed to give products, likely due to disfavored formation of four- and seven-membered cyclic transition states. The reaction also proceeds under outdoor solar light.

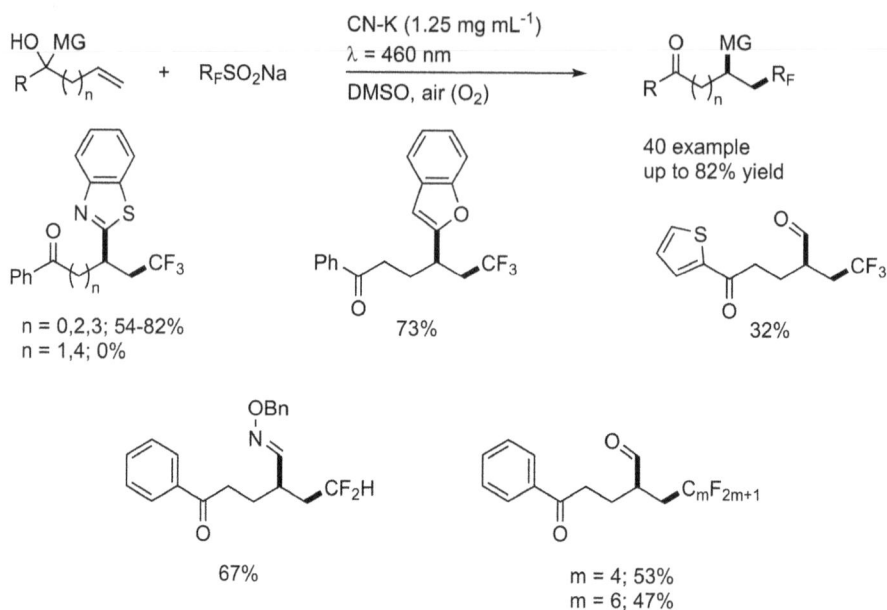

Fig. 8.6: Perfluoroalkylation-distal functionalization of alkenes with ionic carbon nitride CN-K. Selected structures of products are shown.

Carbon nitride CN-K reduces ester, based on *N*-hydroxyphthalimide (NHPI), as shown in Fig. 8.7 via PET to the corresponding C-centered radical, which abstracts hydrogen atom from acetonitrile and generates $^{\bullet}CH_2CN$ [14]. The addition of the $^{\bullet}CH_2CN$ radical to the alkene, followed by cyclization and back transfer of the electron and H^+ at the photocatalyst turnover step, delivers a series of indolines, oxindoles, isoquinolinones and isoquinolinediones. Further derivatization of structures is performed by employing reactivity of the pendant CN-group.

8.3.2 Bifunctionalization of organic compounds

As a result of band edges alignment that are suitable both for the reduction and oxidation of a broad range of organic compounds, carbon nitride photocatalysis offers a very practical approach to bifunctionalize organic compounds, employing their innate reactivity. Thus, Ghosh, König et al. performed bromination and trifluoromethylation of electron-rich 1,3,5-trimethoxybenzene, performed in a sequential fashion (Fig. 8.8) [8]. Once trimethoxybenzene was consumed, KBr was added to the reaction mixture and illumination was continued, which afforded 2-bromo-1,3,5-trimethoxy-4-(trifluoromethyl)benzene with 74% yield. The same product was also obtained in 41% yield from trimethoxybenzene by swapping the order of CF_3SO_2Na and KBr addition.

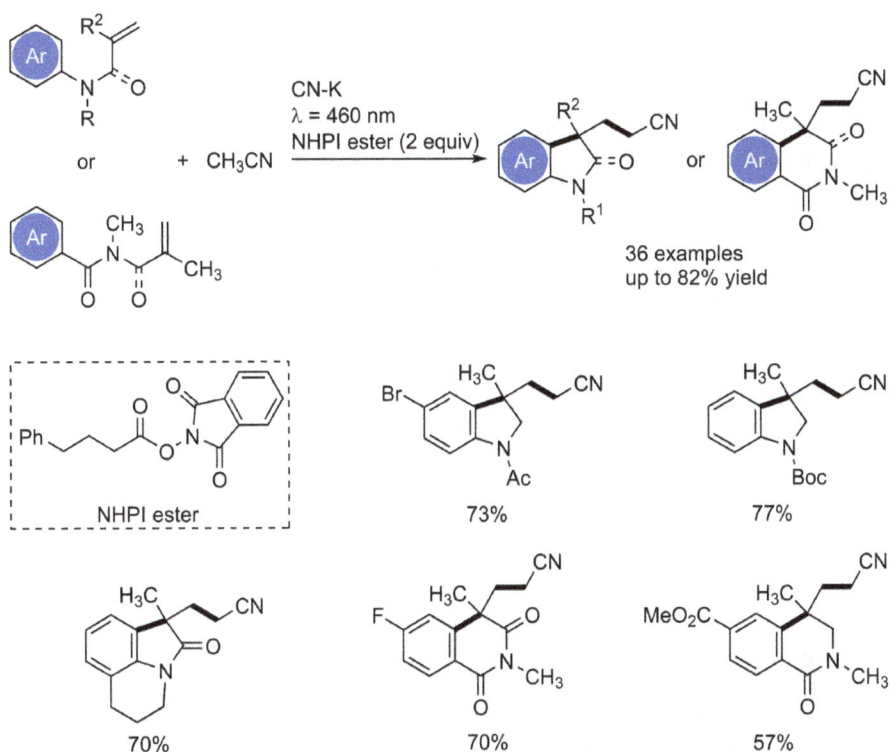

Fig. 8.7: Cyanomethylarylation mediated by carbon nitride CN-K. Structures of selected products are shown.

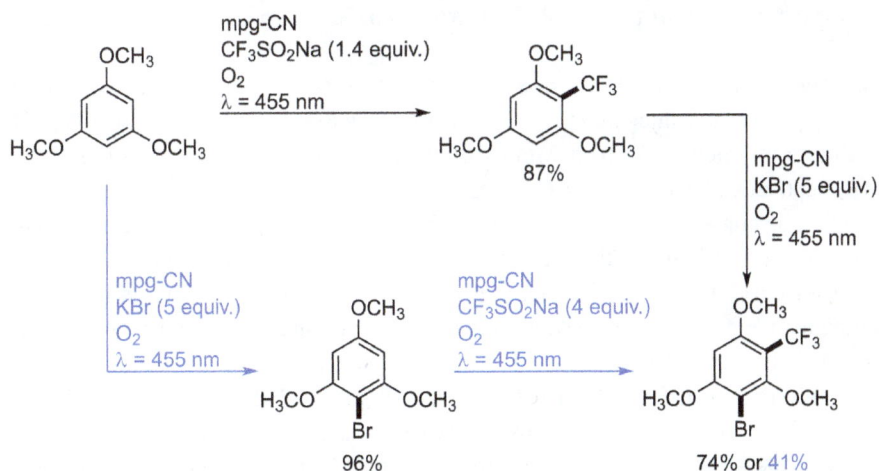

Fig. 8.8: Sequential bifunctionalization of 1,3,5-trimethoxybenzene by mpg-CN.

On the other hand, mpg-CN enables simultaneous installation of two functional groups – either –Br and –CH(CO$_2$Et)$_2$ or –CF$_3$ and –CF$_2$CO$_2$Et into pyrrole in one pot (Fig. 8.9) [8]. The functional groups are derived either from one reagent – in this case, pyrrole ring is "inserted" into the C–Br bond of diethylbromomalonate; or two reagents – BrCF$_2$CO$_2$Et and CF$_3$SO$_2$Na.

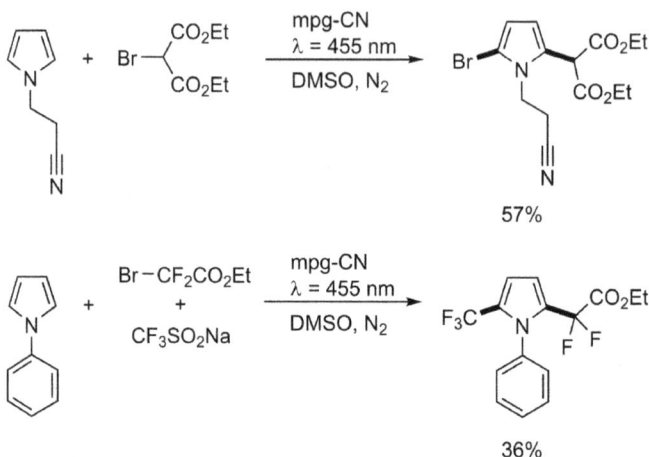

Fig. 8.9: Bifunctionalization of pyrrole derivatives in one pot.

8.3.3 Proton-coupled electron transfer with carbon nitrides

8.3.3.1 Excited-state proton-coupled electron transfer

Basicity of the six-membered nitrogen-containing heterocycles decreases in the order: pyridine (pK_a of the conjugate acid is 5.2) > pyrimidine (pK_a 1.3) > 1,3,5-triazine (pK_a ~ 0) [15]. Due to the mutual electron withdrawing effect of seven sp^2-hybridized nitrogen atoms, heptazine is expected to be a weak base, while its conjugate acid, heptazinium cation, is a strong electrolyte. Nevertheless, heptazine units in Melem, Melem hydrate and Melon-type carbon nitride form an extensive network of hydrogen bonds in solid state. Therefore, it is not surprising that heptazine derivatives also form hydrogen bonding with H-donors, such as water and phenols. Illumination of such a system enables the abstraction of hydrogen atom from the substrate via excited-state proton-coupled electron transfer (ES-PCET) (Fig. 8.10a) [16]. Such ES-PCET is oxidative by its nature, as the substrate molecule loses a hydrogen atom. As a result, the oxidation number of the adjacent atom becomes more positive. The activation barrier depends on the strengths of the complex and decreases, for example, from 0.369 eV to just 0.093 eV for *para*-cyano- and *para*-methoxy-substituted phenols, respectively. Another implication of these findings is that by decreasing the electron density at the heptazine units, a sensitizer (in this case also a photobase) should become capable of breaking even stronger X–H bonds, such as O–H in water, and take a step closer to full water splitting without using sacrificial agents and co-catalysts.

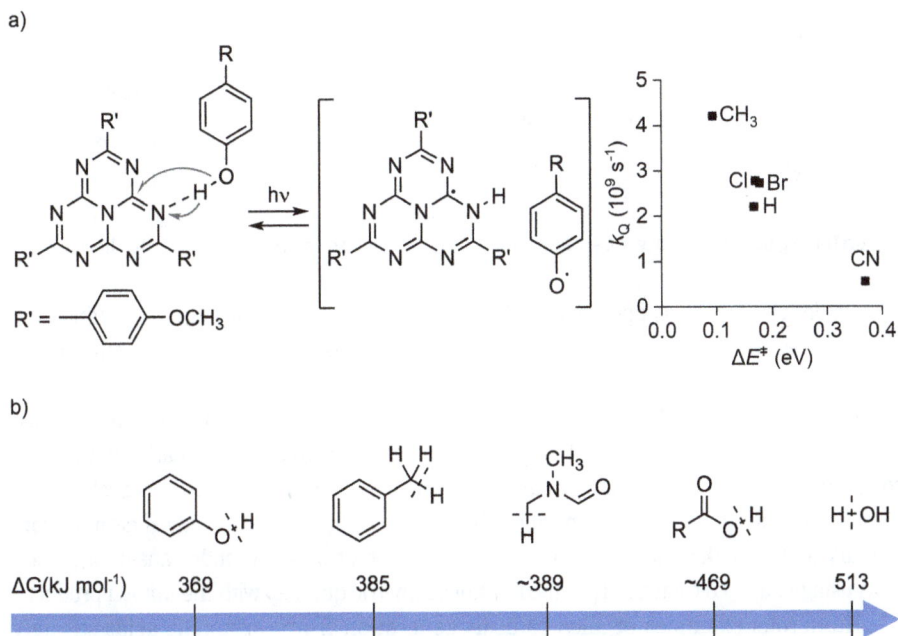

a)

b)

Fig. 8.10: Oxidative ES-PCET in carbon nitride photocatalysis. (a) ES-PCET from phenols to molecular carbon nitrides induced by excitation with light. (b) A scale with X–H bond dissociation free energy values in several classes of organic compounds [17].

When we move from heptazine derivatives (molecular systems) to heptazine-based carbon nitrides (materials), it is reasonable to expect increase of basicity. Due to the cooperative effect of the adjacent heptazine units (similar to "proton sponge"), carbon nitride materials are able to bind protons stronger, compared to heptazine derivatives. Such a feature allows employing carbon nitride materials as photocatalysts for the cleavage of C–H and O–H bonds in several classes of organic compounds via ES-PCET for the subsequent functionalization (Fig. 8.10b). For example, oxidative coupling of resveratrol (a polyphenol) and its analogues by mpg-CN is likely to proceed via initial O–H bond cleavage [18].

Cai et al. applied g-C_3N_4 to mediate oxygenation of benzylic C–H sites in alkylarenes, isochromanes and phthalanes via oxidative ES-PCET as an energetically more facile pathway compared to the stepwise transfer of an electron and a proton (Fig. 8.11) [19].

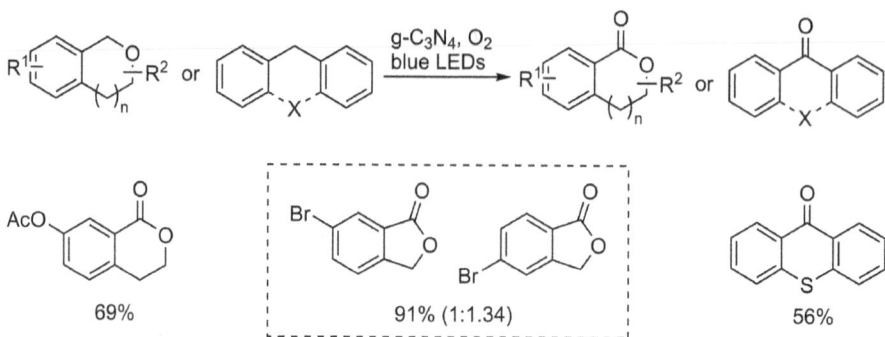

Fig. 8.11: Oxygenation of benzylic C–H sites with g-C$_3$N$_4$. Selected structures of products are shown.

Oxidative ES-PCET is operative in the Minisci reaction between pyridines, quinolines, isoquinolines and compounds possessing labile C–H bonds, next to amide and ether moieties, such as tetrahydrofuran, 1,4-dioxane, 1,3-dioxolane, *N,N*-dimethylformamide, *N,N*-dimethylacetamide and *N*-methylpyrrolidone (Fig. 8.12) [20]. Although O$_2$ from air was used as an electron acceptor, aliphatic alcohols that typically undergo dehydrogenation to carbonyl compounds, gave α-C–H coupling products with isoquinoline. Such reactivity is likely due to the isoquinolimium isoquinolinium cation that traps rapidly C-centered radical derived from primary and secondary alcohols. When the reaction was conducted under anaerobic conditions using Pt as a co-catalyst, H$_2$ formed in stoichiometric quantity with the Minisci product.

Numerous examples of alkyl radicals generation from carboxylic acids strongly support that graphitic carbon nitrides are able to cleave polar O–H bonds, with BDFE as high as ~ 469 kJ mol^{-1} [12, 21]. Finally, due to the highly positive potential of the valence band, +2.2 V versus SHE, and strong basic character, K-PHI can even cleave O–H bond in water (the strongest single O–H bond) via ES-PCET without using a co-catalyst, as long as photogenerated electrons are effectively removed from the conduction band by employing a scavenger [22].

8.3.3.2 Reductive proton-coupled electron transfer

Heptazine derivatives, at least the structure shown in Fig. 8.10a, with electron-donating 4-methoxyphenyl substituents, upon ES-PCET, do not stabilize excessive negative charge to the extent that the heptazinyl radical could survive for days or even weeks. However, under anaerobic conditions, extended structures, such as K-PHI, do form persistent radicals, as deduced from the EPR spectra (Fig. 8.13). Illumination of ionic carbon nitride suspension in aqueous 4-methylbenzyl alcohol as well as its dispersion in benzylamine acetonitrile mixture gives paramagnetic blue or a green state of the material. In order to maintain electrical neutrality, the excessive negative charge must be compensated by the counter ion. Among the counter ions, the most ubiquitous is H$^+$. Observation of

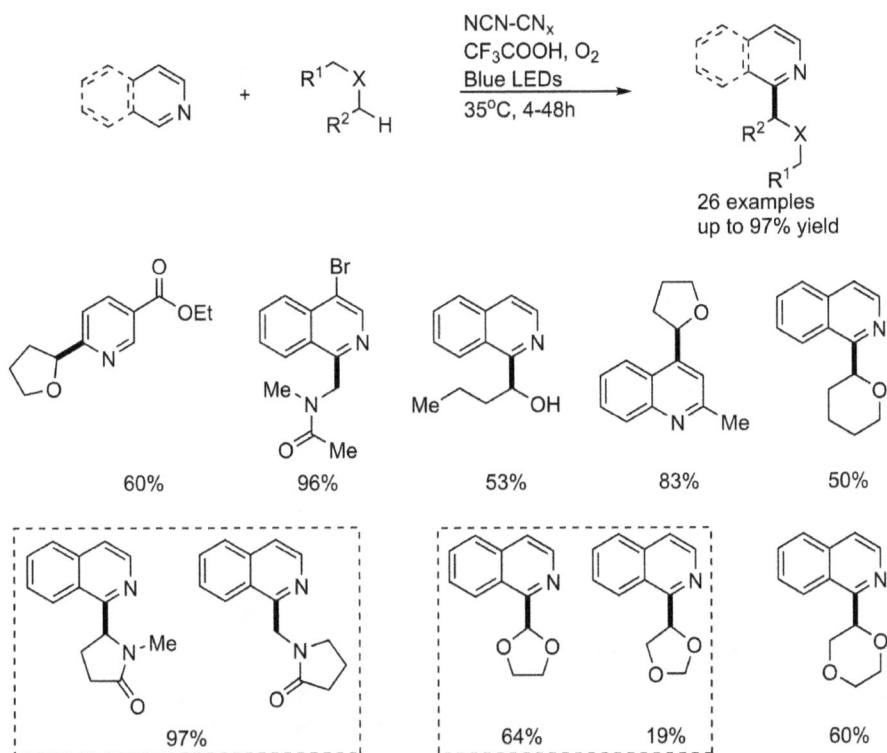

Fig. 8.12: Minisci reaction mediated by NCN-CN$_x$. Selected structures of products are shown.

such photocharged carbon nitrides strongly points at the reductive quenching of carbon nitride excited state under anaerobic conditions. At the same time, photocharged carbon nitrides act as PCET agents in the photocatalyst turnover step. Note that "excited state" is omitted in the context of *reductive* PCET, as transfer of electron and proton from K-PHI(e$^-$/H$^+$) to a substrate is downhill, and taking into account state-of-the art knowledge in this field does not require excitation.

In this chapter, photocharged carbon nitrides are discussed as catalysts – they are added in deficiency versus the reagents, and they are recovered in the catalytic cycle. Application of carbon nitrides as stoichiometric reductants in dark to reduce C = O, C–X (X = Cl, Br, I) bonds and nitro-group[23] or generate H$_2$ upon recombination of the stored electrons and protons by adding a suitable co-catalyst are discussed in Chapter 9.

Similar to oxidative ES-PCET shown in Fig. 8.10 that is essential for the X–H functionalization of organic molecules, reductive PCET also has strong implication in organic synthesis to enable processes, which are otherwise uphill. For example, one-electron reduction of C = O bond in ketones to the corresponding radical anion, typically requires potential more negative than –1.5 V versus SCE (Fig. 8.14) [9]. One of the strategies to facilitate ketone reduction is to activate the carbonyl group with Lewis acid, such as a

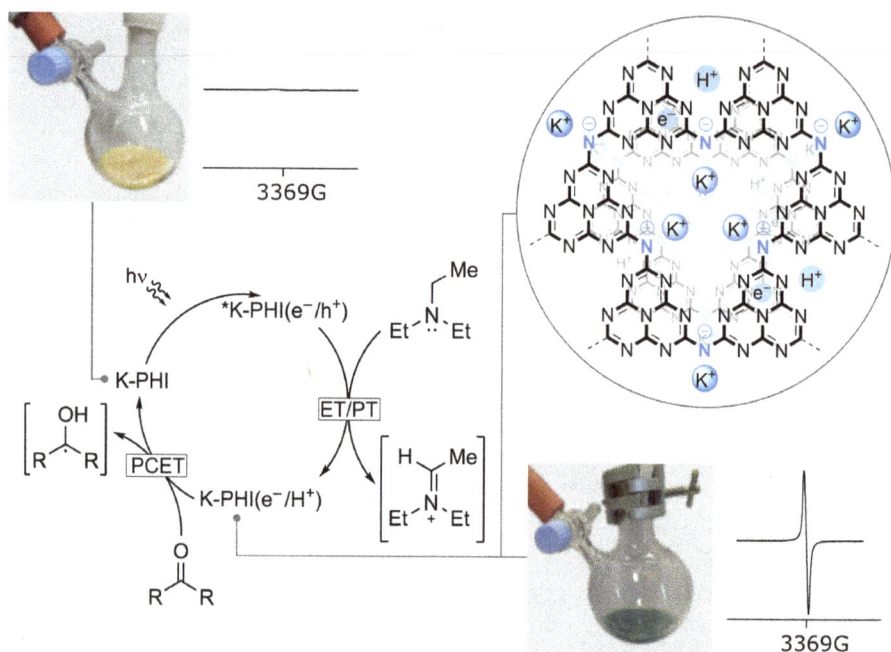

Fig. 8.13: Reductive quenching of K-PHI excited state, appearance of long-lived K-PHI(e⁻/H⁺) specie and its signal at 3,690 G in EPR spectrum. Using excess of Et₃N versus ketone results in the accumulation of K-PHI(e⁻/H⁺) species, which are observed as green solid in the photo. Reduction of carbonyl compound to ketyl radical is shown as an example. Photos are reproduced with permission from [24]. Copyright 2019 John Wiley & Sons, Inc.

proton or Li^+. Reduction of ketone to ketyl radical, thus proceeds via PCET and requires a less negative potential.

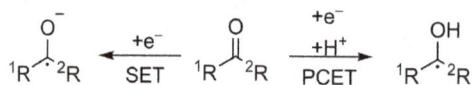

SET – challenging:
$E < -1.5V$
Charged radical

PCET – facile:
$E > -1.5V$
Neutral radical

Fig. 8.14: Reduction of ketone to the radical anion and ketyl radical via SET and PCET, respectively.

Due to the presence of two reactive sites in α,β-unsaturated ketones, C = C and C = O, this class of organic molecules possess rich chemistry that may be elegantly used to construct more complex molecules. Kurpil et al. found that ketyl radicals generated in situ by K-PHI under net-reductive conditions readily undergo addition to the C = C bond of the second enone molecule and yield cyclopentanoles (Fig. 8.15) [25].

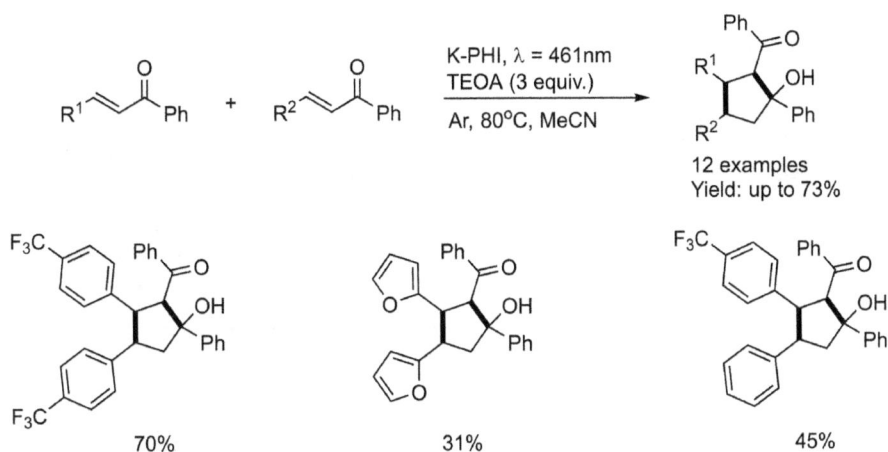

Fig. 8.15: Cyclodimerization of enones over K-PHI. Selected structures of cyclopentanols are shown.

Similarly, PCET is operative in the synthesis of *N*-fused pyrroles from enones and tetrahydroisoquinolines (Fig. 8.16) [26]. Interesting that without K-PHI and light irradiation, a classical Michael addition of tetrahydroisoquinoline to the β-carbon atom of the enone takes place, while the adduct is not converted into the *N*-fused pyrrole upon the subsequent addition of K-PHI and light irradiation. Such results underline the peculiar role of photocatalysis to enable reaction pathways via PCET. Upon excitation with UV light, *N*-fused pyrroles fluoresce, with the maximum at 400–450 nm, while external quantum efficiency reaches 24%.

Our group found that by replacing MeCN with chloroform, which is more reactive under net-reductive conditions, K-PHI(e^-/H^+) enables formation of dichloromethyl radical, $CHCl_2^•$, which is added to the β-carbon atom of the enone (Fig. 8.17) [27]. This reaction, a type of Giese radical addition, affords masked 1,4-dicarbonyl compounds, which could be easily 'deprotected' upon alkaline hydrolysis.

Cai et al. found that *para*-quinone methides exhibit peculiar reactivity, when combined with carbon nitride CN-K (Fig. 8.18) [28]. Under redox neutral conditions, carbon nitride generates alkyl and acyl radicals from carboxylic and α-ketoacids, respectively, which are then coupled with diarylmethyl radical, derived from *para*-quinone methides, upon PCET. Cyclic voltammetry revealed shift of $E_{p1/2}$ from –1.18 V for *para*-quinone methide to –1.03 V versus SCE, when measurements were carried out in the presence of CF_3CH_2OH. These results imply involvement of CF_3CH_2OH as a proton donor that facilitates reduction of *para*-quinone methide to the corresponding radical via PCET.

Under similar conditions, carbon nitride CN-K enables decarboxylative coupling of 2-(4-Methoxyphenyl)acetic acid and 2-oxo-2-phenylacetic acid with activated alkenes (Fig. 8.19).

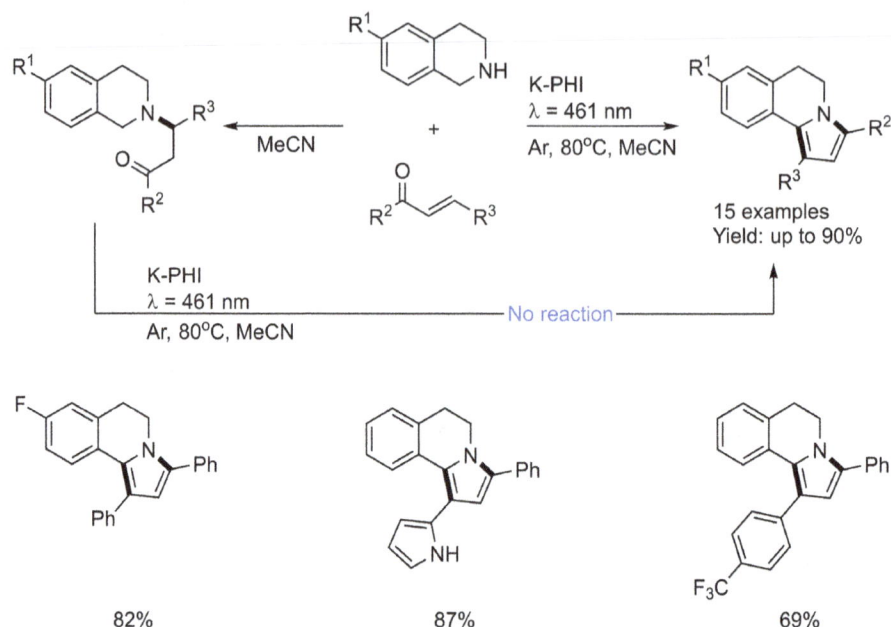

Fig. 8.16: Synthesis of *N*-fused pyrroles from enones and tetrahydroisoquinolines. Selected structures of products are shown.

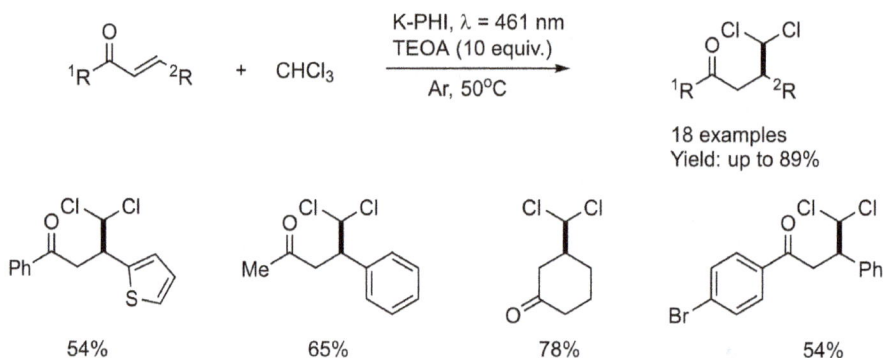

Fig. 8.17: Synthesis of γ,γ-dichloroketones from enones. Selected structures of products are shown.

Under net-reductive conditions, using sodium formate as a sacrificial electron donor, CN-K mediates reductive dimerization of *para*-quinone methides (Fig. 8.20) [28].

All reactions highlighted above are based on $1e^-/1\,H^+$ reduction of the unsaturated bonds, which could, in general, be also mediated by molecular sensitizers. One of the greatest advantages of K-PHI(e^-/H^+) and, in general, photocharged semiconductors,

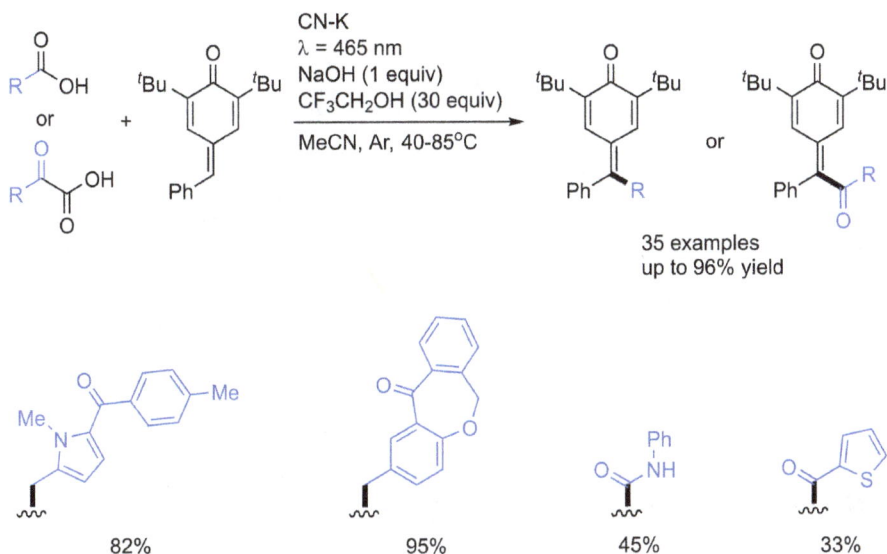

Fig. 8.18: Redox neutral decarboxylative coupling of carboxylic and α-ketoacids with *para*-quinone methides. Selected moieties that were transferred from carboxylic and α-ketoacids are shown.

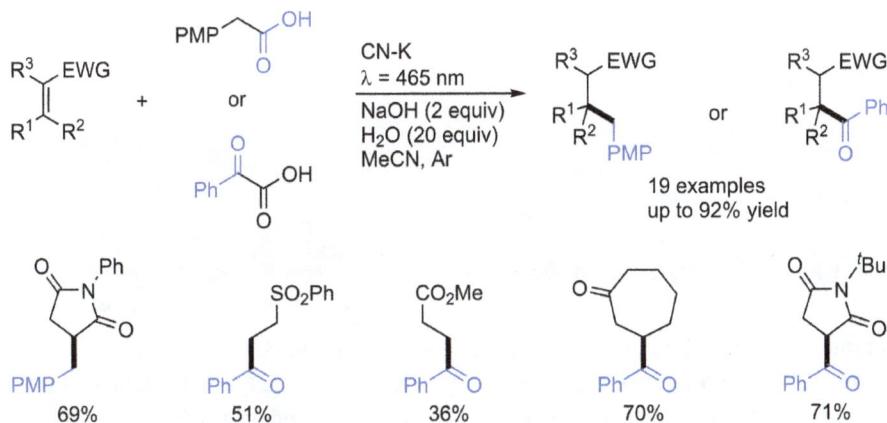

Fig. 8.19: Redox neutral decarboxylative alkylation and acylation of activated alkenes. Selected structures of products are shown.

compared to molecular reductants, however, is their ability to deliver multiple electrons at nearly the same potential [29, 30]. Reduction of nitrobenzene is known to be a complex reaction that may terminate at the step of nitrozobenzene (2e$^-$/2 H$^+$ reduction product), *N*-hydroxyaniline (4e$^-$/4 H$^+$ reduction product) [31] as well as diazo- and azoxy-compounds [32]. However, when reduction of nitro compounds is performed in deep eutectic solvent (DES) made of ammonium formate and glycerol, K-PHI yields

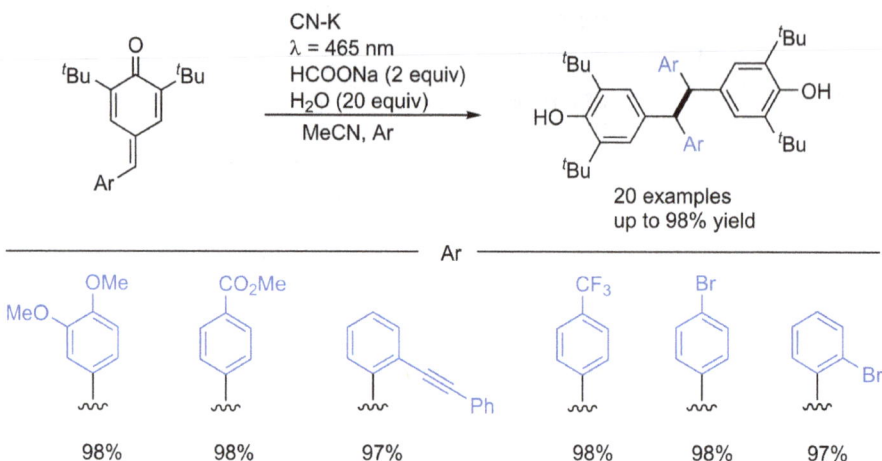

Fig. 8.20: Reductive dimerization of *para*-quinone methides.

exclusively aniline (the product of $6e^-/6$ H^+ reduction of nitrobenzene), while other unsaturated products are not formed (Fig. 8.21) [33]. By switching from ammonium formate: glycerol DES to that composed of ammonium formate: glycolic acid in situ formylation of aniline is achieved in one pot.

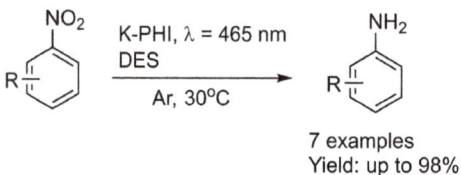

Fig. 8.21: Reduction of nitro compounds to aniline using PCET from photocharged K-PHI(e^-/H^+).

Mazzanti et al. found that photocharged K-PHI(e^-/H^+) reduces nitrobenzene to aniline in dark [23]. Such a feature and remarkable selectivity toward a fully reduced product, which otherwise is possible by using H_2 at Pt or Pd, underlines the similarity between photocharged K-PHI(e^-/H^+) and noble metals.

8.3.4 Elemental sulfur as a reagent in carbon nitride photocatalysis

Carbonization of organic compounds is accompanied by the evolution of small thermodynamically stable molecules, such as CO_2, water, N_2. According to the concept proposed by Antonietti and Oschatz, carbon materials may be divided into noble and non-noble materials, based on the potential of their valence band (Fig. 8.22) [34]. In such a notation, carbon materials that are derived from electron-rich precursors, such as hydrocarbons and lignin, are non-noble – elimination of water upon pyrolysis sets the oxidation potential of such materials slightly below the potential of proton reduction. Such carbon-based materials are electron rich (reductive). Therefore, they readily burn in oxygen atmosphere.

Contrarily, carbon materials that are prepared from electron-deficient precursors, such as azoles, azines and azolium salts, are noble (oxidizing). Oxidation potential of such precursors is more positive than the oxidation potential of water. Therefore, elimination of water during their pyrolysis shifts the oxidation potential of carbons to even more positive values. Noble carbons are electron deficient. Similar to noble metals, such as gold, they are stable against oxidation.

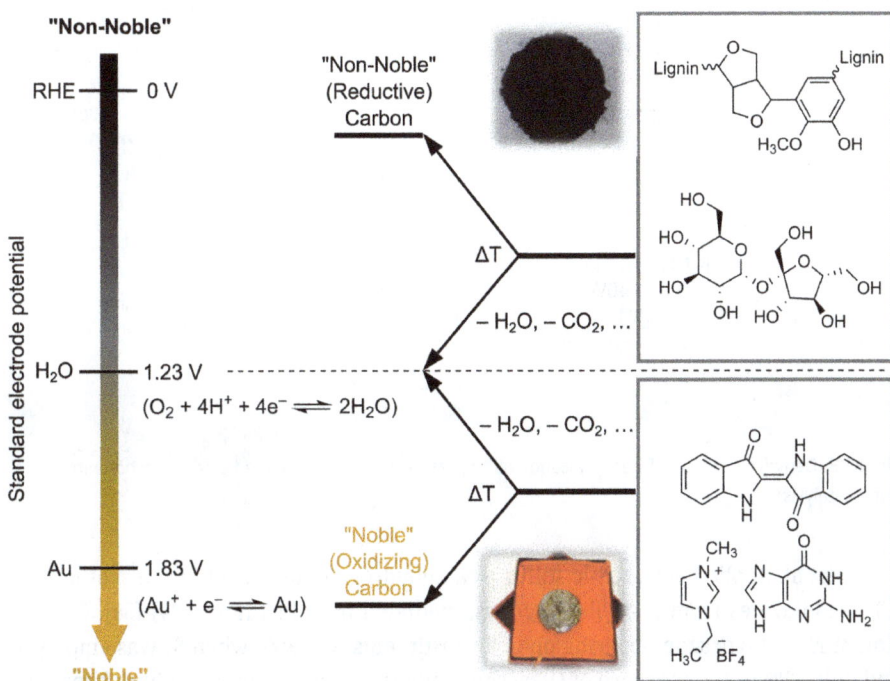

Fig. 8.22: Noble and non-noble carbons (reproduced with permission from [34]. Copyright 2018 John Wiley & Sons, Inc.). Image of silvery shining solid of nitrogen-doped carbon, prepared by calcining ionic liquid at 1,000 °C is shown in the image (reproduced with permission from [35]. Copyright 2013 John Wiley & Sons, Inc.).

From this standpoint, carbon nitrides belong to noble carbons – they are prepared from melamine, which is electron-deficient heterocycle, while the valence band potential of carbon nitrides is more positive than the water oxidation potential. Such a feature makes carbon nitrides stable against the oxidizing species and electrophilic radicals generated in photocatalysis, and also allows coupling carbon nitride photocatalysis with reagents that are seldom used in combination with transition metals.

Elemental sulfur (or octaatomic sulfur, S_8), which in the periodic table stands just below O_2, is an overlooked electron acceptor and reagent in photocatalysis, likely because of its strong corrosive effect and foul odor of its reduction product, H_2S. However, compared to H_2O_2, H_2S is less prone to trigger chain radical processes, which stems from higher BDFE of H–S, 392 kJ mol^{-1} [17], compared to O–O bond, 183 kJ mol^{-1} [36]. Indeed, in photocatalysis, O_2 through its reduction product, H_2O_2, serves as a source of highly oxidative HO$^\bullet$ radicals, generation of which in the reaction mixture unavoidably decreases selectivity. Overoxidation of aromatic aldehydes to carboxylic acids is a common problem in photocatalysis. Savateev et al. succeeded to suppress this side reaction by using S_8 instead of O_2 as the electron acceptor (Fig. 8.23) [37].

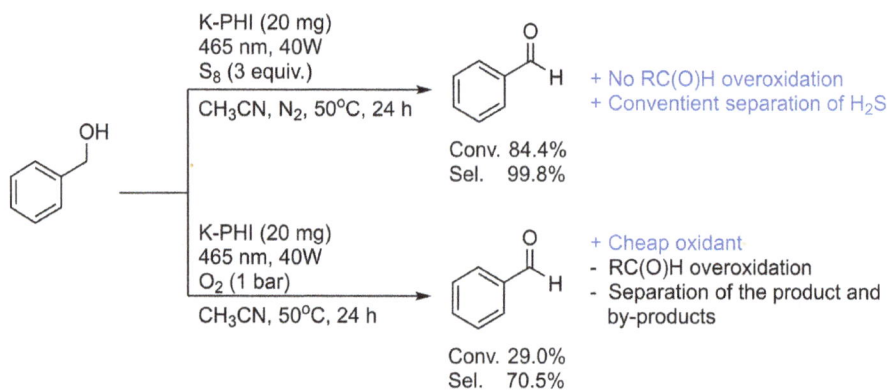

Fig. 8.23: Dehydrogenation of benzyl alcohol in the presence of O_2 and S_8, mediated by carbon nitride photocatalysis.

Kurpil et al. applied octaatomic sulfur as a dehydrogenating agent in the synthesis of 1,3,4-oxadiazoles from *N*-acylhydrazones, mediated by K-PHI (Fig. 8.24) [38]. Conversion of *N*-acylhydrazone-bearing phenyl substituents was 86% when S_8 was employed and only 47% in the presence of O_2, accompanied by a significant amount of benzaldehyde as the by-product that originates from N = C bond cleavage upon nucleophilic attack of superoxide radical.

Elemental sulfur is a promising 100% atom-efficient alternative to common sulfurating agents such as Lawessons reagent, in which sulfur content is only 32%. Savateev et al. enabled benzylic C–H photocatalytic oxidation of alkylarenes in the presence of S_8 to obtain dibenzyldisulfides (Fig. 8.25) [11].

Oxidation of benzylic amines, mediated by mpg-CN, in the presence of O_2 yields imines (Fig. 8.26) [39]. When the reaction is conducted in the presence of S_8 and in strictly anaerobic conditions, the products of amines coupling are thioamides [40]. Thioamides from two different benzylic amines are also accessible via this method.

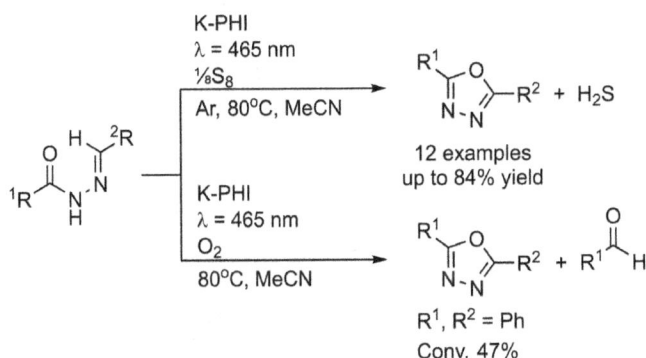

Fig. 8.24: Synthesis of 1,3,4-oxadiazoles by the dehydrogenation of *N*-acylhydrazones, mediated by K-PHI in the presence of octaatomic sulfur.

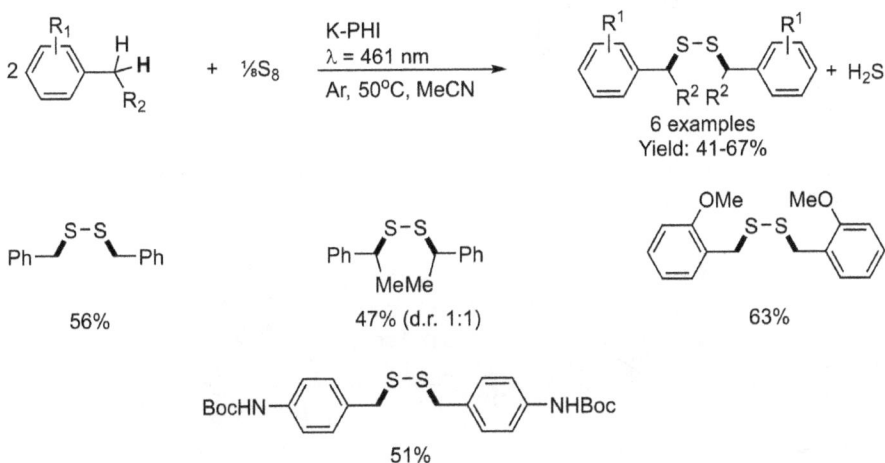

Fig. 8.25: Synthesis of dibenzyldisulfides from alkylarenes and octaatomic sulfur, mediated by K-PHI photocatalyst.

Su et al. applied polymeric carbon nitride in combination with CuI to synthesize a series of thio- and selenoethers, while chalcogene atoms are derived from octaatomic sulfur and elemental selenium (Fig. 8.27) [41].

8.3.5 Triplet excited state and energy transfer in carbon nitride photocatalysis

Similar to small organic molecules and organic semiconductors, irradiation of carbon nitrides with UV–vis light generates their singlet and triplet excited states. Radiative relaxation of the singlet excited state back to the singlet ground state, fluorescence, is

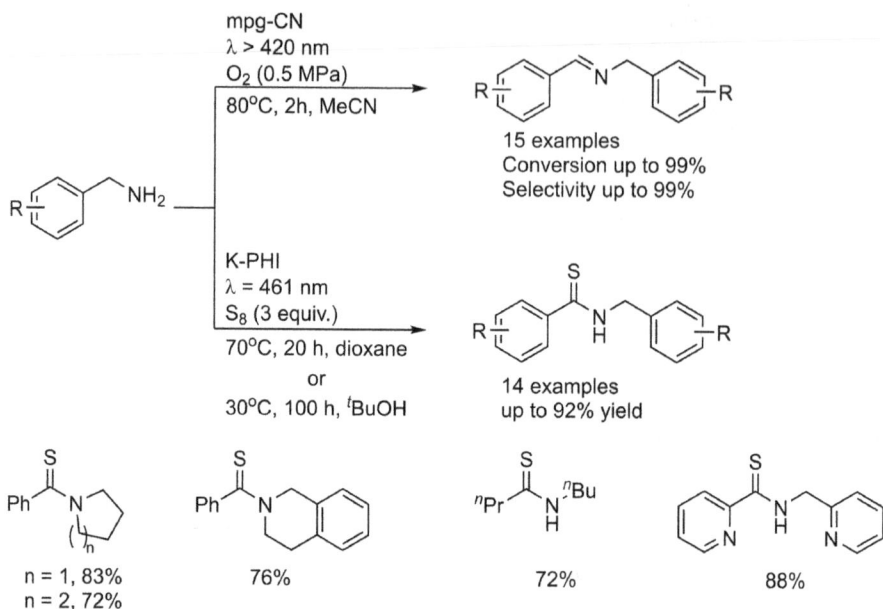

Fig. 8.26: Synthesis of imines and thioamides from benzylic amines, mediated by carbon nitride photocatalysis. Selected structures of thioamides are shown.

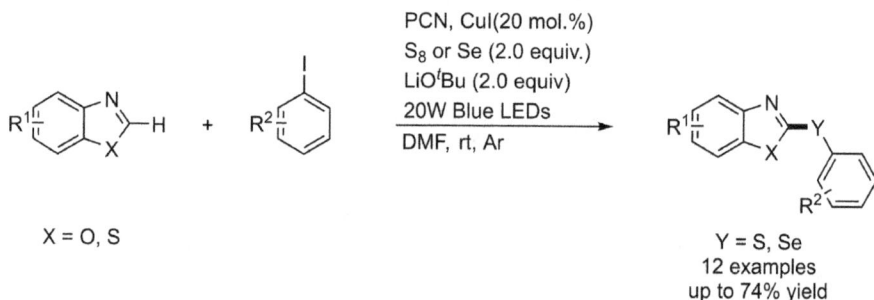

Fig. 8.27: Dual Cu/carbon nitride photocatalytic synthesis of thio- and selenoethers using S_8 and Se.

a spin-allowed process. Therefore, the lifetime of the singlet excited state of carbon nitride, such as K-PHI, as well as many organic dyes is limited to hundreds of ps or up to tens of ns (Fig. 8.28). Successful PET between a reagent and a short-lived carbon nitride singlet excited state requires the reagent to be located within the distance of electron transfer, prior light excitation. In this case, diffusion as a potential limiting factor for PET is effectively eliminated. Note that carbon nitride particles discussed in this chapter have a diameter of hundreds and thousands of nanometers. Therefore, compared to small organic molecules in dispersion, they behave as static objects.

Intersystem crossing (ISC) produces emissive or non-emissive triplet states. Radiative relaxation of a triplet excited state back to singlet ground state, phosphorescence, is a spin-forbidden process. Therefore, the lifetime of carbon nitrides triplet excited state often exceeds microseconds. Such long-lived triplet excited states are essential for processes that are limited by the diffusion of a reagent to the carbon nitride surface or kinetically sluggish processes. On the other hand, triplet states may engage in energy transfer, which implies transfer of the photogenerated electron and the hole. As such, redox neutral reactions often proceed via energy transfer.

During the ISC, a sensitizer loses a fraction of its energy, which is defined by the singlet-triplet energy gap, ΔE_{ST} – triplet states, in general, are more stable compared to the singlet excited states. For example, triplet excited state of dicyanoanthracene is 1.09 eV lower in energy compared to its singlet excited state, which implies that certain thermodynamically challenging reactions are not accessible for the triplet state of this sensitizer.

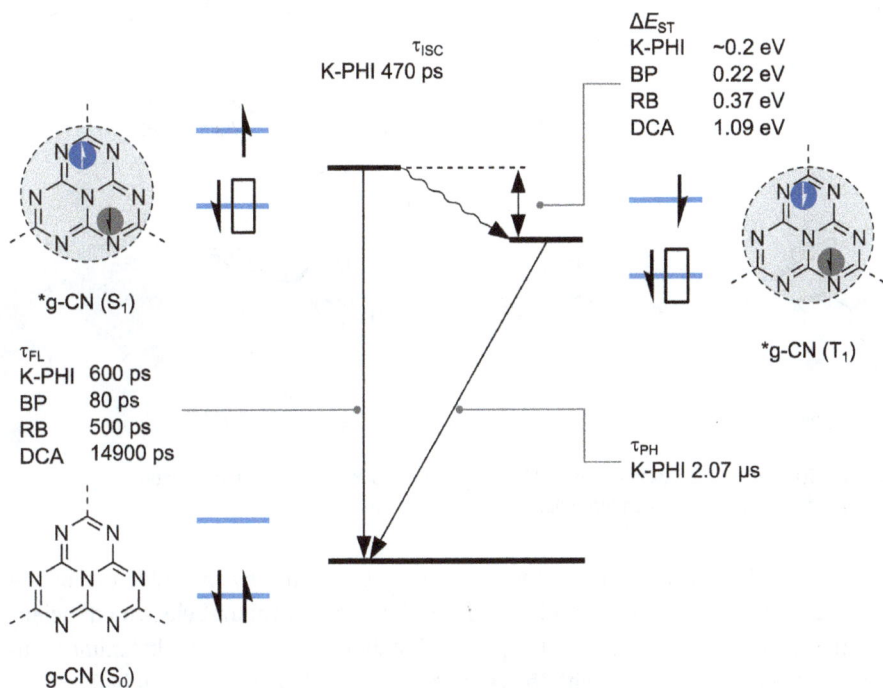

Fig. 8.28: Jablonski diagram summarizes the singlet and triplet excited state lifetimes, time of ISC, and ΔE_{ST} values for molecular sensitizers. BP, benzophenone; RB, rose Bengal; DCA, dicyanoanthracene [5] and K-PHI [42]. Exciton in K-PHI is schematically shown to be localized at a single heptazine unit.

Due to abundance, oxygen is the most commonly used electron acceptor and acceptor of energy in photocatalysis. Typically, superoxide radical $O_2^{\cdot-}$ is generated upon PET from the carbon nitride excited state in the presence of sacrificial electron donors. Long-lived triplet state of K-PHI is engaged in energy transfer to produce singlet oxygen, $^1O_2$, which was unambiguously detected via its phosphorescence at 1,280 nm [42]. Note that an alternative pathway for $^1O_2$ generation exists, which is based on the one-electron reduction of O_2 that gives $O_2^{\cdot-}$, followed by oxidation of the latter by the photogenerated hole [43]. Therefore, often, $^1O_2$ and $O_2^{\cdot-}$ co-exist in the reaction mixture.

Regardless of the $^1O_2$ generation pathway, it is a likely intermediate that enables the generation of nitrile oxides from oximes, which then participate in [3 + 2]-cycloaddition to $C \equiv N$ and $C = C$ bonds, leading to the formation of oxadiazoles-1,2,4 and isoxazoles (Fig. 8.29).

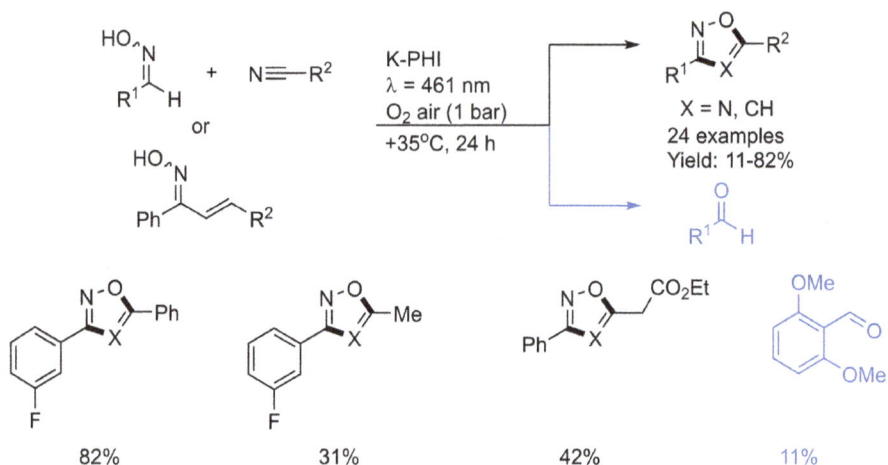

Fig. 8.29: Synthesis of 1,2,4-oxadiazoles and isoxazoles assisted by K-PHI. Selected structures of oxadiazoles and aldehyde that undergo deoxymation are shown.

Selectivity of 1,2,4-oxadiazole synthesis is strongly affected by the structure of the oxime. Decent yields have been obtained for electron-neutral and electron-deficient substrates, while electron rich and aliphatic aldoximes, and especially ketoximes, undergo deoxymation reaction, with the recovery of the carbonyl compound [44].

Long-lived carbon nitride triplet excited state is also essential for redox neutral reactions. Thus, cyanamide-modified carbon nitride ($^{NCN}CN_x$) triplet state having a lifetime of 1.77 µs was used in the hydroamidation reaction (Fig. 8.30) [45].

In this example, the multisite PCET from the substrate yields carbamyl radical – the electron quenches the photogenerated hole in $^{NCN}CN_x$, while the base, tributylmethylammonium dibutylphosphate, receives H$^+$. The overall efficiency of the process depends on the effective coupling of the photocatalytic B/BH$^+$ and HAT cycles. The formation of a

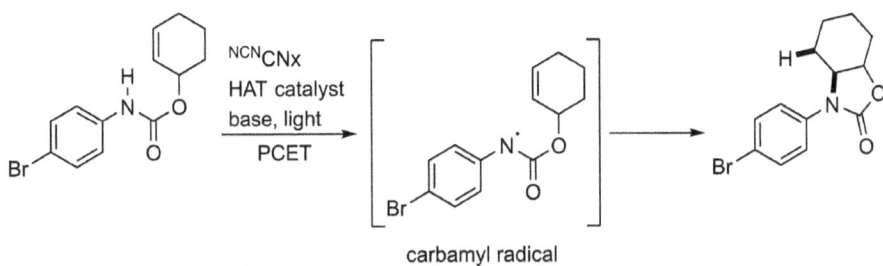

Fig. 8.30: Hydroamidation reaction mediated by triplet excited state of carbon nitride.

long-lived carbon nitride radical anion provides a temporary buffer for the storage of an electron, which the carbon nitride then passes to the HAT-catalyst, MesS$^{\bullet}$ radical. Albeit the yield of the product was twice lower in the absence of the base, this experiment gives additional piece of evidence that the basicity of $^{NCN}CN_x$ is sufficient to mediate reactions via oxidative ES-PCET.

Optimization of the carbon nitride structure toward a faster ISC, lower ΔE_{ST} and longer lifetime of the triplet excited state are the targets to achieve in the next generation of high-performing carbon nitride photocatalysts for application in redox-neutral reactions.

8.3.6 Photons as reagents – chromoselective photocatalysis

Wavelength (λ, nm) and the corresponding energy (E, eV) of a photon are connected via the Planck-Einstein relation:

$$E = h\nu = \frac{hc}{\lambda \cdot 10^{-9} \times 1.6 \cdot 10^{-19}} \sim \frac{1,240}{\lambda} \tag{8.9}$$

where c is the speed of light in vacuum, 3×10^8 m s^{-1}; h is the Planck constant, 6.63×10^{-34} m^2 kg s^{-1}; ν is the frequency of the electromagnetic wave, Hz. Wavelength of photons and the corresponding energy are shown in Fig. 8.31.

The energy carried by a single microwave photon is orders of magnitude lower compared to the IR photon, while IR photons in turn is less energetic than the photons in visible and UV range of the electromagnetic spectrum. Upon absorption, microwave, IR and UV–vis photons induce different changes in the structure of a molecule. Thus, microwave photons enable rotational transitions, IR photons – vibrational transitions, while UV–vis photons generate electronically excited state of a molecule and even enable homolysis of chemical bonds. For example, photolysis of water O–H bonds is achieved by irradiation with far UV light, while cleavage of S–S bond in disulfides is facile even under green light.

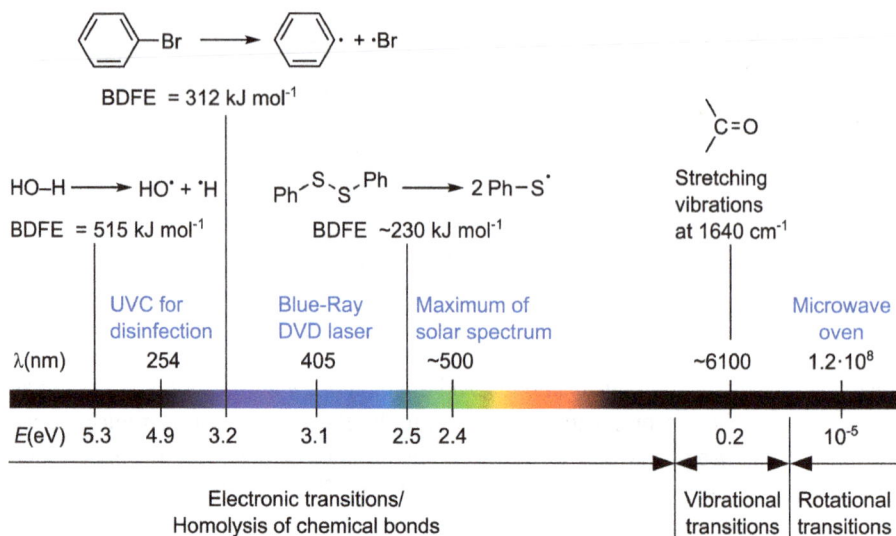

Fig. 8.31: Electromagnetic spectrum with wavelength, energy of photons generated by certain sources and energy required to enable certain processes labeled on the scale.

Typically, the quantum efficiency of a photocatalytic and photochemical processes, which are not free-radical chain reactions, is significantly lower than 100%. Therefore, photons are "added" into the reaction mixture in superstoichiometric quantity versus molecular reagents. Given that the composition of a reaction mixture strongly affects selectivity of a dark chemical reaction, selectivity of a *photochemical reaction* is influenced by the energy (color) of photons that chemists shine onto the reactor [46]. Similarly, photocatalysts can enable a specific reaction pathway from a set of potentially possible, for a given combination of chemical reagents, by choosing photons of a specific wavelength (Fig. 8.32a) [47]. Such a mechanism of tuning the outcome of a photocatalytic reaction is called *chromoselectivity*.

The mechanism behind the chromoselective behavior of heptazine-based carbon nitrides is explained in Fig. 8.32b-d. For carbon nitrides, the steep absorption band typical for semiconductors is observed at ~ 460 nm, which is similar to aromatic structures – this is ascribed to π–π* transitions. Heptazine-based carbon nitrides possess plenty of (sp^2)N-lone pairs. Therefore, often an additional band of lower intensity is observed. The shoulder corresponds to n–π* transitions that starts in the visible or near-IR range of the electromagnetic spectrum and gradually merges with the primary absorption band. In heptazine-based carbon nitride, free of defects, n–π* transitions are virtually absent, as seen in Fig. 8.32b (black curve and pale yellow powder in inset). The shoulder becomes increasingly pronounced when the density of defects increases. Note the evolution of the absorption band at ~ 600 nm – the color of the samples change from yellow to light brown.

Figure 8.32c (top) shows a fragment of "ideal" heptazine-based carbon nitride. In such a structure, n–π* transitions are effectively suppressed due to the nitrogen lone pairs being perpendicular to the plane of the conjugated system – the dihedral angle between the adjacent heptazine units is 0°. It was discussed in Chapter 2 (Fig. 2.14c) that distortion of the system from perfectly planar improves the overlap between nitrogen lone pair and the LUMO (Fig. 8.32c, bottom), which increases the magnitude of the otherwise prohibited n–π* transitions. Note that the discussed "defective" structure was chosen to illustrate the principle. However, it is neither the only nor the most plausible source of defects in carbon nitrides. Substitution of some nitrogen atoms in the heptazine unit by carbon (see a series of cyclazines in Chapter 2) or other heteroelements (doping) as well as missing carbon and nitrogen atoms (called carbon and nitrogen vacancies respectively) promote n–π* transitions. The band structure of a "defective" carbon nitride is shown in Fig. 8.32d. Photons of higher energy excite electrons from the valence band to the conduction band (π–π* transitions), while photons of lower energy are only capable of promoting an electron from intraband states (IBS, formed by the non-bonding molecular orbitals) to the conduction band (n–π* transitions). While the reductive power of the photogenerated electrons is the same in both cases, holes resting in the valence band are stronger oxidants compared to the holes in the IBS. This feature of carbon nitrides is the origin of their chromoselective behavior.

A carbon nitride CN-OA-m that possesses IBS was employed in photo-chemo-biocatalytic cascades (Fig. 8.33) [50]. The oxidation power of the holes generated in the IBS, upon CN-OA-m excitation with 528 nm photons, is not sufficient to oxidize ethylbenzene directly. Instead, methanol as a sacrificial electron donor effectively quenches the holes. The photogenerated electrons then reduce O_2 to H_2O_2, which is used by an unspecific peroxygenase (UPO) from A. aegerita (AaeUPO) acting as chiral catalyst to convert ethylbenzene into (R)-phenylethanol. On the other hand, excitation of CN-OA-m with 455 nm photons generates holes in the valence band. The oxidation power of photogenerated holes in such a case is sufficient to convert ethylbenzene into acetophenone. In this case, O_2 serves as a sacrificial agent that consumes photogenerated electrons and therefore drives oxidation of the substrate. Acetophenone is reduced in situ by alcohol dehydrogenase (ADH-A) from Rhodococcus ruber in the presence of NAD^+ as a cofactor.

S-Thiophenolacetate is a molecule with a rather complex photoredox chemistry. Markushyna et al. found that despite a comparable photon flux density of 0.05–0.22 μmol cm^{-2} s^{-1}, the selectivity of S-thioacetate photocatalytic oxidation by K-PHI depends on the wavelength of the incident photons (Fig. 8.34) [51]. Thus, under 365 and 410 nm photons, S-thioacetate is converted primarily into arylchloride. Under 465 nm, S-thioacetate yields sulfonylchloride. Under 525 and 635 nm, it is converted into diphenyldisulfide. The reason for such chromoselectivity is the dependence of the yield of the reactive of oxygen species, namely singlet oxygen and superoxide radical, on the wavelength of photons. Thus, green and red photons enable excitation of an

a)

b)

c)

n–π* suppressed

n–π* promoted

d)

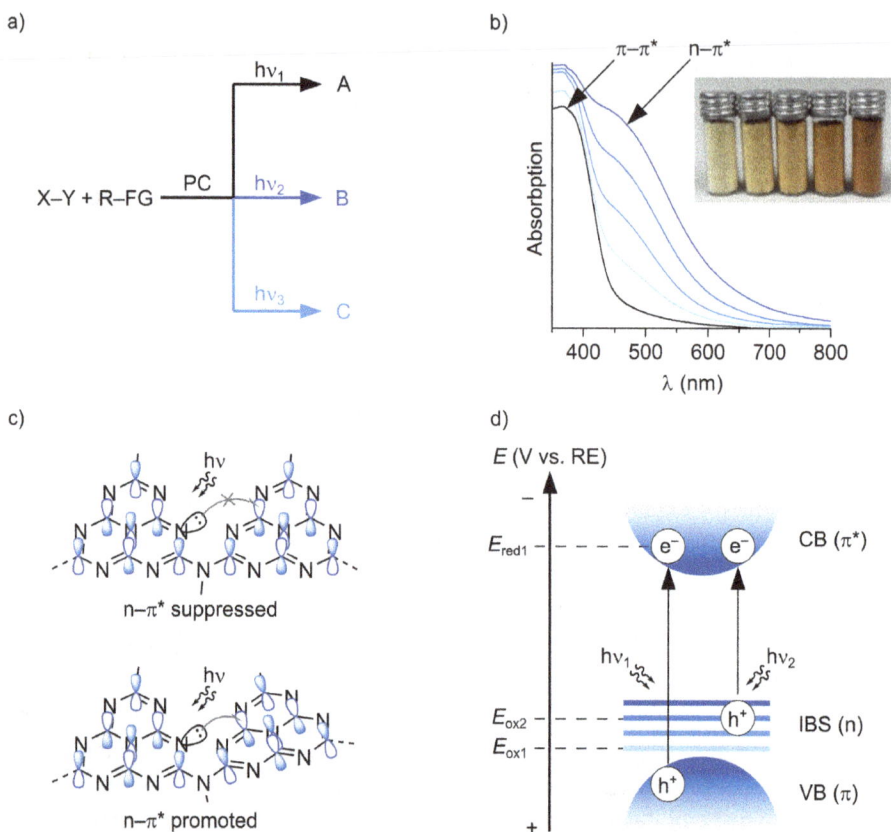

Fig. 8.32: Origin of chromoselective photocatalytic behavior of carbon nitrides. (a) Schematic illustration of the concept of chromoselective photocatalysis. (b) UV–vis absorption spectra of carbon nitrides, with the density of IBS changing from low (light blue) to high (dark blue) and the appearance of carbon nitride powders in inset. (c) Schematic representation of n–π* transitions in "defective" (bottom) and otherwise prohibited n–π* transitions in "ideal" heptazine-based carbon nitride (top). Blue p-type orbitals represent the conduction band (LUMO) [48]. (d) Band diagram of a semiconductor possessing IBS. Excitation of electron from the valence band to conduction band (π–π* transitions) and from the IBS to the conduction band (n–π* transitions) with photons of higher and lower energy respectively. Panel (b) is reproduced with permission from [49] (License CC BY 3.0. Copyright 2017, the Royal Society of Chemistry).

electron from the IBS to the conduction band, while the excited state then facilitates $^1O_2$ sensitization. On the other hand, high-energy blue and UV photons enable electron excitation from the valence band to the conduction band, and fast (<0.4 ns) charge separation. The photogenerated electrons reduce O_2 to the superoxide radical. By replacing HCl with NH_4Cl, ethylammonium- or n-butylammonium chlorides, the corresponding sulfamides are accessible via this method.

AaeUPO (25 nM)
CN-OA-m (2 mg mL^{-1})
MeOH (1% v/v)
Tricine buffer (pH 7.5, 100 mM)
λ = 528 nm
(1330 µmol photons m^{-2} s^{-1})
24 h, 30 °C

OH
| Me

(R)-isomer
7.1 mM
71%, ee = 99%

Me

CN-OA-m (2 mg mL^{-1})
MeOH (1% v/v)
KPi buffer (pH 7.5, 100 mM)
λ = 455 nm
(1400 µmol photons m^{-2} s^{-1})
24 h, 30 °C

O
‖ Me

not isolated

ADH A (17.5 µg mL^{-1})
NAD$^+$ (0.1 mM)
iPrOH (300 mM)
12 h, 30 °C

OH
| Me

(S)-isomer
2.5 mM
30%, ee = 93%
one pot,
stepwise

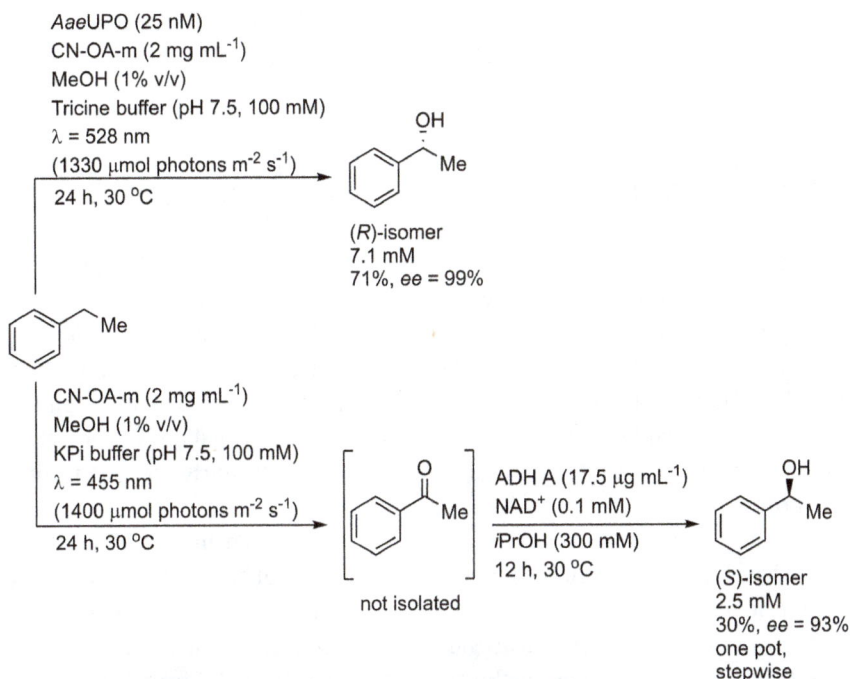

Fig. 8.33: Chromoselective synthesis of (R)- and (S)-2-phenylethanol from ethylbenzene employing photo-chemo-biocatalytic cascade reaction.

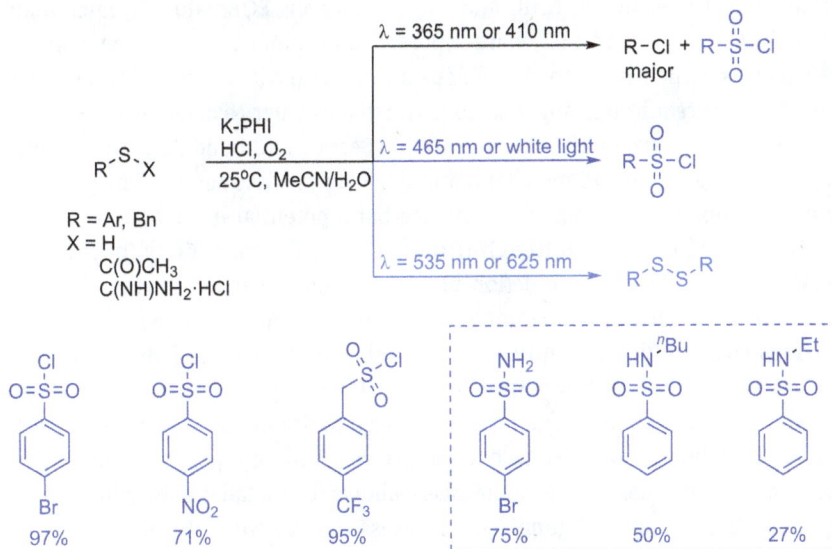

λ = 365 nm or 410 nm

R–Cl + R–S–Cl
major

K-PHI
HCl, O$_2$
25°C, MeCN/H$_2$O

R$^{\cdot}$S$^{\cdot}$X

R = Ar, Bn
X = H
C(O)CH$_3$
C(NH)NH$_2$·HCl

λ = 465 nm or white light

R–S–Cl

λ = 535 nm or 625 nm

R$^{\cdot}$S$^{\cdot}$S$^{\cdot}$R

Cl
O=S=O
|
Br
97%

Cl
O=S=O
|
NO$_2$
71%

O
‖–Cl
S
‖
O
|
CF$_3$
95%

NH$_2$
O=S=O
|
Br
75%

^{n}Bu
HN
O=S=O

50%

Et
HN
O=S=O

27%

Fig. 8.34: Chromoselective transformations of thiophenol derivatives.

The scope of arylsulfonylchlorides may be expanded further by employing aryl-diazonium salts and SO_2 under anaerobic conditions, instead of thiophenol derivatives [52].

8.3.7 Dual-transition metal–carbon nitride photocatalysis

The importance of palladium-catalyzed cross-coupling reactions for synthetic organic chemistry cannot be overstated. It allows constructing complex molecules from simple building blocks via convergent approach. However, the high cost of palladium motivates chemists to develop alternative protocols that are based on earth-abundant metals such as Ni, Co, Cu and Fe. Ni is a harder Lewis acid and more nucleophilic. Therefore, oxidative addition (OA) is more facile (Fig. 8.35). Aryl halides bearing electron-withdrawing groups undergo OA to Ni(0) species particularly well, as in this case, the aromatic ring is reduced. Among arylhalides of similar structure, OA becomes more facile when moving in the periodic table from chlorine to iodine. However, reductive elimination from Ni(II) species, which is the final step, for example, in C–O bond formation, as shown in Fig. 8.35, is endergonic [53]. In addition, in the case of Ni(II)-alkoxy complexes bearing hydrogen atom in β-position to the metal site, elimination of hydride is the dominant pathway. On the other hand, RE from Ni(III) complex is exergonic. The photocatalytic approach is particularly appealing to generate Ni(III) species in situ as it allows the conducting reaction without the stoichiometric oxidant. In Fig. 8.35, the catalytic cycle for C–O cross-coupling is shown as an example, and it implies resting states of Ni(0) and Ni(II). However, EQE values greater than unity, namely 2.2 ± 0.2 and 2.4 ± 0.1, were reported in C–N and C–O in Ni metallaphotoredox catalysis, which point to the self-sustaining Ni(I)/Ni(III) cycle [54]. In other words, photons are required in substoichiometric quantity per Ni catalytic cycle.

The properties of carbon nitrides discussed in Sections 8.3.1 and 8.3.5 are also elegantly applied in the context of metallaphotoredox catalysis, as supported by plentiful reports on this topic. For example, the valence band potential in carbon nitrides is sufficiently positive to oxidize Ni(II) to Ni(III), $E_{ox} = +0.41$ V versus SCE in (bpy)NiArBr [55]. On the other hand, the conduction band potential in carbon nitrides is sufficiently negative to reduce, for example, Ni(II) to Ni(I) in (dtbbpy)NiCl$_2$ ($E_{red} \sim -0.85$ V versus NHE), which is required to close the catalytic cycle [54]. Long-lived triplet states are engaged in energy transfer to Ni(II) complex to generate the excited Ni(II)* specie, which is an alternative pathway used to destabilize Ni(II) species and promote RE. Finally, the nitrogen-rich structure resembling a polypyridine ligand stabilizes metal single sites and, in some cases, allows for metallaphotoredox catalysis without adding the extra ligand. The progress in dual Ni/photocatalytic cross-coupling reactions is summarized later.

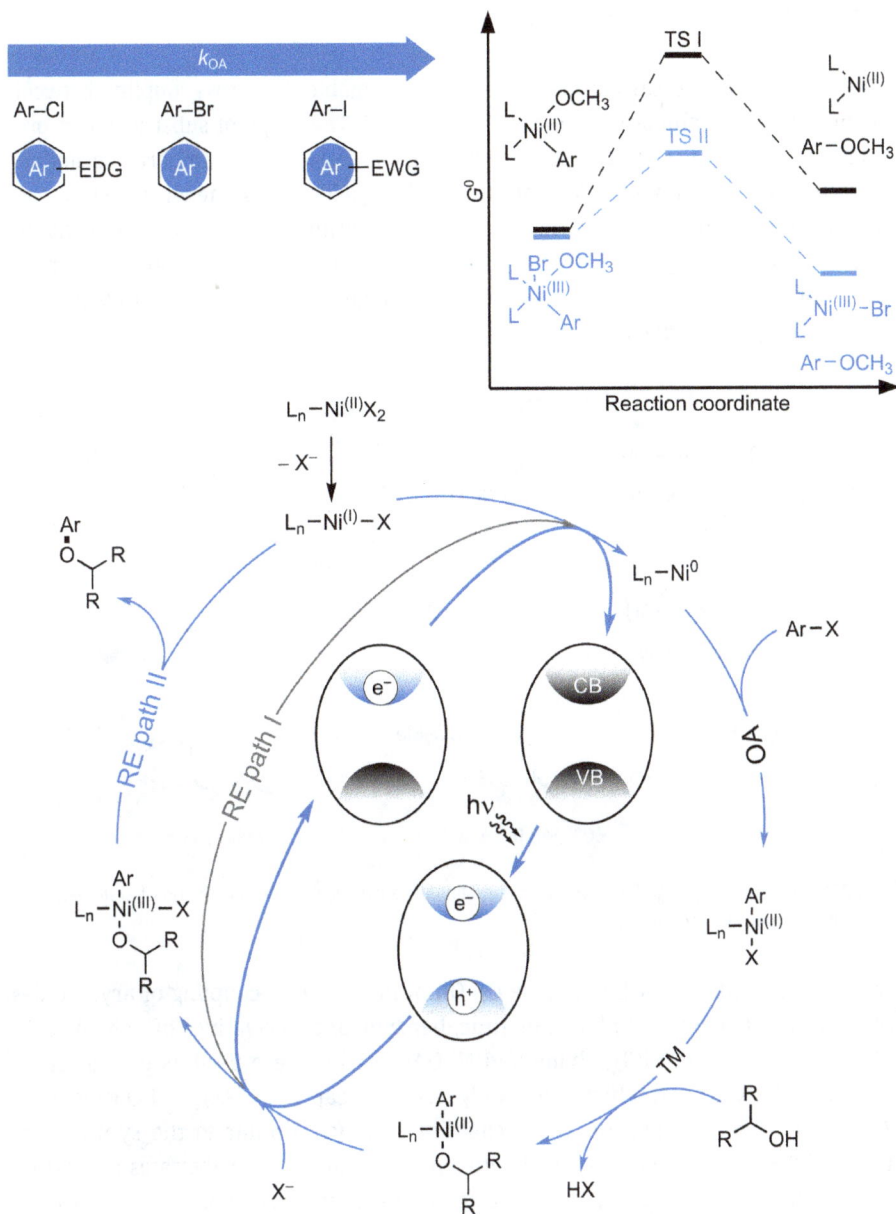

Fig. 8.35: A general mechanism of C–O cross-coupling. Blue lines show dual Ni/photocatalytic pathway that includes RE from Ni(III) complex (downhill). Gray line highlights dual Ni/photocatalytic pathway, in which RE occurs from Ni(II) complex (uphill).

8.3.7.1 C–O and C–S cross-couplings

Cavedon et al. applied carbon nitride CN-OA-m to enable C–O cross-coupling between aryl bromides and aliphatic alcohols (Fig. 8.36) [56]. The scope of substrates is represented mainly by primary and secondary alcohols. Sterically encumbered secondary and tertiary alcohols gave coupling products in low yields, while phenols failed to react due to their low nucleophilicity. In a competing experiment using 2-mercaptoethanol, the C–S cross-coupling product was formed exclusively. Tertiary and aromatic thiols give coupling products due to the involvement of highly reactive thiyl radicals that are added to the Ni(I) intermediate in the catalytic cycle.

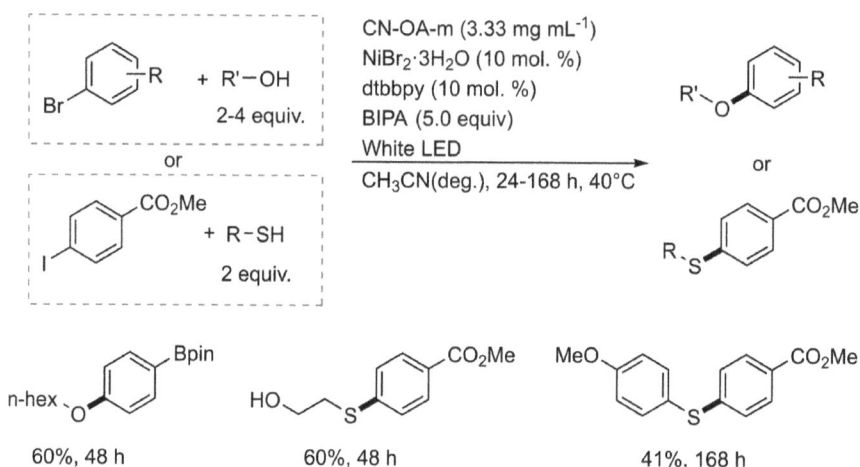

Fig. 8.36: C–O cross-coupling of arylbromides and aliphatic alcohols; C–S cross-coupling of aryliodides with aliphatic thiols and arylthiols.

Pieber et al. employed carbon nitride CN-OA-m in C–O cross-coupling of aryl iodides with carboxylic acids – aliphatic, olefinic, benzylic and derivatives of benzoic acid [57]. In this example, $NiCl_2 \cdot$ glyme and $Ni(OAc)_2 \cdot 4H_2O$ were used as pre-catalysts. While selectivity is overall higher for $NiCl_2 \cdot$ glyme, cheaper $Ni(OAc)_2 \cdot 4H_2O$ gave up to 22% of the aryliodide-acetate anion coupling product. Similar to the synthesis of ethers (Fig. 8.37), the methods enable effective coupling of the substrates possessing electron-withdrawing groups, while electron-rich aryliodides demonstrate lower reactivity. Illumination with quasi monochromatic blue and green light instead of white LED slows down the reaction, but it does not alter the selectivity.

CN-OA-m (3.33 mg mL^{-1})
NiCl$_2$·glyme or Ni(OAc)$_2$ (10 mol. %)
dtbppy (10 mol. %)
BIPA (3.0 equiv) or (5 equiv)
White LED
DMSO(deg.), 14-168 h, 40°C

94% or 75%   91% or 80%   78% or 68%

93% or 77%   89% or 59%

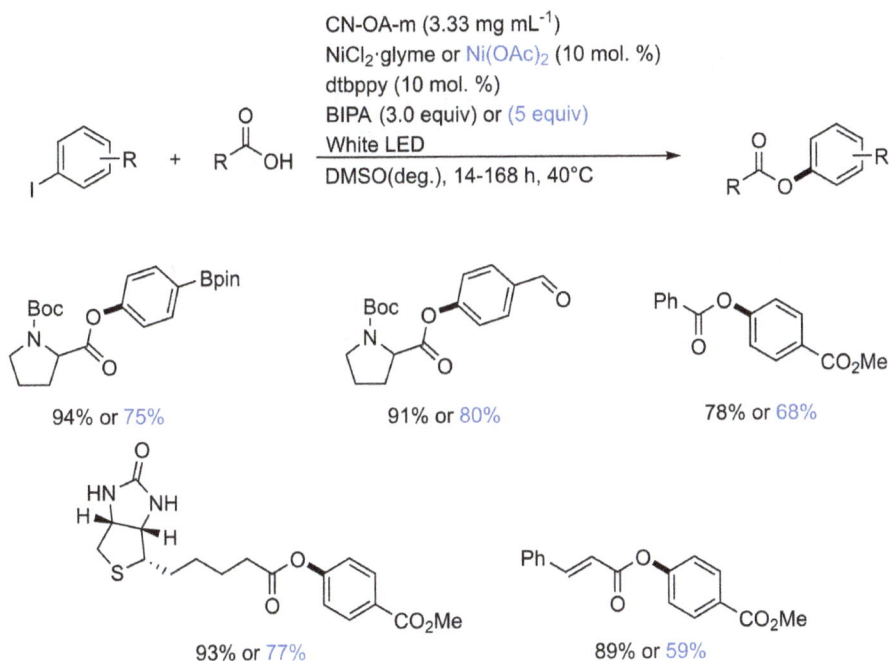

Fig. 8.37: C–O cross-coupling of aryliodides and carboxylic acids.

8.3.7.2 C–N cross-coupling

Ghosh, König et al. found that unlike C–O cross-coupling, coupling of electron-deficient arylhalides with primary and secondary aliphatic amines, aromatic amines and sulfonamides that are mediated by Ni/mpg-CN do not require an external ligand due to a nucleophile or a base, which typically is a tertiary amine, to serve as a ligand (Fig. 8.38) [8].

In addition to the scope of coupling partners demonstrated above, carbon nitride CN-OA-m enables cross-coupling of electron-rich arylbromides with less nucleophilic *n*-butylamine (Fig. 8.39) [58]. Giesbertz et al. achieved this by adding 7-methyl-1,5,7-triazabicyclo[4.4.0]dec-5-ene (MTBD) into the reaction mixture, which activates low-valent Ni-species toward OA. Still, the coupling of electron-rich arylchlorides present a significant synthetic challenge compared to the more reactive bromides.

Beyond the convenience in handling and recyclability, the presence of IBS in the band structure of carbon nitride CN-OA-m has important implication in the dual Ni/photocatalytic C–N cross-coupling as well. Thus, illumination of the reaction mixture with 450 nm photons leads to deposition of Ni black, while changing the light source to 520 nm photons allows overcoming such problem (Fig. 8.40) [58]. The idea behind this approach is to decrease the rate of RE by storing the Ni catalyst in the resting state II, instead of the low-valent Ni species, such as resting state I, which is prone to

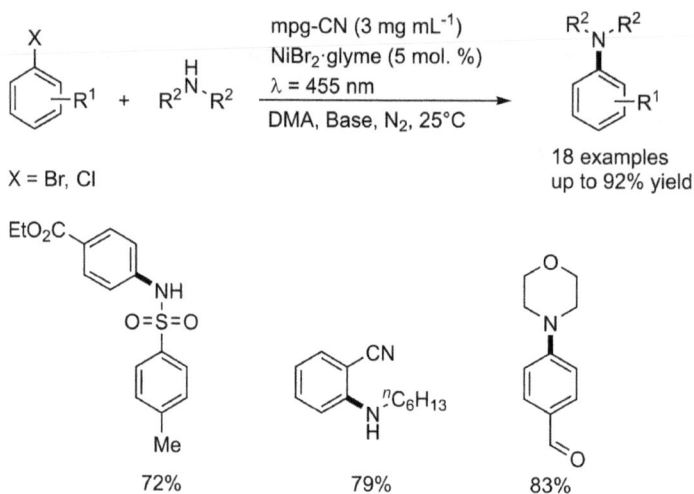

Fig. 8.38: C–N cross-coupling of arylhalides with NH-nucleophiles mediated by Ni/mpg-CN dual photocatalysis.

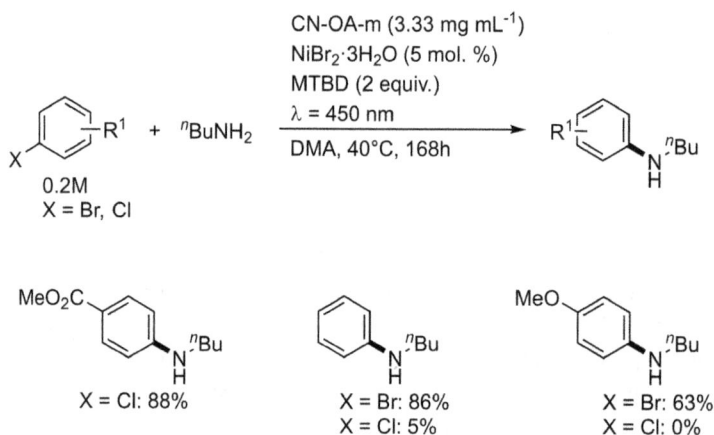

Fig. 8.39: C–N cross-coupling of arylhalides and n-butylamine.

form nickel nanoparticles and the subsequent deposition of nickel-black. Similar effect is achieved under 450 nm photons by increasing the substrate concentration from 0.2 M to 1.2 M.

8.3.7.3 C–C cross-coupling

A two-electron transmetallation step (Fig. 8.41) in C–N and C–O cross-coupling is, in general, facile due to coupling partners, amines, alcohols and carboxylate anion, being

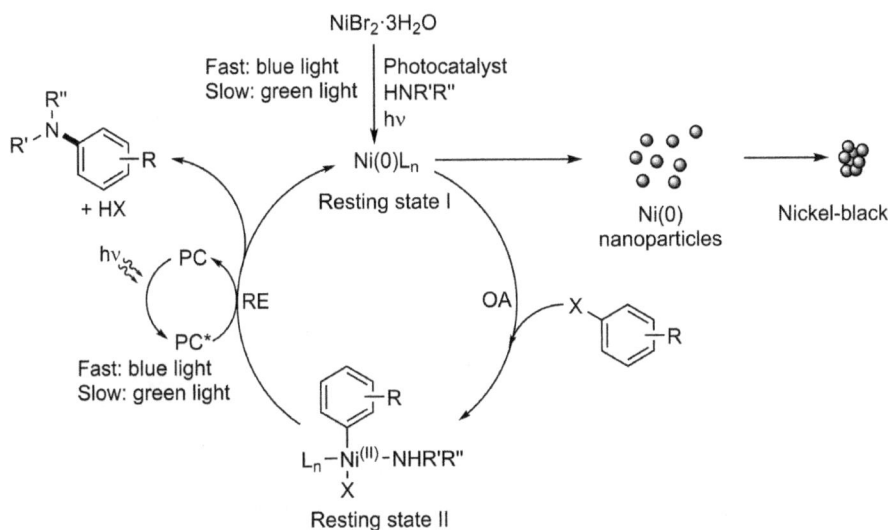

Fig. 8.40: Chromoselective behavior of carbon nitride that is employed to suppress the counterproductive generation of nickel-black and decrease the rate of RE.

sufficiently nucleophilic to substitute the halide atom in the metal coordination sphere. On the other hand, nucleophilicity of, for example, boronic ethers which serve as a source of aryl moieties in Suzuki cross-coupling is low (Fig. 8.41). Because of this, the activation energy of the two-electron transmetalation step is high, which makes it the rate-limiting step of most Suzuki cross-couplings. To overcome the activation energy barrier, traditional C–C cross-couplings require high temperature and stoichiometric quantity of a base. An alternative, one-electron transmetalation, includes generation of a reactive radical upon one-electron oxidation of precursors, such as trifluoroborates [59], carboxylic acids [60], bis(chatecolato)silicates [61], Hantzsch esters [62], sulfinate salts [63], etc.

Khamrai et al. applied mpg-CN to enable C–C cross-coupling of arylhalides with allyl- and benzyl trifluoroborates through single electron transmetalation (Fig. 8.42) [64]. Unlike C–O and C–N coupling, arylbromides bearing different electron-donating and electron-withdrawing groups as well as electron-rich hetarylbromides participate in the cross-coupling reaction with potassium benzyl trifluoroborate. In addition, electron-deficient heteroaryl*chlorides* are coupled with potassium benzyl trifluoro borate. mpg-CN also mediates the generation of allylic radicals and their coupling with electron-deficient arylbromides.

Alternative precursor of alkyl radicals that is used in dual Ni/mpg-CN photocatalysis is bis(chatecolato)silicates (Fig. 8.43) [65]. Various aromatic and heteroaromatic halides, primarily electron deficient, as well as vinylbromides were coupled with bis(chatecolato)silicates.

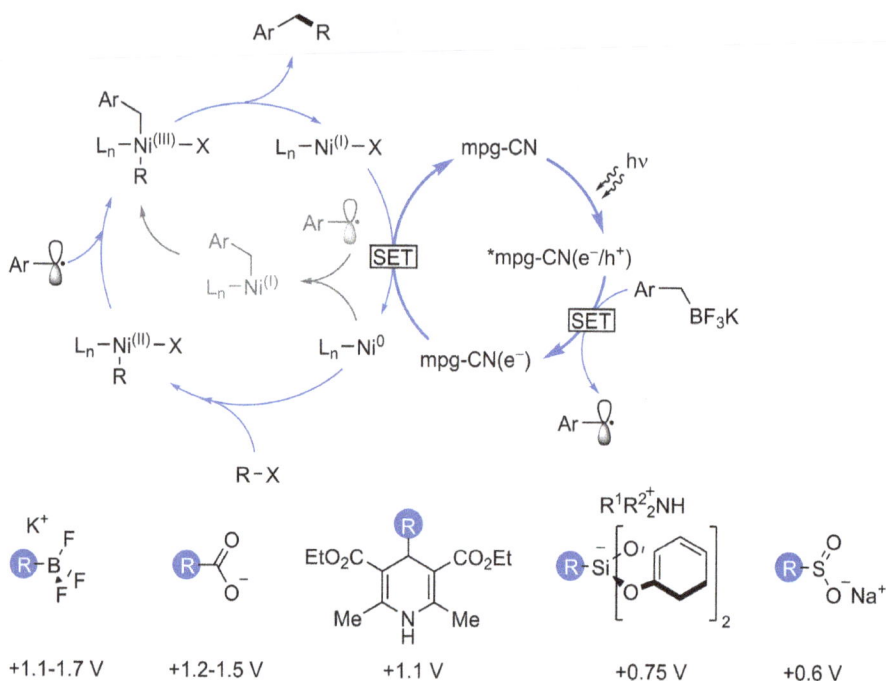

Fig. 8.41: Two-electron transmetalation of Pd-catalyst versus single-electron transmetalation of Ni-catalyst (adapted with permission from [59]. Copyright 2019 John Wiley & Sons, Inc.). Oxidation potentials of radical precursors are shown under the structures (in V vs. SCE).

C–H bonds are the most abundant in organic molecules. Their activation eliminates the need for prefunctionalization. Das et al. combined mpg-CN and Ni-catalyst to enable arylation of C–H bond in β-position to nitrogen in dialkylamides (Fig. 8.44) [66]. The protocol allows for a wide range of aryl- and hetaryl bromides and chlorides to be coupled with N,N-dialkylamides, including tetramethylurea and N-methyl-2-pyrrolidone. A range of functional groups, such as HO-, amide-, carboxylic-, sulfonamide and boronic acid were tolerated. Furthermore, late-stage functionalization of organic compounds was accomplished. Collectively, these aspects point to an exceptionally high selectivity of amides functionalization by Ni/mpg-CN dual photocatalysis.

Although the mechanism of this transformation proceeding via SET, similar to that shown in Fig. 8.35, sound piece of evidence for energy transfer pathway was obtained by illuminating the preformed (bpy)Ni(II)(2-CH$_3$C$_6$H$_4$)Br complex II and DMA (Fig. 8.45). According to the proposed mechanism, OA of 2-bromotoluene to Ni(0)-complex I gives intermediate II. Energy transfer from *mpg-CN(e$^-$/h$^+$) to Ni(II)-complex II generates electronically excited Ni(II)-complex III. Dissociation of Ni–X bond in the complex III releases a halogene atom, which abstracts hydrogen from N-alkylamide. Ni(I)-complex V traps the alkyl radical and gives Ni(II)-complex VI. RE from Ni(II)-complex

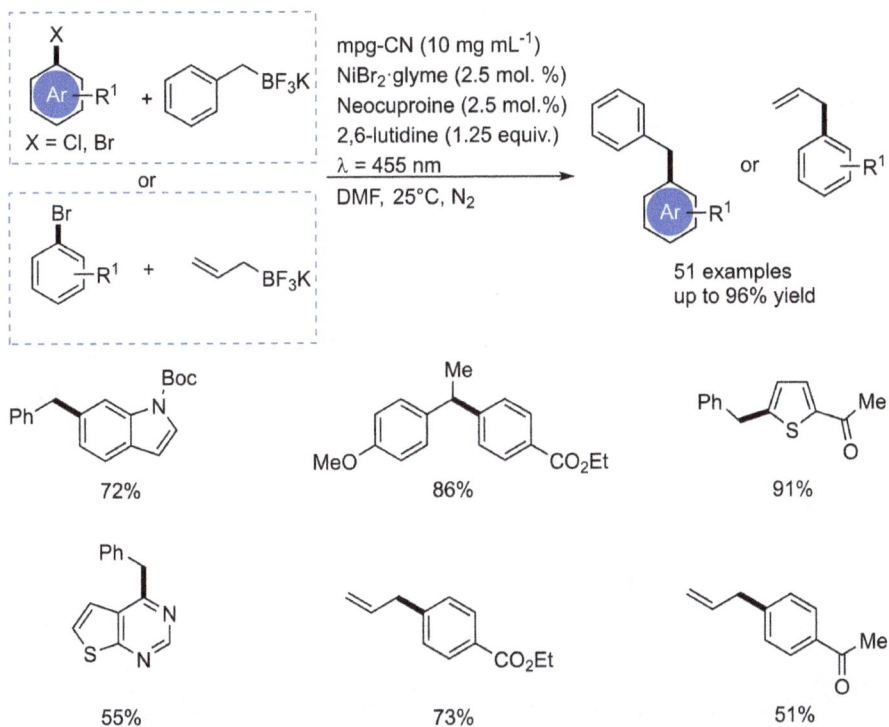

Fig. 8.42: Scope of C–C cross-coupling reaction between potassium benzyl- and allyltrifluoroborates and (het)arylhalides.

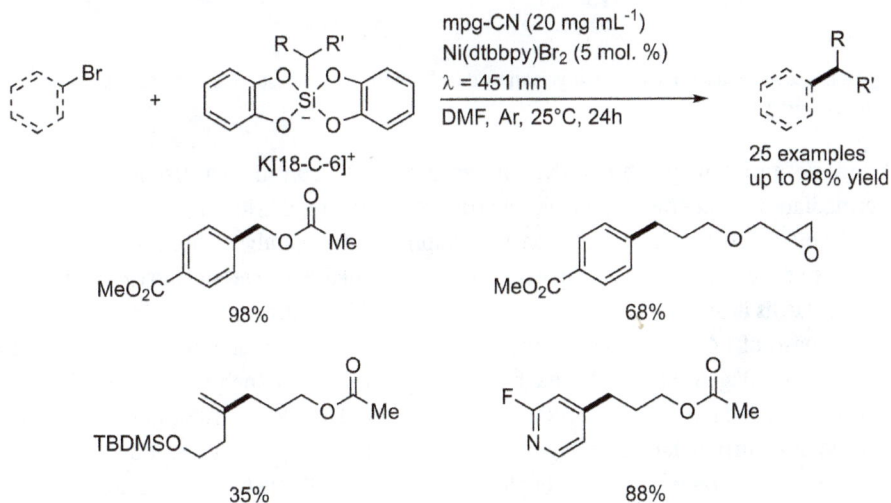

Fig. 8.43: Scope of C–C cross-coupling of bis(chatecolato)silicates and arylbromides, mediated by Ni/mpg-CN dual photocatalysis. Structures of selected products are shown.

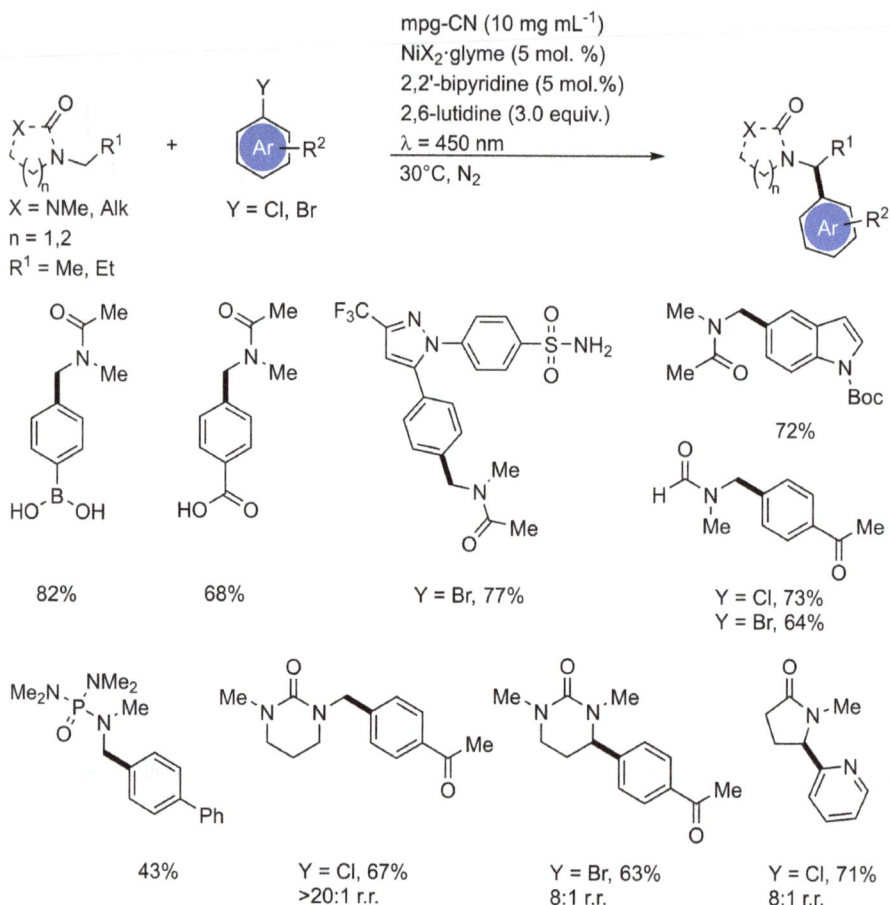

Fig. 8.44: A scope of alkylamides C–H arylation by dual Ni/mpg-CN photocatalysis. Selected structures of products are shown.

VI does not proceed spontaneously, but becomes facile from the electronically excited intermediate VII, generated upon energy transfer from *mpg-CN(e⁻/h⁺).

Although this section focuses on metallaphotoredox catalysis, the role of Ni-salts and salts of other transition complexes, when combined with carbon nitride, might be broader. Lewis acids, when coupled with mpg-CN, facilitate Mizoroki-Heck cross-coupling and synthesis of 1,4-dicarbonyl compounds from vinylacetate or styrene derivatives and bromoalkenes (Fig. 8.46) [67]. Although mpg-CN gives the coupling product, the addition of Lewis acid, such as Cu(OTf)$_2$, In(OTf)$_3$, Sc(OTf)$_3$, Zn(OTf)$_2$, Y(OTf)$_3$, AlCl$_3$ and, especially, NiBr$_2$·glyme, further increases the yield. The recovered mpg-CN, without adding fresh portion of NiBr$_2$·glyme, gave the coupling product with 74% yield, which is comparable to that using a combination of mpg-CN and NiBr$_2$·glyme (86%).

entry	light source	yield
1.	340 nm	17%
2.	450 nm	20%
3.	dark	ND

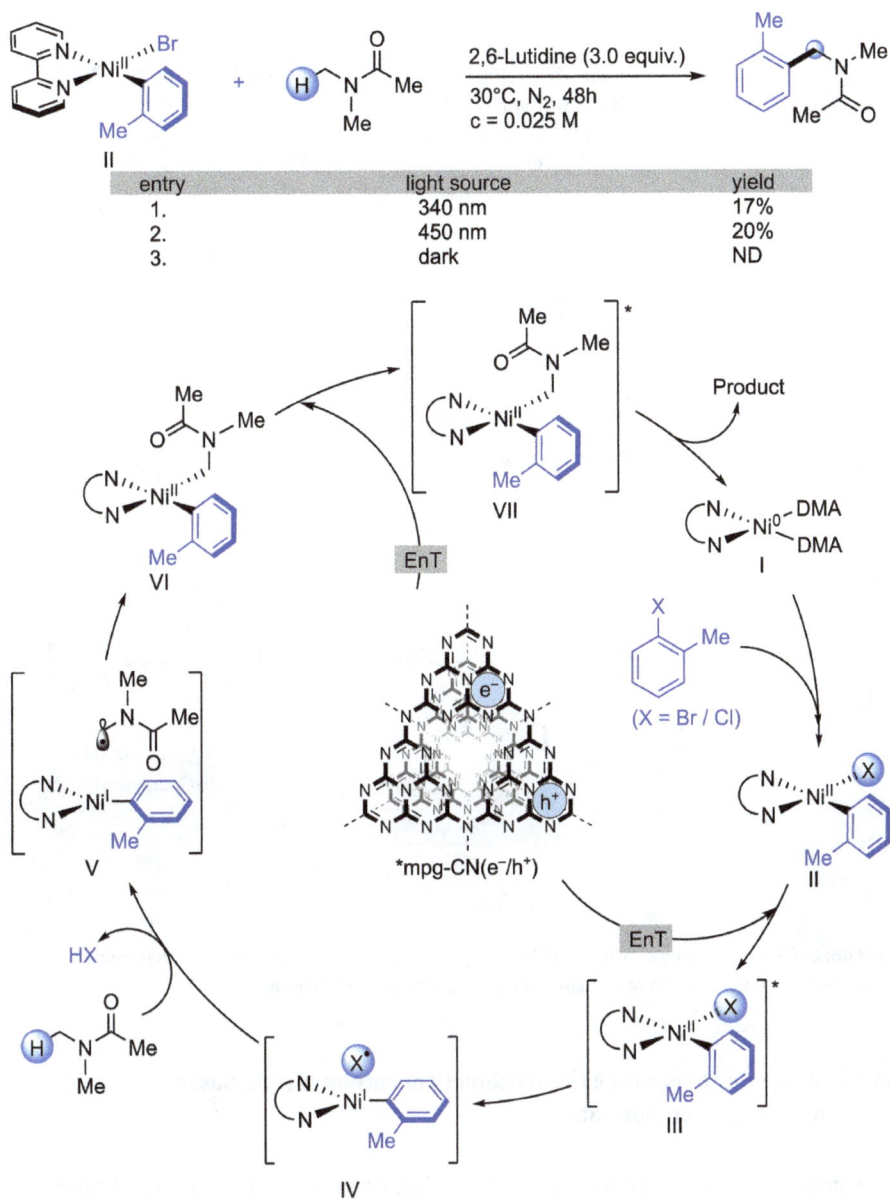

Fig. 8.45: A proposed mechanism of alkylamides C–H arylation by dual Ni/mpg-CN photocatalysis that is based on energy transfer from mpg-CN excited state.

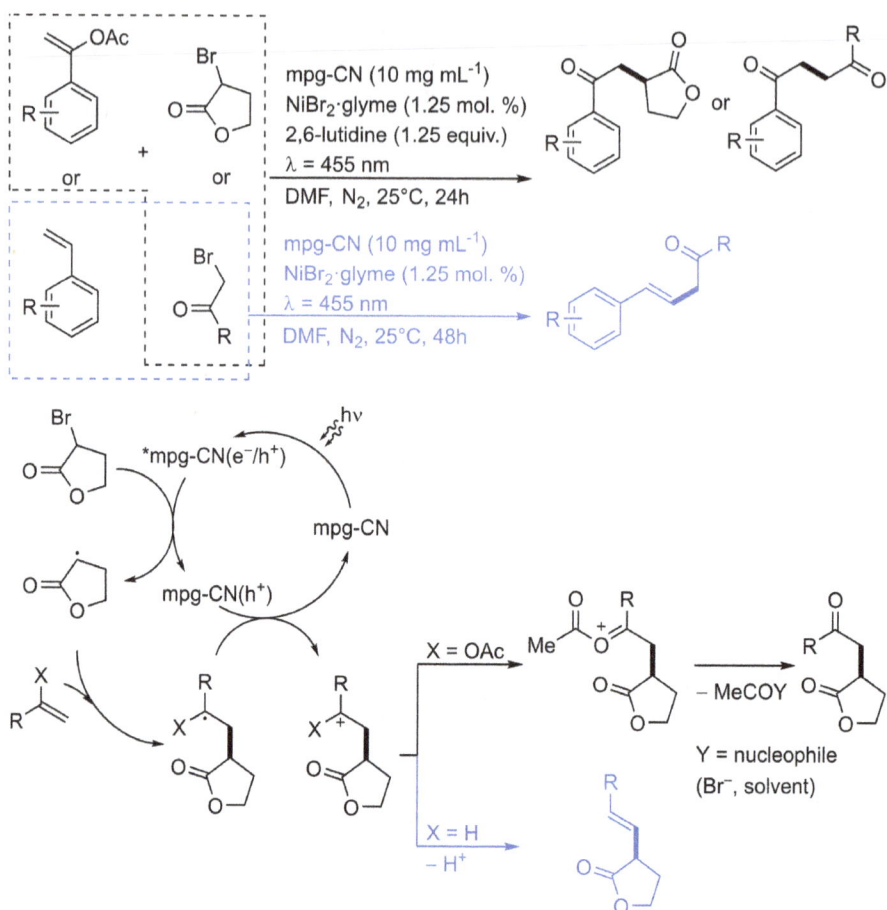

Fig. 8.46: C–C cross-coupling of alkylbromides and vinylacetates or styrene derivatives that yield 1,4-dicarbonyl compounds and Mizoroki-Heck coupling product, respectively.

8.3.7.4 Strategies to improve the reusability of carbon nitride-based metallaphotoredox catalysts

In examples discussed in Sections 8.3.7.1–8.3.7.3, carbon nitrides recovered after the photocatalytic reaction contain up to 12.6 wt% of Ni (Tab. 8.1).

The difference may be explained by a type of cross-coupling reaction – overall, greater amount of Ni is deposited in C–N cross-coupling, in which aliphatic amine is add as a base, but could also act as hydrogen donor, facilitating reduction of Ni(II) to Ni(0). However, the difference may also be explained by the structures of ionic carbon nitrides, such as CN-OA-m, used in C–N cross-coupling, and the covalent one, mpg-CN, used in C–C cross-couplings. The ionic nature of CN-OA-m is likely to facilitate the

Tab. 8.1: Amount of Ni (wt%) detected by ICP in carbon nitride materials recovered after the metallaphotoredox catalysis.

Entry	Type of cross-coupling	Coupling partners	Carbon nitride	Ni content (wt%)	Reference
1	C–N	Pyrrolidine, 1-bromo-4-fluorobenzene	CN-OA-m	12.6[a] 3.6[b]	[58]
2	C–O	Methyl 4-iodobenzoate, Boc-Pro-OH	CN-OA-m	1.4	[57]
3	C–C	Bis(chatecolatosilicate), 4-bromobenzotrifluoride	mpg-CN	0.146 ± 0.002	[65]
4	C–C	N,N-Dimethylacetamide, 4-bromobenzonitrile	mpg-CN	0.07 ± 0.013	[66]
5	C–C	1-(p-Tolyl) vinyl acetate, α-bromo-γ-butyrolactone	mpg-CN	0.0268 ± 0.00033	[67]

[a]Photocatalysis at 450 nm.
[b]Photocatalysis at 520 nm.

formation of a highly reactive coordinatively unsaturated species, such as 14e⁻ Ni-complex, while the surface chemistry of covalent mpg-CN is able to stabilize 16e⁻ Ni-species (Fig. 8.47), which are less prone to form nickel-black [67].

Except for few examples [67], the deposited Ni is catalytically inactive. Therefore, fresh Ni precatalyst is required in the next round of carbon nitride use. Several approaches have been proposed in dual Ni/carbon nitride photocatalysis to deposit catalytically active Ni species at the surface of the semiconductor, with the intent to improve its reusability.

Reischauer et al. grafted bipyridine ligands that are equipped with phosphonic acid groups to the surface of carbon nitride CN-OA-m, followed by the complexation of NiBr$_2 \cdot$3H$_2$O, which provides a material with 1 mol% NiBr$_2 \cdot$3H$_2$O and 3 mol% dpbpy (Fig. 8.48) [68]. The material retains its activity in C–S cross-coupling after ten cycles, without leaching of Ni and dpbpy ligand. However, this heterogeneous material is not applicable in C–O cross-coupling of carboxylic acids and arylhaldies, which is likely due to the detrimental effect of phosphonic groups. Optimization of ligand structure is thus necessary to expand the scope of reactions enabled by this catalyst.

The multitude of (sp^2)N atoms in carbon nitrides provides sites for immobilization of Ni atoms [69]. In other words, the carbon nitride framework can serve as a polydentate ligand in metallaphotoredox catalysis. C–O cross-coupling of electron-deficient arylbromides with aliphatic alcohols, which were also used as solvent, was accomplished by Ni-mpg-CN carrying 4.14 ± 0.99 wt% Ni (Fig. 8.49) [70]. Compared to C–O cross-coupling of arylhalides with aliphatic alcohols, mediated by carbon nitride CN-OA-m,

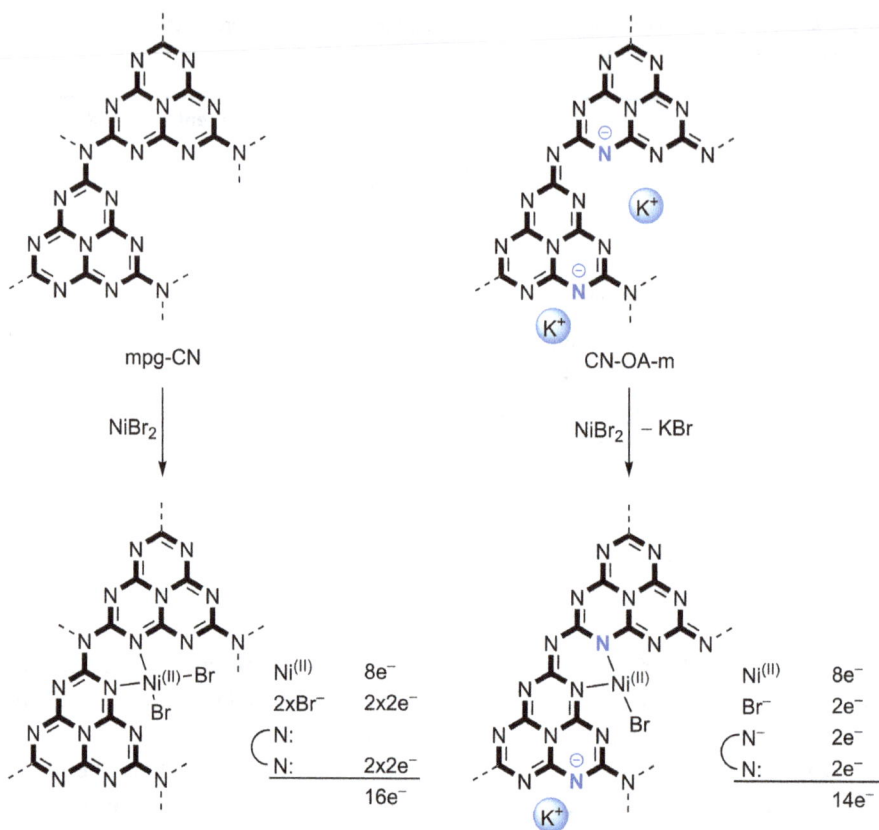

Fig. 8.47: Tentative schematic structures of 16e⁻ and 14e⁻ Ni-complexes with carbon nitride framework serving a polydentate ligand.

which requires up to 10 mol% of NiBr$_2$·glyme versus arylhalide (Fig. 8.36), more than 100 times lower amount of Ni is used in this case.

In a series of arylhalides of similar electronic structure, reactivity follows the order Ar–I > Ar–Br > Ar–Cl (Fig. 8.49). Electron-deficient arylbromides and iodides give ethers with higher yields, while electron-rich substrates are significantly less reactive, which is reflected in the longer time and incomplete conversion. Therefore, despite a higher degree of Ni utilization, compared to dual Ni/carbon nitride photocatalysis, in this example, the integrated Ni-mpg-CN photocatalysis does not allow solving the problem of low rate of OA for electron-rich arylhalides. However, imidazole and, to some extent, other nitrogen-containing heterocycles, when added to the reaction mixture, can serve as auxiliary ligands that are likely to facilitate switching of Ni between several oxidation states, and as a result, improve the performance of such integrated Ni-mpg-CN photocatalysts. Unlike the example shown in Fig. 8.36 [56], phenol gave a coupling product with 4-bromoacetophenone, with 20% yield. This example

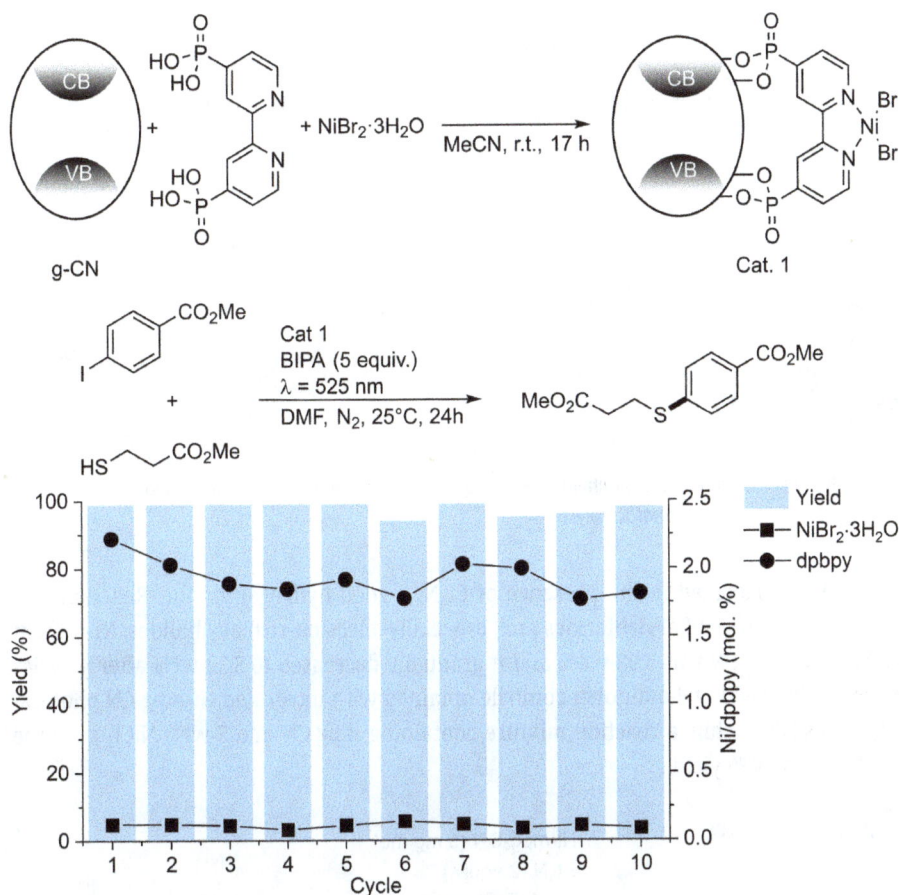

Fig. 8.48: Preparation of carbon nitride metallaphotoredox catalyst by grafting bipyridine ligands, and application in C–S cross-coupling.

points to higher activity of the photocatalytic system when Ni is immobilized directly at carbon nitride. In addition, further optimization of the carbon nitride structure for activating low-valent Ni species may enhance activity of the photocatalyst so that even extremely challenging coupling of electron-rich arylchlorides with weak nucleophiles could become possible.

Reisner et al. accomplished the synthesis of anilines from aryl halides and sodium azide using the integrated Ni-mpg-CN photocatalyst (Fig. 8.50) [71]. Reactivity pattern, which is similar to that discussed above, was observed for a series of arylhalides. Thus, electron-deficient arylbromides gave primary anilines in high yields. Although aryliodides are expected to undergo OA faster than arylbromides, the yield of the corresponding anilines was lower due to formation of the hydrocarbon via formal reduction of C–I bond. This process is mode facile compared to C–O cross-coupling (Fig. 8.49), as the

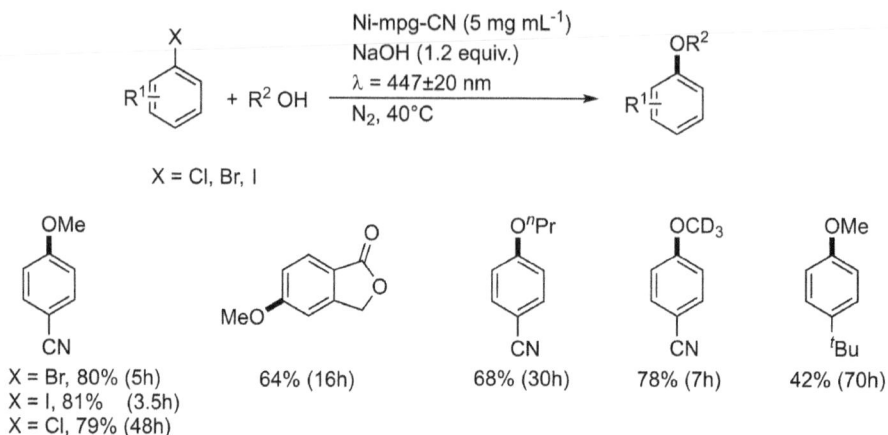

Fig. 8.49: C–O cross-coupling of arylhaldies and aliphatic alcohols, mediated by integrated heterogeneous Ni-mpg-CN photocatalyst.

reaction is performed in the presence of Et$_3$N, a good hydrogen donor. Reaction was sluggish in the case of arylchlorides and, especially, electron-rich arylhalides. Ni content in fresh Ni-mpg-CN was 6.98 wt.% and it gradually decreased to 5.39 wt% after 4 cycles of reuse. The yield of 4-aminobenzonitrile obtained with integrated Ni-mpg-CN photocatalyst was 88%, while a reaction mixture containing mpg-CN and 5 wt% NiCl$_2$ gave the product with 63% yield.

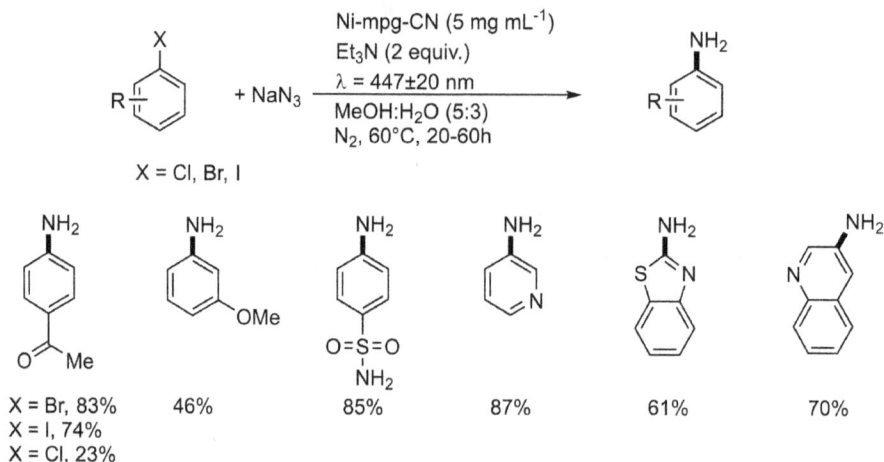

Fig. 8.50: Synthesis of primary anilines from arylhalides and sodium azide, mediated by Ni-mpg-CN.

8.4 Summary and outlook

With the shortage of raw materials, the transition from fossils-based economy to that based on renewables is unavoidable to keep the pace of progress. While synthesis of raw materials for the chemical industry from biomass still suffers from poor selectivity and therefore requires separation of the components from complex mixture, using sunlight as the energy source instead of combustibles to convert molecules via photochemistry is a more mature approach, at least in terms of selectivity and the scope of reactions. Solar energy is ultimately a nondepletable source in comparison to the lifespan of human civilization. More than a century ago, in 1912, Giacomo Ciamician, in his visionary work, emphasized that *"On the arid lands there will spring up industrial colonies without smoke and without smokestacks; forests of glass tubes will extend over the plants and glass buildings will rise everywhere; inside of these will take place the photochemical processes that hitherto have been the guarded secret of the plants, but that will have been mastered by human industry which will know how to make them bear even more abundant fruit than nature, for nature is not in a hurry and mankind is."* [72, 73], While photochemistry itself enables a range of chemical transformations, photocatalysis may direct the reaction according to a different pathway. In order to realize photocatalysis on a large industrial scale, we need a robust, functional and cheap photocatalyst. Robustness of graphitic carbon nitrides was highlighted at least several times in this book.

While the functionality of carbon nitrides in preparative organic chemistry is the primary topic of this chapter, readers are invited one more time to analyze carbon nitride photocatalysis that starts with oxidation of toluene (Fig. 8.51). In the presence of O_2 or S_8, carbon nitrides yield benzaldehyde and dibenzyldisulfide – the first-generation products of photocatalysis. These basic compounds may be employed directly as reagents in the follow-up photocatalytic transformations, such as oxidation of benzaldehyde to benzoic acid. However, using simple non-photocatalytic transformations, they may be converted into a series of substrates, enones, benzaldehyde oxime, acylhydrazones and thiouronium salt. Carbon nitride photocatalysis converts these substrates into the products of the second generation.

Provided that reagents are derived from renewable sources and considering the availability of precursors for carbon nitride synthesis: urea – for covalent carbon nitrides or urea and sea water (3.5 wt% NaCl solution) – for sodium poly(heptazine imide), this class of materials may indeed fulfill the prophesy of Giacomo Ciamician.

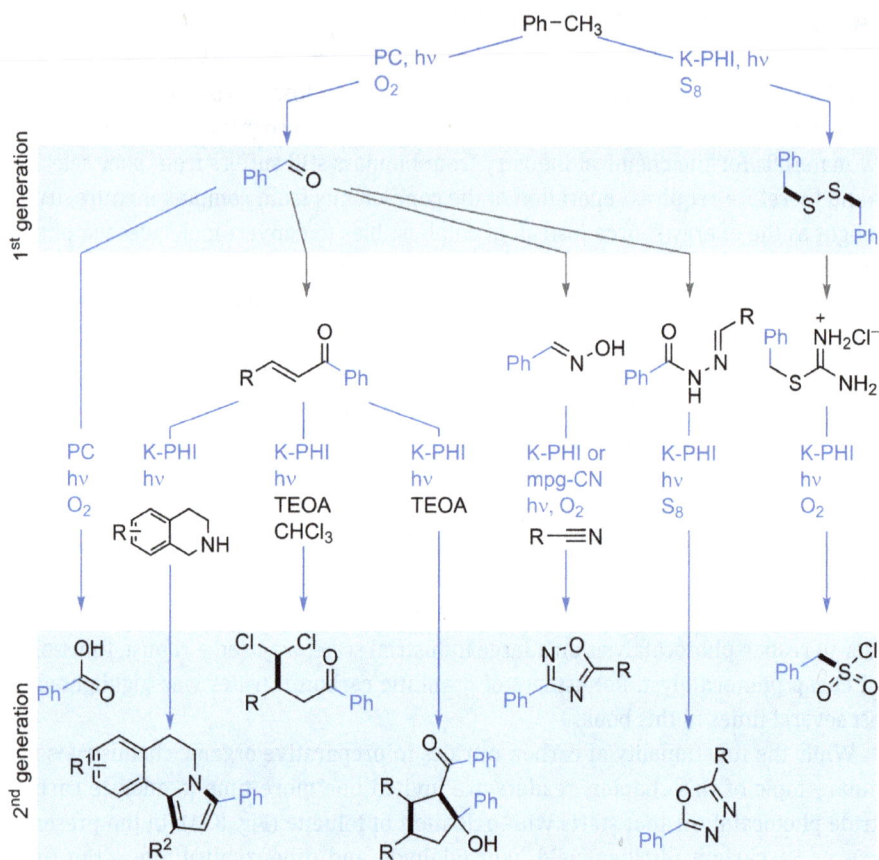

Fig. 8.51: Valorization of toluene into various organic compounds. Gray lines correspond to non-photocatalytic steps, blue – photocatalysis mediated by carbon nitrides.

References

[1]   Thermochemical data from, https://webbook.nist.gov/chemistry/.
[2]   Osterloh FE. Photocatalysis versus photosynthesis: A sensitivity analysis of devices for solar energy conversion and chemical transformations. ACS Energy Lett 2017, 2(2), 445–53.
[3]   It is assumed that HOMO potential is equal to the oxidation potential of a molecule, while its LUMO potential is equal to reduction potential.
[4]   Rehm D, Weller A. Kinetik und Mechanismus der Elektronübertragung bei der Fluoreszenzlöschung in Acetonitril. Ber Bunsen-Ges Phys Chem 1969, 73(8-9), 834–39.
[5]   Romero NA, Nicewicz DA. Organic photoredox catalysis. Chem Rev 2016, 116(17), 10075–166.
[6]   Liu P, Liu W, Li C-J. Catalyst-free and redox-neutral innate trifluoromethylation and alkylation of aromatics enabled by light. J Am Chem Soc 2017, 139(40), 14315–21.

[7]   Baar M, Blechert S. Graphitic carbon nitride polymer as a recyclable photoredox catalyst for fluoroalkylation of arenes. Chem – Eur J 2015, 21(2), 526–30.
[8]   Ghosh I, Khamrai J, Savateev A, Shlapakov N, Antonietti M, König B. Organic semiconductor photocatalyst can bifunctionalize arenes and heteroarenes. Science 2019, 365(6451), 360–66.
[9]   Roth HG, Romero NA, Nicewicz DA. Experimental and calculated electrochemical potentials of common organic molecules for applications to single-electron redox chemistry. Synlett 2016, 27(05), 714–23.
[10] Pavlishchuk VV, Addison AW. Conversion constants for redox potentials measured versus different reference electrodes in acetonitrile solutions at 25°C. Inorg Chim Acta 2000, 298(1), 97–102.
[11] Savateev A, Kurpil B, Mishchenko A, Zhang G, Antonietti M. A "waiting" carbon nitride radical anion: A charge storage material and key intermediate in direct C–H thiolation of methylarenes using elemental sulfur as the "S"-source. Chem Sci 2018, 9(14), 3584–91.
[12] Cai Y, Tang Y, Fan L, Lefebvre Q, Hou H, Rueping M. Heterogeneous visible-light photoredox catalysis with graphitic carbon nitride for α-aminoalkyl radical additions, allylations, and heteroarylations. ACS Catal 2018, 8(10), 9471–76.
[13] He Y, Dan X, Tang Y, Yang Q, Wang W, Cai Y. Semi-heterogeneous photocatalytic fluoroalkylation-distal functionalization of unactivated alkenes with R_FSO_2Na under air atmosphere. Green Chem 2021, 23(23), 9577–82.
[14] Pan G, Yang Q, Wang W, Tang Y, Cai Y. Heterogeneous photocatalytic cyanomethylarylation of alkenes with acetonitrile: Synthesis of diverse nitrogenous heterocyclic compounds. Beilstein J Org Chem 2021, 17, 1171–80.
[15] Joule JA, Mills K. Heterocyclic chemistry. John Wiley & Sons, 2010.
[16] Rabe EJ, Corp KL, Huang X, Ehrmaier J, Flores RG, Estes SL, et al. Barrierless heptazine-driven excited state proton-coupled electron transfer: Implications for controlling photochemistry of carbon nitrides and aza-arenes. J Phys Chem C 2019, 123(49), 29580–88.
[17] Warren JJ, Tronic TA, Mayer JM. Thermochemistry of proton-coupled electron transfer reagents and its implications. Chem Rev 2010, 110(12), 6961–7001.
[18] Song T, Zhou B, Peng G-W, Zhang Q-B, Wu L-Z, Liu Q, et al. Aerobic oxidative coupling of resveratrol and its analogues by visible light using mesoporous graphitic carbon nitride (mpg-C_3N_4) as a bioinspired catalyst. Chem – Eur J 2014, 20(3), 678–82.
[19] Geng P, Tang Y, Pan G, Wang W, Hu J, Cai Y. A g-C_3N_4-based heterogeneous photocatalyst for visible light mediated aerobic benzylic C–H oxygenations. Green Chem 2019, 21(22), 6116–22.
[20] Vijeta A, Reisner E. Carbon nitride as a heterogeneous visible-light photocatalyst for the Minisci reaction and coupling to H_2 production. Chem Commun 2019, 55(93), 14007–10.
[21] Poletti L, Ragno D, Bortolini O, Presini F, Pesciaioli F, Carli S, et al. Photoredox cross-dehydrogenative coupling of N-aryl glycines mediated by mesoporous graphitic carbon nitride: An environmentally friendly approach to the synthesis of non-proteinogenic α-amino acids (NPAAs) decorated with indoles. J Org Chem 2022, 87(12), 7826–37.
[22] Savateev A, Pronkin S, Epping JD, Willinger MG, Wolff C, Neher D, et al. Potassium poly(heptazine imides) from aminotetrazoles: Shifting band gaps of carbon nitride-like materials for more efficient solar hydrogen and oxygen evolution. ChemCatChem 2017, 9(1), 167–74.
[23] Mazzanti S, Schritt C, ten Brummelhuis K, Antonietti M, Savateev A. Multisite PCET with photocharged carbon nitride in dark. Exploration 2021, 1(3), 20210063.
[24] Savateev A, Antonietti M, Ionic Carbon Nitrides in Solar Hydrogen Production and Organic Synthesis: Exciting Chemistry and Economic Advantages, ChemCatChem 2019, 11(24), 6166–76.
[25] Kurpil B, Markushyna Y, Savateev A. Visible-light-driven reductive (cyclo)dimerization of chalcones over heterogeneous carbon nitride photocatalyst. ACS Catal 2019, 9(2), 1531–38.

[26] Kurpil B, Otte K, Mishchenko A, Lamagni P, Lipiński W, Lock N, et al. Carbon nitride photocatalyzes regioselective aminium radical addition to the carbonyl bond and yields N-fused pyrroles. Nat Commun 2019, 10(1), 945.

[27] Mazzanti S, Kurpil B, Pieber B, Antonietti M, Savateev A. Dichloromethylation of enones by carbon nitride photocatalysis. Nat Commun 2020, 11(1), 1387.

[28] Yang Q, Pan G, Wei J, Wang W, Tang Y, Cai Y. Remarkable activity of potassium-modified carbon nitride for heterogeneous photocatalytic decarboxylative alkyl/acyl radical addition and reductive dimerization of para-quinone methides. ACS Sustainable Chem Eng 2021, 9(5), 2367–77.

[29] Castillo-Lora J, Delley MF, Laga SM, Mayer JM. Two-electron–two-proton transfer from colloidal ZnO and TiO_2 nanoparticles to molecular substrates. J Phys Chem Lett 2020, 11(18), 7687–91.

[30] Savateev O. Photocharging of semiconductor materials: Database, quantitative data analysis, and application in organic synthesis. Adv Energy Mater 2022, 12(21), 2200352.

[31] Pei L, Tan H, Liu M, Wang R, Gu X, Ke X, et al. Hydroxyl-group-modified polymeric carbon nitride with the highly selective hydrogenation of nitrobenzene to N-phenylhydroxylamine under visible light. Green Chem 2021, 23(10), 3612–22.

[32] Dai Y, Li C, Shen Y, Lim T, Xu J, Li Y, et al. Light-tuned selective photosynthesis of azo- and azoxy-aromatics using graphitic C_3N_4. Nat Commun 2018, 9(1), 60.

[33] Markushyna Y, Völkel A, Savateev A, Antonietti M, Filonenko S. One-pot photocalalytic reductive formylation of nitroarenes *via* multielectron transfer by carbon nitride in functional eutectic medium. J Catal 2019, 380, 186–94.

[34] Antonietti M, Oschatz M. The concept of "Noble, heteroatom-doped carbons," their directed synthesis by electronic band control of carbonization, and applications in catalysis and energy materials. Adv Mater 2018, 30(21), 1706836.

[35] Fellinger T-P, Thomas A, Yuan J, Antonietti M. 25th anniversary article: "Cooking carbon with salt": Carbon materials and carbonaceous frameworks from ionic liquids and poly(ionic liquid)s. Adv Mater 2013, 25(41), 5838–55.

[36] Bach RD, Ayala PY, Schlegel HB. A reassessment of the bond dissociation energies of peroxides. An Ab Initio Study J Am Chem Soc 1996, 118(50), 12758–65.

[37] Savateev A, Dontsova D, Kurpil B, Antonietti M. Highly crystalline poly(heptazine imides) by mechanochemical synthesis for photooxidation of various organic substrates using an intriguing electron acceptor – Elemental sulfur. J Catal 2017, 350, 203–11.

[38] Kurpil B, Otte K, Antonietti M, Savateev A. Photooxidation of N-acylhydrazones to 1,3,4-oxadiazoles catalyzed by heterogeneous visible-light-active carbon nitride semiconductor. Appl Catal: B 2018, 228, 97–102.

[39] Su F, Mathew SC, Möhlmann L, Antonietti M, Wang X, Blechert S. Aerobic oxidative coupling of amines by carbon nitride photocatalysis with visible light. Angew Chem Int Ed 2011, 50(3), 657–60.

[40] Kurpil B, Kumru B, Heil T, Antonietti M, Savateev A. Carbon nitride creates thioamides in high yields by the photocatalytic Kindler reaction. Green Chem 2018, 20(4), 838–42.

[41] Zhang Z, Xu Y, Zhang Q, Fang S, Sun H, Ou W, et al. Semi-heterogeneous photo-Cu-dual-catalytic cross-coupling reactions using polymeric carbon nitrides. Sci Bull 2022, 67(1), 71–78.

[42] Savateev A, Tarakina NV, Strauss V, Hussain T, ten Brummelhuis K, Sánchez Vadillo JM, et al. Potassium poly(heptazine imide): Transition metal-free solid-state triplet sensitizer in cascade energy transfer and [3+2]-cycloadditions. Angew Chem Int Ed 2020, 59(35), 15061–68.

[43] Daimon T, Hirakawa T, Kitazawa M, Suetake J, Nosaka Y. Formation of singlet molecular oxygen associated with the formation of superoxide radicals in aqueous suspensions of TiO_2 photocatalysts. Appl Catal A 2008, 340(2), 169–75.

[44] Galushchinskiy A, ten Brummelhuis K, Antonietti M, Savateev A. Insights into the mechanism of energy transfer with poly(heptazine imide)s in a deoximation reaction. ChemPhotoChem 2021, 5(11), 1020–25.

[45] Rieth AJ, Qin Y, Martindale BCM, Nocera DG. Long-lived triplet excited state in a heterogeneous modified carbon nitride photocatalyst. J Am Chem Soc 2021, 143(12), 4646–52.

[46] Protti S, Ravelli D, Fagnoni M. Wavelength dependence and wavelength selectivity in photochemical reactions. Photochem Photobiol Sci 2019, 18(9), 2094–101.

[47] Markushyna Y, Savateev A. Light as a tool in organic photocatalysis: Multi-photon excitation and chromoselective reactions. Eur J Org Chem 2022, 2022(24), e202200026.

[48] Ehrmaier J, Rabe EJ, Pristash SR, Corp KL, Schlenker CW, Sobolewski AL, et al. Singlet–triplet inversion in heptazine and in polymeric carbon nitrides. J Phys Chem A 2019, 123(38), 8099–108.

[49] Zhang G, Savateev A, Zhao Y, Li L, Antonietti M. Advancing the n → π* electron transition of carbon nitride nanotubes for H_2 photosynthesis. J Mater Chem A 2017, 5(25), 12723–28.

[50] Schmermund L, Reischauer S, Bierbaumer S, Winkler CK, Diaz-Rodriguez A, Edwards LJ, et al. Chromoselective photocatalysis enables stereocomplementary biocatalytic pathways. Angew Chem Int Ed 2021, 60(13), 6965–69.

[51] Markushyna Y, Schüßlbauer CM, Ullrich T, Guldi DM, Antonietti M, Savateev A. Chromoselective synthesis of sulfonyl chlorides and sulfonamides with potassium poly(heptazine imide) photocatalyst. Angew Chem Int Ed 2021, 60(37), 20543–50.

[52] Markushyna Y, Antonietti M, Savateev A. Synthesis of sulfonyl chlorides from aryldiazonium salts mediated by a heterogeneous potassium poly(heptazine imide) photocatalyst. ACS Org Inorg Au 2022, 2(2), 153–58.

[53] Zhu C, Yue H, Jia J, Rueping M. Nickel-catalyzed C-heteroatom cross-coupling reactions under mild conditions *via* facilitated reductive elimination. Angew Chem Int Ed 2021, 60(33), 17810–31.

[54] Qin Y, Martindale BCM, Sun R, Rieth AJ, Nocera DG. Solar-driven tandem photoredox nickel-catalysed cross-coupling using modified carbon nitride. Chem Sci 2020, 11(28), 7456–61.

[55] Zhu C, Yue H, Nikolaienko P, Rueping M. Merging electrolysis and nickel catalysis in redox neutral cross-coupling reactions: Experiment and computation for electrochemically induced C–P and C–Se bonds formation. CCS Chem 2020, 2(2), 179–90.

[56] Cavedon C, Madani A, Seeberger PH, Pieber B. Semiheterogeneous dual nickel/photocatalytic (thio) etherification using carbon nitrides. Org Lett 2019, 21(13), 5331–34.

[57] Pieber B, Malik JA, Cavedon C, Gisbertz S, Savateev A, Cruz D, et al. Semi-heterogeneous dual nickel/photocatalysis using carbon nitrides: Esterification of carboxylic acids with aryl halides. Angew Chem Int Ed 2019, 58(28), 9575–80.

[58] Gisbertz S, Reischauer S, Pieber B. Overcoming limitations in dual photoredox/nickel-catalysed C–N cross-couplings due to catalyst deactivation. Nat Catal 2020, 3(8), 611–20.

[59] Milligan JA, Phelan JP, Badir SO, Molander GA. Angew Chem Int Ed 2019, 58, 6152–63.

[60] Zuo Z, Ahneman DT, Chu L, Terrett JA, Doyle AG, MacMillan DWC. Merging photoredox with nickel catalysis: Coupling of α-carboxyl sp^3-carbons with aryl halides. Science 2014, 345(6195), 437–40.

[61] Corcé V, Chamoreau L-M, Derat E, Goddard J-P, Ollivier C, Fensterbank L. Silicates as latent alkyl radical precursors: Visible-light photocatalytic oxidation of hypervalent bis-catecholato silicon compounds. Angew Chem Int Ed 2015, 54(39), 11414–18.

[62] de Assis FF, Huang X, Akiyama M, Pilli RA, Meggers E. Visible-light-activated catalytic enantioselective β-alkylation of α,β-unsaturated 2-acyl imidazoles using Hantzsch esters as radical reservoirs. J Org Chem 2018, 83(18), 10922–32.

[63] Knauber T, Chandrasekaran R, Tucker JW, Chen JM, Reese M, Rankic DA, et al. Ru/Ni dual catalytic desulfinative photoredox Csp^2–Csp^3 cross-coupling of alkyl sulfinate salts and aryl halides. Org Lett 2017, 19(24), 6566–69.

[64] Khamrai J, Ghosh I, Savateev A, Antonietti M, König B. Photo-Ni-dual-catalytic C(sp^2)–C(sp^3) cross-coupling reactions with mesoporous graphitic carbon nitride as a heterogeneous organic semiconductor photocatalyst. ACS Catal 2020, 10(6), 3526–32.

[65] Schirmer TE, Abdellaoui M, Savateev A, Ollivier C, Antonietti M, Fensterbank L, et al. Mesoporous graphitic carbon nitride as a heterogeneous organic photocatalyst in the dual catalytic arylation of alkyl bis(catecholato)silicates. Org Lett 2022, 24(13), 2483–87.

[66] Das S, Murugesan K, Villegas Rodríguez GJ, Kaur J, Barham JP, Savateev A, et al. Photocatalytic (het) arylation of C(sp^3)–H bonds with carbon nitride. ACS Catal 2021, 11(3), 1593–603.

[67] Khamrai J, Das S, Savateev A, Antonietti M, König B. Mizoroki–Heck type reactions and synthesis of 1,4-dicarbonyl compounds by heterogeneous organic semiconductor photocatalysis. Green Chem 2021, 23(5), 2017–24.

[68] Reischauer S, Pieber B. Recyclable, bifunctional metallaphotocatalysts for C–S cross-coupling reactions. ChemPhotoChem 2021, 5(8), 716–20.

[69] Zhao X, Deng C, Meng D, Ji H, Chen C, Song W, et al. Nickel-coordinated carbon nitride as a metallaphotoredox platform for the cross-coupling of aryl halides with alcohols. ACS Catal 2020, 10(24), 15178–85.

[70] Vijeta A, Casadevall C, Roy S, Reisner E. Visible-light promoted C–O bond formation with an integrated carbon nitride–nickel heterogeneous photocatalyst. Angew Chem Int Ed 2021, 60(15), 8494–99.

[71] Vijeta A, Casadevall C, Reisner E. An integrated carbon nitride-nickel photocatalyst for the amination of aryl halides using sodium azide. Angew Chem Int Ed 2022, 134, e202203176.

[72] Ciamician G. The photochemistry of the future. Science 1912, 36(926), 385–94.

[73] Albini A, Fagnoni M. 1908: Giacomo Ciamician and the concept of green chemistry. ChemSusChem 2008, 1(1-2), 63–66.

Filip Podjaski, Vincent W.-h. Lau and Bettina V. Lotsch*

Chapter 9
Photocharging carbon nitrides: from fundamental properties to applications combining solar energy conversion and storage

9.1 Introduction

Among the photoresponsive materials of the carbon nitride family, only the two-dimensional PHI (poly(heptazine imide)) has, to date, been reported to exhibit the exceptional ability to stabilize photogenerated electrons, following reductive quenching [1]. Since these electrons remain trapped well after irradiation has ceased and are highly reductive, this phenomenon opens a new paradigm in energy storage, particularly for time-delayed solar fuel production following the methodology illustrated in Fig. 9.1. Here, band gap irradiation of M-PHI (M being the respective countercation) generates electron–hole pairs [2–4], of which the photoholes are quenched in the presence of a suitable reducing agent. The remaining photo-electrons persist as long-lived anion radicals akin to small polarons and are macroscopically evident by a color change from the yellow material to a blue hue in aqueous conditions. Provided that the surrounding is free of electron acceptors such as dissolved oxygen gas, the trapped charges can persist for days (albeit with progressively decreasing concentration) and may later be applied to initiate reduction reactions, such as reduction of protons to H_2 as a chemical fuel. Various labels have been used to describe this phenomenon: the initial reports coined the terms "trapped electrons" and "dark photocatalysis," alluding to the light-dependent and light-independent (dark) reactions of natural photosynthesis in terms of temporally separating the two half-reactions. Others used the term "illumination-driven electron accumulation in semiconductors" (abbreviated to IDEAS) to explicitly describe the photocarrier dynamics (see later sections), or phrases such as "waiting

Note: Filip Podjaski and Vincent W.-h. Lau are contributed equally.

*Corresponding author: Bettina V. Lotsch,** Nanochemistry Department, Max Planck Institute for Solid State Research, Heisenbergstrasse 1, 70569 Stuttgart, Germany; Department of Chemistry, University of Munich (LMU), Butenandtstrasse 5-13, 81377 München, Germany, e-mail: b.lotsch@fkf.mpg.de
Filip Podjaski, Nanochemistry Department, Max Planck Institute for Solid State Research, Heisenbergstrasse 1, 70569 Stuttgart, Germany; Department of Chemistry, Imperial College London, 82 Wood Lane, W12 0BZ London, United Kingdom
Vincent W.-h. Lau, Department of Chemistry, National Cheng Kung University, Tainan 701, Taiwan

https://doi.org/10.1515/9783110746976-009

radical anion" to reference the delayed usage of the electrons after photogeneration [5–7]. Irrespective of the exact nomenclature, as a phenomenon that has been observed only quite recently (relative to the history of the carbon nitride family of materials), there is still much to explore regarding the origin and mechanism of light-induced charge trapping in PHI-type carbon nitrides. In particular, the close coupling between photogenerated electronic and ionic charge carriers, as observed in PHI and discussed below, opens up the essentially uncharted territory of optoionic materials, which bridge the gap between solar energy conversion and energy storage [8].

A better understanding of how this stabilization effect can be controlled through structural tailoring of PHI or optimizing its operational environment may ultimately broaden its scope of applicability in solar energy conversion and storage. Hence, this chapter will first summarize our present understanding of the structural and spectroscopic characteristics of PHI that enable this phenomenon, prior to presenting potential applications that deliberately exploit the long lifetime and reductive nature of the radicals.

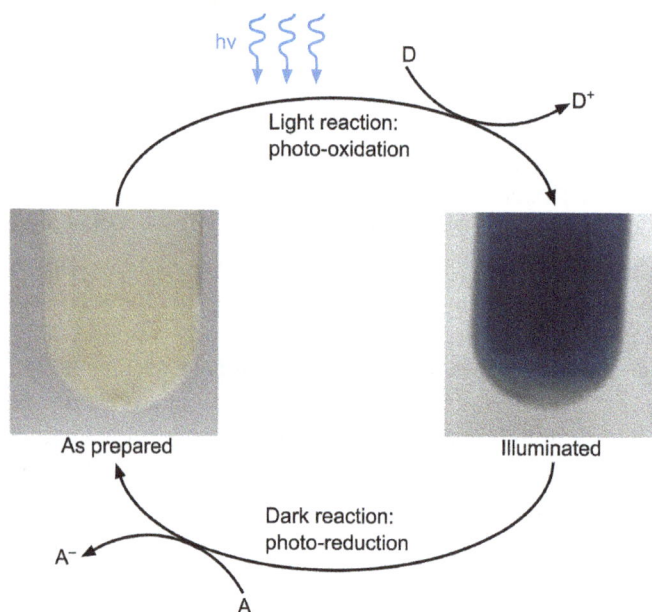

Fig. 9.1: Schematic of the electron storage process in a PHI suspension: irradiation of the yellow PHI with a reducing agent (i.e., electron donor, represented by D in the figure) initiates a photooxidation reaction (D → D⁺), yielding the color change to blue as characteristic of trapped electrons, which can then be used to drive photoreduction of a specific electron acceptor (represented in the figure as A → A⁻), well after irradiation has ceased (photographs reproduced from reference [3] with permission from the American Chemical Society).

9.2 Fundamentals of charge trapping in PHI

9.2.1 Structural features of PHI and general properties related to electron stabilization

PHI is synthesized under ionothermal conditions, i.e., in salt melts rather than in solid–gas reactions known to form 1D carbon nitrides such as Melem oligomers or Melon [3]. While PHI has been synthesized using various precursors (e.g., Melon, di-cyandiamide, and 5-aminotetrazole) in different types of mostly potassium-based salt melts (e.g., chloride, hydroxide and thiocyanate), light-induced electron trapping is observed, irrespective of the synthesis route, suggesting that this phenomenon is fairly robust and tolerant to small structural variations [3, 7, 9, 10]. Nevertheless, we will show below that the nature of the (hydrated) metal ions residing in the pores of PHI have a profound influence on the charge stabilizing ability of PHI. The local and long-range structure of PHI as depicted in Fig. 9.2 was solved from X-ray scattering and electron diffraction methods, and corroborated by infrared (IR) and solid-state nuclear magnetic resonance (NMR) spectroscopies [3]. PHI is distinct from the archetype carbon nitride, Melon, in that it is a truly two-dimensional (2D) material analogous to poly(triazine imide) (PTI), but comprising heptazine building blocks. Each heptazine is connected to three neighboring heptazine units via the three imide groups on its 2-, 5-,8-positions, forming a trigonal arrangement of heptazine units in the 2D backbone. Furthermore, PHI possesses different anionic functional groups; these include the imides bridging two heptazine units, and cyanamide groups that exist as crystal termination or as defects in its interior, as evidenced by its spectroscopic signatures (IR $\approx 2,200$ cm^{-1} and a ^{13}C solid state MAS-NMR signal at ≈ 120 ppm) [3, 11]. These anionic groups are charge-balanced by hydrated potassium or other pore fitting cations, which originate from the synthesis salt melt and reside within the PHI layers inside the pores generated by the trigonal arrangement of heptazine units. The cations in its interior can be exchanged when PHI is immersed in an aqueous salt solution, provided that the ions have the hydrodynamic radius to fit into the heptazine network [4, 12, 13]. The cations are hydrated and displaced towards the carbon nitride backbone rather than residing in the middle of the pores, thus suggesting the formation of ion-dipole interactions with the nitrogen atoms of the heptazine within the plane as well as in neighboring layers, additional to the ionic and hydrogen bonding interactions with the imide and cyanamide groups. Several sites within the pores are occupied by ca. 7 water molecules, which are largely disordered. As a result, interlayer stacking has contributions from both ionic, hydrogen bonding and π-π interactions of the aromatic subunits, which rationalizes both the structure-directing influence of the hydrated cations in terms of stacking polytypes and stacking faults, and also the narrower interlayer separation as compared to the non-ionic Melon, as evidenced by their diffraction patterns.

Prior to discussing the origin of the electron stabilization in relation to PHI's structure, some general properties of these photogenerated and stabilized charges will be

a)

b)

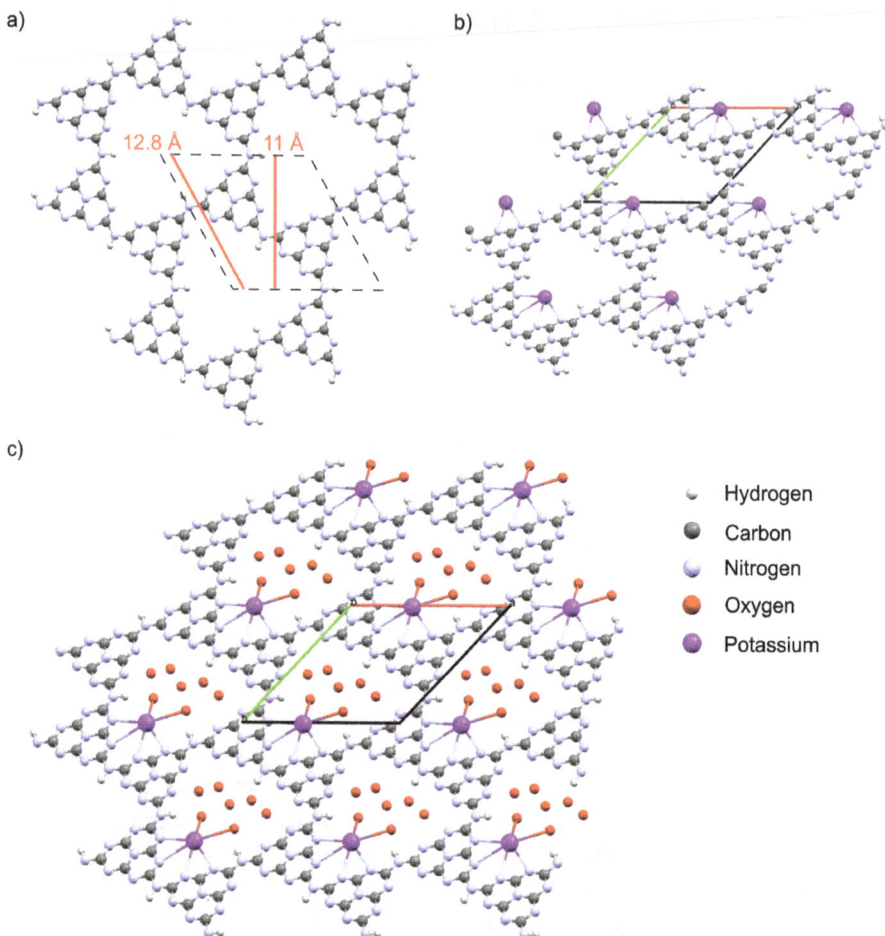

c)

- Hydrogen
- Carbon
- Nitrogen
- Oxygen
- Potassium

Fig. 9.2: Structure models of (a) H-PHI where the bridging imides are protonated. (b) K-PHI where water molecules incorporated in the pores have been omitted for clarity. Legends for models: dark gray – carbon; blue – nitrogen; magenta – potassium. (c) Structure model with the positions of K^+ and water (shown without hydrogen), refined by pair-distribution function analysis using the structural model obtained by Rietveld refinement of the K-PHI diffraction data (reproduced from reference [3] with permission from the American Chemical Society).

described. These long-lived radicals are formed upon illumination in the presence of a suitable reducing agent, and are macroscopically visible in terms of material color change, from PHI's native yellow to a bluish hue due to a broad absorption feature centered at approximately 680 nm in its UV–Vis absorbance or reflection spectrum. In its W-band CW electron paramagnetic resonance spectrum, one main signal with a g-value near 2.003 is indicative of heptazinyl-centered π-radicals; anisotropic broadening and a distribution of relaxation times suggests the co-existence of different π-radicals with slight structural differences (i.e., delocalization, size of the framework, defect density, etc.) owing to the structural heterogeneity of the sample [2]. From an electrochemical perspective, the trapped electrons have a redox potential of -600 mV vs. Ag/AgCl or more negative in neutral conditions, as measured by open-circuit photopotential (OCP) and verified by the pH-independent redox indicator, methyl viologen (MV^{2+}, $E° = -430$ mV vs. NHE, normal hydrogen electrode), demonstrating that they are sufficiently reductive to initiate a number of reactions including the hydrogen evolution reaction (HER) [2, 13]. PHI's capacity to store these electrons can also be quantified by titration with a suitable electron acceptor. Quantification of the amount of H_2 evolved (reduction of H^+ via platinum nanoparticles as cocatalyst) or amount of methyl viologen MV^+ (reduction of MV^{2+}) suggests roughly one electron is stored over every 4–9 heptazine units [2, 5]. This broad range may suggest that the efficacy of the stabilization phenomenon is sensitive to various factors, including the properties of the PHI itself as well as its environment, as detailed below.

The current understanding attributes this stabilization phenomenon to PHI's ionic composition, which provides the charge-screening via the hydrated cations to effectively stabilize the photo-electrons, presumably in conjunction with the (partial) conjugation of the heptazine units in an extended 2D arrangement. The evidence for the role of the ionic components is that this stabilization phenomenon is largely absent when the potassium in PHI is cation-exchanged for protons, but can manifest when the H^+-PHI is immersed in a (K^+-containing) electrolyte, upon which the K-PHI structure is restored through H/K exchange [3]. The molecular cation $[NH_4]^+$, as well as other pore-fitting ions also appear to induce this stabilization phenomenon [4, 14], as long as the cations have a smaller hydrodynamic radius to fit into the 1D channel pores of M-PHI. Moreover, PHI has been reported to exhibit different UV–Vis absorbance and photoluminescence signatures and, in turn, photocatalytic activity when immersed in different electrolytes [15]. Here, K^+ and Rb^+ (and increasing concentrations thereof) were most effective in increasing the absorbance and photocatalytic activity while suppressing photoluminescence, though the trends relating to the anion of the electrolyte were less clear. It should be noted that, in this study, an electrolyte dependency for photocatalytic activity was also observed for Melon-type carbon nitride but without as drastic differences as observed with PHI, thus suggesting an effect of the ions on the (photo) carrier dynamics.

While the presence of ionic species is vital for the stabilization of the photogenerated electrons, the mobility of the ionic charges also appears to play a key role [4]. In fact, the alkali / alkaline earth metal cations within PHI were found to be mobile within the pores, with their local mobility and long-range conductivity increasing with increasing relative humidity, up to $\geq 10^{-6}$ S cm^{-1} for Na-PHI. The observed correlation between ion mobility, charge carrier lifetimes as probed by transient photoluminescence spectroscopy (on the ps-ns timescale) and photocatalytic activity, suggests that cation-mediated dielectric screening of the photogenerated electrons through ion rearrangement on both short and long (beyond microseconds) timescales is relevant for photocatalysis [4, 16–18]. In line with these findings, divalent or large ions that do not fit into

PHI's pores lead to a decrease in photocatalytic activity for HER [4]. While such light-induced charge carrier stabilization through ionic rearrangement is difficult to track directly via time-resolved photoluminescence (PL) or in transient absorption spectroscopy (TAS) – topics that will be discussed in the next section – the coupling of optoelectronic and ionic charge carriers in PHI suggests that the dielectric and dynamic nature of the carbon nitride "lattice" (including the pore content), plays a much more fundamental role in photocatalysis than previously thought.

9.2.2 Spectroscopic characterization of PHI

To understand how these light-induced trapped electrons are formed and stabilized, various research groups have applied different types of spectroscopies including time resolved methods (e.g., TAS and time-resolved PL). Note that, although the electron stabilization phenomenon has been observed for PHI synthesized by various routes, direct comparison of spectra from different publications may not be possible due to minute structural differences (e.g., type and density of structural defects). The subsequent discussions will, thus, focus primarily on key spectral features. In the steady state (Fig. 9.3a), the UV–Vis absorption of PHI has an absorption threshold of around 2.64–2.76 eV (450–470 nm) with a tail or a shoulder band at higher wavelengths, and a photoluminescence band at around 2.45–2.75 (450–505 nm) [9, 19]. In Fig. 9.3b for spectra collected at 1–10 ps and 1,000 ps after the pump pulse, photo-excitation leads to two features within several picoseconds, as seen in a photo-induced absorption band centered at around 680 nm and a broad absorption extending into the infrared (>1,100 nm). These two features are discussed separately, below.

The feature at 680 nm was assigned to electrons accumulated in PHI, and the decay kinetics of this signal has a strong dependence on the presence of a reducing agent. As seen in Fig. 9.3c, upon addition of the donor 4-methylbenzyl alcohol, its decay accelerates from 3 ns without the donor to 10 ps (lifetime given here refers to when the signal intensity is reduced by 50%). This accelerated decay was attributed to electrons accumulation in the material (considered as n-doping by the authors), which are thought to recombine through a charge trapping/de-trapping process [20]. However, TAA (transient absorption anisotropy) spectroscopy shows that dipole moment orientation, which already is randomized with a decay time of 1.1 ps, has no dependency on electron donor up to the maximum measurement time of 2 ns. The initial rapid orientation loss of the dipole moment was attributed to either the cation responding to the light-induced charge density shift or to fast structural reorientation of surface functional groups, the cyanamide and/or oxygen-bearing groups [21]. Regarding the absence of donor dependency, the authors of this study suggest that the electrons and/or the charge in the reduced functional groups (reduced by the electron donor) are localized to within individual chromophoric units, without diffusion into neighboring heptazines, at least within the first 100 ps. Despite the accelerated decay,

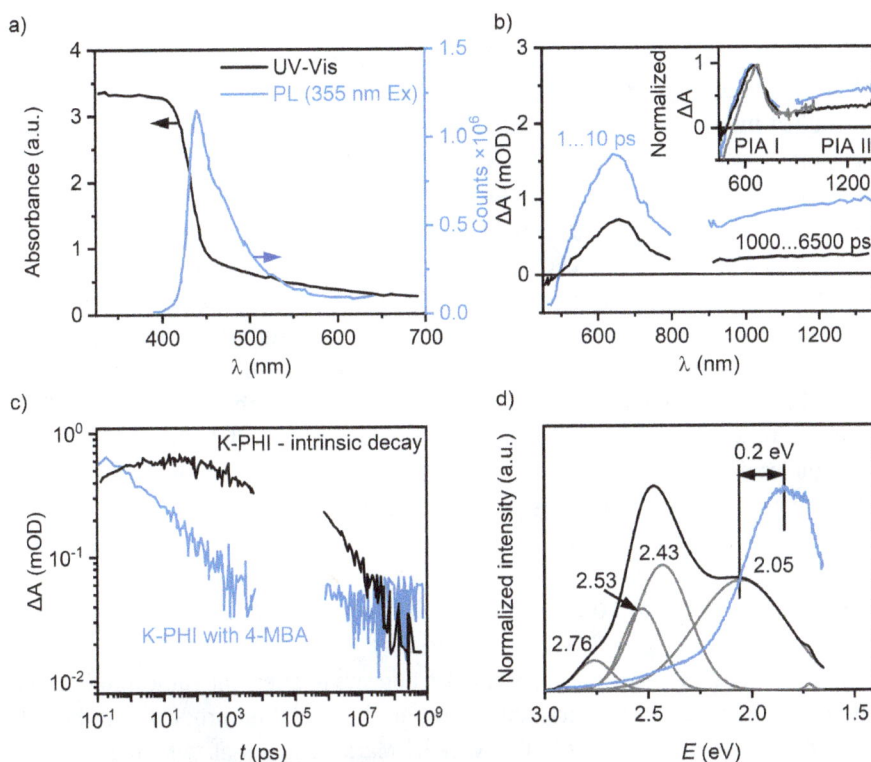

Fig. 9.3: (a) Steady-state solid UV–Vis and photoluminescence spectra of PHI. (b) Transient absorption spectra averaged in the 1–10 ps and 1,000–6,500 ps timescale, with the corresponding normalized spectra shown in the inset. (c) Evolution of the 680 nm signal seen in panel 'b' with time as dependent on the presence of the donor, 4-methylbenzyl alcohol (4-MBA) leading to charge accumulation [3]. (d) Deconvoluted fluorescence spectrum in black and the phosphorescence spectrum in the blue plot (10 ms delay), both using 360 nm excitation (reproduced from reference [20] (panels (a)–(c)) and reference [19] (panel (d)) with permission from the American Chemical Society and Wiley VCH).

TAS shows that around 10% of the initially photogenerated electrons survive to 6 ns and beyond to millisecond scale, presumably resulting in the blue radicals macroscopically observed [20]. Measurements from the TAA study show a similar result, where significant randomization of the dipole moment in the 100s ps-ns timescale indicates charge transport out of the individual heptazine units. Such charge diffusion is likely based on a hopping mechanism mediated by trap states, which the authors associate with the aforementioned functional groups [21]. Delocalization of the radical over several heptazines is also consistent with two-dimensional EPR spectroscopy on small heptazine-based model systems without stacking, where the experimental spectrum can be fitted with the simulated one modelled by PHI terminated by the anionic cyanamide [2]. Electrons accumulated in the trap states are extracted (e.g., by an electron mediator or cocatalyst) on a much longer timescales: a nickel phosphine molecular catalyst for HER scavenges the electrons in the second timescale, which is rather slow and may explain the persistence of this photogenerated radical.

The second absorption feature in the infrared decays within 1–10 ps and has not been studied in as much detail [20, 22]. However, one assignment is that this feature refers to intersystem crossing to form a triplet state [19]. Phosphorescence associated with this triplet was detected after 10 ms delay in PL spectroscopy, yielding a signal at 1.85 eV (Fig. 9.3d). When compared to the fluorescence peak at 2.05 eV, this would give a singlet-triplet gap of around 200 meV, although this calculation assumed the intersystem crossing started in an interband state, rather than from the conduction band edge minimum. As further evidence of a triplet state, the authors demonstrated that PHI can produce singlet oxygen as detected by the characteristic $^1O_2$ phosphorescence signal at 1,270 nm, a finding that led to a number of studies exploring the use of PHI-generated singlet oxygen for organic transformations [19, 23]. While these results may suggest the radical anion is in a triplet state, how it is related to both the absorption bands at 680 nm and at >1,100 nm is yet to be resolved.

9.2.3 Dependency of radical formation on donor

Though less discussed, the reducing agent also appears to affect the radical formation, as there may be stabilizing interactions between the oxidation product and the PHI anion radical. As a case in point in the work by Markushyna et al., who used benzylamine as the electron donor in acetonitrile (i.e., nonaqueous conditions), PHI was found to store more radicals when irradiated under a CO_2 atmosphere rather than argon, as assessed by the depth of color change, which was greenish and not blue in this nonaqueous case, and corroborated by titration with MV^{2+} [5]. Characterization of the reaction mixture shows that benzylamine reacts with CO_2 to form the corresponding carbamic acid, suggesting this to be the true reducing agent, although it is not immediately obvious why the acid is a better reducing agent than the corresponding amine for generating the radical anion in these conditions. However, it was deduced that the radical anion was charge-balanced by benzylammonium from the observation that quenching the radical with silver triflate yielded both metallic silver and benzylammonium triflate. This may suggest that the efficiency of radical formation is dependent on a number of features, including the donor's redox potential and its intermolecular interactions with PHI, as well as the solvent used. Presently available data on the donor dependency in radical formation are collected in Tab. 9.1, although direct comparison is difficult due to differences in experimental conditions and the synthetic procedure for PHI.

Tab. 9.1: Charge storage properties of reported PHI materials in the presence of different donors.

Electron donor	Radical formed (μmol g^{-1})	PHI synthesis conditions	Reference
Methanol (10 vol %, aqueous)	32.2*	PHI colloid by synthesis from melamine in KOH/NaOH salt melt	[24]
4-Methylbenzyl alcohol (10 μM, aqueous)	354[†]	Synthesis from reacting Melon in KSCN salt melt	[2]
Benzylamine (0.2 mmol, CH$_3$CN solution) in argon	701*	5-Aminotetrazole in LiCl–KCl salt melt	[5]
Benzylamine (0.2 mmol, CH$_3$CN solution) in CO$_2$	957*	5-Aminotetrazole in LiCl–KCl salt melt	[5]
Methanol (10 vol %, aqueous)	11.7*	Mel-PHI[‡] from K-PHI with dicyandiamide and ammonium chloride	[14]
Triethanolamine (TeOA, 10 vol % aqueous)	30.1*	Mel-PHI[‡] from K-PHI with dicyandiamide and ammonium chloride	[14]

*Calculated from MV^{2+} titration.
[†]Calculated from H$_2$ evolved.
[‡]PHI with cyanamide terminal groups condensed into melamine.

9.3 Applications exploiting the long-lived photoelectrons

PHI has been employed in standard photocatalysis for solar fuel production and organic transformations, the term standard here referring to the photoredox reactions being performed immediately without relying on the long-lived state. As these topics have been reviewed already [17, 25, 26], this chapter focuses exclusively on applications that exploit the long-lived property of this highly reducing species. Application of PHI for various photoelectrochemical devices also reveals mechanistic aspects pertaining to the electron stabilization, from which one can derive a better understanding of the overall phenomenon.

9.3.1 Toward all-in-one direct solar batteries

Given the fluctuating nature of solar energy, there is an increased need to store electricity generated by photovoltaics (PV), especially on short and intermediate time-scales. To do so, different solar energy storage technologies can be employed. On the one hand, the production of solar fuels is intensely discussed and studied as a solution for mid- and long-term reversible energy storage and fuel production, which is relevant for the transportation and other energy sectors. However, the conversion of electrical to chemical energy (and backwards) comes at high energetic costs and requires additional infrastructure, as envisaged for the hydrogen economy. On the other hand, concepts for the direct electrochemical storage of solar energy within so-called solar battery devices are being developed, relying on more or less direct photoelectrochemical energy storage (PES). In principle, three approaches are being discussed at different stages of technological maturity [27]. First, PV panels can be coupled directly to batteries or capacitors that are being charged by the PV module, and discharged on demand. This technology is relatively mature, but expensive to be used due to high costs of batteries. Second, a photoelectrode can be used to drive photoreactions and reduce or oxidize adjacent electrodes, which store the charges. This typically occurs in a three-electrode system. The energy stored on the adjacent material can then be discharged on demand. While this technology is being studied employing various materials, it currently has little technological relevance. A third approach is employing direct solar batteries consisting of bifunctional materials combining light absorption and charge storage properties. Here, a solar battery material absorbs light and intrinsically stores one kind of charge (electrons or holes). The countercharge is shuttled to a counter electrode, forming a full battery (two-electrode device) [27]. PHI is a perfect candidate for such direct solar battery devices due to its intrinsic solar energy storage ability, as will be discussed in the following [13, 28].

Figure 9.4a depicts a proof-of-concept photoanode of such a two-electrode solar battery system. PHI nanosheets were deposited on conducting glass as electron storage photoelectrode, immersed in a phosphate electrolyte containing a chemical reducing agent (4-methylbenzyl alcohol) to simulate hole quenching (electrons supplied by the cathode via a redox mediator) during the cell photocharging process [13]. Note that though water can in principle also act as weak electron donor, the alcohol is the stronger reducing agent that is irreversibly oxidized [29]. In this application, PHI's high intrinsic overpotential for both the hydrogen and oxygen evolution reactions is beneficial for avoiding the hazardous H_2 and O_2 mixture being formed. At the same time, it ensures prolonged charge trapping on the carbon nitride backbone in the absence of a cocatalyst, making it suitable for applications in aqueous solar batteries.

In this two-electrode solar battery configuration, PHI acts as the battery (photo) anode. Employing conditions optimized for charge storage in terms of PHI loading and experimental parameters, illumination leading to band gap excitation generates an immediate open-circuit potential (OCP), i.e., the half-cell voltage, of −600 mV vs. Ag/AgCl that becomes more negative toward −800 mV vs. Ag/AgCl with longer irradiation time. These accumulated electrons can be discharged subsequently (Fig. 9.4b). The apparent capacity after solar charging is limited by the materials' poor conductivity, especially in the discharged state, and by self-discharge. In this case, self-discharge may be accelerated by the presence of oxygen leaking into the cell, or by hydrogen evolution through

a)

b)

c)

d)

e)

f)

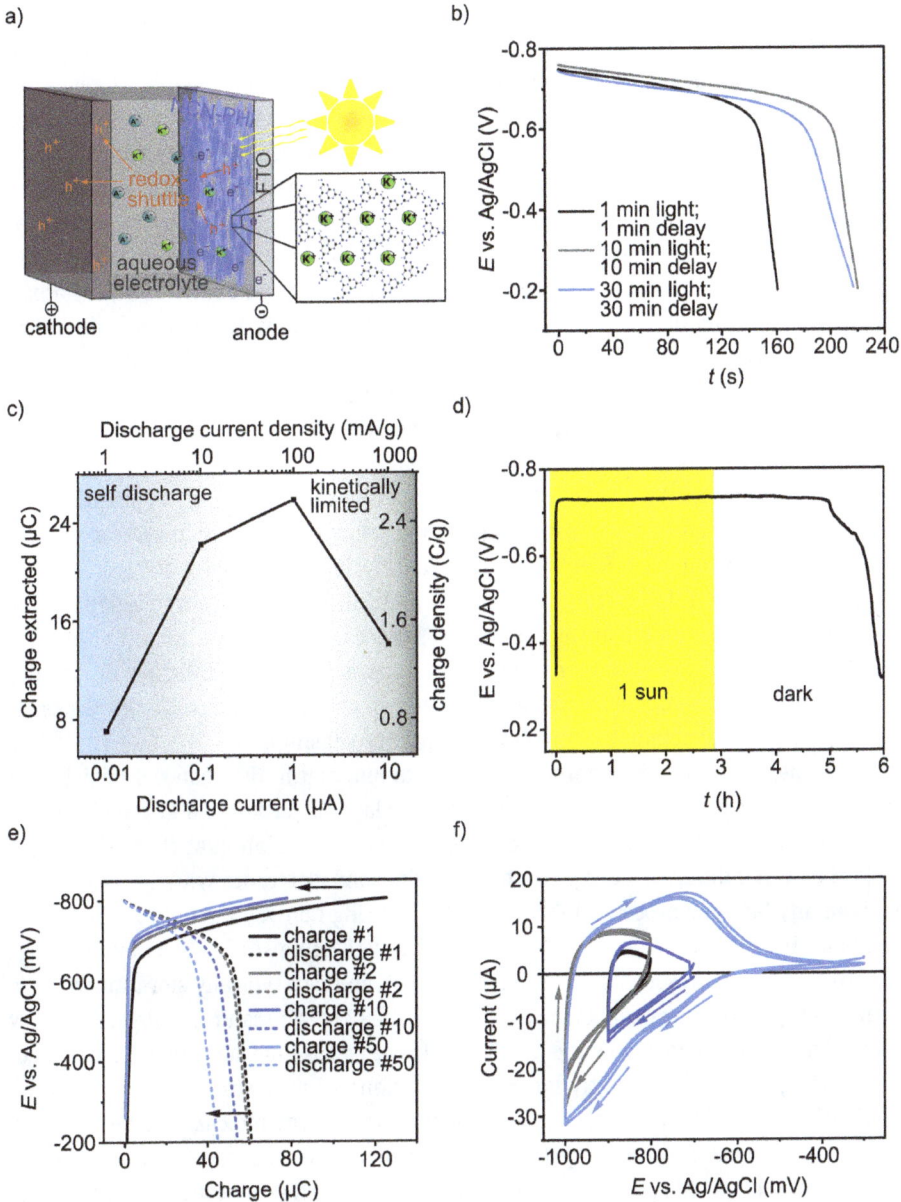

Fig. 9.4: (a) Schematic of a solar battery comprising a thin film of PHI over conducting glass substrate (fluorinated tin oxide, FTO) as the photoanode. (b) Discharge curves with different irradiation duration and delay between light off and electric discharge. (c) Capacity of the material after illumination in presence of a donor, as a function of discharge rate, highlighting kinetic losses at fast discharge and self-discharge contributions at slow discharge. (d) Open-circuit potential (OCP) stability during and after photocharging. (e) Purely electrical galvanostatic charge–discharge curves showing PHI's ability to act as electrically charged battery electrode in water, in addition to light-induced applications. (f) Cyclic

water reduction via the substrate, which is feasible considering the half-cell potential. As shown in Figure 9.4c, the optimum discharge current is 100 mA g^{-1}, a balance between capacity loss through self-discharge and low capacity extracted due to kinetic limitations. The optimal conditions, however, are strongly dependent on the electrode fabrication and PHI loading. As a demonstration of its ability to store charge over long duration, the generated OCP can remain steady for two hours even after illumination has ceased (Fig. 9.4d). Other than solar charging, the cell can also be charged purely by electricity, as illustrated in Figure 9.4e and f. As seen in these figures, the cell can be electrically charged/discharged over many cycles, thus demonstrating the stability of PHI as a battery electrode material over the long term, attaining coulombic efficiencies in the range 75–85% when charged to –800 mV versus Ag/AgCl. Here, capacity decrease over the long term was attributed to mechanical detachment of the PHI nanosheets from the conductive substrate, with the PHI possibly being dislodged by H$_2$ bubbles produced by HER from the substrate, rather than intrinsic material instability.

Investigations of the reaction kinetics in the half-cells reveal the mechanistic aspects of the electron stabilization phenomenon, which may rationalize some of the cell performance metrics. Scan-rate-dependent cyclic voltammetry (CV) reveals an asymmetry between the charging (reduction) and subsequent discharge kinetics. This asymmetry, mainly visible at high areal loadings, has two explanations that may be mutually inclusive: the first being kinetic limitations in the ion diffusion during the charge storage process, and the second, a change in conductivity at different states-of-charge (vide infra, Section 9.3.5 on photomemristive sensing).

The charge storage mechanism is not easy to pinpoint in this material, which may have two modes of charge storage: electric double layer capacitance and pseudocapacitance. Briefly, double layer capacitance refers to the accumulation of charge at the interface between different phases (in this case, the solid electrodes with the electrolyte) without any faradaic process. On the other hand, using Conway's definition, pseudocapacitance involves faradaic reaction(s) associated with some surficial phenomena (e.g., electro-sorption or absorption of charged species), and can typically store much more charge than double layer capacitance [30]. From the CVs, fast scan rates appear to show double layer capacitance to be the dominant charge storage mechanism, whereas pseudocapacitance has an evidently larger contribution at lower scan rates (negative of –600 mV vs. Ag/AgCl); pseudocapacitive behavior in this regime can also be observed by electrochemical impedance spectroscopy and by the fact that the capacity increases negative of this specific potential. In general, however, further investigations on the nature and kinetics of the charge storage process are needed.

Fig. 9.4 (continued)

voltammograms (scanned at 50 mV s^{-1}) showing the increased charge density in the pseudocapacitive region negative of –600 mV vs. Ag/AgCl (reproduced from reference [13] with permission from Wiley VCH).

Regarding the second explanation – conductivity variation with state-of-charge–impedance spectra of the electrode with and without illumination (i.e., at the ground state potential and in the photoreduced state at approx. –800 mV vs. Ag/AgCl) showed a decrease in overall charge transport resistance when the material is being reduced. Hence, charging increases the conductivity (akin to doping) [31], and discharging creates an interface of low conductivity with the substrate, affecting the CV symmetry and influencing discharge kinetics. Galvanostatic direct current measurements outside any electrolyte reservoir with and without illumination further indicate an increase in electronic conductivity upon illumination by a factor of approx. 40. Such behavior is typically observed in classical semiconductors where the concentration of free charge carriers increases upon illumination. For photocharging, this property is very beneficial. Later sections of this chapter will elaborate in greater detail on how the carrier dynamics within PHI are influenced by ions that can fit within its pores, as demonstrated by various methods (NMR, EIS, etc.) [4].

Results from this proof-of-concept also illustrate a few challenges in improving a PHI-based solar battery in terms of capacity and mechanical stability. On a material level, improving the storage density to the level of one electron per 9–10 heptazine units or beyond is essential to increasing the intrinsic capacity of PHI, while mitigating self-discharge would ensure that the electrons are stored for longer, after light/electric charging. At a device level, the weak adherence of the PHI thin film to the conducting substrate shortens its operational lifetime due to eventual detachment. Furthermore, a low mass loading of PHI was employed, which, while facilitating charge transfer from the substrate to the poorly conducting PHI, limits the overall device capacity. While being general problems of such bifunctional materials [27], these issues can possibly be mitigated by forming appropriate composites with conducting and cohesive additives, so as to improve charge transfer and electrode integrity, which, in turn, would improve overall device performance.

9.3.2 Delayed solar fuel production

As an alternative to storing solar energy in an electrochemical storage device, it can be converted into chemical fuels, of which hydrogen evolution is the prime example due to its suitable redox potential and relative ease of production (as compared to hydrocarbons from CO_2 reduction). Since PHI itself has a large intrinsic overpotential for proton reduction, a dedicated electrocatalyst is necessary to trigger H_2 formation. This property, in tandem with the unique photocharging ability of PHI described above, has recently opened up a new field in carbon nitride research for which the term "dark photocatalysis" has been coined [2, 22]: Akin to natural photosynthesis where the light and dark reactions (light harvesting and carbon fixation, respectively), are separated in space and time, light absorption and catalysis can be temporally separated on PHI, allowing the production of solar fuels in the dark. Time-delayed solar fuel or electricity production, thus, productively engages the intermittency of solar irradiation by extending solar fuel or electricity production into phases of low or zero solar irradiation, which can, in turn, be used as an energy buffer system, i.e., to reduce the load on the grid in times of high energy demand [1, 2, 13, 27, 32].

From a practical perspective, the controlled detrapping of electrons accumulated on PHI during photocharging is a vital step in the process of "dark photocatalysis." Within aqueous suspensions of PHI particles, most of the reported electrocatalysts are molecular or nanoparticulate, so as to maximize the collision frequency, and therefore, electron transfer probability/kinetics from the PHI to the electrocatalyst. A nickel phosphine molecular electrocatalyst and aqueous platinum colloids have been reported as such HER electrocatalysts, and application of either proceeds as graphically summarized in Figure 9.5a,b and described as follows [2, 22]. An oxygen-free PHI suspension in aqueous solutions containing salts and a suitable reducing agent is irradiated, generating the long-lived radical state characterized by the suspension's color change. Afterwards, the electrocatalyst can be introduced into the suspension whenever H_2 is desired, and the PHI is restored to its original state and cycled through these steps again. For the photoreaction system in this study, irradiation for 2 h is necessary to maximize H_2 production from 20 mg of PHI, evolving nearly 3 µmol H_2 when the electrocatalyst is added immediately after switching the light off. This corresponds to roughly one electron to every 9 heptazines. As shown in Figure 9.5c, if the time gap between light off and electrocatalyst addition is extended to between 1 and 6 h, the amount of H_2 evolved decreases to around 1.5 µmol H_2, and extending this time gap further leads to decreased H_2 evolved approaching zero. One origin of this decay may be quenching of the stabilized photogenerated charges by oxygen leaking into the imperfectly sealed photoreactor, but also internal loss processes leading to charge decay are possible, including the reductive formation of other chemicals [33].

In evaluating the use of the trapped, light-induced electrons for "dark photocatalysis," further insights into the properties of PHI as related to this stabilization phenomenon have been acquired. From Figure 9.5c, the amount of hydrogen evolved may act as a proxy for the radical population, or strictly speaking, the population of radicals with sufficient reducing potential for hydrogen evolution. Here, two decay kinetics in the hour timescale are observed, consistent also with how EPR (electron paramagnetic resonance) signal intensity evolved over time for PHI in sealed tubes, which would eliminate extrinsic effects such as oxygen leakage. In the first, a large portion of the radical population is lost in the immediate hours after illumination ceased. Afterwards, population decay has slowed; while EPR shows a paramagnetic signal (of half the initial intensity) beyond 16 h after illumination ceased, no "dark hydrogen" evolved beyond 12 h, either due to quenching by oxygen leakage into the imperfectly sealed photoreactor (extrinsic effect) or loss of the electrons' potential energy (intrinsic effect). The intrinsic explanation appears consistent with the relaxation mechanism proposed by Durrant and coworkers, which was originally developed for Melon-type carbon nitrides, but may be partially applicable also to PHI [20, 31, 34, 35]. In their model, charge trap states tailing into the interband region of Melon are modelled with a Boltzmann-like distribution, where electrons relax into

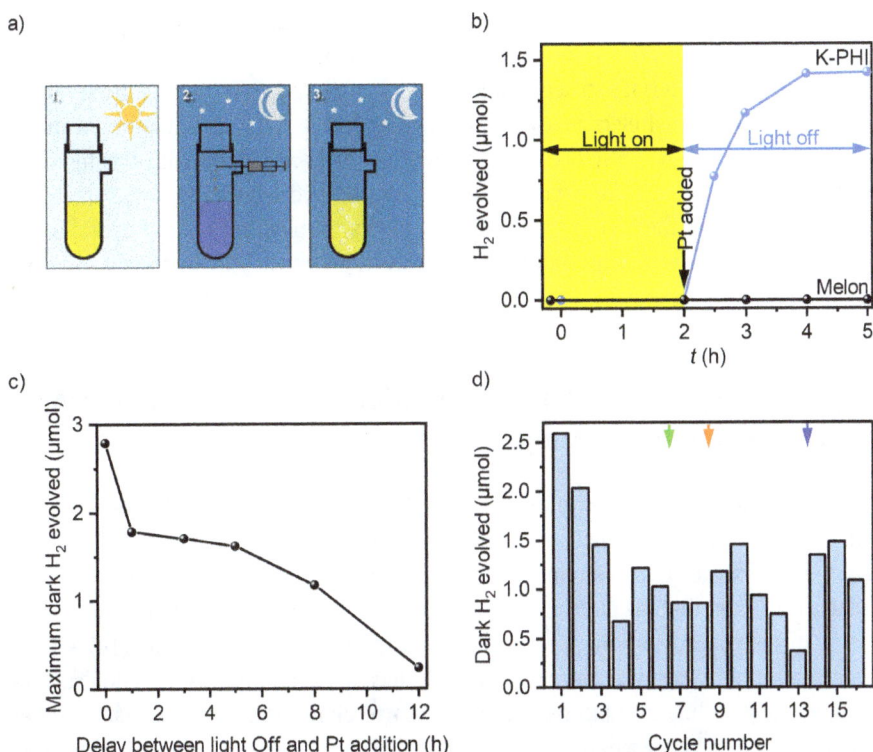

Fig. 9.5: (a) Graphical summary of the delayed H_2 evolution showing (1) photocharging of a deaerated PHI suspension containing a sacrificial electron donor; (2) injecting hydrogen evolution electrocatalyst after illumination has ceased; and (3) H_2 evolution. (b) Hydrogen evolution with time illustrating decoupling of irradiation and H_2 evolution. (c) Plot of hydrogen evolved as a function of delay time between ceasing illumination and injection of platinum colloid as H_2 evolution electrocatalyst. (d) Repeated cycling of the dark H_2 evolution (2 h irradiation), where the green arrow indicates replenishment of electron donor (4-methylbenzyl alcohol), the orange arrow addition of Pt colloid, and blue arrow addition of both electron donor and Pt colloid (reproduced from reference [2] with permission from Wiley VCH).

traps with progressively lower energy. Upon photocharging of PHI, however, the traps are filled and accumulated electrons can also be considered free charge carriers. However, upon partial charge decay, these electrons may fill traps, ultimately being rendered insufficiently energetic for HER. While this model would be consistent with the observed amount of dark H_2 evolved in Figure 9.5c and how the EPR signal intensity changes with time after illumination ceased, it highlights some routes for further research, especially in terms of identifying the molecular equivalence of these trap states and controlling the energetics of these states to maintain the charge's reducing strength. When PHI is repeatedly used for the "dark hydrogen" experiment (Fig. 9.5d), the total amount of hydrogen evolved declined by a third from the first cycle to the third cycle; in subsequent cycles of this experiment, hydrogen evolution never exceeded ≈ 1.5 μmol. This may be rationalized by consumption of electron donor for H_2 evolution with cycling, surface clogging by the donor, or structural modifications upon washing and/or ion exchange. The latter can slightly modify the material's structure and ionic nature, which is essential for the stabilization of photogenerated electrons, as mentioned above [3, 4, 11]. Deactivation of the Pt colloid electrocatalyst may also account for the declining H_2 evolution with cycling, since addition of fresh Pt colloid for cycle 9 and 14 increased hydrogen evolution, somewhat. Long-term stability of the PHI for "dark photocatalysis" in terms of time of operation or turnover needs to be further ascertained, especially in preserving the material's ionic nature and structure.

9.3.3 Summary of electron accumulation in PHI for solar energy conversion and storage

The phenomenon of electron accumulation following a photooxidation involves two characteristics of PHI: its photoactivity as is known from carbon nitride photocatalysis and its ability to stabilize the electrons remaining after hole extraction for a duration beyond the diurnal cycle. Given that the electrons are generated by light and stored in a highly reducing state, this effect provides the means to store solar energy in a readily convertible form for later use, either directly as electricity or for reductive chemical transformation such as HER. Proofs-of-principle for both have been presented above as integrated solar batteries and "dark" photocatalysis where, even after cessation of illumination, the PHI photoelectrode maintains its voltage after charging, while aqueous PHI suspension can evolve H_2 as a fuel chemical. Although the amount of charge (and therefore, solar energy) storable is not yet competitive to those of modern battery materials, PHI as a multifunctional material capable of harvesting and storing solar energy may find niche applications that require a portable, yet (solar) rechargeable, power source. One such niche use could be for powering micro-devices remotely, as detailed in the following section. Hence, further developments to increase the charge storage capacity and maintaining their reductive strength will be required to enhance the applicability of PHI in energy-related technologies.

9.3.4 PHI as part of light-driven microdevices

As demonstrated earlier, PHI can store solar energy as highly reducing electrons for later use in chemical conversion or extracted as electric current. As illustrated in the following section, PHI micron-scale particles can also act as light-driven microswimmers, which, owing to their unique charge-trapping ability, can be photocharged to power nanoscale vehicles also in the dark, a property that has been termed "solar battery swimming," and fulfill many other functions that are unprecedented in the field. In the following, the term "microrobots" will be used for devices capable of performing multiple functions, partially autonomously.

9.3.4.1 Janus particle microswimmers

Microswimmers are micron-scale machines that are propelled in liquids, usually powered by an external supply of energy. In many cases, this external energy source may take the form of magnetic field for use with microswimmers having magnetic components, ultrasonic actuation, and conversion of chemical fuels present in the liquid [21, 36–40]. In the latter case, besides possible bubble propulsion, the source of propulsion force is the built-up of a concentration gradient of solute species (ionic or nonionic) across the surface or sides of the microswimmer. This gradient is counteracted by fluid flow of solute species across the particle surface, giving rise to motion. If internal charge transfer between the two hemispheres of the particle is involved, balanced by fluid flow of countercharges across the particle surface, the motion is called self-electrophoretic.

In cases where the surface reactions are induced by light, light-driven microswimmers can be realized. Hence, thrust can be generated from solar energy, which is beneficial from a practical perspective, as it does not require sophisticated external machinery to control the movement of small particles. Since photocatalysis offers energy for uphill thermodynamic reactions to create concentration gradients around the microswimmer, also water or biological species, which are stable under natural conditions, can be used with light-driven microswimmers, making this light-driven approach attractive, especially in biological contexts.

Light-driven microswimmers that are propelled by photocatalytic reactions are typically comprised of a semiconductor part that absorbs light to drive reactions on the particle surface. Directional thrust to propelling the microswimmer is generated, provided that the photo-driven reactions take place asymmetrically on the microswimmer particle. One realization of such a system is a Janus particle with metal or insulator-capped semiconductor hemispheres.

The first demonstration of light-driven microswimmers based on carbon nitride was reported in 2017, using spherical Melon microparticles (1 to 3 µm) with randomly distributed Pt nanoparticles on its surface [41]. Upon band gap illumination and in the presence of ethanol as sacrificial agent, controllable back and forth motion can be observed, although the exact mechanism of propulsion and charge transfer processes remained vague. Subsequently, Sridhar *et al.* studied the propulsion mechanism of PHI-based Janus-type microswimmers [29].

To form asymmetric Janus-type particles, size-separated PHI micron particles were deposited onto a substrate, and a thin layer of a metal or oxide was evaporated onto the exposed surface, yielding a defined capping structure, as shown in Fig. 9.6a,b. Insights into the propulsion mechanism of these microswimmers were obtained using different capping materials (Pt, Au and SiO_2) as well as chemical fuels (water, methanol (MeOH), H_2O_2) acting as electron donors and acceptors (Fig. 9.6c-f) [29]. In pure water, under ambient conditions (i.e., containing dissolved O_2), the oxidation of water is the only possible reaction for photogenerated holes. At the same time, the reduction reaction – the most facile being the reduction of oxygen – leads mainly to the formation of radical oxygen species (ROS) as confirmed by electrochemistry and gas chromatography for PHI- and TiO$_2$-based Janus microparticles. Due to the excellent electron-accepting properties of dissolved O$_2$, efficient propulsion from the PHI hemisphere is observed even with the passive capping material, SiO$_2$. If a donor such as MeOH is added, however, the efficiency of hole extraction is enhanced, translating into increased propulsion speed. (Fig. 9.6c,d).

If the Janus particles are capped with Au or Pt, which act as electron sinks, the photo-electrons generated by PHI can be extracted more efficiently by these metals, which are also efficient as electrocatalysts in performing HER or ROS formation. Hence, photoinduced electrons generated on the PHI side of the particles also flow towards the metal-capped side where they engage in reduction reactions at the metal surface, while the photoinduced holes on the PHI side are transferred to the surface to engage in oxidation reactions. In addition to the self-diffusiophoretic propulsion mechanism for the SiO$_2$ caps, with the metal caps, a self-electrophoretic propulsion component contributes. Since thrust and propulsion speed are related to the rate of charge separation and surface catalysis, these experiments also confirm that efficient charge separation, transport and extraction are important factors for efficient photocatalysis on PHI. Under high fluence or intensity, the intrinsic accumulation of electrons on PHI may lead to increased recombination and, thus, reduce the rate of free charge carriers being produced, thereby limiting photocatalysis and, in the case of these light-powered PHI-based microswimmers, also limiting their overall propulsion [20, 42, 43].

Of the fuels studied in this publication, the one most commonly used for the propulsion of microswimmers (solely driven chemically or with light, as is the case here) is H$_2$O$_2$ [37, 38, 40]. Even at concentrations as low as 0.05%, light-induced swimming is significantly increased in comparison to MeOH (Fig. 9.6e), since this fuel is easily reduced and oxidized by charge transfer reactions. (Photo-)electrochemical studies with this fuel confirmed that the photo- and dark currents are more pronounced and the system is insensitive to the presence of oxygen because the redox of H$_2$O$_2$ is favored both kinetically and thermodynamically. The propulsion mechanism using H$_2$O$_2$ is as described above: the SiO$_2$-capped particles are propelled by a self-diffusiophoretis from photoredox reaction on the PHI hemisphere, while propulsion for the Au-/Pt-capped particles is based on a self-electrophoretic mechanism (Fig. 9.6f). Interestingly, however, the Au-capped Janus microswimmers are propelled faster than the Pt-capped ones, which are propelled even slower than SiO$_2$-PHI at 0.5% H$_2$O$_2$, a trend opposite to the case in pure water or in aqueous methanol. One rationale is that H$_2$O$_2$ can decompose even without light on the metal cap, with reaction rates faster on Pt than on Au [43, 44], thereby creating a counterforce that slows down particle motion through light-independent chemical reactions. While, at first sight, this reactivity may seem "adverse," it can nevertheless be exploited to enable photocharging to realize *solar battery swimming*, as outlined in the next section.

a)

b)

c)

d)

e)

f)

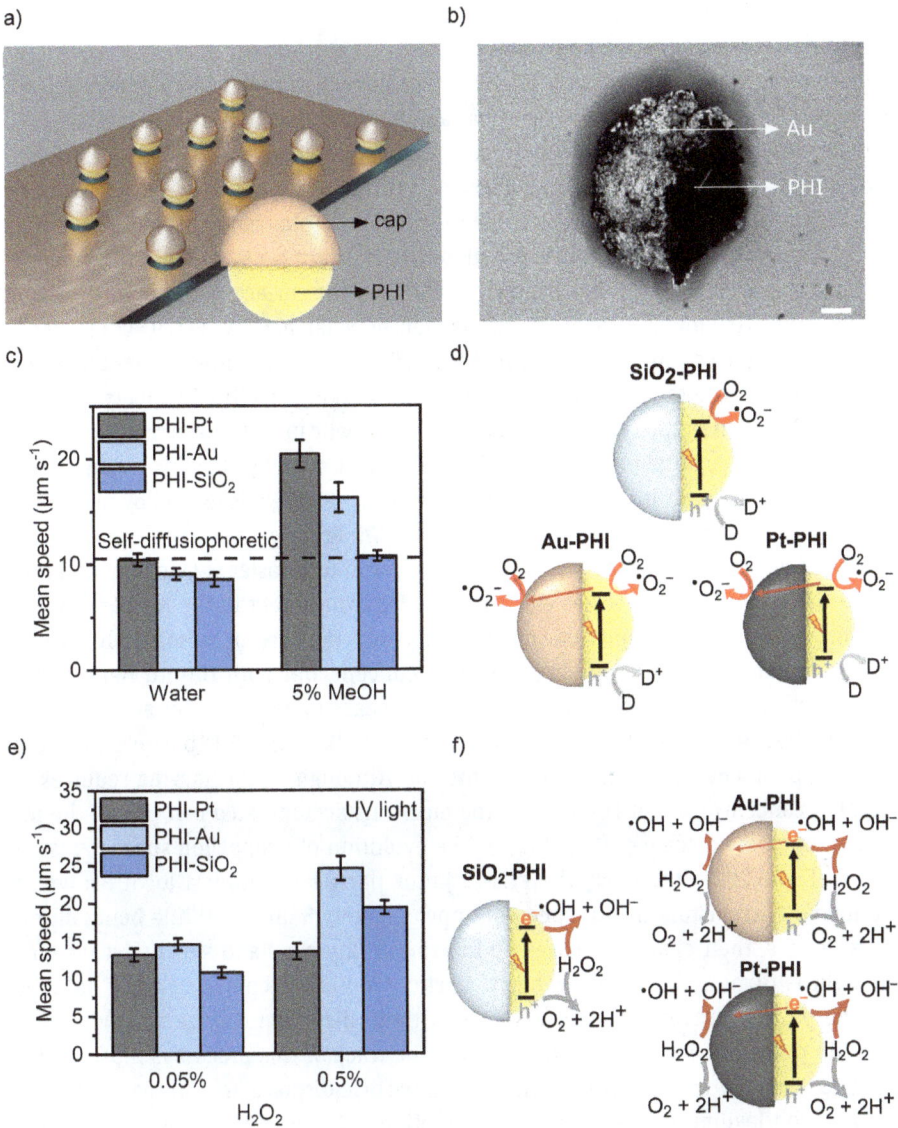

Fig. 9.6: Light-driven propulsion of photocatalytic PHI Janus-microparticles. (a) Fabrication principle for PHI-based Janus microswimmers by evaporation of the capping material. (b) Au-capped PHI microswimmers imaged by SEM (scale bar: 400 nm). (c) Mean speed of propulsion under UV illumination in pure water and upon addition of 5% MeOH for PHI microswimmers capped with Pt, Au or SiO$_2$. (d) Illustration of the proposed light-induced surface reactions and charge transport mechanisms. (e) Janus microswimmer propulsion speeds in water with H$_2$O$_2$ added. (f) Redox mechanisms proposed for explaining the propulsion trends shown in panel (e) using H$_2$O$_2$ as fuel (reproduced from reference [29] with permission from National Academy of Sciences (NAS)).

9.3.4.2 Solar battery swimming

Photocharging is a prerequisite to enable "solar battery swimming" by photocatalysis. However, at ambient conditions, electrons cannot accumulate in PHI since they are continuously used for photocatalytic propulsion the moment they are generated. Hence, to enable solar battery swimming, the charging process of the microswimmers must be faster than the decay/consumption rate of the photogenerated electrons in PHI. For this, accumulation of electrons within the bulk of the PHI particle is not only highly desirable but also feasible, since the charge percolation to the surface is rather sluggish and in line with PHI's low conductivity (see section on solar battery, Fig. 9.4c) [4, 13]. To make charging (electron accumulation) of the PHI microparticle interior feasible, a possible route is to use the metal cap to assist photocharging of PHI by supporting redox reactions. The concept here is to generate electrons that can be transferred through the metal cap to quench the photogenerated holes on PHI, thereby allowing accumulation of photoelectrons in its interior. One realization of this strategy is to employ metal caps in conjunction with H_2O_2 as a redox fuel (see Fig. 9.7). From previous works on the propulsion of Au-Pt Janus particles with H_2O_2 as chemical fuel, faster H_2O_2 oxidation kinetics on the Pt hemisphere causes electron transfer from the Pt to the Au hemisphere, resulting in the self-electrophoretic mechanism that is the basis of these purely metallic microswimmers [43, 44]. Applying the same concept, but with the Au hemisphere being replaced by PHI, H_2O_2 oxidation on Pt appears to generate net electrons that are transferred to quench the photoholes in the bulk of the PHI particles, thereby enabling photocharging of the particle interior. Although photocharging requires an elevated concentration of H_2O_2 (>1%), the electrons accumulated can propel the microswimmers by discharge (Fig. 9.7a,b). The evolution of propulsion speed is visualized in Fig. 9.7c. In the dark, the PHI-Pt Janus particles exhibit a local Brownian motion with an instantaneous speed of approximately 5 μm s^{-1}. While being illuminated for 5 s, their speed increases up to around 12 μm s^{-1} and slowly decays once the light is turned off. Simultaneously, particle motion becomes strongly ballistic, i.e., the particle travels a longer distance without direction change than would be possible purely by local Brownian motion. The trace of this process is illustrated in Fig. 9.7d: the gray trace represents the (small) particle displacement in the dark; upon illumination leading to charging, particle motion is increased as shown in the blue trace, and ballistic particle motion continues after illumination has ceased as in the orange plot, indicative of solar battery swimming. Due to the increased speed and ballistic motion, the total displacement is also significantly increased (Fig. 9.7e). One particularly remarkable feature of this solar battery swimming is the duration of this process: increasing the illumination time significantly increases the duration of solar battery swimming, where ballistic microswimmer propulsion after light has ceased can continue for up to 30 min, with only 30 s of photocharging (Fig. 9.7f).

The exact mechanism of the solar battery propulsion is not entirely understood. Some open questions include: (i) which side of the Janus-type particle is dominant in driving the catalytic reaction and (ii) what is the origin of the nonlinearity between photocharging time and propulsion duration. One possibility to account for the non-linearity is that, with short charging, the subsequent discharge rate driving catalytic propulsion is overall insufficient, either in terms of kinetics and/or energetics, to drive any discernible ballistic motion.

As demonstrated in this proof-of-concept, the combination of photocatalysis translating into mechanical force and solar charge storage illustrates the potential of PHI to act as the solar-powered propulsion component. One possible application of such swimmers includes degradation of pollutants, where pollutant molecules are simultaneously decomposed and act as propulsion fuel [45–47]. In this case, a random motion being enhanced for its displacement by ballistic propulsion is the desirable method, which capped PHI microswimmers already exhibit. Another example is the use of PHI as a micro-vehicle for payload drop-off and delivery, as will be elaborated in the next section. Further development of these swimmers should replace this noble electrocatalyst and toxic fuel to broaden their scope of applicability. Improving process efficiency for solar energy conversion and energy storage is desirable though not immediately necessary, as the envisaged applications do not require long operational lifetime and can be recharged in a straightforward manner (e.g., by illumination). However, a more (near-) infrared light powered charging mechanism would be beneficial for applications deeper inside (living) tissue. Nevertheless, the comparatively simplistic design of these swimmers, not requiring any sophisticated structural control, underscores the benefits of PHI. It is particularly advantageous in the context of micro-/nano-machineries, where identifying suitable components for energy harvesting, propulsion and energy storage and connection thereof is technically nearly impossible at the moment, and very difficult to upscale. In these contexts for specific and short term use, and where external (wireless) refueling is desired and easily possible, this simplicity may be much more relevant than overall solar energy harvesting efficiency [27].

9.3.4.3 PHI microswimmer applications in high salinity and biological conditions

For applications in environmental, biological and medical contexts, the microswimmers introduced in the earlier section and related devices must have multiple functionalities. Hence, in addition to self-propulsion, the microdevice should be able to perform a function in a specific location as triggered by a stimulus, one example being cargo pickup and drop-off at a desired location. Realizing multiple functionalities requires sophisticated microstructure design to integrate the (photochemical) propulsion unit with the cargo delivery functional component. Additional criteria include compatibility to natural or biological conditions, especially for medical applications, so that the micro-device is nontoxic and can be operated within biological media [38, 48]. This compatibility is especially critical for the light-driven swimmers in the previous section, since the propulsion through ionic solutions (e.g., biological media, natural saline solutions) is

a)

Pt-PHI: photocharging

b)

Pt-PHI: discharge

c)

d)

e)

f)

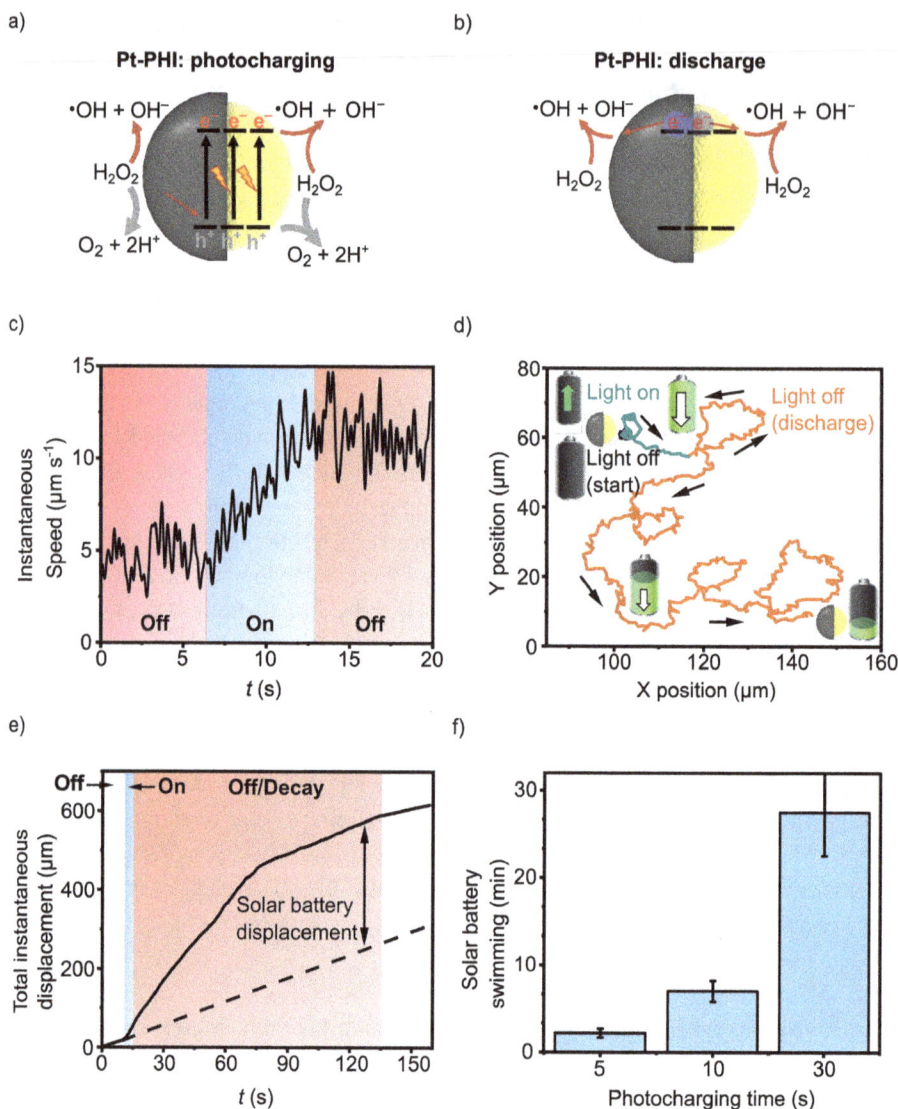

Fig. 9.7: Solar battery swimming enabled by cooperative effects in Janus microswimmers. (a) Proposed mechanism of charge accumulation in Pt-PHI microswimmers, where electrons generated from H_2O_2 decomposition on Pt quench the photogenerated holes in PHI. (b) Subsequent propulsion after charging (i.e., the discharge process) is then driven by the electrons accumulated on PHI. (c) Instantaneous speed of Pt-PHI microswimmers, being low in the dark, increasing continuously under illumination, and staying elevated after illumination has ceased. (d) Particle trajectory prior to illumination (0–10 s in the dark, gray trace), under illumination (10–15 s, blue trace), and during solar battery swimming (from 15 to 120 s, orange trace). (e) Integrated total instantaneous displacement for the *solar battery swimming* shown in panel (c). (f) Solar battery swimming time increasing with photocharging time (reproduced from reference [29] with permission from AAAS).

typically made impossible due to the presence of ions adversely affecting electrophoretic or diffusiophoretic propulsion mechanisms. Specifically, since the microswimmers' propulsion originates from the electric field generated around them through photocatalytic reactions, having the ions present in the swimming medium typically screens this generated electric field, eliminating this propulsion force. Even with sophisticated microstructure designs aimed at minimizing this screening effect, microswimmer propulsion is typically possible only when the solution is in the low millimolar salt concentration, well below the 100 mM range found in biological systems [38, 49]. However, microswimmers comprising carbon nitride materials, especially PHI, have been demonstrated to overcome this issue, presumably due to their ability to interact with ions in their surrounding [29]. One illustration of this ability has been reported by Antonietti and coworkers, who prepared solar-driven ion pumps using Melon-type carbon nitride. In detail, they prepared a membrane comprising carbon nanotubes decorated with Melon which, when illuminated, can transport ions against a strong concentration gradient, as enabled by electrostatic interactions between the photogenerated charges on the Melon and the ions [50, 51]. The ionic interaction as demonstrated is especially accentuated in PHI due to its 2D structure and its structural porosity, and can be exploited for developing microswimmers capable of swimming through ionic media propelled by photocatalytic reactions, as illustrated below.

As initially reported, spherical C_3N_4 microswimmers decorated with Pt nanoparticles did not exhibit light-driven propulsion within ionic solutions (tested at 100 mM), which confirms the propulsion mechanism to be based on self-diffusiophoresis [41, 49]. However, carbon nitride microswimmers comprising Melon and PHI have been demonstrated to be capable of swimming through ionic media [52]. In initial experiments, small particles of these two carbon nitrides were prepared by size selection through sonication and stepwise centrifugation, and tested for propulsion without any metal caps or cocatalyst. Here, the microparticles could be efficiently propelled by UV and visible light within not only pure water, but also in biologically relevant buffers, such as DMEM (Dulbecco's modified Eagle's medium) and DPBS (Dulbecco's phosphate buffered saline), both at concentrations of ≈ 160 mM, as well as even in diluted blood solution (Fig. 9.8a). To compare the propulsion through ionic media, the metric EI50 number was introduced, which is defined as the salt concentration required for reducing the propulsion speed by 50% [49, 53]. While in previous reports even optimized swimmers based on nanorod morphology had EI50 values in the range of 1–10 mM, the texturally porous Melon-based swimmers have EI50 values of around 100 mM, while EI50 for PHI-based swimmers is over 1 M, as shown in Fig. 9.8c. Of note, significant propulsion of PHI-based swimmers can still be observed even within a salt concentration of 5 M, which is practically saturated. The effectiveness of the carbon nitrides as the light-driven propulsion unit is attributed to the strong photocatalytic activity to generate an ionic gradient outside the particle, as well as to the ability to generate an optoionic interaction through the particle volume, which would have otherwise eliminated propulsion in materials to be considered as solid spheres, not being permeable to ions (Fig. 9.8b). In particular, for PHI, with its stronger photocatalytic activity (compared to Melon), its structural porosity through which cations can be transported and ion-mediated photocharge stabilization, microswimmers based on this material have an enhanced tolerance to ionic solutions in terms of propulsion. However, also for Melon, where textural porosity coupled with some ionic interactions and good photocatalytic activity are present, sustained propulsion in the presence of ions remains possible.

In the course of experiments on pure carbon nitride-based microswimmers, an important feature was observed: the ability to control the swimming direction purely by light. Specifically, the microswimmers move towards the source of light, which suggests positive phototaxis to be at play, which is desirable as a method to direct the microswimmers toward the targeted location. Otherwise, in the absence of phototaxis, the microswimmers would move only randomly (though this may be advantageous as a search function in, say, autonomous micromachines for random sampling). Using light for directional control can also provide a method of assembling and collectively directing a swarm of microswimmers that have been randomly distributed over a volume of solution to a given point (e.g., using focused light), thus facilitating recovery of the microswimmers after they have performed their function for possible reuse. Phototaxis and collective motion have also been observed for Pt-decorated C_3N_4 microswimmers as described above [41], although only when being contained in salt-free ethanol solution. The origin of this phototactic response is related to the microswimmers' zeta potential [54], the effective electric potential at the surface of the particles, which is negative for both Melon (in neutral and alkaline pH) and PHI (at all pH). The symmetry breaking required for directional propulsion is generated by the light itself: the part being illuminated is photocatalytically activated, while the opposite part is being shadowed so that no photocatalytic reaction takes place. Hence, even without strict morphological control (the PHI microparticles used in this work were prepared by ultrasonication and have poor shape control), this phototactic property is present, and the shadowing effect may also explain the motion of C_3N_4 microparticles with randomly distributed Pt nanoparticulate cocatalyst [41, 55, 56].

9.3.4.4 PHI microswimmers for smart drug transport and drop-off

To broaden the scope of applications of the motion-controlled microswimmers, they must be able to perform multiple functions, such as cargo uptake and drop-off. For this example of function to be applied in a biological or medical context, the microswimmers must be compatible to cells. This is the case for PHI-based microswimmers: in studies with human fibroplast, when HT-29 and SKBR3 cell lines were exposed to PHI at 30 µg mL^{-1} in the dark for 24 h, no immune response was elicited, indicative of biocompatibility. Under UV illumination, cell decay is observed within 30 min, whereas under illumination with visible light (415 nm) the cells remain viable for 30 min, which is sufficient time for light-driven applications. This biocompatibility can be attributed to the chemical stability and

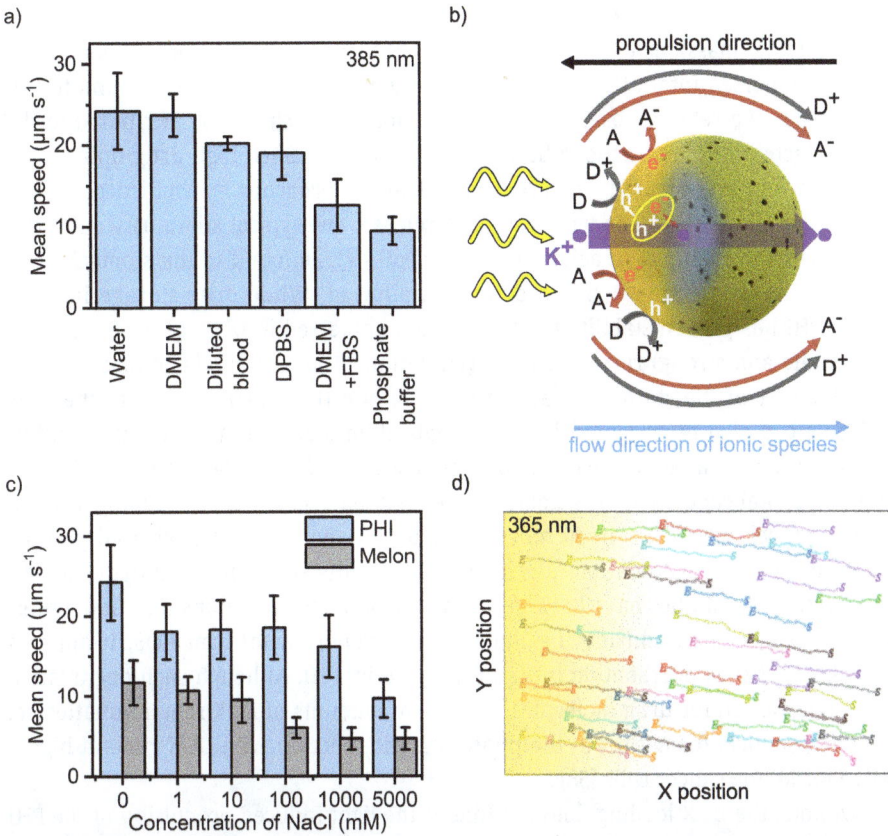

Fig. 9.8: Carbon nitride (CN$_x$) microswimmers in biological and ionic media. (a) Propulsion speed of PHI microparticles under UV illumination in water, biological buffers such as diluted blood and phosphate buffer (see main text for definition of abbreviations used). (b) Light-driven symmetry breaking and propulsion in the presence of ions, presumably enabled by ion flux through the particle, and photo-ionic interactions. (c) Propulsion speed of PHI- and Melon-type CN$_x$ microswimmers under UV illumination, decreasing with increasing concentrations of NaCl. (d) Positive phototaxis of PHI microswimmers when illuminated from the (left) side. For the trace representing particle trajectory, S indicates the starting position, and E, the end position (reproduced from reference [52] with permission from AAAS).

organic nature of PHI: it is inert to cellular interactions and does not contain metals or other potentially pathogenic/antigenic molecular structure that would otherwise elicit an immune response [48]. Although full biocompatibility in terms of in vivo studies remains to be confirmed, PHI is a promising material for developing functional microswimmers in various biomedical contexts due to its biocompatibility, tolerance to ionic solutions and biological media, as well as responsiveness to visible light. A proof-of-concept demonstrating its potential is detailed later.

In biomedical application, one desirable function for microrobots is stable cargo uptake and triggered release. PHI has been demonstrated to soak up the anticancer drug, Doxorubicin (DOX), from the biological buffer DPBS into its pores with a loading of up to 185% (mass), as evidenced by UV–Vis absorption spectroscopy. When subsequently immersed into a DOX-free medium in the dark, the DOX stays bound to PHI with no passive release observed for over a month, which can be attributed to the strong electrostatic interactions between DOX and PHI. Such high drug uptake and stability are unprecedented, and are usually being approached by sophisticated drug capping structures [57]. For triggered drug release, one typical stimulus would be a pH change, for example, in acidic conditions (pH < 4), as found in the stomach or in the gastrointestinal tract when a drug is swallowed. While drug storage in DOX-loaded PHI has proven stability at pH 6.7, when immersed in DPBS of pH 3.5, a burst (immediate and strong) release is observed within 10 min intervals, offloading a cumulative 65% of the storage capacity within 60 min (Fig. 9.9a). Release of the drug molecule can also be triggered by visible-light illumination in the biological buffer DMEM. Also here, a burst release is observed based on UV–Vis spectroscopy – around 14% of the total capacity was estimated to be released every 10 min, reaching a cumulative total release of 35% in 30 min (Fig. 9.9b). An analysis of the released product revealed a partial degradation of DOX, probably induced by photocatalytic conversion. Since these studies have been performed in ambient conditions (i.e., in the presence of oxygen) and considering that PHI was reported to form ROS, it appears plausible that DOX degradation is photocatalyzed by PHI. Although such degradation is not uncommon for drug molecules, significant amounts of DOX remain unaffected, and the products formed from the photocatalytic reaction with DOX remain highly active for killing cancer cells [58].

Of note, the DOX loading does not hinder the light-induced propulsion of the PHI microswimmers significantly, slowing it down only slightly, thereby confirming the propulsion mechanism explanations made earlier. While being of benefit to be able to drive both effects by light only, in the future, a partial decoupling of both actions might be beneficial to only release the drug where required, if the microswimmer transport takes long.

When changing to oxygen-poor conditions as found naturally in the environment of cancer cells, the color of the suspension changed to blue under 415 nm illumination (Fig. 9.9c) as evidence of photocharging of the DOX-loaded PHI. Simultaneously, DOX and related products are released significantly faster (Fig. 9.9b, red line). In this case, the distribution of DOX degradation products is different from that in the presence of oxygen, thereby supporting the assumption that drug degradation is related to the photocatalytic formation of ROS [59]. Moreover, the amount of drug released has increased in every release step when compared to under ambient conditions, reaching nearly 65% after 30 min, implicating photocharging as a process that modifies the electrostatic interactions between PHI as carrier and DOX as cargo, resulting in better drug release. The process is illustrated in Fig. 9.9d. As proof-of-concept with DOX-loaded PHI microparticles in the vicinity of SKBR3 cancer cells, illumination under 415 nm light for 20 min indeed triggered sufficient release of DOX to be taken up by the cells, causing their subsequent death (Fig. 9.9e). Meanwhile, PHI keeps adhering to the cells, at least partially.

9.3.4.5 Carbon nitride microswimmer summary

As illustrated above, PHI has various properties resolving several challenges encountered in the development of (light-driven) microswimmers, namely: (i) tolerance to ionic/biological media for propulsion; (ii) phototaxis; (iii) biological compatibility even under illumination, (iv) large capacity for loading guest (drug) molecules including a high stability against passive release; and (v) triggered and hence, responsive release with stimuli such as pH change or light. Coupled with the well-known advantages of graphitic carbon nitride in photocatalysis – structural and/or textural porosity, chemical stability, high photocatalytic activity – the proof-of-concept studies above demonstrate PHI to be a promising component in the future design of multifunctional microswimmers and microrobotics, particularly for biomedical applications. Further, PHI can respond to its environment through a change of color and electrostatic/intermolecular interactions, implying its applicability in sensing, which may be harnessed as an active trigger to perform some of the functions mentioned above, in an unprecedented semiautonomous fashion. This feature is key to the development of smart robotics, often put in context to "neuromorphic" sensing and action triggering, akin to neurons transmitting information to trigger actions [8, 60]. Motivated by potential applications in these emerging technologies, especially for microrobotics, sensorial and energy applications, the next section focuses on the applicability of PHI as novel sensor type and memory device for information storage. This implies its use as autonomous particle like in the case of microswimmers, and as wired electrodes, akin to the solar battery design.

9.3.5 Photomemristive sensing by charge accumulation in PHI

9.3.5.1 Basic principle: direct sensing and solar battery properties

Harnessing the photocharging ability of PHI, Gouder *et al.* have recently developed a PHI-based photomemristive sensing platform [61]. The idea behind the concept is illustrated in Fig. 9.10. To explain, we first focus on direct (amperometric) sensing for the analyte interaction, while photomemristive sensing properties, relying on the measurement of changes in a range of physical quantities upon light induced charge accumulation, including photovoltage, photocurrent, resistance, impedance, absorption, and photoluminescence, are described as follows.

When PHI is illuminated with energies above the band gap (450 nm or less), electron–hole pairs are generated in the material. As the analytes are oxidized by the

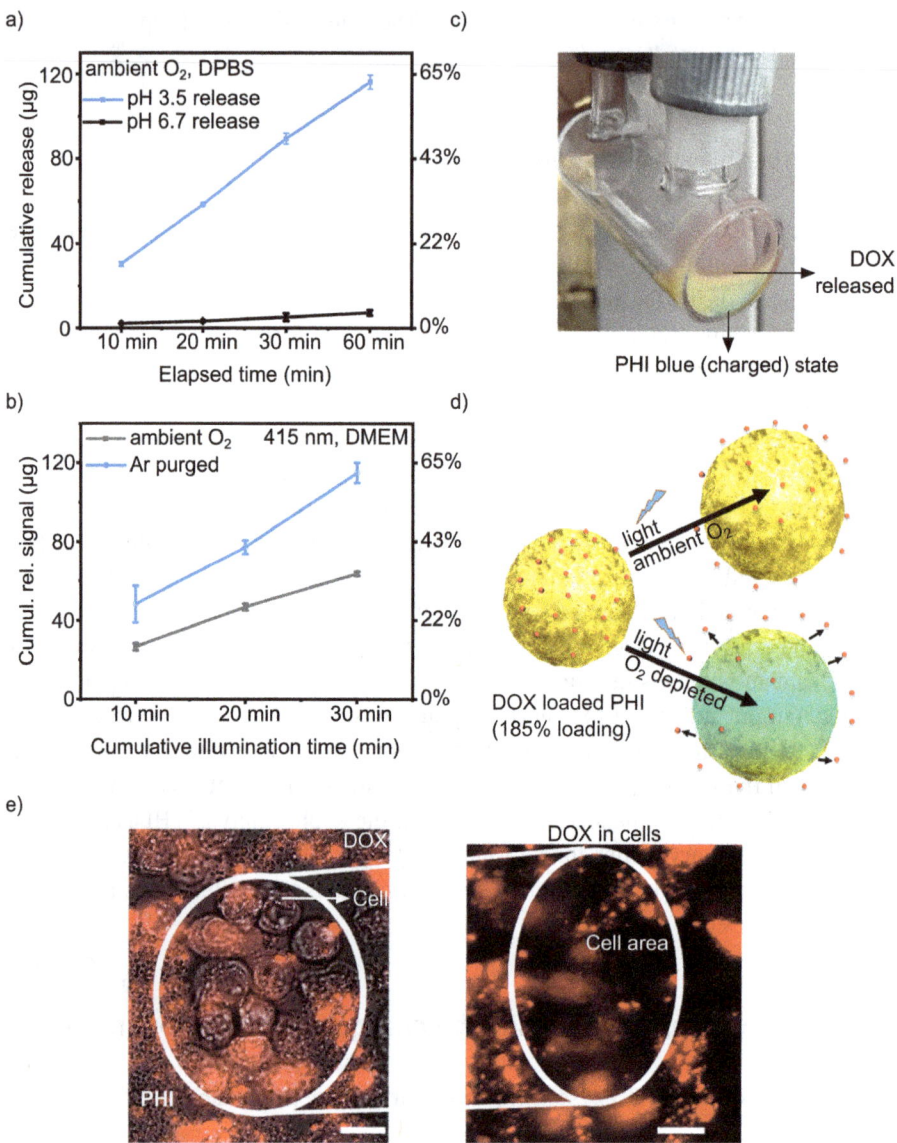

Fig. 9.9: PHI as responsive drug delivery shuttle. (a) Cumulative DOX release from preloaded PHI particles in DPBS triggered by a change in pH to acidic conditions. (b) DOX release equivalent measurement in DMEM triggered by visible-light illumination, which is more efficient in oxygen poor conditions. (c) Photograph of a suspension of DOX-loaded PHI after illumination for release, showing the blue (photocharged) PHI and red DOX released. (d) Illustration of the environmentally sensitive, smart release properties enabled by PHI's conditioned photocharging ability. (e) Optical microscopy and fluorescence image (overlaid) showing SKBR3 cancer cells (circled area). After 20 min of 415 nm illumination, DOX uptake by the cells is evidenced by fluorescence (red), leading to cell death. Scale bar: 10 μm (reproduced from reference [52] with permission from AAAS).

holes, this oxidation rate is mirrored by the generation of trapped electrons, which can then be used to quantify the concentration of analytes through traditional photocurrent measurements. Provided that the photocurrent is not lost through other channels like O_2 in solution, its magnitude is directly proportional to the analyte oxidation rate, which itself is correlated to the analyte concentration. Of note, there is no need for additional signal transducers or reactants in the analyte medium to generate the readout signal, as commonly used in earlier designs of carbon-nitride-based sensors [62]. Due to the low-lying valence band of PHI (+2.2 V vs. NHE), a broad range of analytes can be sensed, such as sugars (glucose, fructose, lactose and maltose), which, in the case of glucose, can be detected down to 11 µM in the setup reported. Physiologically relevant species such as ascorbic acid and uric acid can also be detected, as can many different alcohols including the ones typically used as electron donors in photocatalysis (methanol, ethanol, 2-propanol, 1-hexanol, triethanolamine (TEOA), or 4-methylbenzyl alcohol). The authors highlight that this direct sensing also enables validation of donor strengths and reaction rates for photocatalysis, independent of the fate of the electrons, since both charge carrier extraction rates can be limiting, in principle [4, 11, 14].

Based on this direct sensing interaction through analyte oxidation, charge accumulation on PHI was studied as a function of the analyte and other conditions. This light-induced "memory" sensing interaction after analyte interaction is termed photomemristive sensing. In the absence of an electron extraction bias, PHI is photocharged, with the photocharging being accompanied by modification of its photophysical properties, of which the most visual is the formation of the "blue state" (Fig. 9.1 and 9.10c) [13]. To acquire detailed insights into the sensing mechanism based on changes in PHI's properties, further experimentations were conducted using glucose as the test analyte, given its importance in the medical field (e.g., blood sugar quantification). The following is a summary of the "wired" (i.e., electrochemical) sensor functionalities of PHI and its "wireless" or "remote" application for sensing as (autonomous) particles in suspension, relating the charge accumulation phenomenon to the change in material optical properties.

9.3.5.2 Wired sensor functionalities

The photophysical and electrochemical properties of PHI are changed upon stable charge accumulation in PHI following analyte photooxidation, in the absence of discharge channels. By depositing PHI onto a suitable conductive substrate, such as FTO, these altered properties can be probed by a potentiostat. Upon analyte photooxidation with 30 s illumination of AM1.5 G light, the electrochemical potential increases as in the case for solar batteries (Fig. 9.11a), rising slowly from 0 V to –400 mV versus Ag/AgCl when the glucose concentration is low (50 µM). However, the potential increases sharply if the glucose concentration surpasses 1 mM, and then starts to plateau at approximately –600 mV vs. Ag/AgCl, becoming even more negative slowly over time. Hence, for a given illumination time, the electrochemical potential of the PHI sensor

a)

b)

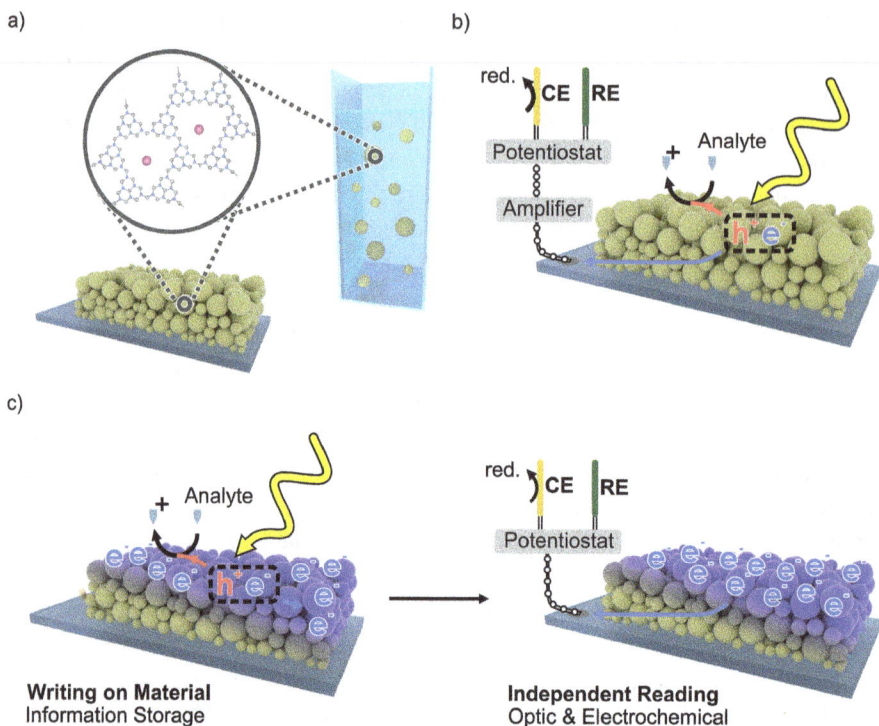

c)

Writing on Material
Information Storage

Independent Reading
Optic & Electrochemical

Fig. 9.10: Sensing by light-induced interactions with PHI. (a) PHI as (photomemristive) sensor platform when used as thin film photoelectrode, or as autonomous, dispersed particle for remote sensing. (b) Operating principle of direct photoelectrochemical sensing with PHI thin films, illustrating the facile design without further additives or signal transducers. This panel shows the photooxidation of the analyte and measurement of the resultant photocurrent with a potentiostat as amperometric readout scheme. (c) Illustration of memory sensing exploiting the phenomenon of charge accumulation in PHI, reflecting the amount of light-induced interaction with the analyte. Right side of panel shows subsequent readout by optical and electrochemical methods, based on the change in photophysical and electrochemical properties of PHI from charge accumulation (reproduced from reference [61] with permission from RSC).

depends directly on the analyte concentration. A readout of the potential is possible during illumination (at OCP), or afterwards (Fig. 9.11b). This type of analytical technique is called *potentiometric sensing*. The fast increase in potential up to −600 mV vs. Ag/AgCl is attributed to a greater surface capacitive response in this voltage domain, which is more suited for detecting analytes of low concentrations. Photocharging negative of −600 mV has also faradaic contributions, going hand in hand with redox events that require intercalation of ions into the bulk of PHI. In this regime, the specific differential capacity is increased, and with it, photocharging is less pronounced in terms of voltage change in the very negative potential window.

The second physical quantity modified by the amount of photocharging (or "photo-doping" as referred to by Beranek, Durrant and coworkers [31]) is the material's conductivity (*impedimetric sensing*). Although the nature of charge transport in PHI is not yet fully clear as it has strong ionic and some electronic contributions [4], the absolute value of the impedance (measured at OCP conditions) is monotonously correlated with the amount of charging, which itself is correlated with the analyte concentration (for a fixed illumination time), as shown in Fig. 9.11c. With increasing analyte concentration, more charges are accumulated, and the overall magnitude of impedance decreases (Fig. 9.11d). The conductivity changes are more pronounced with decreasing frequency and best discernible at 0.1 Hz, pointing to slow (ionic) processes dominating AC charge transport. This method can be used at single frequencies up to 100 Hz. Beyond, a distinction between the sensor responses for different analyte concentrations becomes difficult for the setup reported. This versatility in frequency ranges, however, conveys an important feature of the sensor: adaptable sensitivity ranges within one method. Depending on the frequency chosen, a faster measurement with less detailed concentration information is possible, whereas slower measurements (0.1 Hz, equivalent to 10 s) give higher accuracy and more detailed information. Of note, these two electrochemical measurement techniques do not interfere with each other since the charging state of the sensor is not modified (i.e., the methods are considered noninvasive). Hence, the measurements can be carried out sequentially, and even some time after photooxidation of the analyte (exploiting the longevity of the charged state), yielding results that can be corroborated. In terms of device limitations, the sensitivity of this *impedimetric* method is mainly restricted by the overall poor conductivity of PHI. However, this disadvantage may be overcome by decreasing the film thickness to enhance charge transport, or blending PHI with conductive additives such as graphene, though one needs to ensure that such modifications do not impact the homogeneous substrate coverage or affect the stability of the electron accumulation.

The third method of accessing the analyte concentration is based on coulometry by discharging the electrons accumulated in the sensor (*coulometric sensing*) through the application of a potential more positive than the OCP after photocharging (e.g., 0 V vs. Ag/AgCl, see Fig. 9.11e). This process is, in principle, akin to discharging a solar battery electrode, or to photochemical titration, to quantify the electrons extracted from photooxidation [5, 13]. For a given illumination time, the amount of charges stored on the sensor increased monotonously with concentrations between 0 and 50 mM. Of note, this discharge also resets the sensor and enables subsequent reuse; note that this reset procedure also applies to the impedimetric or potentiometric measurement method described above. This sensing technique is, however, significantly slower than the two methods described above, since a full discharge typically takes several minutes.

By modifying the illumination time, the amount of charges accumulated and, hence, the absolute sensitivity range can be easily tuned without further modification of the sensor (or the arrangement of the setup). Of note, tuning the absolute sensitivity range by increasing the illumination time may also have an impact on the dynamic range (orders of magnitude of analyte being accessible for a given measurement protocol). However, photocharging is highly nonlinear and self-limiting, as highlighted by the authors and explained in the previous sections. As the amount of electrons being stored on the material increases, the rate of further electron storage (and the analyte oxidation rate) decreases, especially in the faradaic charging regime negative of −600 mV vs. Ag/AgCl, in the case reported here. Such a nonlinear photocharging behavior is typical for light-charged materials [1] and is the reason for a more complex fitting equation being required for using the whole sensing range, independent of the method. However, the authors found that a single type of equation, taking into account the capacitive and faradaic regime independently, can be applied to all sensing cases and all ranges.

9.3.5.3 Wireless sensing

As highlighted earlier, in the case of PHI microswimmers, PHI can be used as a sensor for electron donors when dispersed in a medium. In the absence of wiring for electrochemical-based sensing of the analyte, sensing based on optical readout can be carried out, instead (Fig. 9.12a). As described earlier, the color of PHI changes from its original yellow hue with an absorption threshold at 450 nm to blue, due to the emergence of a broad absorption band that extends into the infrared region (Fig. 9.12b). The strongest band of this absorption spectrum was accurately determined to be 672 nm by absorptance measurements of the PHI suspension, using an integrating sphere. To perform such *colorimetric* sensing in a degassed electrolyte, the suspension is again illuminated for a fixed time, and the analyte concentration is directly related to the integral of the absorption band, or the absorptance at a single wavelength. This colorimetric sensing mechanism can, in principle, be applied to films of PHI, but is more challenging to implement, due to strong scattering of the material (and periphery) on the wired sensing platform.

Another optical property that is altered by photocharging is the PL emission yield. Exciting the pristine material with UV light yields a broad PL signal centered at around 460 nm (Fig. 9.12c and d), the intensity of which is attenuated with increasing photocharging. In other words, one can draw a direct relationship between the PL signal intensity and the extent of photocharge as dependent on the donor concentration for a given illumination time. As with the colorimetric-based sensing above, the whole PL spectrum can be recorded and integrated for higher accuracy, and faster measurements are possible by probing the PL intensity at a single wavelength. Also, this *fluorimetric sensing* technique can be applied to thin films of PHI, if desired. To reset the particle sensor, it is sufficient to purge the suspension with oxygen to discharge the system. After subsequent degassing (and washing to exchange the medium, if desired), the material can be reused easily. Both fluorimetric and colorimetric sensing can be combined to obtain two sets of complementary data, which can be used for

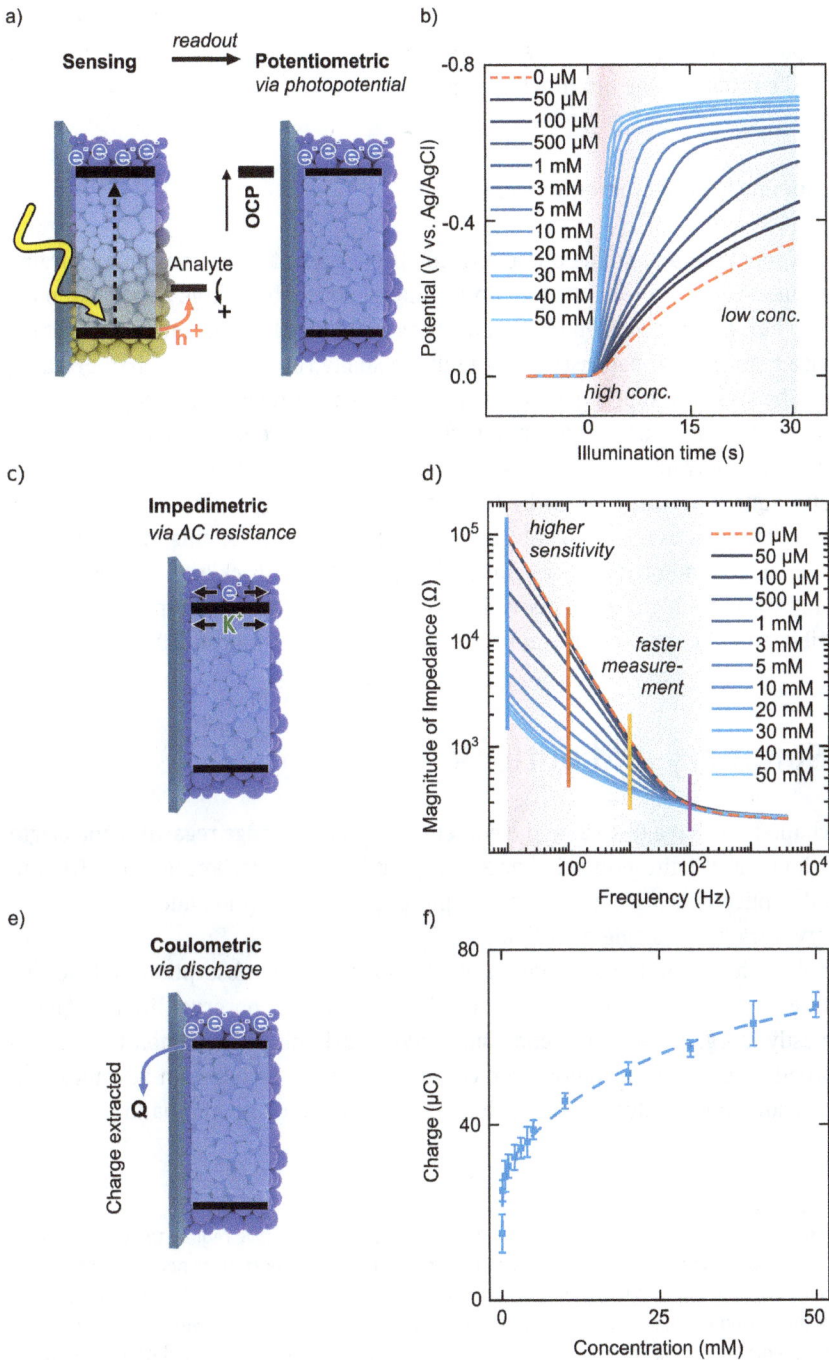

Fig. 9.11: Sensing based on light-induced modification of PHI properties. (a, b) Potentiometric sensing: for a given illumination time, the OCP of the PHI film electrode shifts negatively with increasing donor

data validation, especially in the case that one method is more reliable than another within a particular range of analyte concentration. Also here, the absolute sensitivity range can be tuned via the illumination time.

9.3.5.4 Summary of photomemristive sensing

The use of modified and stable optoelectronic properties of PHI upon photocharging provides an easy means to memorize donor interactions. The charge accumulation is assisted by ions and depends on the flux of charges. Owing also to its memory function, this device resembles a memristor with multiple stable resistance (i.e., charge) states, which can be accessed to perform multivariate or even neuromorphic computation. Finally, the accumulation of charges and, hence, energy storage by a "solar battery sensor" enables not only information storage and interfacing with electric circuits, but also drives subsequent actions directly by the accumulated energy. Such responsive function, often discussed in the context of neuromorphic signal processing, bodes well for the development of autonomous systems realizing a tight feedback loop between detection and reaction, as highlighted above in the example of light-driven microswimmers that exhibit stimuli-responsive, diagnostic and therapeutic (theranostic) functions.

9.4 Summary and outlook

In this chapter, we have described the current state of knowledge regarding the origin and mechanism of light-induced charge trapping in the 2D carbon nitride PHI, and discussed applications arising from the coupling of light absorption and electrochemical energy storage in a single material.

Although the property combination of simultaneous light absorption and electrochemical energy storage is not unique to PHI and has been observed in a couple of other, mostly inorganic materials and composites [1, 27], this phenomenon in PHI enables a broad scope of applicability when combined with the well-known advantages of carbon nitride photocatalysts, namely, having suitable band gap and band potentials,

---

Fig. 9.11 (continued)
concentration. Two regimes must be taken into account: a surface capacitive regime to −600 mV vs. Ag/AgCl and a more pseudocapacitive regime with higher charge density at more negative potentials. (c, d) Impedimetric sensing: As more charges accumulate, the magnitude of impedance decreases as a result of increasing charge carrier concentration (photodoping). Readout can even be performed at a single frequency. (e, f) Coulometric sensing: after charge accumulation following photooxidation of the analyte, a bias is applied to discharge the sensor. The cumulative (integrated) charge extracted from this measurement is monotonously related to the analyte concentration. Simultaneously, this measurement resets the sensor (reproduced from reference [61] with permission from RSC).

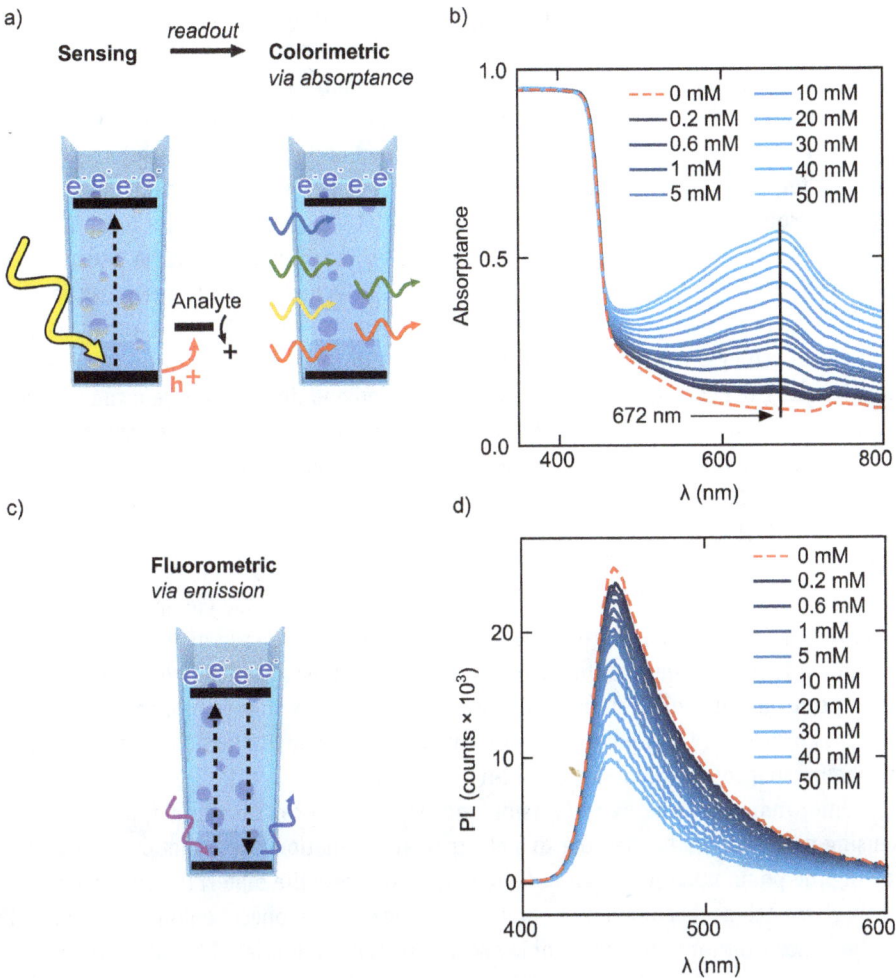

Fig. 9.12: Optical photomemristive sensing using suspended PHI particles. (a) The continuous formation of a broad absorption band with a maximum at 672 nm during photocharging due to a color change in the material. (b) The amount of charging is quantified by absorptance measurements, where signal intensity increases with increasing analyte concentration for a fixed illumination time. The measurement method is referred to as *colorimetric*. Readout can also be performed at a single wavelength.
(c) Fluorometric sensing: analyte quantification by measuring the attenuation of the photoluminescence (PL) signal after photocharging. (d) Spectra showing decreasing intensity of the PL signal with increased photocharging at increasing analyte concentrations due to charge accumulation affecting radiative emission. For short measurement time, quantification of the PL signal can be done on a single emission wavelength. In both cases, readout is carried out remotely, circumventing the electric connection required for electrochemical methods (reproduced from reference [61] with permission from RSC).

excellent chemically stability, biocompatibility, earth-abundance and partial processability into microparticulate suspensions and films. Intrinsically, the structural porosity of PHI enables intercalation of mobile (hydrated) cations, which were found to be conducive to this electron stabilization phenomenon via charge screening interactions. Given that these electrons can be accumulated and can maintain their reductive strength for tens of hours or more, generating such electrons by photooxidation of suitable electron donors is equivalent to the storage of solar energy in the form of high potential electrons, thus bridging the gap between solar energy conversion and storage in a single material. A special characteristic of PHI is its high intrinsic overpotential for water reduction, enabling charge accumulation negative of RHE, which, otherwise, is difficult to achieve in aqueous conditions. Only with these energetics and surface energy barriers, the chemical energy conversion on demand described herein can be realized. Similarly, the negative storage potential is of benefit for aqueous (solar) batteries, as their operation voltage on the anode side can be increased beyond the water reduction potential. Methods of solar energy conversion and storage as presented here include using PHI as battery photoelectrode (recovering stored energy as electricity), in "dark photocatalysis" for chemical transformations (recovering solar energy as chemical fuels), and as propulsion component in microscale devices such as microswimmers (recovery as kinetic energy from chemical reactions). In the case of microswimmers, additional functionalities can be installed, exploiting other beneficial characteristics of PHI. Specifically, its (textural) porosity enables high capacity storage of a drug molecule, which is prevented from passive release by intermolecular interactions with PHI, and its photoresponsiveness allows for light-triggered drop-off.

Other than solar energy conversion and storage, PHI can also find applications as sensing platforms by harnessing this electron accumulation phenomenon which, following the photooxidation of an analyte chemical, alters the material's optoelectronic properties such as its conductivity and color. Since such property changes are related to the concentration of the oxidizable species, information related to the analyte concentration can be stored within PHI and read out by electrochemical and spectroscopic techniques. Realization of multiple stable conductivity states in PHI that are triggered in the presence of light and persist in the dark imparts it with photomemristive properties, which may broaden the application scope of this material to circuit components in neuromorphic computation devices for (semiautonomous) decision-making. This particular application would be especially attractive when combined with cargo-transporting microswimmers, where decision related to payload drop-off may be made autonomously, or combined with sensing platforms, where further actions are automatically taken in the presence of specific analytes (i.e., analytes acting as stimuli to trigger further processes).

Considering the broad scope of applications as demonstrated by the proof-of-concepts presented here, improving the performance of various devices is underpinned by a better control of the electron trapping process in PHI, especially in terms of increasing the electron storage capacity, maintaining their reductive strength for longer duration and magnitude of the change in optoelectronic or (opto-)ionic properties. These may be realized through suitable modification of the underlying PHI structure, as guided by a better fundamental understanding of the factors giving rise to this phenomenon, especially in controlling the dynamics of solvated cations within the structural pores. Further afield, these insights in PHI may be adapted to the development of other materials and technologies, particularly for photo(electro)chemically active materials, in terms of controlling the photocarrier stabilization dynamics, which is one of the limiting factors in photon-to-charge carrier conversion efficiency. Advances in both the fundamental understanding and technological development related to the electron accumulation phenomenon, in PHI as well as in other (in)organic materials, are anticipated to lead to a wide range of functional devices with potential applications in sectors spanning energy conversion and storage, chemical analysis, micro-machineries, environmental remediation and biomedical systems.

References

[1]   Savateev O. Photocharging of semiconductor materials: Database, quantitative data analysis, and application in organic synthesis, Adv Energy Mater 2022, 12(21), 2200352.

[2]   Lau VWH, Klose D, Kasap H, Podjaski F, Pignie MC, Reisner E, et al. Dark photocatalysis: Storage of solar energy in carbon nitride for time-delayed hydrogen generation. Angew Chem Int Ed 2017, 56(2), 510–14.

[3]   Schlomberg H, Kröger J, Savasci G, Terban MW, Bette S, Moudrakovski I, et al. Structural insights into poly(Heptazine Imides): A light-storing carbon nitride material for dark photocatalysis. Chem, Mater 2019, 31(18), 7478–86.

[4]   Kröger J, Podjaski F, Savasci G, Moudrakovski I, Jiménez-Solano A, Terban MW, et al. Conductivity mechanism in ionic 2D carbon nitrides: From hydrated ion motion to enhanced photocatalysis. Adv Mater 2022, 34(7), 2107061.

[5]   Markushyna Y, Lamagni P, Teutloff C, Catalano J, Lock N, Zhang G, et al. Green radicals of potassium poly(heptazine imide) using light and benzylamine. J Mater Chem A 2019, 7(43), 24771–75.

[6]   Zeng Z, Quan X, Yu H, Chen S, Zhang Y, Zhao H, et al. Carbon nitride with electron storage property: Enhanced exciton dissociation for high-efficient photocatalysis. Appl Catal: B 2018, 236, 99–106.

[7]   Savateev A, Kurpil B, Mishchenko A, Zhang G, Antonietti M. A "waiting" carbon nitride radical anion: A charge storage material and key intermediate in direct C–H thiolation of methylarenes using elemental sulfur as the "S"-source, Chem Sci 2018, 9(14), 3584–91.

[8]   Podjaski F, Lotsch BV. Optoelectronics meets optoionics: Light storing carbon nitrides and beyond, Adv Energy Mater 2021, 11(4), 2003049.

[9]   Lau VW-H, Moudrakovski I, Botari T, Weinberger S, Mesch MB, Duppel V, et al. Rational design of carbon nitride photocatalysts by identification of cyanamide defects as catalytically relevant sites. Nat Commun 2016, 7(1), 12165.

[10] Krivtsov I, Mitoraj D, Adler C, Ilkaeva M, Sardo M, Mafra L, et al. Water-soluble polymeric carbon nitride colloidal nanoparticles for highly selective quasi-homogeneous photocatalysis. Angew Chem Int Ed 2020, 59(1), 487–95.

[11] Kröger J, Jiménez-Solano A, Savasci G, Lau VWh, Duppel V, Moudrakovski I, et al. Morphology control in 2D carbon nitrides: Impact of particle size on optoelectronic properties and photocatalysis. Adv Funct Mater 2021, 31(28), 2102468.

[12] Savateev A, Pronkin S, Willinger MG, Antonietti M, Dontsova D. Towards organic zeolites and inclusion catalysts: Heptazine imide salts can exchange metal cations in the solid state, Chem – Asian J 2017, 12(13), 1517–22.

[13] Podjaski F, Kröger J, Lotsch BV. Toward an aqueous solar battery: Direct electrochemical storage of solar energy in carbon nitrides, Adv Mater 2018, 30(9), 1705477.

[14] Kröger J, Jiménez-Solano A, Savasci G, Rovó P, Moudrakovski I, Küster K, et al. Interfacial engineering for improved photocatalysis in a charge storing 2D carbon nitride: Melamine functionalized poly(heptazine imide). Adv Energy Mater 2021, 11(6), 2003016.

[15] Li X, Bartlett SA, Hook JM, Sergeyev I, Clatworthy EB, Masters AF, et al. Salt-enhanced photocatalytic hydrogen production from water with carbon nitride nanorod photocatalysts: Cation and pH dependence. J Mater Chem A 2019, 7(32), 18987–95.

[16] Wang Y, Vogel A, Sachs M, Sprick RS, Wilbraham L, Moniz SJA, et al. Current understanding and challenges of solar-driven hydrogen generation using polymeric photocatalysts. Nat Energy 2019, 4(9), 746–60.

[17] Banerjee T, Podjaski F, Kröger J, Biswal BP, Lotsch BV. Polymer photocatalysts for solar-to-chemical energy conversion, Nat Rev Mater 2021, 6(2), 168–90.

[18] Kosco J, Gonzalez-Carrero S, Howells CT, Fei T, Dong Y, Sougrat R, et al. Generation of long-lived charges in organic semiconductor heterojunction nanoparticles for efficient photocatalytic hydrogen evolution. Nat Energy 2022, 7(4), 340–51.

[19] Savateev A, Tarakina NV, Strauss V, Hussain T, ten Brummelhuis K, Sánchez Vadillo JM, et al. Potassium poly(heptazine Imide): Transition metal-free solid-state triplet sensitizer in cascade energy transfer and [3+2]-cycloadditions. Angew Chem Int Ed 2020, 59(35), 15061–68.

[20] Yang W, Godin R, Kasap H, Moss B, Dong Y, Hillman SAJ, et al. Electron accumulation induces efficiency bottleneck for hydrogen production in carbon nitride photocatalysts. J Am Chem Soc 2019, 141(28), 11219–29.

[21] Li C, Adler C, Krivtsov I, Mitoraj D, Leiter R, Kaiser U, et al. Ultrafast anisotropic exciton dynamics in a water-soluble ionic carbon nitride photocatalyst. Chem Commun 2021, 57(82), 10739–42.

[22] Kasap H, Caputo CA, Martindale BCM, Godin R, Lau VW-H, Lotsch BV, et al. Solar-driven reduction of aqueous protons coupled to selective alcohol oxidation with a carbon nitride–molecular Ni catalyst system. J Am Chem Soc 2016, 138(29), 9183–92.

[23] Rieth AJ, Qin Y, Martindale BCM, Nocera DG. Long-lived triplet excited state in a heterogeneous modified carbon nitride photocatalyst, J Am Chem Soc 2021, 143(12), 4646–52.

[24] Li C, Hofmeister E, Krivtsov I, Mitoraj D, Adler C, Beranek R, et al. Photodriven charge accumulation and carrier dynamics in a water-soluble carbon nitride photocatalyst. ChemSusChem 2021, 14(7), 1728–36.

[25] Savateev A, Antonietti M. Ionic carbon nitrides in solar hydrogen production and organic synthesis: Exciting chemistry and economic advantages, Chemcatchem 2019, 11(24), 6166–76.

[26] Lau VW-H, Lotsch BV. A Tour-guide through carbon nitride-land: Structure- and dimensionality-dependent properties for photo(electro)chemical energy conversion and storage, Adv Energy Mater 2022, 12(4), 2101078.

[27] Lv J, Xie J, Mohamed AGA, Zhang X, Wang Y. Photoelectrochemical energy storage materials: Design principles and functional devices towards direct solar to electrochemical energy storage, Chem Soc Rev 2022, 51(4), 1511–28.

[28] Gouder A, Podjaski F, Jiménez-Solano A, Kröger J, Wang Y, Lotsch B. V. Energy Environ Sci 2023, DOI: 10.1039/D2EE03409C.

[29] Sridhar V, Podjaski F, Kröger J, Jiménez-Solano A, Park B-W, Lotsch BV, et al. Carbon nitride-based light-driven microswimmers with intrinsic photocharging ability. Proc Natl Acad Sci U S A 2020, 117 (40), 24748–56.

[30] Conway BE. Electrochemical Supercapacitors. New York, Springer, 1999.

[31] Adler C, Selim S, Krivtsov I, Li C, Mitoraj D, Dietzek B, et al. Photodoping and fast charge extraction in ionic carbon nitride photoanodes. Adv Funct Mater 2021, 31(45), 2105369.

[32] Loh JYY, Kherani NP, Ozin GA. Persistent CO_2 photocatalysis for solar fuels in the dark, Nat Sustain 2021, 4(6), 466–73.

[33] Ou H, Tang C, Chen X, Zhou M, Wang X. Solvated electrons for photochemistry syntheses using conjugated carbon nitride polymers, ACS Catal 2019, 9(4), 2949–55.

[34] Godin R, Wang Y, Zwijnenburg MA, Tang J, Durrant JR. Time-resolved spectroscopic investigation of charge trapping in carbon nitrides photocatalysts for hydrogen generation, J Am Chem Soc 2017, 139(14), 5216–24.

[35] Godin R, Durrant JR. Dynamics of photoconversion processes: The energetic cost of lifetime gain in photosynthetic and photovoltaic systems, Chem Soc Rev 2021, 50(23), 13372–409.

[36] Fernández-Medina M, Ramos-Docampo MA, Hovorka O, Salgueiriño V, Städler B. Recent advances in nano- and micromotors, Adv Funct Mater 2020, 30(12), 1908283.

[37] Šípová-Jungová H, Andrén D, Jones S, Käll M. Nanoscale inorganic motors driven by light: Principles, realizations, and opportunities, Chem Rev 2020, 120(1), 269–87.

[38] Wang J, Xiong Z, Zheng J, Zhan X, Tang J. Light-driven micro/nanomotor for promising biomedical tools: Principle, challenge, and prospect, Acc Chem Res 2018, 51(9), 1957–65.

[39] Eskandarloo H, Kierulf A, Abbaspourrad A. Light-harvesting synthetic nano- and micromotors: A review, Nanoscale 2017, 9(34), 12218–30.

[40] Dong R, Cai Y, Yang Y, Gao W, Ren B. Photocatalytic micro/nanomotors: From construction to applications, Acc Chem Res 2018, 51(9), 1940–47.

[41] Ye Z, Sun Y, Zhang H, Song B, Dong B. A phototactic micromotor based on platinum nanoparticle decorated carbon nitride, Nanoscale 2017, 9(46), 18516–22.

[42] Kasap H, Godin R, Jeay-Bizot C, Achilleos DS, Fang X, Durrant JR, et al. Interfacial engineering of a carbon nitride–graphene oxide–molecular Ni catalyst hybrid for enhanced photocatalytic activity. ACS Catal 2018, 8(8), 6914–26.

[43] Moran JL, Posner JD. Phoretic self-propulsion, Annu Rev Fluid Mech 2017, 49(1), 511–40.

[44] Moran JL, Wheat PM, Posner JD. Locomotion of electrocatalytic nanomotors due to reaction induced charge autoelectrophoresis, Phys Rev E 2010, 81(6), 065302.

[45] Kong L, Mayorga-Martinez CC, Guan J, Pumera M. Photocatalytic micromotors activated by UV to visible light for environmental remediation, micropumps, reversible assembly, transportation, and biomimicry, Small 2020, 16(27), 1903179.

[46] Urso M, Pumera M. Nano/microplastics capture and degradation by autonomous nano/ microrobots: A perspective, Adv Funct Mater 2022, 32(20), 2112120.

[47] Hermanová S, Pumera M. Micromachines for microplastics treatment. ACS Nanosci Au 2022, 2, 225–32.

[48] Ussia M, Pumera M. Towards micromachine intelligence: Potential of polymers, Chem Soc Rev 2022, 51(5), 1558–72.

[49] Zhan X, Wang J, Xiong Z, Zhang X, Zhou Y, Zheng J, et al. Enhanced ion tolerance of electrokinetic locomotion in polyelectrolyte-coated microswimmer. Nat Commun 2019, 10(1), 3921.

[50] Xiao K, Chen L, Chen R, Heil T, Lemus SDC, Fan F, et al. Artificial light-driven ion pump for photoelectric energy conversion. Nat Commun 2019, 10(1), 74.

[51] Xiao K, Tu B, Chen L, Heil T, Wen L, Jiang L, et al. Photo-driven ion transport for a photodetector based on an asymmetric carbon nitride nanotube membrane. Angew Chem Int Ed Engl 2019, 58(36), 12574–79.

[52] Sridhar V, Podjaski F, Alapan Y, Kröger J, Grunenberg L, Kishore V, et al. Light-driven carbon nitride microswimmers with propulsion in biological and ionic media and responsive on-demand drug delivery. Sci Rob 2022, 7(62), eabm1421.

[53] Wei M, Zhou C, Tang J, Wang W. Catalytic micromotors moving near polyelectrolyte-modified substrates: The roles of surface charges, morphology, and released ions, ACS Appl Mater Interfaces 2018, 10(3), 2249–52.

[54] Dai B, Wang J, Xiong Z, Zhan X, Dai W, Li C-C, et al. Programmable artificial phototactic microswimmer. Nat Nanotechnol 2016, 11(12), 1087–92.

[55] Uspal WE. Theory of light-activated catalytic Janus particles, J Chem Phys 2019, 150(11), 114903.

[56] Singh DP, Uspal WE, Popescu MN, Wilson LG, Fischer P. Photogravitactic microswimmers, Adv Funct Mater 2018, 28(25), 1706660.

[57] Gao Y, Chen Y, Ji X, He X, Yin Q, Zhang Z, et al. Controlled intracellular release of doxorubicin in multidrug-resistant cancer cells by tuning the shell-pore sizes of mesoporous silica nanoparticles. ACS Nano 2011, 5(12), 9788–98.

[58] Calza P, Medana C, Sarro M, Rosato V, Aigotti R, Baiocchi C, et al. Photocatalytic degradation of selected anticancer drugs and identification of their transformation products in water by liquid chromatography–high resolution mass spectrometry. J Chromatogr A 2014, 1362, 135–44.

[59] Akram MW, Raziq F, Fakhar-e-alam M, Aziz MH, Alimgeer KS, Atif M, et al. Tailoring of Au-TiO2 nanoparticles conjugated with doxorubicin for their synergistic response and photodynamic therapy applications. J Photochem Photobiol A 2019, 384, 112040.

[60] Wan C, Xiao K, Angelin A, Antonietti M, Chen X. The Rise of bioinspired ionotronics, Adv Intell Syst 2019, 1(7), 1900073.

[61] Gouder A, Jiménez-Solano A, Vargas-Barbosa NM, Podjaski F, Lotsch BV. Photomemristive sensing *via* charge storage in 2D carbon nitrides. Mater Horiz 2022, 9, 1866–77.

[62] Xavier MM, Nair PR, Mathew S. Emerging trends in sensors based on carbon nitride materials, Analyst 2019, 144(5), 1475–91.

Nobuhiko Mitoma* and Takuzo Aida*

Chapter 10
Graphitic carbon nitride thin films: synthesis, properties, actuators and electronic devices

10.1 Introduction

The production of hydrogen as a clean energy source using nontoxic and abundant elements and visible light is an important strategy for realization of a sustainable society. Graphitic carbon nitride (GCN) is a substance of great interest because it contains no metals, is composed only of carbon and nitrogen, which are abundant in the earth, and shows photocatalytic activity in the photodegradation of water [1]. Many GCN materials have been obtained as powders, and research on them and nanocomposites has developed greatly [2]. However, what if GCN could be obtained in film form? Films are an important group of materials, along with fibers, because they can be used for layering, folding, wrapping and separating objects. Films are important in terms of not only applications but also basic science. For example, the discovery and understanding of conductive polymers would not have been possible without the synthesis of polyacetylene in film form. The first polyacetylene was synthesized by Natta et al. and attracted the interest of many researchers, but the research did not progress much because only insoluble and nonmelting black powder was initially obtained. However, the situation drastically changed when Shirakawa et al. reported the synthesis of polyacetylene in film form [3]. When doped, polyacetylene films have a metallic luster, show high conductivity and are known today as conductive polymers. Previously, organic materials were regarded as insulators, but research on polyacetylene overturned this common knowledge. Later, research on conductive polyacetylene led to the development of organic electronic devices such as transistors, light-emitting diodes (LEDs) and capacitors. Without the realization of polyacetylene films, we would not have recognized the importance of organic semiconductor electronics. In addition to conductivity, polyacetylene films have been successfully observed to have

Acknowledgments: This work was supported by JSPS KAKENHI grant numbers JP21K04840 and JP18H05260. We thank Dr. Oleksandr Savateev (Max Planck Institute of Colloids and Interfaces) for fruitful comments on the chapter.

*Corresponding author: Nobuhiko Mitoma, RIKEN Center for Emergent Matter Science, Saitama, Japan; Department of Chemistry and Biotechnology, The University of Tokyo, Tokyo, Japan, e-mail: nobuhiko.mitoma@riken.jp
*Corresponding author: Takuzo Aida, RIKEN Center for Emergent Matter Science, Saitama, Japan; Department of Chemistry and Biotechnology, The University of Tokyo, Tokyo, Japan, e-mail: aida@macro.t.u-tokyo.ac.jp

https://doi.org/10.1515/9783110746976-010

clear IR and Raman spectra, and their chemical structure has been elucidated [4, 5], making them applicable to a larger scope of analytical methods than powder. Despite the numerous advantages of films, research on GCN films is still in its infancy [6, 7].

GCN powder can be obtained by simply heating nitrogen-containing precursors such as cyanamide, dicyandiamide, and melamine, and attempts have been made to elucidate the polymerization mechanism. Jürgens et al. obtained GCN intermediates, analyzed their crystal structures and found that Melem is a stable intermediate [8]. Since Melem can be synthesized from various precursors, there are various theories on the mechanism of its formation. May proposed a mechanism for thermal decomposition of melamine to cyanamide, which reacts with another melamine to form Melem [9]. Dicyandiamide has been reported to react with melamine to form Melem [10]. Polyaddition and polycondensation reactions for the transformation to the intermediate Melem have been found to occur at approximately 390 °C, and desorption of ammonia and further polycondensation occur at approximately 520 °C. In addition, GCN is a semiconductor that shows catalytic activity in the photolysis of water. The energy level of the GCN conduction band is higher than that of TiO_2, which translates into higher reducing ability [1, 11]. GCN is a semiconductor with very useful properties, but when considering its applications in devices, processing it from powder to film is very difficult because it neither dissolves in any solvent nor melts. Film materials are suitable for applications such as optics, electronics and mechanics and have potential that cannot be realized with powder materials.

In a typical film coating technique, a suspension of GCN is applied onto a substrate [12]. Chen et al. produced GCN films with fewer cracks and higher uniformity by applying pressure during film coating, but there was no difference in the structure or vibration spectrum compared to GCN powder [13]. This indicates that the orientation of the GCN film domains is not aligned but random. In 2016, our group reported that self-supporting, uniform and highly oriented film samples could be obtained by vapor deposition polymerization (VDP) [14]. The films have different vibrational spectra from those of conventional powders, and they also exhibit actuation behavior via adsorption and desorption of small amounts of water vapor in the air. In the next and subsequent sections, a series of new highly oriented GCN film materials will be discussed.

10.2 GCN film synthesis using vapor deposition techniques

There are several methods to synthesize a film. If the polymer is soluble in a solvent, then a thin film can be obtained by spin casting or ink jetting. For thermoplastic polymers, extrusion molding or hot pressing can be used to produce a thin film. Since the structural order of polymers changes depending on the film formation method, their physical properties are closely related to the sample preparation method. However,

depending on the solubility and melting behavior of the polymer, the sample preparation methods are limited.

VDP is a method of forming films via sublimation of precursor monomers followed by chemical reaction on a target substrate. Compared to the solution casting
method, VDP enables homogeneous film formation and film thickness control. An important point regarding VDP is that it can form films even when the polymer has a
rigid structure or a large molecular weight that makes dissolution in solvents difficult.
A typical example of a film made by VDP is parylene, which is an insulator with a
high dielectric constant and used as an electronic material because it is highly resis-
tant to corrosion and moisture. Transparent and flexible film can be fabricated from
this material [15, 16]. The precursor of parylene is [2.2]paracyclophane, which is converted to para-xylene by applying high temperature in vacuum followed by radical
polymerization on the target substrate [17]. The two precursors can also be combined
to perform VDP. For example, polyimide films can be synthesized from diamine and
acid anhydride monomers [18], and polyurea films can be synthesized from diamine
and diisocyanate monomers [19].

Our group synthesized the first free-standing highly oriented GCN film by VDP
[14]. Several structures have been proposed for carbon nitrides, including α, β and
cubic structures [20]. In the standard state, the graphitic phase, i.e., GCN is the most
stable structure [2, 21]. However, until recently, the structure of GCN has remained
elusive. The basic structure of GCN is proposed to consist of 2,5,8-triamino-tri-s-
triazine (heptazine) units. However, its insolubility in solvents limits the structural
identification methods, and the experimental results previously reported show that
the catalytic, electronic and optical properties of GCN differ among research groups
[22, 23]. This suggests that the structure of GCN can differ depending on the sample
preparation method and equipment. However, even if the structure of GCN is different,
it is expected to be a sheet-like material due to hydrogen bonding between polymer
chains and a periodic layer-to-layer distance that has been observed in X-ray diffraction
experiments [2, 22]. Our group focused on guanidine salts as molecular precursors for
the synthesis of GCN films. Guanidine salts are highly soluble in water and form stable
cations due to charge delocalization in aqueous solution. The guanidinium ion is known
to be an analog of dicyandiamide, a typical precursor of GCN, and the melaminium ion
[24], a GCN intermediate, can be obtained from guanidine hydrochloride. Synthesis of
GCN powder from guanidinium ions was also reported to be possible [25]. The structure
of the obtained GCN is expected to be controlled by changing the counteranion of guanidine. As a result of investigating various counteranions, our group found that guanidinium carbonate was the most suitable precursor for this purpose (Fig. 10.1a).

The GCN film synthesis method developed by our group is described below. The
guanidinium carbonate (Gdm$_2$CO$_3$) was ground well and placed at the bottom of a test
tube. The target substrate was placed in the center of the test tube, covered with aluminum foil, and heated to 550 °C (Fig. 10.1b). Guanidinium carbonate polymerized as previously reported, yielding GCN powder at the bottom of the test tube [2]. At the same

a)

Guanidinium carbonate
(Gdm$_2$CO$_3$)

NH$_3$, H$_2$O, CO$_2$

1) 25 $\longrightarrow$ 550°C (10°C min^{-1})
2) 550°C (30 min)

Repeating unit
(heptazine)

b)

Growth side

Substrate side

Glass target   Al foil cap

Gdm$_2$CO$_3$

c)   d)   e)

1.0 cm

RIKEN

1.0 µm   1.0 µm

Growth side   Cross-section

2.0 mm

Growth
side

Substrate
side

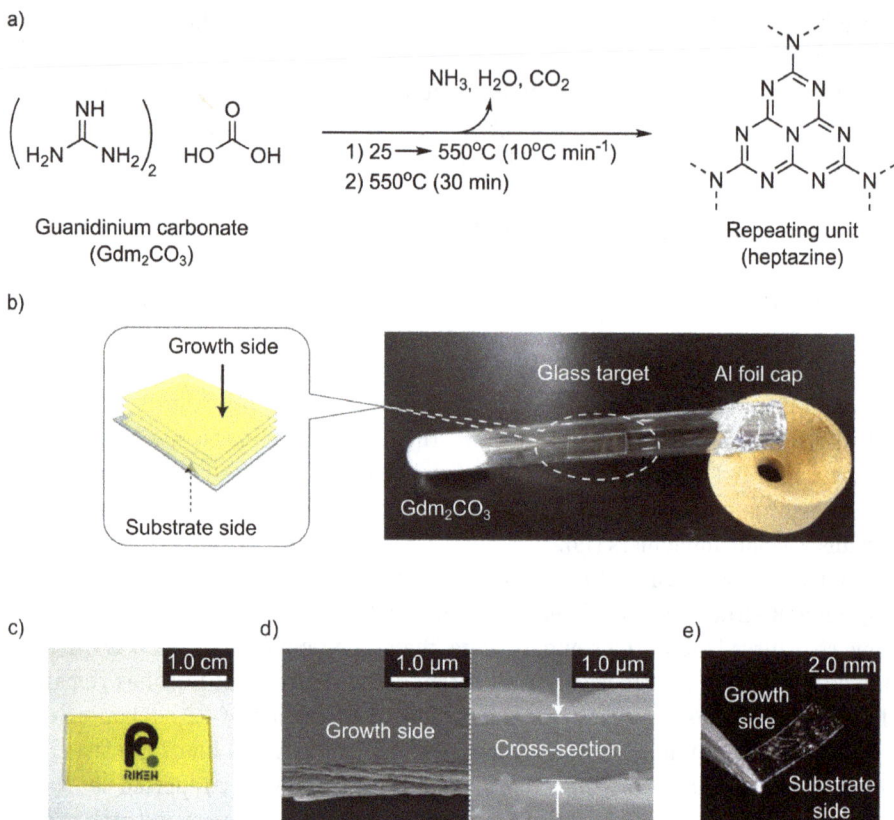

Fig. 10.1: (a) Scheme of the synthesis of GCN consisting of heptazine as a repeating unit. (b) Schematic representation of the VDP experimental setup used to synthesize GCN films on a target substrate from Gdm$_2$CO$_3$ as a starting compound. (c, d) Photograph (c) and scanning electron micrographs (d) of a GCN film (top (left) and side (right) views) obtained by VDP on a glass substrate. (e) Optical micrograph of a GCN film peeled off from a glass substrate after VDP (reproduced with permission from reference [14]. Copyright 2016, Macmillan Publishers Ltd.).

time, by using this method, GCN films could be deposited on glass, silicon, fluorine tin oxide substrates, carbon fibers and other heat-resistant materials. A 30-min GCN film deposition at 550 °C is shown in Figure 10.1c. A uniform yellow transparent film was produced on the glass substrate. The resulting GCN film was observed by electron microscopy to have a crack-free surface, ordered layered structure and a thickness of approximately 0.9 µm (Fig. 10.1d). The thickness of the GCN film could be varied from 0.1 to 1.8 µm by changing the amount of guanidinium carbonate as a precursor and the heating time. Even thicker films could be synthesized, but the transparency was lost. By immersion of the substrate in water at 70 °C, 0.8 µm-thick self-standing and free-of-wrinkles GCN film was obtained (Fig. 10.1e), which is a clear difference from previous reports.

Inspired by our VDP method, Cai et al. obtained a GCN film by physical vapor transport (PVT) [26]. First, GCN powder was obtained by heating melamine to 550 °C in air. The obtained GCN powder was placed in a quartz tube under an Ar flow of 100 mL min^{-1} and heated at temperature ramp of 7.5 °C min^{-1} to 720 °C and kept at this temperature for 4 h. A translucent pale yellow film was formed at the opposite end of the quartz tube, heated at approximately 200–240 °C. The film could be removed from the tube wall by cooling it to room temperature and immersing the quartz tube in water. The GCN powder residue after heating had more structural defects than that before heating, which is explained by release of Melem vapor upon heating at high temperature. Similar to our film prepared by VDP, this film also showed actuation in response to moisture. Furthermore, the film showed tough mechanical properties.

Giusto et al. obtained GCN films by a two-zone chemical vapor deposition (CVD) method consisting of a low-temperature section and a high-temperature section [27]. The precursor melamine and the target substrate were placed in a quartz tube. The pressure in the quartz tube was reduced to 10 Pa, and 50 sccm N_2 carrier gas was introduced. The target substrate installed downstream was heated to 550 °C beforehand, and the precursor melamine installed upstream was heated to 300 °C for 30 min, with a temperature ramp of 10 °C min^{-1} to obtain a GCN film on the substrate. The resulting film was a transparent and bendable polymer material. The refractive index of typical polymer materials is 1.33–1.70, but the GCN film prepared by Giusto et al. had larger refractive index of 2.32–2.43, which is expected to be useful for optical devices. Liu et al. also recently reported that a 4-inch wafer-scale GCN film could be obtained by a similar CVD method [28]. The surface roughness of the obtained GCN film was 0.63 nm, which was a significant improvement over the previously reported best value of 12 nm [29]. Regardless of the precursor used or the configuration of the furnace, thermal polymerization that yields GCN occurs at 550 °C. Increasing the amount of precursor or increasing the deposition time can increase the thickness of the resulting film.

10.3 Analysis of GCN films

The GCN films obtained by VDP or CVD methods have similarities with and differences from GCN powders that are synthesized previously. X-ray diffraction, Fourier-transform infrared (FTIR) spectroscopy and X-ray photoelectron spectroscopy (XPS) were used to analyze the GCN films. Figure 10.2 shows the measurement results.

The two-dimensional wide-angle X-ray diffraction (WAXD) pattern of the film with almost horizontal X-ray incidence is shown in the inset of Figure 10.2a. In this case, the periodic structure can be observed only in the direction perpendicular to the film surface, indicating that the π-plane orientation is very high. A one-dimensional plot of the scattering intensity is shown in Figure 10.2a, where a single peak is observed, and the

a)

b)

c)

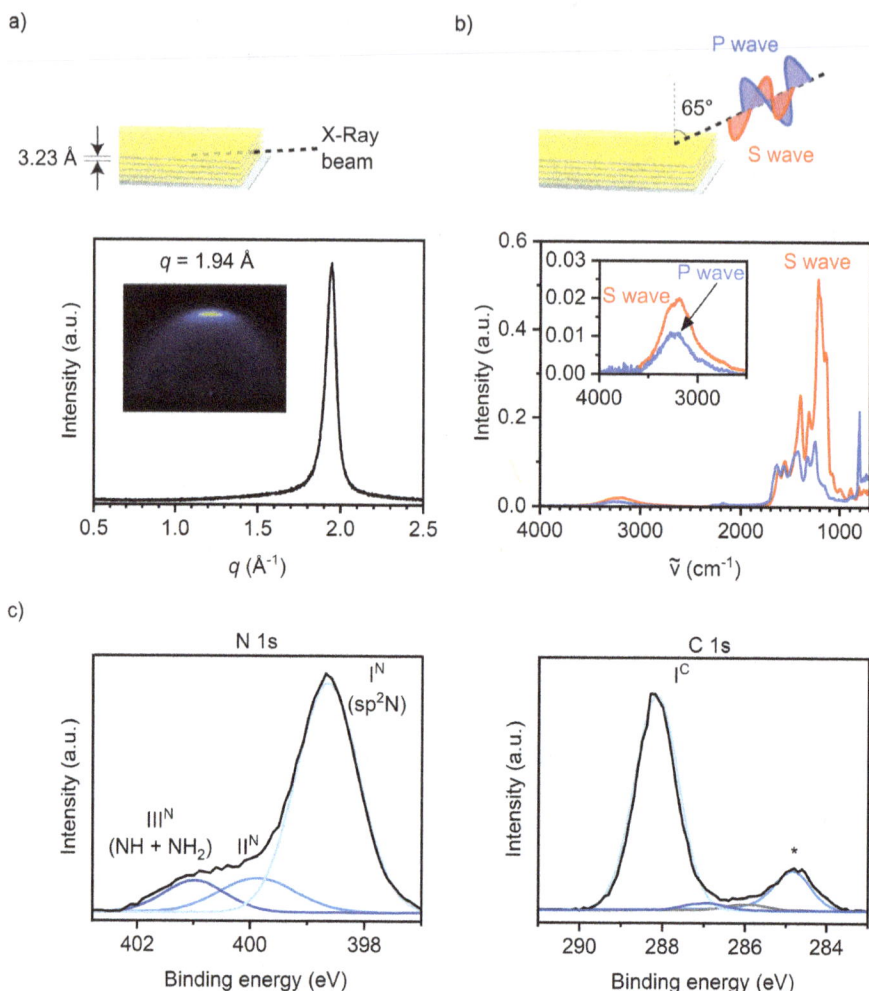

Fig. 10.2: (a) Edge-view 2D-WAXD image of a GCN film obtained by VDP on a glass substrate and its 1D-WAXD pattern obtained along the out-of-plane direction. (b) Polarized FTIR spectra, obtained using s-polarized (S wave) and p-polarized (P wave) light of a GCN film (peeled off from a glass substrate after VDP). The sample was attached to the stainless-steel substrate via a carbon tape. The inset shows N–H vibrational bands of the unreacted amino groups in a GCN film. (c) N 1s and C 1s XPS spectra of a GCN film on a glass substrate (reproduced with permission from reference [14]. Copyright 2016, Macmillan Publishers Ltd.).

scattering wavenumber indicates that the periodicity of the structure of the GCN film is approximately 0.32 nm, which is in good agreement with the value of the interlayer distance previously reported for nanosheet powders.

Three characteristic bands can be observed in the FTIR spectrum of GCN: a band at $3,000–3,400 \ cm^{-1}$ originating from $CH_x–$ and $NH_x–$groups, a band at $2,200 \ cm^{-1}$ originating from the sp-hybridized C and N atoms in the CN–groups and a band at 1,000–2,000 cm^{-1} originating from the π-conjugation of sp^2-hybridized C and N atoms [30]. The remaining N–H vibrational mode in the GCN film is observed as a broad band between 3,000 and 3,400 cm^{-1} (Fig. 10.2b), which is particularly sensitive to s-polarized light and is due to the presence of unreacted amino groups oriented parallel to the GCN plane. The band at 2,200 cm^{-1} originates from the cyano group and is not present in the perfectly condensed heptazine-based network. Therefore, this mode can be used as an indication of structural defects in GCN. The sharp peak at 800 cm^{-1} is the out-of-plane vibrational mode of the amino group and is detected well under p-polarized IR light (Fig. 10.2b). Compared to melamine, a molecule with three amino groups attached to the triazine backbone, the sharp peak at 800 cm^{-1} in GCN is less pronounced due to the out-of-plane vibration mode being suppressed by film formation.

The chemical bonding state of elements in the GCN film can also be assessed based on XPS results. Deconvolution of the N 1s region shows that I^N peak originates from sp^2-hybridized N atoms of N–C = N moieties and II^N peak originates from sp^3-hybridized N atoms. These data, together with the I^C peak in the C 1s region assigned to the sp^2-hybridized C atoms of C–N = C moieties, confirm that GCN is composed of Melem units. III^N peaks originating from NH– or NH$_2$–groups at the Melem ends [2, 31] are also observed in the N 1s region, indicating that GCN is composed of di- and tribranched Melem units. The C/N ratio deduced from the elemental analysis of the GCN film is 0.68, suggesting that it has more terminal amino groups than the ideal g-C$_3$N$_4$, which is consistent with the XPS results. In addition, the peak in the C 1s spectrum marked with the asterisk corresponds to adventitious carbon from the air [32, 33]. The XPS N 1s spectrum in Figure 10.2c was obtained for GCN films deposited at glass substrates. No change in the spectral shape was observed when the films were formed on silicon or fluorine-doped tin oxide substrates.

Carbon nitride has been theoretically predicted to be a superhard material [34]. Employing nanoindentation method, our group measured the Young's modulus (E_r) and hardness (H) of the film to be 12 GPa and 1 GPa, respectively. Compared to typical polymeric materials, such as polyethylene (E_r = 2.8 GPa and H = 0.045 GPa), polycarbonate (E_r = 6.8 GPa and H = 0.26 GPa), polymethyl methacrylate (E_r = 8.6 GPa and H = 0.39 GPa), GCN films are less deformable and stiffer.

Cai et al. synthesized GCN films with thickness greater than 10 μm and investigated their toughness [26]. For a 3 cm-wide film, the tensile strength reached 0.1 GPa, which is 300 times greater than the value of 0.35 MPa, typical for mammalian skeletal muscle. The GCN film is mechanically tough and can support objects with weights 25,000 times greater than its own weight (Fig. 10.3).

Elemental analysis showed that 32 mol% of hydrogen was present in the GCN film prepared by Cai et al. In general, higher degree of GCN disorder results in the presence of greater amount of hydrogen [22], which may compromise its mechanical properties. Thus, reducing the number of defects and, as a result, amount of hydrogen in GCN is desirable to obtain tougher films.

a)

b)

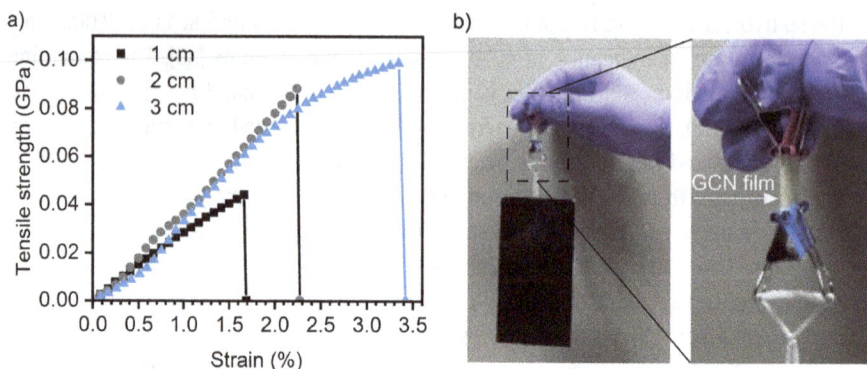

Fig. 10.3: (a) Typical stress–strain curves of GCN films with different widths. (b) Photograph of a strip of a GCN film (1.5 cm × 1.2 cm × 10 μm) with a smartphone hanging from it (reproduced with permission from reference [26]. Copyright 2019, American Chemical Society).

10.4 Actuators

Actuators play an important role in wearable devices, rehabilitation robotics, prosthetics, etc. In GCN films, an amino group located parallel to the π-plane can strongly trap water through hydrogen bonding. The free-standing GCN film synthesized by our group functions as a humidity-responsive autonomous actuator, i.e., the film bends as water is desorbed from GCN and straightens as water is adsorbed on GCN [14]. This can be confirmed by changing the relative humidity of the air from 54% to 18% (Fig. 10.4a). Since the quantum yield of the GCN film luminescence is very low (approximately 0.1%), most of the incident energy is converted into heat. Thus, water can be desorbed by irradiating the GCN film with light shorter than 400 nm. Very fast actuation behavior was observed when the GCN film was irradiated with UV light in air (Fig. 10.4b). For example, when the GCN film was irradiated with UV light (365 nm, 200 mW cm^{-2}) from a distance of 2.0 cm, it curled up within 50 ms. When the film was irradiated with 1,064 nm light, the light was absorbed by water molecules and converted into molecular vibrations. Thus, the actuation behavior via desorption of water from the GCN film when irradiated with 1,064 nm light was also unambiguously confirmed.

The GCN film is capable of fast actuation and it is so light that jump may be induced by illumination (Fig. 10.4c). In this experiment, a 1.3 μm-thick GCN film measuring 2 × 2 mm^2 and weighing 10 μg was used. When the growth side of the GCN film was placed on a graphite substrate and the film was irradiated with 365 nm light, the film jumped 10 mm from the substrate, which is 1,000 times higher than its own thickness. A time lag of approximately one second between the light irradiation and the jump was noted. It is explained by the fact that the energy is released only after the frictional force between the GCN film upon bending and the substrate is generated.

a)

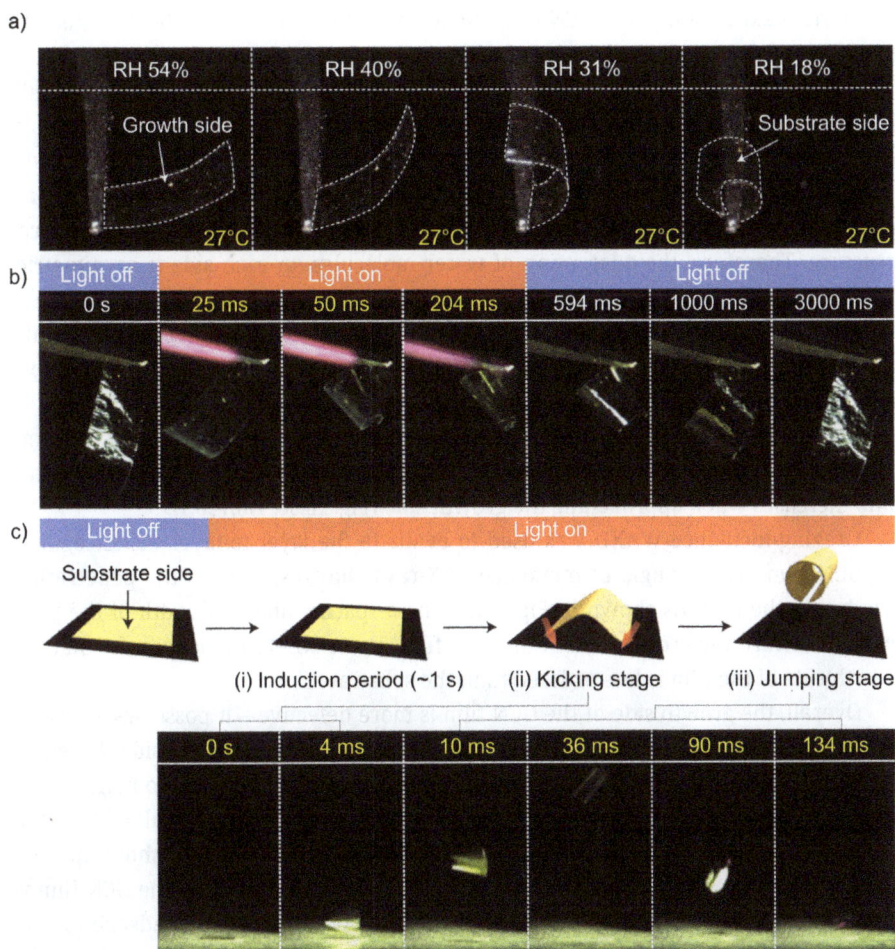

b)

c)

Fig. 10.4: (a) Photographs of a GCN film peeled off from a glass substrate after VDP and suspended under different relative humidity (RH) values at 27 °C. (b) High-speed snapshots of the motions of a GCN film in response to turning on and turning off UV (λ = 365 nm) irradiation. (c) High-speed snapshots of the motions of a GCN film upon exposure to UV light. All of the films were obtained by VDP on a glass substrate and then peeled off (reproduced with permission from reference [14]. Copyright 2016, Macmillan Publishers Ltd.).

This time lag does not exist when the actuation is performed in air and the GCN film is not supported by a substrate.

The GCN film is very durable and stable. The process of irradiating a GCN film with UV light for 0.1 s at an interval of 1.9 s was repeated for over 5 h. This allowed the GCN film to be bent and straightened more than 10,000 times, and no degradation of the sample was observed. In a glove box with a relative humidity of 0.0002%, the GCN film curled up and did not show any response to UV irradiation. It implies that the thermal expansion of the GCN film due to UV irradiation is negligible and that the actuation behavior is caused by the adsorption and desorption of water.

A film bends when the degrees of expansion and contraction differ on each side (Fig. 10.5a). During growth in the VDP process, the substrate side of the film is in constant contact with the underlying substrate. In contrast, the opposite side (growth side) is exposed to the source gas. As a result, the growth conditions are different and asymmetric in the thickness direction, which result in slightly different chemical structure. Therefore, the adsorption of water molecules on each side is asymmetric, resulting in differences in the degrees of expansion and contraction.

To obtain the chemical composition profile of the GCN film along the growth direction, XPS measurements were performed at five points of diagonally sliced cross section of the sample (Fig. 10.5c, Tab. 10.1).

The plot of the integrated intensity ratio of unreacted sp^3–hybridized N atoms (I_{NH} and I_{NH2}) to sp^2-hybridized N atoms (I_{sp2N}) against film depth shows that this ratio increases when approaching the growth side (Fig. 10.5c, Tab. 10.1).

Grazing-incidence WAXD was used to evaluate the layer-to-layer distance, d, in the GCN film; as the angle of incidence of X-rays changes, so does the penetration length into the film. As shown in Fig. 10.5d, the d-spacing and half-width of the peak increase toward the growth side of the GCN film. These observations indicate that the growth side of the film is less ordered than the substrate side.

Overall, the growth side of the GCN film is more defective – it possesses a greater number of unreacted NH– and NH$_2$–groups compared to the substrate side. Melem, a derivative of heptazine, is known to form cocrystals with water [35]. Two neighboring Melem units are cross-linked by water molecules through intermolecular hydrogen bonds of amino groups (Fig. 10.5b). The FTIR results in Figure 10.2b show that the amino groups are oriented in the in-plane direction. Therefore, when the GCN film is placed under moist conditions, a large number of water molecules adsorb on the growth side rather than on the substrate side, resulting in film bending. Most of the GCN film is hydrophobic, and water molecules are selectively adsorbed on structural defects such as unreacted amino groups. Water molecules are adsorbed and desorbed mainly on the surface layer of the GCN film, resulting in fast response and actuation behavior with a small amount of water.

Considering the practical application of GCN films, those with high mechanical strength are desirable. Cai et al. used GCN powder as a precursor and fabricated a tough, film that was several tens of micrometers thick [26] by the PVT method. In this film, actuation is also induced by water adsorption and desorption, but the film stretches at high temperature and under light exposure, and bends at low temperature and in dark conditions (Fig. 10.6a). Such behavior is opposite to that of the GCN thin film reported by our group (Fig. 10.5b), which, nevertheless, may be explained as follows. As with the film chemical structure prepared by VDP method, a greater number of structural defects on the growth side compared to the substrate side were reported by Cai et al. However, under humid conditions, the growth side was proposed

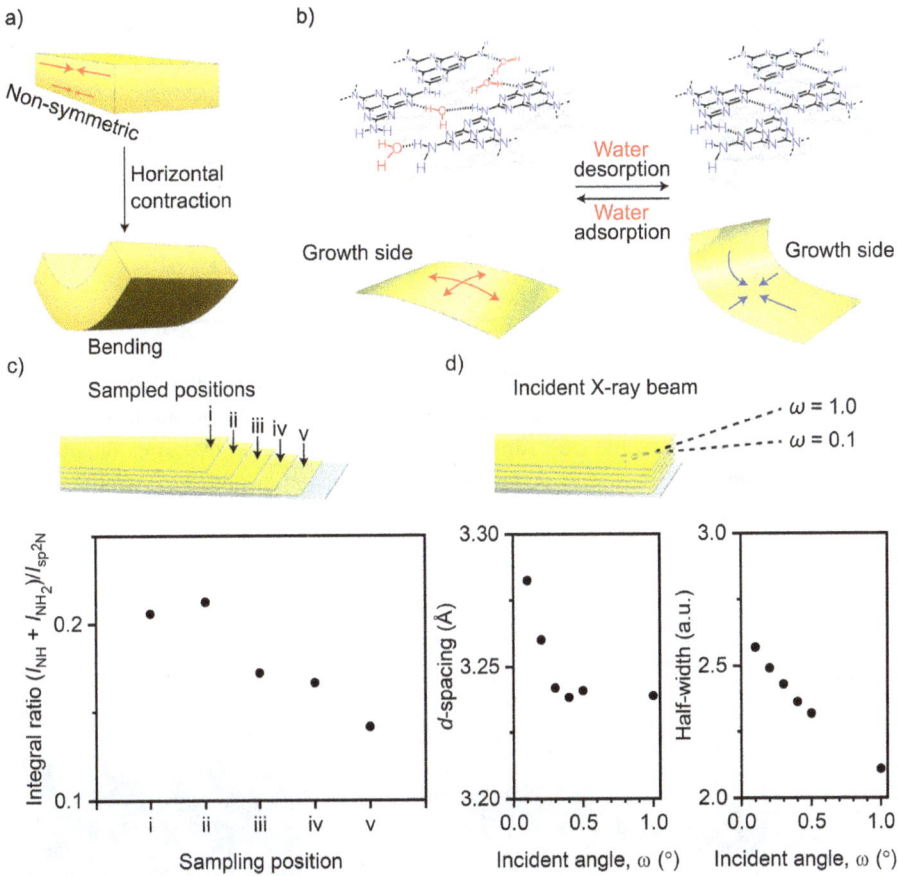

Fig. 10.5: (a) Schematic representations of a nonsymmetric film along the thickness direction that bends upon anisotropic contraction. (b) Schematic illustration of a possible mechanism for the efficient conversion of water desorption/adsorption events into the bending and straightening motions of a GCN film. Unreacted amino groups on the growth side of the GCN film are hydrogen-bonded along the film plane and allow its curled shape to be locked (right). When water is adsorbed on the film surface, these hydrogen bonds are reorganized by incorporation of water molecules such that the film is relaxed and straightens (left). (c, d) Depth profiles evaluated by XPS and grazing-incidence WAXD of a GCN film obtained by VDP on a glass substrate. (c) XPS peak integral ratios of unreacted sp^3 nitrogen (I_{NH} and I_{NH2}) to sp^2 nitrogen (I_{sp2N}) at different depths. The GCN film sample was diagonally sliced, as illustrated by the cartoon. (d) d-spacings and half-widths due to the periodic sheet distance evaluated by 1D-WAXD at different incident angles of the X-ray beam relative to the GCN film. All of the films were obtained by VDP on a glass substrate and then peeled off (reproduced with permission from reference [14]. Copyright 2016, Macmillan Publishers Ltd.).

to adsorb more water, causing the film to expand and bend toward the substrate side. What makes substantial difference in the behavior of the GCN films is the amount of adsorbed water. Our group found that 680 ng of water was adsorbed and desorbed on a 250 μg film, which is approximately 0.3 wt% of the total mass of the film. In contrast,

Tab. 10.1: XPS depth profiles of a GCN film on a glass substrate. Reprinted with permission from Reference [14] (Copyright 2016, Macmillan Publishers Ltd.).

No.	Peak position (sp² N), eV	Peak position (NH + NH₂), eV	Peak area I_{sp2N}	Peak area $I_{NH} + I_{NH2}$	Integral ratio $(I_{NH} + I_{NH2})/I_{sp2N}$
i	398.6	400.8	102	21	0.206
ii	398.7	401.0	80	17	0.213
iii	398.6	400.9	58	10	0.172
iv	398.7	400.9	43	7	0.167
v	398.5	401.0	113	16	0.142

Cai et al. found that a 5 mg GCN film prepared by PVT method adsorbs 20 wt% of water, which is a much greater amount. This can be explained taking into account different morphologies of the GCN films. The GCN film prepared by Cai et al. is translucent, has rough surface and thus, has a larger surface area (Fig. 10.6) compared to the film prepared by our group.

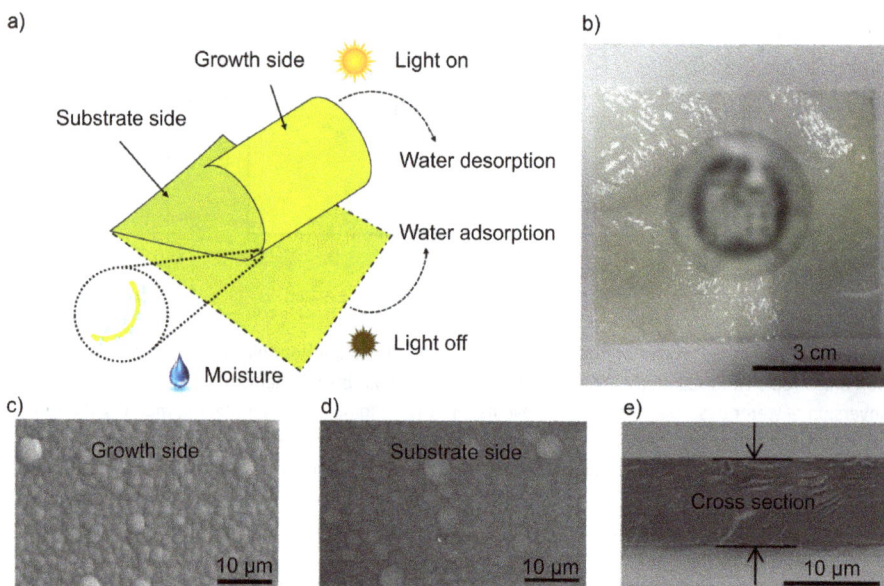

Fig. 10.6: (a) Schematic illustration of the GCN film deformation induced by the adsorption/desorption of water molecules in dark and upon illumination, respectively. (b) Photo of a synthesized GCN film covering a logo. (c) Scanning electron micrographs of the growth side, (d) substrate side, and (e) cross section of a GCN film with a thickness of ca. 8 μm (reproduced with permission from reference [26]. Copyright 2019, American Chemical Society).

Cai et al. suggested that the GCN film prepared by PVT method may be used as a component of smart curtain that uses solar energy. In a demonstration experiment, one end of a rectangular GCN film was fixed to the window frame, and the substrate side faced the outside of the window. When exposed to 460 mW cm^{-2} sunlight, the GCN film is in an extended state, but in the dark, the film bends (Fig. 10.7a and b). This film can be used as a smart curtain that spontaneously rolls up in the middle of the night or on cloudy days and spontaneously covers the window when natural sunlight shines on it.

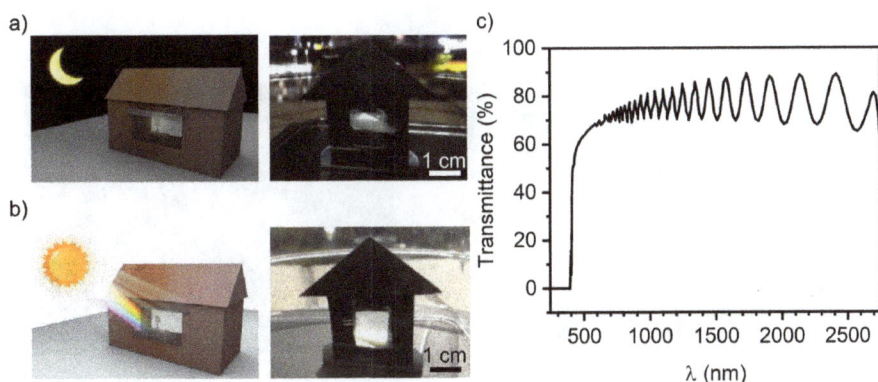

Fig. 10.7: (a) Smart curtain in the dark and (b) upon irradiation with natural sunlight. The experiment was performed at an ambient temperature of 25 °C at night (a) and at 42 °C on a sunny day (b). (c) Transmission spectra of the GCN film versus the wavelength (reproduced with permission from reference [26]. Copyright 2019, American Chemical Society).

The smart curtains reported earlier completely block the incident light when covering the window [36, 37], while the smart curtain based on GCN film blocks only UV light and allows visible light to pass through (Fig. 10.7c). Therefore, the GCN film can be used as a smart curtain that can maintain the brightness of the room while eliminating harmful UV light.

10.5 Electronic devices

The application of GCN in optical and electronic devices is a new research area unique to films. Although actuation behavior has been realized with free-standing films, optical and electronic elements do not necessarily have to be free standing, and research has been conducted using films formed on supporting substrates.

GCN is a polymeric semiconductor with an energy band gap of approximately 2.7 eV [1]. Modulation of the energy band gap in the visible range has also been studied, and GCN was reported to act as an electron acceptor in solar cells [38]. The emission color of GCN powder materials can be modulated by chemical modification [39]. Xu et al. demonstrated that GCN can be used as an electroluminescent material [40]. They used a cyanuric acid/2,4-diamino-6-phenyl-1,3,5-triazine supramolecular complex as a precursor and deposited GCN films on indium tin oxide (ITO) substrates by heating to 450 °C. Then, Ca was deposited as an electrode, followed by Al protective electrode to fabricate single-layer LEDs (Fig. 10.8a and b). In this structure, electrons are injected from the Ca electrode into the conduction band of GCN, holes are injected from the ITO electrode into the valence band, and light is emitted when the electrons and holes recombine in the GCN film.

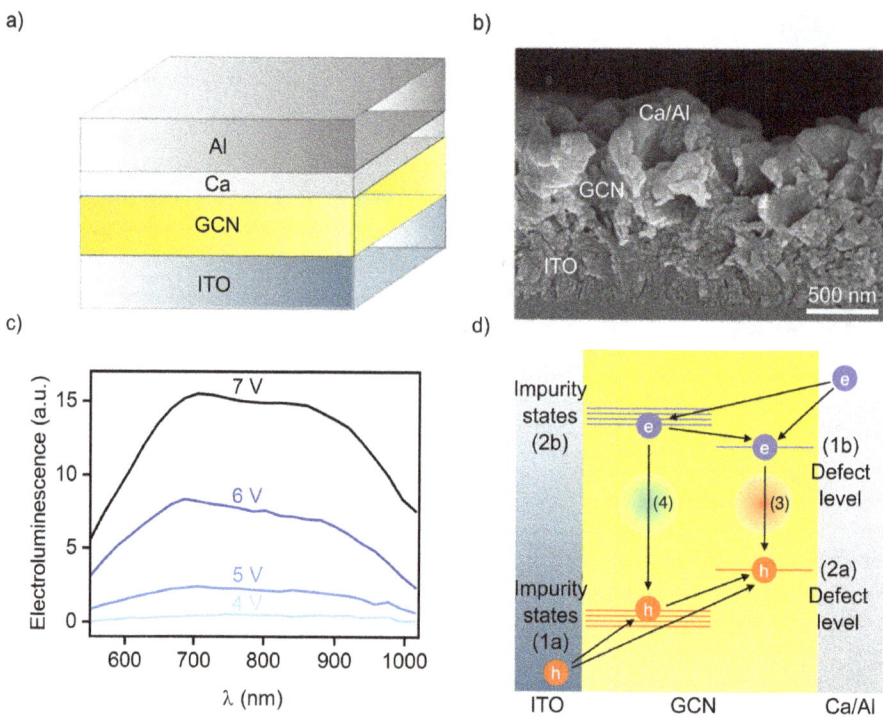

Fig. 10.8: (a) Schematic organic LED (OLED) structure. (b) Cross-sectional scanning electron microscopy image of an OLED device. (c) Electroluminescence (EL) spectra recorded at different applied voltages. (d) Schematic of the proposed mechanism for the photoluminescence (PL) shift and the strong EL redshift, with increasing disorder in GCN (reproduced with permission from reference [40]. Copyright 2015, Wiley-VCH Verlag).

When the synthesized GCN film was irradiated with UV light at 365 nm, green photoluminescence (PL) with a maximum at approximately 550 nm was observed. In contrast, when electrons and holes were injected into GCN by applying an electric field, electroluminescence (EL) in the red to near-IR region with a maximum at approximately 700 nm was observed (Fig. 10.8c). In the EL, a clear redshift was observed compared to the PL, and the intensity of the emission increased with increasing bias voltage applied to the diode, indicating that the emission was not due to chemilumi-nescence but rather due to carrier injection induced by the electric field.

Figure 10.8d shows the process that occurs during the photoemission by GCN. Holes are injected from the ITO electrode into the impurity states (1a) located close to the va-lence band of the GCN, followed by relaxation to the energetically more stable defect level (2a) due to charge carrier transport. Electrons are similarly injected from the Ca electrode into the impurity states (2b) and either relax to the defect level (1b) or are injected directly into (1b). Recombination of the electrons in (1b) and the holes in (2a) is accompanied by the emission of light. The emission is redshifted because the energy difference between the states is smaller than the band gap of GCN (3). In contrast, in the case of PL, a hole in the valence band and an electron in the conduction band or impu-rity states directly combine and emit light (4). As shown here, both electrons and holes can be injected into the GCN depending on the used electrode material. In other words, GCN can be either a p-type semiconductor or an n-type semiconductor.

Since the GCN film is an anisotropic material, its electrical properties depend on the direction. The electrical properties have often been investigated by depositing the GCN on electrodes in electrochemical cells [13]. However, the contribution of not only electron–hole transport in GCN, but also ionic conductivity in the electrolyte makes the discussion of the electron transport mechanism in GCN difficult.

Noda et al. fabricated solid-state electronic devices and found anisotropic electri-cal conductivity in triazine-based GCN films. Thus, the electrical conductivity along the basal plane of triazine-based GCN is 65 times lower than that through the stacked layers, as opposed to graphite [41]. This is explained by the fact that π-electrons are localized in triazine, suppressing in-plane electrical conduction, while hopping elec-trons between layers via π-orbitals is not prohibited.

Similar anisotropic electrical conductivity has been observed in heptazine-based GCN films [42, 43]. Urakami et al. used a CVD method to deposit GCN films with a thickness of 400–500 nm on sapphire substrates, using melamine as a precursor. The GCN films could be peeled off from the sapphire and transferred to any substrate. Using photolithography, they fabricated a device for measuring the out-of-plane con-ductivity (Fig. 10.9a) and a top-gate field-effect transistor for measuring the in-plane conductivity (Fig. 10.9b). The gate insulator used in the field-effect transistor structure was parylene with a thickness of 100 nm. The electrical conductivity in the vertical direction was orders of magnitude higher (10^6) than that in the in-plane direction (Fig. 10.9c). The in-plane conductivity could be modulated by applying a gate voltage (Fig. 10.9d). When no gate voltage is applied, the in-plane direction is insulating. When a negative gate voltage is applied, more carriers are induced, indicating that the GCN in this system acts as a p-type semiconductor. However, since the on/off ratio, which is defined as the current value with a gate voltage of –8 V divided by the cur-rent value without gate voltage, is approximately 20, which is low, further improve-ment of the on/off ratio is desired to use GCN as an electrical switch.

a)

b)

c)

d)

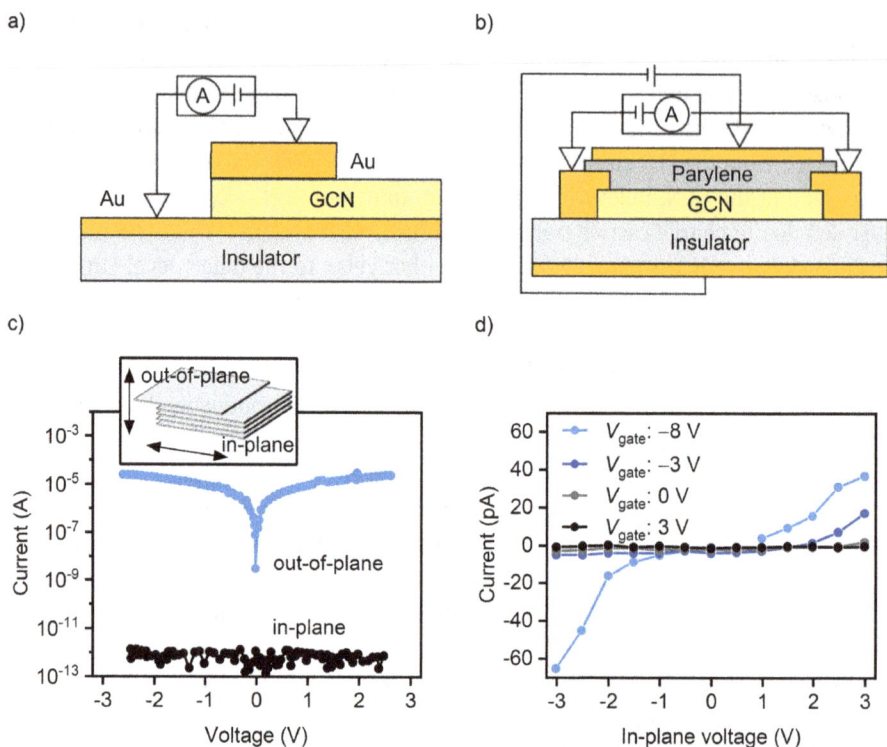

Fig. 10.9: (a) Experimental setup of an I–V measurement for determining the out-of-plane electrical conductivity. (b) Schematic cross section of the GCN top-gate device, which is used to determine the in-plane electrical conductivity. (c) I–V characteristics of a GCN film recorded at different applied bias directions. (d) In-plane I–V characteristics of the GCN film, depending on the gate bias. The gate bias was applied along the out-of-plane direction (reproduced with permission from reference [43]. Copyright 2021, American Institute of Physics).

GCN films can convert electric current into light [39], and the reverse process is also possible. Giusto et al. reported a ratio between the current measured by applying to the GCN film bias voltage of 1.4 V in dark and upon illumination with 465 nm photons (on/off ratio of the photocurrent) to be approximately 1.25, which was obtained by performing measurements in an electrolyte [27]. The out-of-plane direction of the GCN film is more electrically conductive compared to in-plane even in dark. Therefore, it is a challenging task to realize the off state – a state with zero current. Recently, Liu et al. measured the electrical conductivity along the in-plane direction and found that the on/off ratio, with and without UV irradiation, increases by 250 times (Fig. 10.10a and b) [28].

As shown in Figure 10.10c, the on/off ratio of the photocurrent scales linearly with the UV light optical power density, which is an essential characteristic of photodetectors. Moreover, the light and dark currents did not change when the GCN film

a)

b)

c)

d)

e)

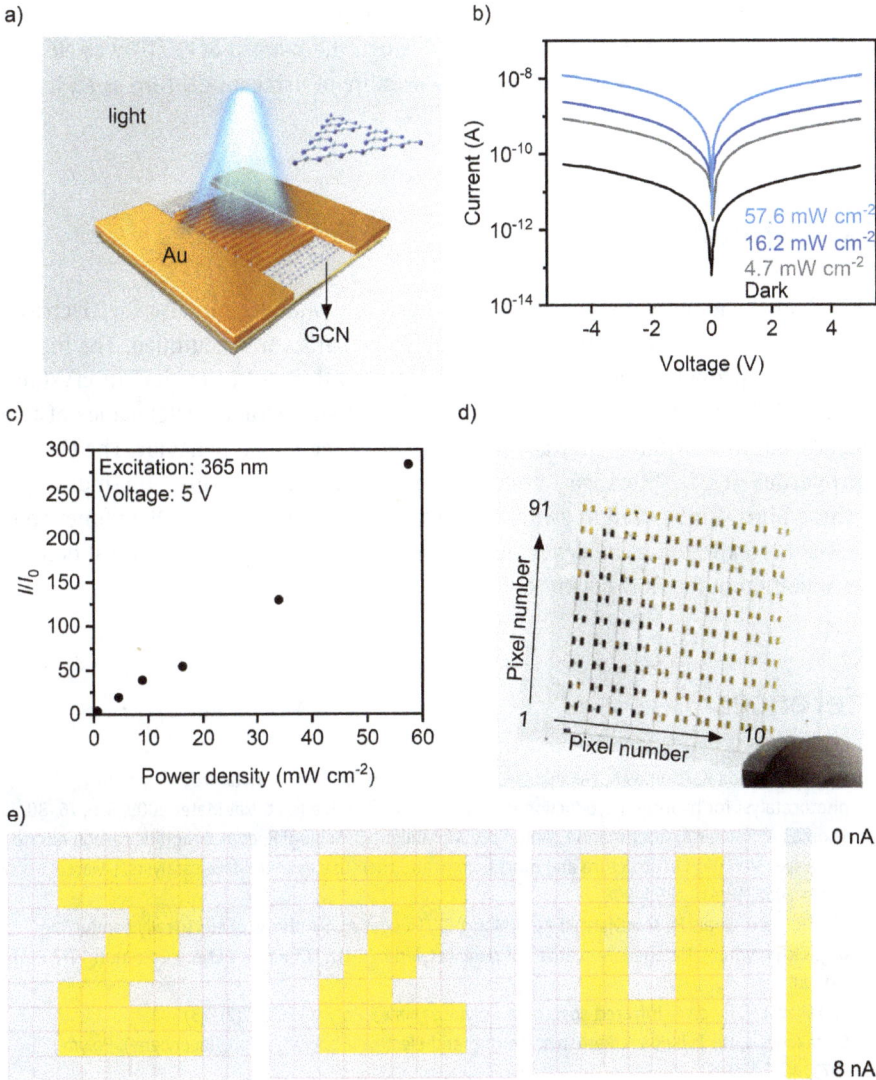

Fig. 10.10: (a) Schematic diagram of the GCN photodetector. (b) I–V curves of the GCN photodetector for different illumination intensities. (c) Dependence of the on/off photocurrent ratio on the illumination intensity. (d) Photograph of the GCN photodetector array on a 4×4 cm^2 GCN film. (e) Image acquired by the GCN photodetector array (reproduced with permission from reference [28]. Copyright 2021, Elsevier Inc.).

was bent from 0° to 150°. The photocurrent did not change even after 1,000 repetitions of the 75° bending test. In addition, the GCN film could be easily transferred onto various substrates and thus be used as a flexible UV photodetector. Figure 10.10d shows a photodetector array composed of 10×10 pixels made of GCN and having the area of 4×4 cm^2 on a flexible substrate. Uniform on and off currents were obtained for all pixels. By recording the location and current values of all pixels, photocurrent imaging could be achieved. UV light irradiation through the pattern of "ZZU" was clearly recognized (Fig. 10.10e), demonstrating the feasibility of using a GCN film as an imaging material.

10.6 Summary

In this chapter, not only the methods of synthesis and analysis, but also the electrical properties of GCN films and their application as actuators were outlined. The highly oriented film provides emergent functions not found in GCN powder. In crystals, structural defects are often considered as features that degrade performance of the materials, but in GCN films, they are the origin of the actuation behavior. The electrical properties of GCN films vary greatly with the orientation of the film, indicating that these films can be used as switching devices and photodetectors. New functional materials are expected to be developed by further controlling the structural defects and orientation in the GCN network.

References

[1]   Wang X, Maeda K, Thomas A, Takanabe K, Xin G, Carlsson JM, et al. A metal-free polymeric photocatalyst for hydrogen production from water under visible light. Nat Mater 2009, 8(1), 76–80.
[2]   Thomas A, Fischer A, Goettmann F, Antonietti M, Müller J-O, Schlögl R, et al. Graphitic carbon nitride materials: Variation of structure and morphology and their use as metal-free catalysts. J Mater Chem 2008, 18(41), 4893–908.
[3]   Shirakawa H, Louis EJ, MacDiarmid AG, Chiang CK, Heeger AJ. Synthesis of electrically conducting organic polymers: Halogen derivatives of polyacetylene, (CH)$_x$. J Chem Soc Chem Commun 1977, (16), 578–80.
[4]   Shirakawa H, Ikeda S. Infrared spectra of poly(acetylene). Polym J 1971, 2(2), 231–44.
[5]   Shirakawa H, Ito T, Ikeda S. Raman scattering and electronic spectra of poly(acetylene). Polym J 1973, 4(4), 460–62.
[6]   Bian J, Huang C, Zhang R-Q. Graphitic carbon nitride film: An emerging star for catalytic and optoelectronic applications. ChemSusChem 2016, 9(19), 2723–35.
[7]   Jia C, Yang L, Zhang Y, Zhang X, Xiao K, Xu J, et al. Graphitic carbon nitride films: Emerging paradigm for versatile applications. ACS Appl Mater Interfaces 2020, 12(48), 53571–91.
[8]   Jürgens B, Irran E, Senker J, Kroll P, Müller H, Schnick W. Melem (2,5,8-triamino-tri-s-triazine), an important intermediate during condensation of melamine rings to graphitic carbon nitride: Synthesis, structure determination by X-ray powder diffractometry, solid-state NMR, and theoretical studies. J Am Chem Soc 2003, 125(34), 10288–300.
[9]   May H. Pyrolysis of melamine. J Appl Chem (London, U K) 1959, 9(6), 340–44.
[10] Hosmane RS, Rossman MA, Leonard NJ. Synthesis and structure of tri-s-triazine. J Am Chem Soc 1982, 104(20), 5497–99.

[11] Chai B, Peng T, Mao J, Li K, Zan L. Graphitic carbon nitride (g-C$_3$N$_4$)–Pt-TiO$_2$ nanocomposite as an efficient photocatalyst for hydrogen production under visible light irradiation. Phys Chem Chem Phys 2012, 14(48), 16745–52.

[12] Zhang Y, Antonietti M. Photocurrent generation by polymeric carbon nitride solids: An initial step towards a novel photovoltaic system. Chem – Asian J 2010, 5(6), 1307–11.

[13] Bu Y, Chen Z, Yu J, Li W. A novel application of g-C$_3$N$_4$ thin film in photoelectrochemical anticorrosion. Electrochim Acta 2013, 88, 294–300.

[14] Arazoe H, Miyajima D, Akaike K, Araoka F, Sato E, Hikima T, et al. An autonomous actuator driven by fluctuations in ambient humidity. Nat Mater 2016, 15(10), 1084–89.

[15] Gorham WF. A New, General synthetic method for the preparation of linear poly-p-xylylenes. J Polym Sci Part A-1: Polym Chem 1966, 4(12), 3027–39.

[16] Lahann J, Klee D, Höcker H. Chemical vapor deposition polymerization of substituted [2.2] paracyclophanes. Macromol Rapid Commun 1998, 19(9), 441–44.

[17] Vaeth KM, Jensen KF. Poly(p-phenylene vinylene) Prepared by chemical vapor deposition: Influence of monomer selection and reaction conditions on film composition and luminescence properties. Macromolecules 1998, 31(20), 6789–93.

[18] Alf ME, Asatekin A, Barr MC, Baxamusa SH, Chelawat H, Ozaydin-Ince G, et al. Chemical vapor deposition of conformal, functional, and responsive polymer films. Adv Mater 2010, 22(18), 1993–2027.

[19] Takahashi Y, Ukishima S, Iijima M, Fukada E. Piezoelectric properties of thin films of aromatic polyurea prepared by vapor deposition polymerization. J Appl Phys 1991, 70(11), 6983–87.

[20] Molina B, Sansores LE. Electronic structure of six phases of C$_3$N$_4$: A theoretical approach. Mod Phys Lett B 1999, 13(06n07), 193–201.

[21] Kroke E. gt-C3N4 – The first stable binary carbon(IV) nitride. Angew Chem Int Ed 2014, 53(42), 11134–36.

[22] Lotsch BV, Döblinger M, Sehnert J, Seyfarth L, Senker J, Oeckler O, et al. Unmasking Melon by a complementary approach employing electron diffraction, solid-state NMR spectroscopy, and theoretical calculations – Structural characterization of a carbon nitride polymer. Chem – Eur J 2007, 13(17), 4969–80.

[23] Jorge AB, Martin DJ, Dhanoa MTS, Rahman AS, Makwana N, Tang J, et al. H$_2$ and O$_2$ evolution from water half-splitting reactions by graphitic carbon nitride materials. J Phys Chem C 2013, 117(14), 7178–85.

[24] Athikomrattanakul U, Promptmas C, Katterle M, Schilde U. An orthorhombic polymorph of melaminium chloride hemihydrate. Acta Crystallogr Sect E: Crystallogr Commun 2007, 63(5), o2154–o6.

[25] Xu J, Wu H-T, Wang X, Xue B, Li Y-X, Cao Y. A new and environmentally benign precursor for the synthesis of mesoporous g-C$_3$N$_4$ with tunable surface area. Phys Chem Chem Phys 2013, 15(13), 4510–17.

[26] Cai Z, Song Z, Guo L. Thermo- and photoresponsive actuators with freestanding carbon nitride films. ACS Appl Mater Interfaces 2019, 11(13), 12770–76.

[27] Giusto P, Cruz D, Heil T, Arazoe H, Lova P, Aida T, et al. Shine bright like a diamond: New light on an old polymeric semiconductor. Adv Mater 2020, 32(10), 1908140.

[28] Liu Z, Wang C, Zhu Z, Lou Q, Shen C, Chen Y, et al. Wafer-scale growth of two-dimensional graphitic carbon nitride films. Matter 2021, 4(5), 1625–38.

[29] Bian J, Li Q, Huang C, Li J, Guo Y, Zaw M, et al. Thermal vapor condensation of uniform graphitic carbon nitride films with remarkable photocurrent density for photoelectrochemical applications. Nano Energy 2015, 15, 353–61.

[30] Rodil SE, Ferrari AC, Robertson J, Muhl S. Infrared spectra of carbon nitride films. Thin Solid Films 2002, 420-421, 122–31.

[31] Zhang J, Zhang M, Zhang G, Wang X. Synthesis of carbon nitride semiconductors in sulfur flux for water photoredox catalysis. ACS Catal 2012, 2(6), 940–48.

[32] Lyth SM, Nabae Y, Moriya S, Kuroki S, Kakimoto M-A, Ozaki J-I, et al. Carbon nitride as a nonprecious catalyst for electrochemical oxygen reduction. J Phys Chem C 2009, 113(47), 20148–51.

[33] Liu J, Zhang T, Wang Z, Dawson G, Chen W. Simple pyrolysis of urea into graphitic carbon nitride with recyclable adsorption and photocatalytic activity. J Mater Chem 2011, 21(38), 14398–401.

[34] Liu AY, Cohen ML. Prediction of new low compressibility solids. Science 1989, 245(4920), 841–42.

[35] Makowski SJ, Köstler P, Schnick W. Formation of a hydrogen-bonded heptazine framework by self-assembly of Melem into a hexagonal channel structure. Chem – Eur J 2012, 18(11), 3248–57.

[36] Zhang X, Yu Z, Wang C, Zarrouk D, Seo J-WT, Cheng JC, et al. Photoactuators and motors based on carbon nanotubes with selective chirality distributions. Nat Commun 2014, 5(1), 2983.

[37] Yamamoto Y, Kanao K, Arie T, Akita S, Takei K. Air Ambient-operated pNIPAM-based flexible actuators stimulated by human body temperature and sunlight. ACS Appl Mater Interfaces 2015, 7(20), 11002–06.

[38] Xu J, Brenner TJK, Chabanne L, Neher D, Antonietti M, Shalom M. Liquid-based growth of polymeric carbon nitride layers and their use in a mesostructured polymer solar cell with V_{oc} exceeding 1 V. J Am Chem Soc 2014, 136(39), 13486–89.

[39] Zhang Y, Pan Q, Chai G, Liang M, Dong G, Zhang Q, et al. Synthesis and luminescence mechanism of multicolor-emitting g-C_3N_4 nanopowders by low temperature thermal condensation of melamine. Sci Rep 2013, 3(1), 1943.

[40] Xu J, Shalom M, Piersimoni F, Antonietti M, Neher D, Brenner TJK. Color-tunable photoluminescence and NIR electroluminescence in carbon nitride thin films and light-emitting diodes. Adv Opt Mater 2015, 3(7), 913–17.

[41] Noda Y, Merschjann C, Tarábek J, Amsalem P, Koch N, Bojdys MJ. Directional charge transport in layered two-dimensional triazine-based graphitic carbon nitride. Angew Chem Int Ed 2019, 58(28), 9394–98.

[42] Takashima K, Urakami N, Hashimoto Y. Electronic transport and device application of crystalline graphitic carbon nitride film. Mater Lett 2020, 281, 128600.

[43] Urakami N, Ogihara K, Futamura H, Takashima K, Hashimoto Y. Demonstration of electronic devices in graphitic carbon nitride crystalline film. AIP Adv 2021, 11(7), 075204.

Yuanyuan Zhang and Jian Liu*

Chapter 11
Thin-film carbon nitride active layers for catalysis, sensing and solar cells

11.1 Introduction

The seminal finding of photocatalytic hydrogen evolution from water using polymeric carbon nitride (PCN, also commonly known as g-C₃N₄) in 2009 has opened up whole new vistas and opportunities in metal-free photocatalytic materials [1]. The success of PCN as particulate photocatalyst makes itself a promising candidate for sustainable chemistry [2]. Due to its unique metal-free nature, visible-light-driven photoresponse and facile synthesis, the past decade has witnessed an explosive growth of PCN-based research in wide fields, spanning from photocatalysis to emerging biosensing, solar cells, nonvolatile memory devices, etc. [3, 4]. However, it is worth to mention that most of the reported work were based on PCN powders.

Actually, efforts to coat PCN onto conductive substrates were initiated in 2009, just shortly after the emergence of particulate PCN-based photocatalytic research. Zhang et al. reported a photoelectrochemical cell using thick PCN coating on indium tin oxide (ITO) and a liquid junction containing a mediating redox couple, thus building a prototype photovoltaic device for converting light energy to electricity [5]. The initial attempt for coating carbon nitride onto the ITO followed the conventional "doctor-blade method", widely used in the dye-sensitized solar cell field, while the specific properties of the PCN (e.g., volatility of precursors, polymerization and deposition properties) were not fully investigated during the film formation process.

The inhomogeneity of the first film was addressed by decreasing the particle size or protonating the PCN powder [5, 6]. However, the inhomogeneous PCN "coatings" and the weak adhesion of the PCN on conductive substrates were intrinsic consequences of textural and grain boundary effects. In 2015, Wang et al. proposed a method to

Acknowledgments: J. Liu thanks Changchao Jia, Gang Lin, Shangfa Pan, Jingwen Bai, Zhenmei Zhang, Lintao Li, Bo Lei for assistance in preparation. J. Liu also thanks the finacial support from the Natural Science Foundation of Shandong Province (ZR2019ZD47, ZR2019JQ05 and ZR2018MB018), the Education Department of Shandong Province (2019KJC006) and the National Natural Science Foundation of China (22175104 and 21802080).

*Corresponding author: Jian Liu, Qingdao Institute of Bioenergy and Bioprocess Technology, Chinese Academy of Sciences, Shandong Energy Institute, Qingdao 266101, P.R. China, e-mail: liujian@qibebt.ac.cn
Yuanyuan Zhang, Qingdao Institute of Bioenergy and Bioprocess Technology, Chinese Academy of Sciences, Shandong Energy Institute, Qingdao 266101, P.R. China

https://doi.org/10.1515/9783110746976-011

prepare a sol "paste" of PCN for film formation by refluxing the suspension in dilute nitric acid solution [7]. In 2013, Xie et al. reported an exfoliation route in water to prepare ultrathin carbon nitride nanosheets from bulk powders [8]. Interestingly, the obtained nanosheets, with lateral size of 70–160 nm, could be vacuum-filtrated onto a cellulose membrane to form large-area films. Moreover, the assembled film could be further transferred to other substrates by dissolving the cellulose in acetone. Such work represented the initial attempt toward the fabrication of high-quality PCN films and also free-standing membranes.

As an ideally infinite extension of nanosheets in two dimensions, thin PCN films are crucial for the development of high-performance energy-related devices. The intrinsically poor processability, low quality and high roughness of the PCN powders prevented their direct application as an active layer in photoelectrochemical devices [9]. New strategies, especially by an in situ synthetic method, using the C, N-containing molecular precursor were high in demand, while the discovery of such film, especially the free-standing counterpart, occurred unintentionally [10, 11]. Actually, it has been long observed that a homogeneous coating on the inner wall of the crucible was formed, complementing the ordinary yellowish powdery substance after thermal condensation. However, it took some time until the PCN film formation, directly from molecular precursors, was systematically investigated. In 2014, Shalom et al. reported a strategy for in situ growing of PCN layers onto the conductive substrate for photoelectrochemical applications, which, however, were still afflicted with grain boundaries and interstitial voids [12]. In 2015, Liu et al. developed a nanoconfinement method to fabricate PCN films directly on conductive substrates and found that the PCN film deposited on the glass substrates could be peeled off to form a free-standing membrane [10]. Bian et al. simplified the film synthesis by directly covering a conductive substrate onto the mouth of the crucible [13]. The PCN coatings obtained by these methods usually showed enhanced photocurrent performance.

The milestone work reported by Aida et al. in 2016 brought PCN films and membranes to a new level. They developed a vapor deposition polymerization method for forming highly oriented crack-free PCN film on the glass substrate by starting from guanidinium carbonate precursor. Such free-standing PCN membrane greatly inspired researchers to construct functional devices while further developing more sophisticated film growth technique [14–16].

As a conjugated polymer semiconductor with a narrow band gap, PCN features unique optical properties (photoluminescence and chemiluminescence). The PCN possesses π-conjugated electronic structure and a rigid C–N plane along the individual layers, suggesting its potential for fluorescence sensing. What makes PCN an ideal candidate for fluorescence sensing is the really extremely stable photoluminescence intensity against photobleaching. By structuring PCN into thin-film active layers, the above mentioned advantages should be kept at the level of sensing device applications [4, 17]. In addition, a thin film, especially the free-standing counterpart, could bring up some unexpected properties. Aida et al. constructed a PCN-film-based actuator, driven by fluctuations in ambient humidity, which could lose the adsorbed water and bend upon heating or light irradiation [11].

Solar cells are devices that use the photovoltaic effect to convert the energy of light directly into electricity, producing electrical charges that can move freely in semiconductors. Currently, the common solar cells are based on silicon, organic polymers and nanocrystalline semiconducting materials, and convert light into electric energy economically. Organic–inorganic lead halide perovskite solar cells are in current focus because of their high conversion efficiency, low cost and easy fabrication. As a π-conjugated material, PCN nanosheets are also suitable for the separation of photogenerated electron–hole pairs. The amines or nitrogen functionalities have shown effective passivation of organic/inorganic halide perovskite materials [18]. It was demonstrated that high-quality PCN active layer via thiazole modification could boost the electronic interface enhancement via suppression of charge recombination [9].

In this chapter, we intend to give a relatively comprehensive review on the fabrication, applications and prospects of such PCN films by exploring their unique properties in photoelectrochemistry, catalysis, sensing, solar cell fields and beyond [4, 17, 19, 20]. In the following context, the synthesis is introduced first, followed by the versatile applications, and ends with an outlook toward future developments of PCN-thin-film-based studies.

11.2 The synthesis of PCN thin films

In a typical synthesis for PCN powder, the molecular precursors, such as melamine ($C_3N_3(NH_2)_3$), cyanamide (CN_2H_2) and dicyandiamide ($C_2N_4H_4$), or even urea (CN_2OH_4) undergo thermolytic condensation. Starting from cyanamide, at temperatures above 47 °C, the cyanamide undergoes melting, and condenses to melamine via dicyandiamide. Heating melamine to 317 °C induces a cascade of condensation reactions, accompanied by the evolution of ammonia. Initially, melamine forms a dimer called melam ($[(H_2N)_2(C_3N_3)]_2 NH$) that rapidly reacts to form a heptazine-based molecule called Melem (triamino-tri-s-triazine, triamino-heptazine, $C_6N_7(NH_2)_3$) above 360 °C. Melem condenses to form 1D chains of amine-linked heptazine units above 520 °C. This poly (amino-imino)heptazine, with the formula $[C_6N_7(NH_2)(NH)]_n$, was named by Liebig's Melon.

Generally, the synthesis of PCN-based films can be divided into post-processing and in situ synthetic routes. The post-processing route refers to the fabrication of the PCN films using a pre-synthesized PCN powdery substance, while the latter describes the direct thermal condensation of the precursors to form the films.

11.2.1 Post-processing from presynthesized powders

Initially, Zhang et al. employed the doctor-blade technique by spreading aqueous slurries of either bulk or mesoporous PCN onto a conductive substrate with a glass rod and using adhesive tapes as spacers (Fig. 11.1a) [5]. In order to obtain high quality of homogeneous films, the particle size of the carbon nitride powdery substance was decreased [6]. More importantly, the dispensability of PCN in the corresponding solvent should be enhanced to improve the processability for film formation [9].

To break the hydrogen bonds between the strand nitrogen atoms and NH/NH_2 groups in PCN, Zhang et al. proposed depolymerization strategy via a strongly oxidizing acid to form a colloidal suspension of PCN [7]. As shown in Fig. 11.1b, a colloidal PCN suspension is obtained from the powder depolymerization by refluxing in concentrated nitric acid at 80 °C. By coating the resultant sol onto the substrate, followed by annealing, thin PCN films featuring highly interconnected porous networks without significant particle boundaries were obtained.

Liu et al. found that PCN nanosheets dispersed in ethanol could be easily spread at the air/water interface due to the Marangoni effect [21]. The gas/liquid interfacial self-assembly was followed by plunging the substrate into the solution for transferring films onto various substrates (Fig. 11.1c) [22]. Furthermore, a PCN/graphene hybrid film electrode could also be fabricated via the same procedure by premixing with graphene nanosheets. Antonietti et al. reported a novel photografting route to increase the dispensability of PCN in organic solvents by attaching the 4-methyl-5-vinylthiazole on the PCN rim [9]. Upon grafting, the dispensability in organic solvents was enhanced. Based on such exceptional organodispersions with well-defined particle thickness and size, transparent and homogeneous film could be readily formed by spray coating or inkjet printing.

Guo et al. reported a secondary thermal condensation method, starting from the as-prepared powdery PCN substance [23]. Accordingly, a yellow PCN film could be found on the inner surface of a quartz tube after "physical vapor deposition" (Fig. 11.1d). It is interesting to notice that the size of the membrane can be up to tens of square centimeters without visible defects, and that the thickness could be adjusted from several to 108 μm.

11.2.2 In situ synthesis from molecular precursor

The in situ synthesis refers to the direct deposition of PCN layers onto the substrates via thermal condensation of molecular precursors. Initially, such efforts were exclusively aimed at high-quality photoelectrodes. By manipulating the interaction between the molecular precursors and the substrate as well as the heating procedures, the growth of high-quality PCN film on the substrate was possible.

In 2014, Shalom et al. adopted a confinement method for short mass transfer by putting the cyanuric acid-melamine supramolecular complex between two substrates,

a)

b)

HNO₃, 353K, 3h

c)

d)

| precursor powder | precursor vapor | PCN film | free-standing PCN film |

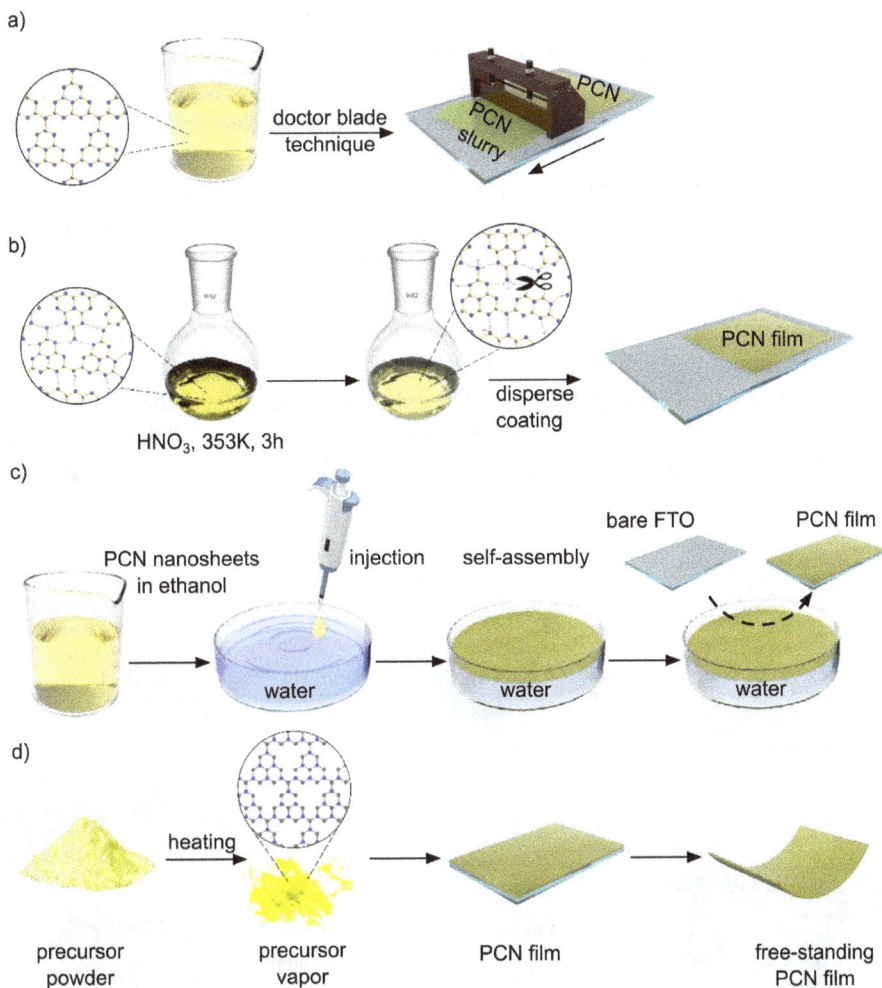

Fig. 11.1: Schematic illustrations of post-processing fabrication techniques. (a) Illustration of the doctor-blade method for fabrication of PCN film [5]. (b) The PCN-sol solution synthesis and the dip/disperse coating technique [7]. (c) Illustration of a self-assembly strategy for PCN film fabrication at water/air interface [22]. (d) Illustration of the secondary evaporation technique for growing PCN films on substrates [23].

and the sandwiched structures were subjected to thermal heating for directly coating PCN on different substrates [12]. Xu and Shalom et al. placed supramolecular precursors in a crucible, completely covering the substrate, and termed such route as "liquid-mediated pathway" to fabricate continuous PCN thin films on substrates (Fig. 11.2a) [24]. The authors claimed that the supramolecular precursors are crucial for such growth, while other conventional precursors such as melamine and dicyandiamide failed to form the continuous film under the given conditions. In 2015, Bian et al. developed a thermal vapor condensation method to deposit a uniform PCN film on the desired substrates, which were acting as a lid of a crucible (Fig. 11.2b). Zhang et al. found out that microwave heating could expedite the synthesis of PCN film on fluorine-doped tin oxide (FTO), while the attachment of the film on substrate could be enhanced. Thus, a continuous film on FTO, with intimate interface, could be obtained via the microwave-assisted condensation of dicyandiamide [25].

For the free-standing PCN membrane, the development occurred differently. In the seminal finding of PCN membrane in 2016, Aida et al. hinted that the study began with a serendipitous discovery that a highly oriented PCN film could be formed upon calcination of guanidinium carbonate (Fig. 11.2c) [11]. By confining the molecular precursor in an unsealed glass tube, the possible intermediates formed during precursor thermal decomposition sublimed, deposited and finally polymerized on the glass substrate to produce the PCN film. Liu et al. placed the cyanamide-filled anodic aluminum oxide (AAO) between two glass slides and also accidently obtained PCN film on the glass substrate instead of AAO surface (Fig. 11.2d) [10]. However, such finding did not

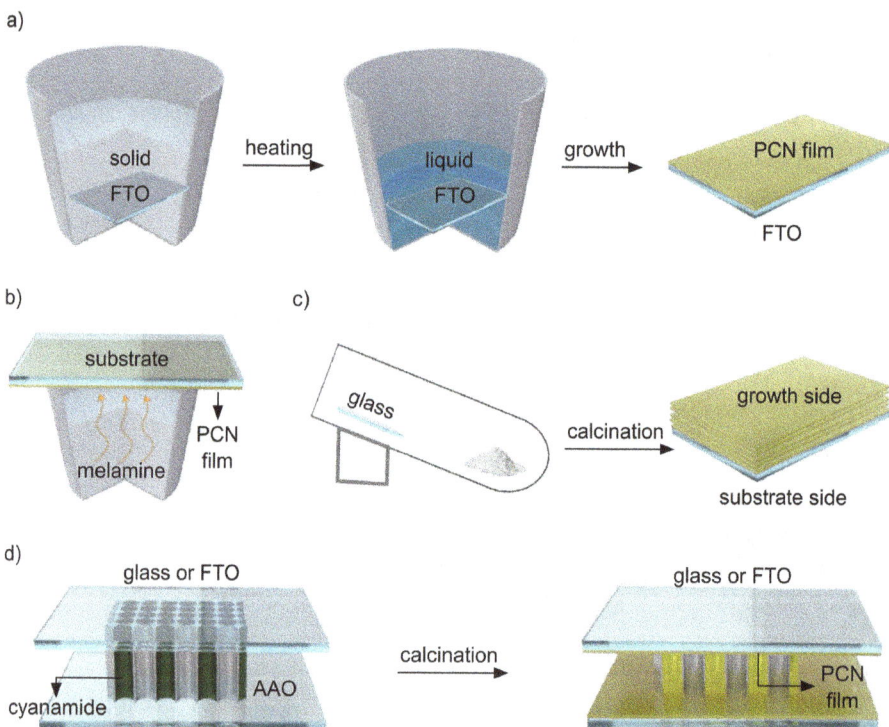

Fig. 11.2: Schematic illustrations for directly growing PCN films. (a) Solid/liquid contact growth modes [24]. (b) Illustration of the thermal vapor condensation procedure for the deposition of PCN films on conductive substrate [13]. (c) Illustration of the vapor deposition polymerization procedure for the deposition of free-standing PCN membrane on substrate [11]. (d) Illustration of fabricating PCN film on substrate using a microcontact-printing-assisted method [10].

attract enough attention initially. Giusto and Antonietti et al. used a technical two-chamber chemical vapor deposition to synthesize a transparent, highly homogeneous PCN thin film over a large area with tunable thickness [15]. Such films possess a highly oriented and conjugated layered structure. A very high intrinsic refractive index (n_D = 2.43) in the visible range, reaching that of diamond for such transparent film was obtained, which could find enormous applications in optoelectronic devices.

It has been realized that the amounts of molecular precursor and synthesis duration could be vital for the quality and thickness of the PCN film [15]. Interestingly, by modulating the local concentration of intermediate species, the surface microstructure of the free-standing membrane can be engineered [26, 27]. More recently, Liu and Gao et al. reported an asymmetric film, with nano-pillars on one side and being smooth on the other side, could be obtained by placing the substrate in the vicinity of the powdery melamine [28]. Liu et al. wrapped the covered crucible with aluminum foil to collect the escaped volatile intermediates. In such circumstances, the film is obtained on both sides of the substrates but with different morphologies. Specifically, the downward side showed conformal coating, while the upward part featured microcluster arrays and in-between nanosized bulges [29].

11.3 Thin-film-based applications

This section summarizes new applications of PCN films or membranes covering photoelectrocatalysis, sensing, and energy storage and conversion fields.

11.3.1 Photoelectrocatalysis

Since the groundbreaking discovery of photoelectrochemical water splitting on rutile TiO_2 electrodes by Fujishima and Honda in 1972, research on solar-to-fuel conversion has attracted extensive attention worldwide [30–32]. Featuring an appropriate band gap, abundant composition element, good thermal stability and nontoxicity, metal-free PCN film was frequently employed as photoelectrode in photoelectrochemical systems [19]. Generally, the exploration of the photoelectrochemical properties of PCN requires the creation of an intimate contact to a conductive substrate. However, pristine PCN materials exhibit a quite rapid recombination rate of its photoinduced electron–hole pairs, which suppresses photocurrent output to some extent.

In early studies, Zhang et al. started photoelectrochemical investigations of PCN electrodes [5, 6]. Conventional "doctor blading" and sintering by using as-prepared PCN powder were employed to prepare such film electrodes. However, the performance of the obtained PCN electrodes was not encouraging due to nonoptimized structures. The particle size of the PCN powder decreases through protonation and

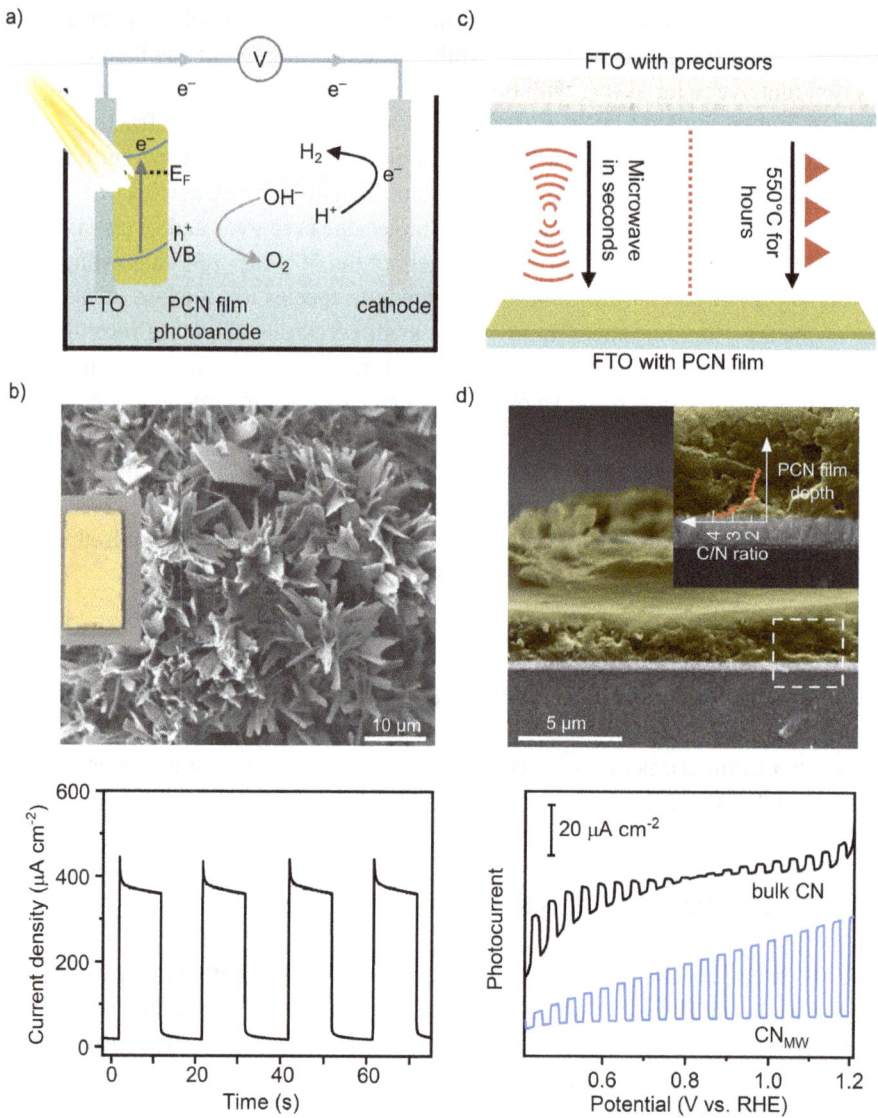

Fig. 11.3: Photoelectrochemical applications. (a) Schematic illustration of PCN film photoanode-based PEC cell. (b) SEM image of PCN film obtained from calcination of the directly grown CN monomer layers, and photocurrent density curve of such PCN photoanode [33]. (c) Schematic illustration of conventional heating and microwave heating methods for the synthesis of PCN films on FTO. (d) The cross-sectional SEM image of PCN film from microwave-assisted condensation, and the linear sweep voltammetry curves under chopped light [25].

deprotonation procedures. A homogeneous "sol" solution was formed after depolymerization of the bulk PCN powder in HNO_3 at elevated temperatures, as reported by Wang et al. [7]. The photocurrent density obtained with the sol-processed electrode could be up to 8 μA cm^{-2}, while the photoelectrode prepared using doctor-blade technique exhibited a current density of only 1.2 μA cm^{-2}. Such attempt could alleviate textural and grain boundary effects in PCN electrodes to some extent; however, the intrinsically aggregation problem of the nanoparticle aggregates was not tackled.

The PCN film quality can be greatly improved by in situ synthetic method due to the enhanced adhesion strength between the film and the substrate. By manipulating the interaction between the vapor of the intermediates and the substrates, Liu et al. and Zhang et al. reported a confinement method for growing the PCN active layers directly on the conductive FTO [10, 13]. Accordingly, an enhanced photoelectrochemical response, compared to the post-processing counterparts, was recorded. Under illumination, the transfer of electrons from the conduction band of PCN to the electrode could also give rise to an anodic photocurrent when the holes on the valence band are consumed by electron donors (water or triethanolamine) solubilized in electrolyte.

Shalom et al. found that the seeded crystallization of precursors on the FTO substrate could be crucial for PCN film formation [34]. The film acted as photoanode and exhibited a low onset potential of 0.25 V_{RHE}. In the absence of any sacrificial reagents, an impressive photocurrent density of 116 μA cm^{-2} at 1.23 V_{RHE} was recorded in an alkaline solution. Recently, Shalom et al. reported a novel method for the direct growth of uniform PCN layers on conductive glass [33]. By depositing thiourea on FTO while introducing melamine vapor during layer growth, a highly uniform PCN layer could be engineered upon calcination at 550 °C (Fig. 11.3a). For PEC applications, the best PCN photoanode demonstrated a benchmark-setting photocurrent density of 353 μA cm^{-2} (51% Faradaic efficiency for oxygen) (Fig. 11.3b) and an incident photo-to-current conversion efficiency above 12% at 450 nm at 1.23 V versus RHE in an alkaline solution.

Concerning the time-consuming synthesis using the conventional electric furnace, Zhang et al. employed microwave heating for facile growth of PCN film onto FTO (Fig. 11.3c) [25]. During the rapid condensation (several seconds), some highly reactive C, N-containing intermediates accumulated and reacted with FTO to form a robust linkage. On the contrary, under conventional heating using electronic furnace (~several hours), the highly reactive intermediates were of lower concentration and faced difficulties to establish an effective interaction with FTO. Benefiting from the high energy transfer efficiency of microwave heating, ultrafast growth of a compact PCN film (CN_{MW}) on FTO could be achieved (Fig. 11.3d upper). For PEC applications, the photocurrent of CN_{MW} under the chopped light irradiation was prompt, steady, and stable, and could reach 38 μA cm^{-2} at 1.23 V_{RHE} (Fig. 11.3d lower). The result was attributed to the improved charge mobility and the gradient carbon-rich PCN texture.

As is well known, light absorption and charge separation at the PCN/electrolyte interface play important roles in PEC performance. As reported by Wang et al., by integrating carbon nanotubes into the PCN thin film for improving the conductivity, the photocurrent density could be further enhanced to 17 $\mu A\ cm^{-2}$ [7]. In addition, the introduction of metal/metal oxide hosts into PCN films could promote the charge-carrier transfer rate [35]. As reported by Liu et al., embedding of Co^{2+} within the PCN film electrode doubles the photocurrent and lowers the water oxidation overpotential of a photoelectrochemical device [36]. Furthermore, graphdiyne, a material with high hole transport mobility, was used by the Lu group to interact with PCN to construct a PCN/graphdiyne heterojunction [37]. The composite film was employed for photoelectrocatalytic water splitting and it exhibited high efficiency. The excellent photogenerated carrier separation performance of PCN/graphdiyne photocathode was sevenfold that of PCN. A high photocurrent density of – 98 $\mu A\ cm^{-2}$ was recorded by the composite photocathode, which is threefold higher than that of the PCN photocathode. Further engineering of PCN-based composite film with high quality could contribute to its wide utilization in photoelectrochemical and electronic devices [38]. However, the lack of deeper understanding toward photophysical and photochemical properties of a PCN thin-film layer impedes attempts to further improve its photoelectrochemical performance.

In the absence of light irradiation, PCN films can also be used as electrode for the electrocatalytic hydrogen evolution reaction. A metal-free hybrid catalyst composed of PCN and graphene was reported by Zheng et al, and the sample exhibited an unexpected hydrogen evolution reaction activity (overpotential of ~ 240 mV to achieve a 10 mA cm^{-2} HER current density and a Tafel slope of 51.5 mV dec^{-1}) [39]. Inefficient charge injection restricts the films' catalytic efficiency, and in situ direct growth of PCN film on the conducting substrate would help to alleviate this problem. A highly ordered carbon nitride array on the electrode reported by Shalom et al. was used for electrocatalytic hydrogen evolution reaction [12]. The catalytic current is recorded at a low overpotential (vs RHE) of approximately 0.25 and 0.1 V for pH 6.9 and pH 13.1, respectively. The activity of the PCN/TiO_2 composite film electrodes was further improved under basic conditions, showing a low overpotential (ca. 0.1 V) with an outstanding current density (1.3 mA cm^{-2} at a 0.3 V overpotential). The high performance of the PCN film could be ascribed to the surface active sites due to amine termination on the film. The nitrogen atoms of surface amine termination have abundant free electrons, which can adsorb water from the solution via hydrogen bridges. The abundant available hydrogen concentration on the catalysts' surface increased; thus the activation energy required for hydrogen evolution reaction decreased.

Gradually, photoenzymatic catalysis moves into another focus, owing to its excellent catalytic efficiency, stereoselectivity and environmental benignity [40]. Inspired by the natural photosynthesis of plants, light-driven enzyme cooperative catalysis has the advantages of excellent catalytic efficiency, stereoselectivity and environmental benignity. NAD(P)H or its oxidized counterparts (NAD(P)$^+$) used as cofactors are essential in numerous redox reactions, catalyzed by variety of oxidoreductases. In biocatalytic reactions, NADH needs to be regenerated owing to equivalent consumption during enzyme turnover [41].

Based on the extensive studies on photocatalytic NADH regeneration using PCN powder, Liu et al. proposed PCN and PCN/graphene film photoelectrodes for coenzyme regeneration (Fig. 11.4a) [22, 41–43]. Water was used as the electron donor, and a rhodium complex $[Cp^*Rh(bpy)H_2O]^{2+}$ (Cp^* = pentamethylcyclopentadienyl, bpy = 2,2′-bipyridine) acted as an electron and proton mediator, respectively. When electrons transfer from the PCN/graphene photocathode via $[Cp^*Rh(bpy)H_2O]^{2+}$ to NAD^+, an increased cathodic peak current and a potential shift appear. Under the bias of –0.9 V (vs Ag/AgCl) and visible light irradiation, the NADH regeneration yield of PCN/graphene is twice that of PCN using water as the electron donor and $[Cp^*Rh(bpy)H_2O]^{2+}$ as the mediator.

Taking inspiration from the natural photosynthetic scheme, Park et al. constructed a photoelectrochemical system consisting of protonated PCN and carbon nanotube hybrid (CNT/p-PCN) cathode and FeOOH-deposited bismuth vanadate $BiVO_4$ (FeOOH/$BiVO_4$) photoanode (Fig. 11.4b) [44]. In the system, photoexcited electrons are generated from the FeOOH/$BiVO_4$ photoanode. They then move to the CNT/p-PCN hybrid film to reduce the flavin mononucleotide mediator. Subsequently, the reduced flavin mononucleotide could deliver electrons to the old yellow enzyme for the enantioselective conversion of ketoisophorone to (R)-levodione. As a result, the (R)-levodione with an enantiomeric excess of above 83% was synthesized by the CNT/p-PCN composite cathode, while no product was detected by the pristine CNT cathode without using p-PCN. The results demonstrated that p-PCN promotes two-electron transfer and accelerates the flavin mononucleotide mediator reduction when coupled with conductive CNTs. Such work reflects the potential of using such biocatalytic photochemical cells for the synthesis of high-value chemicals using water as an electron donor.

Recently, Norby et al. reported that a PCN film coated on an FTO glass could work for direct electrocatalytic NADH regeneration without using the common Rh complex mediator [45]. Such metal-free regeneration using PCN powder as photocatalyst was firstly reported by Liu et al.; however, the selectivity toward 1,4-NADH was poor [41, 46]. It is debatable that noncovalent π–π interaction between the photoexcited PCN and the NAD^+ can describe the 100% 1,4-NADH regeneration in Norby's work. More detailed characterizations and analyses were necessary for addressing such selectivity concern. Norby et al. then demonstrated that by combining Ta_3N_5 nanotubes as a highly active photoanode, a bio-hybrid photoelectrochemical cell (Fig. 11.4c) with formate dehydrogenase could be constructed. In the bio-hybrid photoelectrochemical cell, a stable aerobic photoelectrochemical CO_2-to-formate reduction at close to 100% Faradaic efficiency and unit selectivity were obtained. The solar-to-fuel efficiency, irradiated with simulated sunlight, is up to 0.063%, approaching that of natural photosynthesis. The main challenge of such an effective biotransformation using solar energy is transferring the photoexcited electrons to or from the redox centers of enzymes [43]. Thus, high-quality PCN films or hybrid films are helpful for the rapid improvement of such coupled energy and catalysis applications.

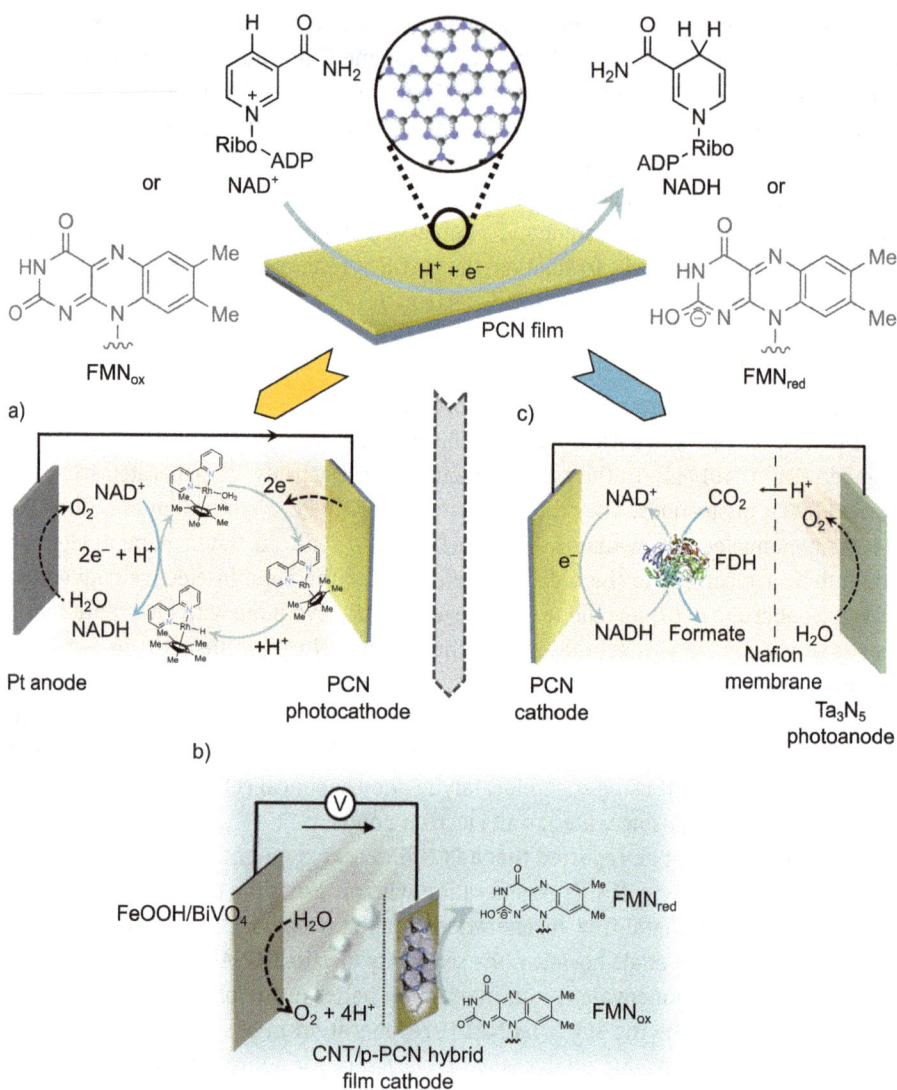

Fig. 11.4: PCN-film-based photoenzyme catalysis. The upper part shows the illustration of the redox relationships between NAD$^+$ (or FMN$_{ox}$) and NADH (or FMN$_{red}$) on a PCN electrode. (a) Illustration of PCN-nanosheets-assembled film photoelectrode for photoelectrochemical NADH regeneration, assisted by homogeneous [Cp*Rh(bpy)H$_2$O]$^{2+}$ [22]. (b) Illustration of electron transfer from FeOOH/BiVO$_4$ to the CNT/p-PCN cathode, followed by the reduction of oxidized flavin mononucleotide [44]. (c) Schematic of the integrated bio-hybrid photoelectrochemical cell comprising PCN photocathode and Ta$_3$N$_5$ photoanode [45].

Continuous-flow chemistry has been extensively used for decades in the chemical industry, owing to the inherent advantages of efficient mass and heat transfer, and simplicity of scale-up. The productivity and selectivity of flow chemistry are enhanced compared to traditional batch reactors. The materials are usually applied in the form of dispersion in liquid medium, which is bound to certain technological limits of applicability. Coating of the photoreactor wall with semiconductor particles is an appealing feature, in combination with flow technology. The merging of flow chemistry with heterogeneous photoredox catalysis reported by Wang et al., was used for the facile production of high-value compounds in a continuous flow reactor under visible light irradiation at room temperature in air [47]. In the flow reactor system, PCN is immobilized onto glass beads and fibers, demonstrating a highly flexible construction possibility for devices of the photocatalytic materials (Fig. 11.5a). The functionalized glass beads or fibers can be easily integrated into a continuous-flow photoreactor with high light penetration, and cyclobutanes can be synthesized in gram scale, with a high yield of 81% [43].

Giusto et al. proposed an innovative approach using carbon nitride thin films, prepared via chemical vapor deposition at different vessel walls, as batch and microfluidic photoreactors (Fig. 11.5b) [16]. The glassware with photocatalytic layer was evaluated in the oxidation of benzyl alcohols under light irradiation. As a result, the film showed at least one order of magnitude higher activity per area unit compared to the process using suspended particles. Interestingly, they recently further extended such PCN film on the glass vial as photoinitiators for radical polymerization [48].

11.3.2 Sensing

Sensors that can selectively detect targeted molecules in complex mixtures are promising tools for improved health monitoring, disease diagnosis, food safety and environment protection. In recent years, PCN materials have drawn much attention in the field of sensors, including ion sensors, biosensors, gas sensors and humidity sensors [4, 49]. As a metal-free polymer semiconductor with a narrow band gap, PCN possesses outstanding optical properties (photoluminescence and chemiluminescence) and in photoelectric conversion (electrochemiluminescence and photoelectrochemistry), which enable multiple signal output modes and various sensors designs. Some researchers explored the interaction between PCN powder or film, with related molecules or metal ions, for sensing [50, 51]. Herein, we summarized the related works on PCN-thin-film-based sensors.

a)

glass bead ⟹ acid etched GB-OH ⟹ NH₂-modified GB-OH ⟹ PCN-coated glass bead

b)

96%
93 min

86%
122 min

100%
108 min

92%
98 min

95%
126 min

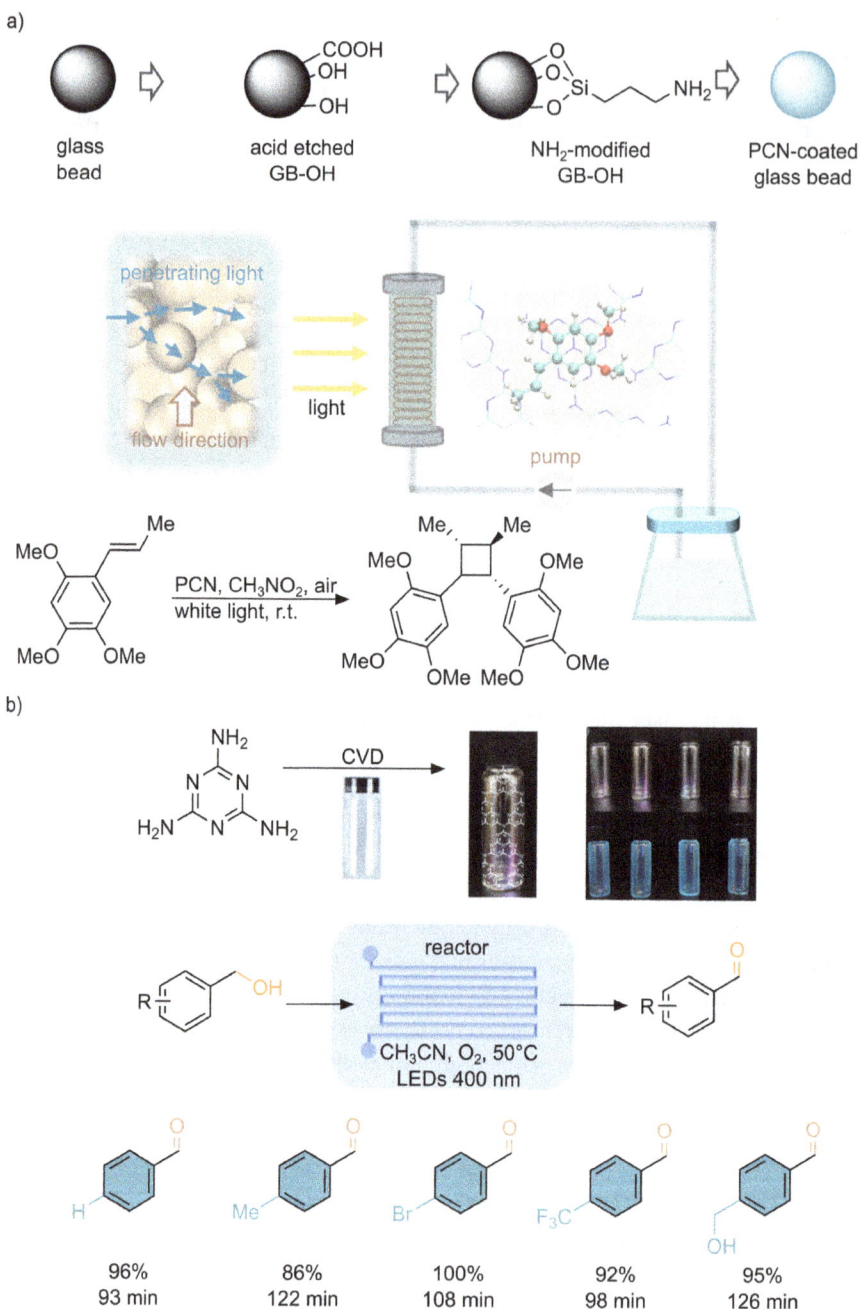

Fig. 11.5: (a) Schematic procedure for coating PCN onto the glass beads surface and illustration of a fixed-bed photoreactor filled with supported PCN for magnosalin production [47]. (b) Schematic representations of the "chemical vapor deposition" process for coating the vessel walls with PCN, which was used as batch and microfluidic reactors for catalytic conversion [16].

11.3.2.1 Metal ion sensors

a)

b)

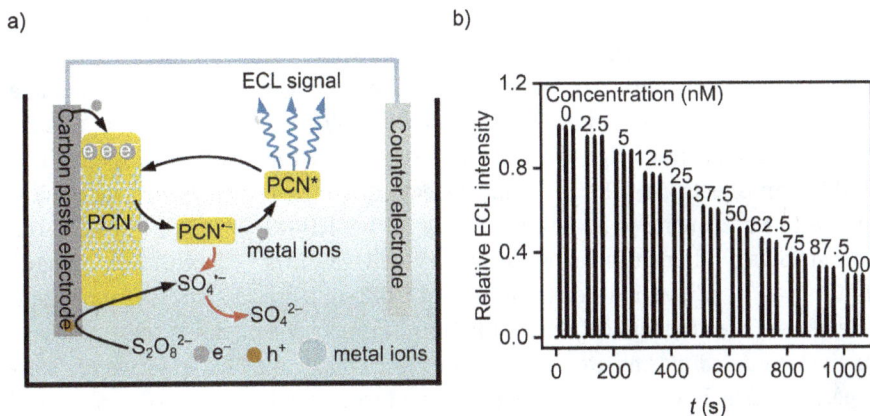

Fig. 11.6: The schematic illustration of the ECL process (a) and the Cu^{2+} concentration-ECL intensity relationship of the PCN-modified carbon paste electrode (b) [52].

Cu^{2+} is an essential element for living organisms but toxic at high concentrations. Li and Zhang et al. constructed a PCN-nanosheet-based electrode on ITO for the purpose of Cu^{2+} detection [53]. Upon illumination, the photogenerated electrons in the conduction band of PCN could be accepted by Cu^{2+} to form Cu^0, which further mediates charge transfer to the surface of Pt electrode. The effect of Cu^{2+} on the photocurrent was found to be concentration-dependent. Furthermore, the ITO/PCN electrode showed high selectivity toward Cu^{2+}, as compared with other metal ions (Mn^{2+}, Cr^{3+}, Co^{2+}, Ni^{2+}, Fe^{3+}, Zn^{2+}, Hg^{2+}, Pb^{2+} and Ag^+). No further explanation was given for such selectivity.

Electrochemiluminescence, as the reverse process of photoelectrochemistry, has been regarded equally important for sensing in recent years. Electrogenerated chemiluminescence is the light emission induced by electron transfer between electronic materials and their co-reactants. Generally speaking, it is easy and cost effective to fabricate an electrochemiluminescence sensor, with a limit of detection at a subnanomolar level. Xiao and Choi et al. reported for the first time the electrochemiluminescence behavior of PCN with $K_2S_2O_8$ as the co-reactant for Cu^{2+} detection [52]. As shown in Fig. 11.6a, electrons from the working electrode (the substrate) are injected to the conduction band of PCN to produce the negatively charged $PCN^{\bullet-}$. Meanwhile, the $S_2O_8^{2-}$ is reduced to $SO_4^{\bullet-}$ and SO_4^{2-} via the conductive substrate. The strong oxidant species ($SO_4^{\bullet-}$) produced from electro-reduction during the cathodic scan subsequently reacts with the reduced form of PCN to produce the excited state PCN^* via electron transfer from $PCN^{\bullet-}$ to $SO_4^{\bullet-}$. Finally, an intense blue emission is obtained when PCN^* decays back to the ground state PCN. The proposed electrogenerated chemiluminescence sensor shows excellent selectivity to Cu^{2+} detection. The suitable redox potential of Cu^{2+}/Cu, located between the conduction and the valence bands of

PCN, allows efficient electron transfer from the negatively charged PCN$^{•-}$ to Cu^{2+}, resulting in quenching of electrochemiluminescence (ECL). The quenching process is also concentration-dependent, with the ECL intensity decreasing along the concentration increase of Cu^{2+} (Fig. 11.6b).

Hg^{2+} is one of the most hazardous pollutants even at very low concentration due to the bioaccumulation and bio-amplification effects. Zhang et al. reported an electrochemical sensor for trace Hg^{2+} detection by using glassy carbon electrodes, modified with PCN nanosheets [54]. The as-obtained electrode showed a strong electrochemical response to Hg^{2+}, owing to its strong affinity to the -NH/-NH$_2$ of ultrathin PCN. The anodic stripping voltametric current is linearly related to Hg^{2+} concentration ranging from 0.1 to 15 μg L^{-1}, with a detection limit of 0.023 μg L^{-1}. The authors claimed that the developed electrochemical sensor for Hg^{2+} in practical samples is comparable to inductively coupled plasma atomic emission spectrometry method, in terms of detection performance; however, it is more simple and cost effective to apply.

11.3.2.2 Biosensing

Biosensor serves the purpose of detecting a biochemical component with a physico-chemical detector. Benefiting from the electrogenerated chemiluminescence behavior of PCN, an electrochemiluminescence biosensor without disturbance by the photoexcitation background could be constructed [52]. Xiao et al. utilized a PCN-based carbon paste electrode as a sensor for rutin detection, a bioactive flavonoid glycoside [55]. The electrogenerated chemiluminescence measurements were carried out in 0.10 mol L^{-1} phosphate buffer solution containing 0.10 mol L^{-1} KCl by cyclic voltammetry. Electro-oxidized PCN can readily react with the intermediate generated from co-reactant triethanolamine, upon electrochemical oxidation, producing excited state PCN, which subsequently decays back to its ground state, emitting strong luminescence, while this luminescence could be quenched by rutin through an energy transfer process.

Zhang et al. developed a "two birds with one stone" strategy for bulk PCN exfoliation via mechanical grinding and also DNA-linking to the exfoliated PCN nanosheets [56]. This DNA biosensor electrode was constructed by drop-casting the functionalized PCN nanosheets dispersion onto the glassy carbon electrode. The resulting electrochemiluminescent biosensor for target DNA exhibited a massively enhanced sensitivity. The authors further extended the PCN-based electrochemiluminescence biosensor to the detection of 8-hydroxy-2′-deoxyguanosine, driven by a mechanism that included competitive catalytic effects and steric hindrances [57]. A hemin/G-quadruplex was assembled on the PCN nanosheets, which were then drop-casted onto the glassy carbon electrode for biosensing. As claimed by the authors, the detection sensitivity on the multiple-mechanism-based electrochemiluminescence was inherently boosted, compared to the single-mechanism-driven biosensing.

11.3.2.3 Humidity sensor

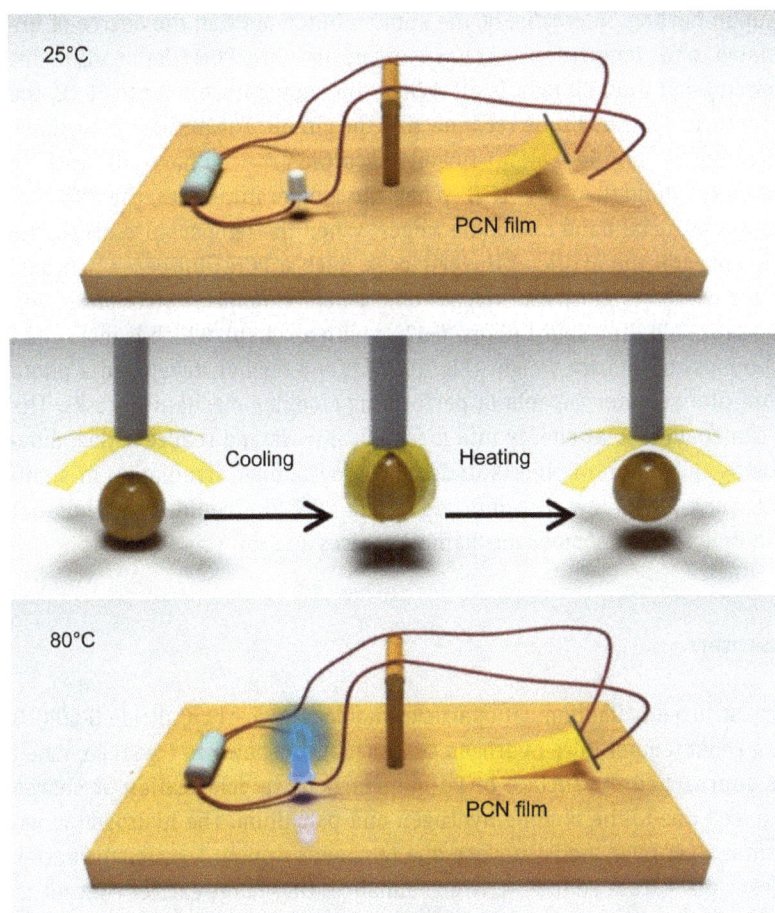

Fig. 11.7: Schematic illustrations of an alarm device for temperature and an artificial hand for grasping and releasing objects using PCN film as finger [23].

Actuators, which can convert environmental energy, including light, electricity, heat and humidity, into mechanical energy, are developed to play important roles in wearable devices, rehabilitation robots, artificial intelligence and so on. Due to their excellent electronic and optical properties as well as their biocompatibility, PCN materials and their membrane derivatives show great potential in actuator applications.

Researchers already developed rather practical PCN film-based actuators that could easily convert heat or light stimuli into sophisticated mechanical work. Guo and coworkers creatively prepared a freestanding PCN film with a thickness of tens of micrometers [23]. The film obtained from the physical vapor deposition of bulk PCN powder underwent macroscopic deformation due to the desorption and adsorption of water upon ambient temperature change. As shown in Fig. 11.7, the PCN film could also work as a simple switch in an alarm device for high temperatures, due to the deformation upon heating. Interestingly, the authors found out that the degree of un-bending is related to the temperature. At low temperatures, the PCN film bends to disconnect the circuit and the LED light is off. When the temperature rises to 80 °C, the unbending film turns on the LED by reconnecting the circuit. The authors moved forward by integrating the film into a four-finger structured artificial hand for grasping and releasing objects (middle of Fig. 11.7). When the temperature rises, the PCN film stretches and the artificial hand opens the finger. When the temperature drops, the film becomes bent and the artificial finger closes. Such a PCN-film-based artificial hand can be engineered to grasp and release objects. For example, a strip of CN film with mass of 5 mg could lift a smart phone (~125 g with a clamp), which is more than 25 000-fold the mass of its own weight (Fig. 10.3b). They further developed a photo responsive thin-film actuator capable of performing complex mechanical tasks. The manipulator can convert solar energy into mechanical work and realize lifting, moving and releasing the action of objects under the programmed irradiation of simulated sunlight. Such facile, low-cost and large-scale PCN film could have potential applications in fields of autonomous mechanical devices and soft robots.

11.3.2.4 Gas sensors

Raghu et al. constructed palladium nanoparticles modified by carbon nitride (Pd/PCN) composite as a room temperature hydrogen sensor [58]. As shown in Fig. 11.8a, when H_2 molecules approached the surface of Pd nanoparticles, a polarization or charge transfer is induced due to the binding hydrogen and palladium. The hydrogen molecules first form a layer over the Pd surface due to chemisorption. A coupled transfer of electrons takes place from Pd to PCN, which enhances the charge carrier concentration and decreases resistance. The variation in the resistance is found to be proportional to the concentration of H_2. The hydrogen-sensing properties of Pd/PCN were studied below the flammability range from 1 to 4 vol% H_2 at room temperature, and 99.8% sensitivity was achieved with a response time of 8.8 s. For the same concentration range of H_2 at 80 °C, the sensor showed 97.4% sensitivity. Here, the sensitivity of the sensor is defined by the equation:

$$\text{Sensitivity} \% = \left[\frac{R_0 - R_g}{R_0} \right] \times 100 \tag{11.1}$$

where R_0 and R_g represent resistance of the film in 100% air and in presence of the gas, respectively. Response time is defined as the time taken to 90% of the total change in sensor response parameter. Recovery time is defined as the time taken to reach 10% of the total change in base resistance.

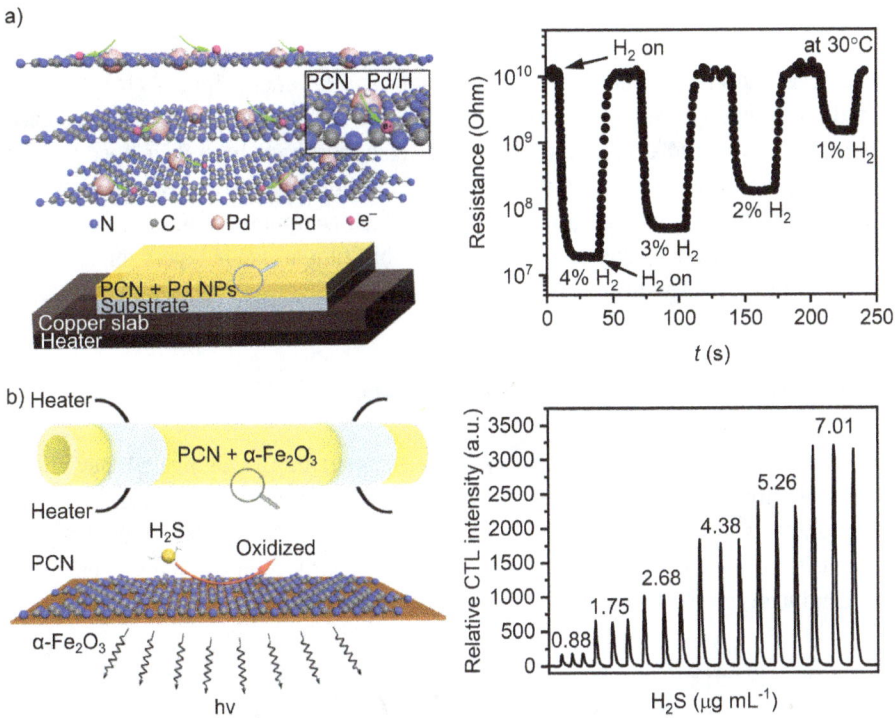

Fig. 11.8: (a) Schematic illustration of hydrogen physisorption and chemisorption over palladium nanoparticles [58]. (b) Schematic diagram of H_2S sensor using α-Fe_2O_3/PCN composites [59].

As a malodorous and toxic by-product of oil refining, mines or food processing industries, hydrogen sulfide (H_2S), even at low concentration, has detrimental effects on the human nerve system. Monitoring the concentration of hydrogen sulfide in the atmosphere has become crucial in a number of occasions. Lv et al. reported a H_2S sensor by utilizing cataluminescence emitted during the oxidation of H_2S on the surface of α-Fe_2O_3/PCN composites coated on a ceramic rod at a proper temperature (Fig. 11.8b) [59]. The cataluminescence could be measured by an ultra-weak luminescence analyzer. The response and recovery time were as fast as 0.1 and 0.6 s at low temperature, respectively. They also found that such a sensor demonstrated high selectivity for H_2S gas without interference from common foreign substances, even at much higher concentration. The high selectivity toward H_2S suggests such a PCN-based composite layer sensor could be of practical value in future environmental monitoring, for example in oil fields and oil refining.

11.3.3 Solar cells and battery

A solar cell is a photovoltaic device that can convert solar energy into electrical energy directly. The representatives of the new photovoltaic technology, such as dye-sensitized solar cells, polymer solar cells and perovskite solar cells, have been developed rapidly in recent years. Due to the unique electronic and optical properties, structuring PCN into thin film was resonating also with the solar cell communities, and a series of inspiring work have been reported by using PCN thin film as active layers for enhancement in device performance.

11.3.3.1 PCN thin film for dye-sensitized solar cells

Dye-sensitized solar cells, firstly developed by O'Regan and Grätzel in 1991, have been intensively investigated owing to the advantages of low materials costs, simplicity of preparation, relatively low production cost, low pollution, and long life [60]. A typical dye-sensitized solar cell contains the following three parts: a photoanode, a sensitized semiconductor, an electrolyte and a counter-electrode [61].

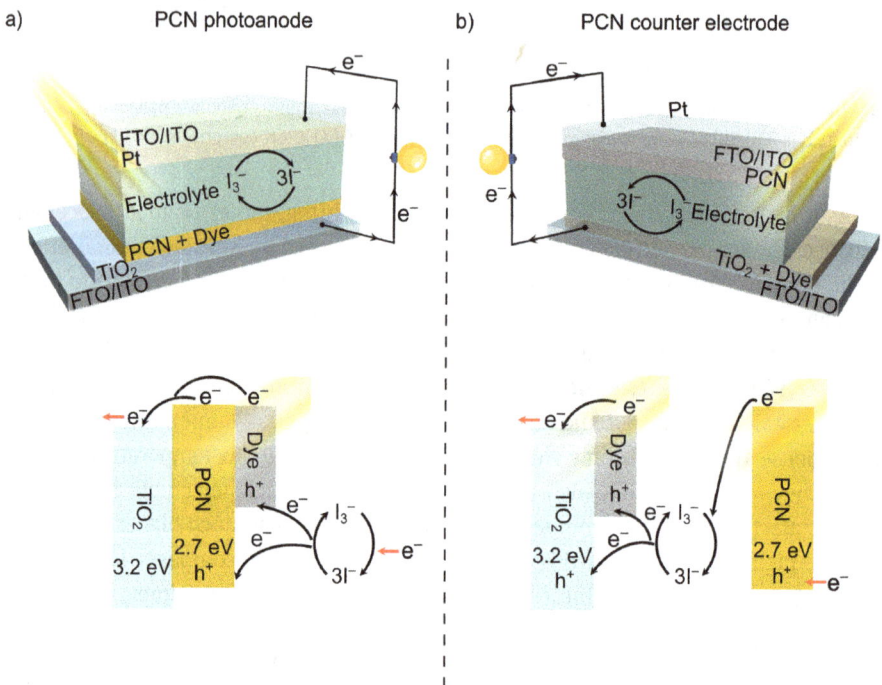

Fig. 11.9: Schematic illustration of the application of PCN composites as multifunctional layer as the (a) photoanode [62] or (b) counter electrode [63] in dye-sensitized solar cells.

As a key part of dye-sensitized solar cells, the photoanode, which can efficiently harvest and utilize solar energy, affords the electron transport between the sensitizer and the external circuit. Yu et al. reported a photoanode of dye-sensitized solar cells using PCN-modified TiO_2 nanosheets, which were fabricated by simply heating the mixture of TiO_2 nanosheets and urea [62]. After modification of the PCN layer, the photoelectric conversion efficiency of the dye-sensitized solar cells increased significantly, approaching 28% enhancement at the optimal addition amount of PCN. As shown in the diagram of Fig. 11.9a, PCN layer plays an important role in improving the solar cell performance as it acts as the blocking layer to suppress the electron backward recombination with the electrolyte. Specifically, PCN has more negative CB position than that of TiO_2; thus the photoinduced electrons from the CB of PCN will transfer to the CB of TiO_2. As a result, the combined structure effectively promotes electron transport and also contributes additional electrons to increase the electron concentration in the photoanodes (Fig. 11.9a lower). Liu et al. used a post-treatment with urea solution to synthesize PCN-coated TiO_2 composites, which was employed as photoanode material in dye-sensitized solar cells. The PCN coating on the TiO_2 surface played again the role of a blocking layer, reducing the recombination rate of electrons and holes. The efficiency of dye-sensitized solar cells increased by 25% when compared with that of pure P25 electrode. Similarly, Yin et al. used PCN-modified TiO_2 nanosheets as photoanodes and Co_9S_8 nanotube arrays as counter electrodes in assembling the dye-sensitized solar cells [64]. The PCN layer acted as a blocking layer to prevent charge recombination at the TiO_2/electrolyte interface efficiently. The smaller electron transport resistance of dye-sensitized solar cell based on TiO_2/PCN photoanodes indicates higher charge transfer ability. A power conversion efficiency of 8.07% was achieved for such a photoanode, much higher than that of the pure TiO_2 counterpart.

The counter electrodes are also indispensable for dye-sensitized solar cells, among which Pt is the most widely used. The development of equivalent alternatives at lower costs is highly requested due to the scarcity and high costs of Pt. High conductivity for charge transport, good electrocatalytic activity for reducing the redox couple and excellent stability should be contained in an optimal counter electrode material. Wang et al. constructed a porous PCN/graphene composite as the counter electrode of dye-sensitized solar cells via a simple hydrothermal method (Fig. 11.9b) [63]. The PCN/graphene composite electrode demonstrated excellent electrocatalytic activity toward I^-/I_3^- and lower charge-transfer resistance, compared with the individual component in the composites. By employing such porous PCN/graphene composite as the counter electrode, a conversion efficiency of 7.13% could be achieved. Similarly, Wang et al. turned to conductive carbon black as the conductive dopant and constructed PCN/conductive carbon black composite via a simple ball-milling process. The Pt-free counter electrode was manufactured by pasting the composite onto the conductive FTO glass. By coupling with the TiO_2 photoanode, a power conversion efficiency of 5.09% could be achieved for such metal-free counter electrode, which is comparable to that of the cell with Pt counter electrode [65].

11.3.3.2 PCN thin film for polymer and perovskite solar cells

Polymer solar cells were once considered a unique and promising alternative to inorganic counterparts due to their renewable, lightweight and low-cost characteristics. With the sudden emergence of perovskite solar cells in 2012 [66], a drastic efficiency increase from 9% to 25.2% was witnessed within 10 years, worldwide [67, 68]. During the rise of perovskite solar cells, polymer solar cells have received lower interest, but were "reloaded" only recently [4, 69]. The driving forces of all such developments are the new materials and design. Due to the above-mentioned many advantages, exploration of PCN, especially thin film layer, in such emerging solar cell layouts is very promising.

Li et al. introduced a post-treated PCN film from exfoliated nanosheets into the polymer solar cell (ITO/PCN/PBDTTT-C:PC$_{71}$BM/MoO$_3$/Ag) as the cathode interfacial layer to facilitate electron collection [70]. The assembled polymer solar cell device with the bulk-heterojunction/ITO interface, modified by the PCN layer, is shown in Fig. 11.10a. Control experiments showed that the PCN film could significantly increase the short circuit current density (from 15.01 to 16.04 mA cm^{-2}) and the open-circuit voltage (from 0.58 to 0.70 V), and improve the fill factor from 42% to 57%. The performance was further confirmed by the external quantum efficiency spectra of the polymer solar cells. The maximum external quantum efficiency values of the device, with and without the PCN film interlayer, are 72% and 64%, respectively. Such work suggests that the post-processed PCN film could act as a new and promising electron transport material for solution-processed organic optoelectronic devices. The PCN film was also used as an acceptor moiety, in combination with poly(3-hexythiophene), and the resultant device exhibited an open circuit voltage exceeding 1 Volt [24]. Interestingly, by doping solution-processable PCN quantum dots into the active layer (Fig. 11.10b), a dramatic efficiency enhancement was observed [71].

PCN thin film can also serve as interfacial electron-transporting layers or hole-blocking layers to improve the performance of perovskite solar cells. Due to the moisture-sensitivity of the organic–inorganic lead halide materials, water is not preferred as the dispersing medium for PCN nanosheets. Such attempt was also discouraged by the poor processability of PCN in organic solvents. In 2019, Antonietti et al. employed a photografting strategy for modifying the PCN with 4-methyl-5-vinylthiazole [72]. Such one-pot modification led to the exceptional organodispersion of PCN–thiazole nanosheets, paving the way for the simple application of a homogeneous film, compatible with the cell buildup. The thiazole-modified PCN film was deposited on top of methylammonium lead iodide (MAPbI$_3$), prior to or after the coating of PC$_{60}$BM layer (Fig. 11.10c, upper and lower panel) [9]. Accordingly, two different architectures (ITO/HTL/MAPbI$_3$/PCN–thiazole/PC$_{60}$BM/AZO/Ag versus ITO/HTL/MAPbI$_3$/PC$_{60}$BM/PCN–thiazole/AZO/Ag) were constructed. The champion device with thiazole-modified carbon nitride realized a 1.09 V open voltage and a current density of 20.17 mA cm^{-2}. The enhancement was attributed to the boosted charge extraction by PCN–thiazole from the

a)

b)

c)

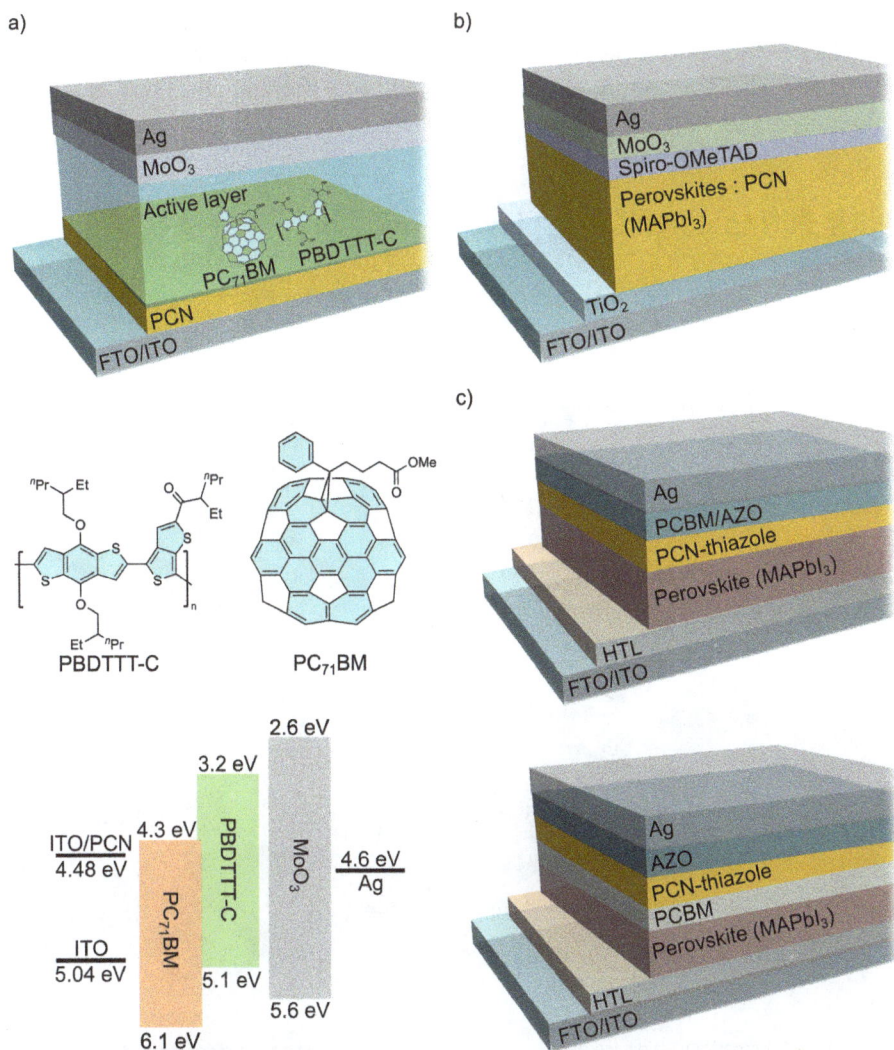

Fig. 11.10: PCN-based solar cell devices. (a) Schematic illustration of the inverted polymer solar cells using PCN nanosheets as cathode interfacial layer (upper), the chemical structures of PBDTTT-C and PC71BM (middle), and the energy level diagram (lower) [70]. (b) Device structure of MAPbI$_3$:PCN-based n-i-p structure perovskite solar cell [18]. (c) Schematic illustration of the inverted perovskite solar cells using PCN–thiazole nanosheets as cathode interfacial layer of MAPbI$_3$-based solar cells, with PCBM coated over PCN-thiazole layer (upper) and perovskite (lower) [9].

perovskite absorber toward the electron transport layer in p-i-n devices. Interestingly, by replacing AZO with PCN–thiazole, a better performance in terms of short-circuit current density was observed, compared to the reference device. By doping the conventional electron transport layer with PCN quantum dots to regulate the interfacial charge dynamics, an enhancement of performance in terms of efficiency and stability could be observed for the perovskite solar cells [73]. Thus, we are sure that the metal-free, cheap, and tunable photophysical PCN will find more applications in a wide range of organic energy devices.

11.3.4 PCN thin film for display

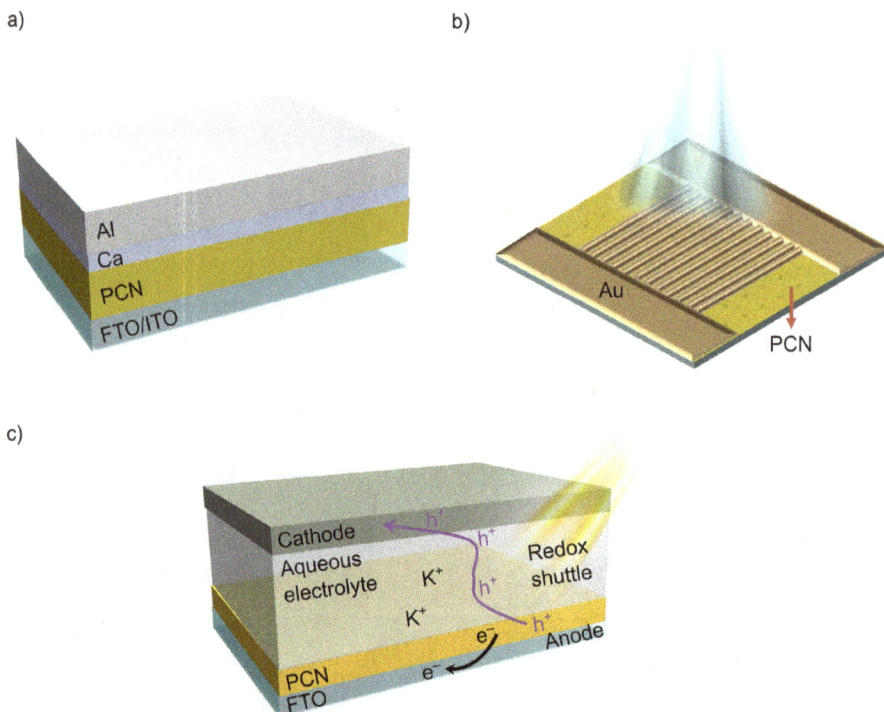

a)

b)

c)

Fig. 11.11: (a) Schematic OLED device based on PCN thin film [74]. (b) Schematic diagram of the photodetector consists of a metal–semiconductor (PCN)–metal structure [77]. (c) A monolithic solar battery based on cyanamide-functionalized polyheptazine imide [75].

Color-tunable photoluminescence in PCN powders has been previously demonstrated [76], promising the potential as a tunable emitter platform. As shown in Fig. 11.11a, Xu and Shalom et al. reported that PCN thin films could serve as emissive layer in single-layer organic light-emitting diode [74]. Electroluminescence generated from prototypical ITO/PCN/Ca/Al organic light-emitting devices, observing a strong redshift of up to 170 nm, compared to the corresponding photoluminescence spectrum. In analogy to conventional polymers, the band gap of PCN can be further tuned through the use of dopants that modify its composition and structure, resulting in color-tunable photoluminescence and near-IR electroluminescence from a thin film. Shan et al. demonstrated a free-standing flexible photodetector prototype by using high-quality PCN films as imaging pixels [77]. As shown in Fig. 11.11b, the photodetector or even flexible photodetector consisting of a metal-semiconductor-metal structure, with an electrode interval of 5 μm, could be demonstrated, benefiting from the outstanding electronic properties and chemical stability of the obtained PCN films.

PCN films can also provide a possibility for on-site storage of solar light in the form of electrical energy. Lotsch et al. integrated the two key functions of energy conversion on a single PCN film: light harvesting and electrical energy storage [75]. They fabricated cyanamide-functionalized poly(heptazine imide) nanosheets by sonicating the pristine powder in deionized water, which were then used to construct electrodes by spin coating. Such PCN layer worked as a photoanode by employing 4-methylbenzyl alcohol as a highly efficient and selective hole quencher. Under illumination, the energy can be harvested by photoreduction of the carbon nitride backbone, and then porous PCN layers can absorb alkali metal ions at the solution interface to store energy as a supercapacitor (Fig. 11.11c). The increasing illumination charging times resulted in the charge storage rising up to 43.7 C g^{-1} (12.1 mA h g^{-1}). Used not only for storing solar energy, the PCN-film-based device can be charged purely electrically, that is, by adopting a negative bias and without adding a sacrificial donor in darkness. The device demonstrates the battery charging and discharging properties, purely electrically, which is consistent with the measurements made after light charging, indicating that reversible electrical charging and discharging is feasible.

11.4 Summary and outlook

In this chapter, recent developments of PCN films in terms of synthesis, properties and emerging applications, especially in photoelectrode applications, electrocatalysis, sensors, batteries, fuel and solar cells and light emitting devices were summarized. PCN thin films usually outperformed powder-based counterparts in most application scenarios; some new are totally beyond people's imagination of the early days. However, studies of PCN films or membranes are still in their infancy, and there are still many challenges to overcome for systematically understanding the film and further applying PCN films in real devices.

Firstly, it is gaining consensus that polymeric carbon nitride studies must move from trial-and-error approaches to one of rational design. This entails understanding film formation at the atomic and molecular level, and characterizing the film with advanced technologies [78]. Exploring the applications of crystalline PCN film could contribute to the trend. Such efforts would also provide great insights for further understanding the performance enhancement behind versatile applications and establish the structure-performance relationship in film-based applications.

Secondly, gaining a fundamental understanding about charge separation and transfer processes in the PCN film as well as powdery substance is crucial for further device developments. Actually, our understanding of these crucial charge transport processes and the underlying knowledge in PCN film is only at the very beginning [79]. The lack of fundamental understanding regarding the photophysics and photochemistry of PCN film impedes our ability to improve their photoelectrochemical performance. A PCN-based photoelectrode gave up to now a maximum photocurrent density of 353 $\mu A\ cm^{-2}$ [33]. However, such photoelectrochemical performance of PCN-based photoelectrode is far lower than state-of-the-art $BiVO_4$- or Si-based photoelectrodes [80, 81]. Further improving the related performance of PCN films rely on the deeper experimental and theoretical understandings toward intrinsic charge separation and transfer properties.

Last but not least, novel methods for synthesizing high-quality PCN film are always desired. The work by Giusto et al. for growing transparent PCN film with the highest refractive index using chemical vapor deposition is very inspiring, suggesting the potential of combining conventional methods with nuanced application [15]. One of the challenges for the use of PCN-based materials as electrocatalysts/electrode is the inherently poor electronic conductivity and the low surface area of bulk PCN powder. Poor conductivity will become even worse with the PCN film being thicker for the sake of enhanced light harvesting, and decoupling of light harvesting and other specific functions in the PCN films could serve such purpose [28].

In summary, with deeper understanding of physical and chemical properties, further exploitations of PCN film in photoelectrochemistry, solar cells, light-emitting diodes and sensors are anticipated. By combining the unique physics and chemistry of PCN and the synthetic advances in the materials field, the vision for developing carbon nitride film-based practical devices will become increasingly clear.

References

[1]   Wang X, Maeda K, Thomas A, Takanabe K, Xin G, Carlsson JM, et al. A metal-free polymeric photocatalyst for hydrogen production from water under visible light. Nat Mater 2009, 8(1), 76–80.

[2]   Zhao D, Wang Y, Dong C-L, Huang Y-C, Chen J, Xue F, et al. Boron-doped nitrogen-deficient carbon nitride-based Z-scheme heterostructures for photocatalytic overall water splitting. Nat Energy 2021, 6(4), 388–97.

[3]   Ong W-J, Tan -L-L, Ng YH, Yong S-T, Chai S-P. Graphitic carbon nitride (g-C_3N_4)-based photocatalysts for artificial photosynthesis and environmental remediation: Are we a step closer to achieving sustainability? Chem Rev 2016, 116(12), 7159–329.

[4]   Liu J, Wang H, Antonietti M. Graphitic carbon nitride "reloaded": Emerging applications beyond (photo)catalysis. Chem Soc Rev 2016, 45(8), 2308–26.

[5]   Zhang Y, Antonietti M. Photocurrent generation by polymeric carbon nitride solids: An initial step towards a novel photovoltaic system. Chem Asian J 2010, 5(6), 1307–11.

[6]   Zhang Y, Thomas A, Antonietti M, Wang X. Activation of carbon nitride solids by protonation: Morphology changes, enhanced ionic conductivity, and photoconduction experiments. J Am Chem Soc 2009, 131(1), 50–51.

[7]   Zhang J, Zhang M, Lin L, Wang X. Sol processing of conjugated carbon nitride powders for thin-film fabrication. Angew Chem Int Ed 2015, 54(21), 6297–301.

[8]   Zhang X, Xie X, Wang H, Zhang J, Pan B, Xie Y. Enhanced photoresponsive ultrathin graphitic-phase C_3N_4 nanosheets for bioimaging. J Am Chem Soc 2013, 135(1), 18–21.

[9]   Cruz D, Garcia Cerrillo J, Kumru B, Li N, Dario Perea J, Schmidt BV, et al. Influence of thiazole-modified carbon nitride nanosheets with feasible electronic properties on inverted perovskite solar cells. J Am Chem Soc 2019, 141(31), 12322–28.

[10]   Liu J, Wang H, Chen ZP, Moehwald H, Fiechter S, van de Krol R, et al. Microcontact-printing-assisted access of graphitic carbon nitride films with favorable textures toward photoelectrochemical application. Adv Mater 2015, 27(4), 712–18.

[11]   Arazoe H, Miyajima D, Akaike K, Araoka F, Sato E, Hikima T, et al. An autonomous actuator driven by fluctuations in ambient humidity. Nat Mater 2016, 15(10), 1084–89.

[12]   Shalom M, Gimenez S, Schipper F, Herraiz-Cardona I, Bisquert J, Antonietti M. Controlled carbon nitride growth on surfaces for hydrogen evolution electrodes. Angew Chem Int Ed 2014, 53(14), 3654–58.

[13]   Bian J, Li Q, Huang C, Li J, Guo Y, Zaw M, et al. Thermal vapor condensation of uniform graphitic carbon nitride films with remarkable photocurrent density for photoelectrochemical applications. Nano Energy 2015, 15, 353–61.

[14]   Xiao K, Chen L, Tu B, Heil T, Wen L, Jiang L, et al. Photo-driven ion transport for photodetector based on asymmetric carbon nitride nanotube membrane. Angew Chem Int Ed 2019, 58(36), 12574–79.

[15]   Giusto P, Cruz D, Heil T, Arazoe H, Lova P, Aida T, et al. Shine bright like a diamond: New light on an old polymeric semiconductor. Adv Mater 2020, 32(10), 1908140.

[16]   Mazzanti S, Manfredi G, Barker AJ, Antonietti M, Savateev A, Giusto P. Carbon nitride thin films as all-in-one technology for photocatalysis. ACS Catal 2021, 11(17), 11109–16.

[17]   Zhou Z, Zhang Y, Shen Y, Liu S, Zhang Y. Molecular engineering of polymeric carbon nitride: Advancing applications from photocatalysis to biosensing and more. Chem Soc Rev 2018, 47(7), 2298–321.

[18]   Jiang -L-L, Wang Z-K, Li M, Zhang C-C, Ye Q-Q, Hu K-H, et al. Passivated perovskite crystallization *via* g-C_3N_4 for high-performance solar cells. Adv Funct Mater 2018, 28(7), 1705875.

[19]   Volokh M, Peng G, Barrio J, Shalom M. Carbon nitride materials for water splitting photoelectrochemical cells. Angew Chem Int Ed 2018, 58(19), 6138–51.

[20]   Yang X, Zhao L, Wang S, Li J, Chi B. Recent progress of g-C_3N_4 applied in solar cells. J Materiomics 2021, 7(4), 728–41.

[21]   Teng C, Lin Y, Tan Y, Liu J, Wang L. Facile assembly of a large-area BNNSs film for oxidation/corrosion-resistant coatings. Adv Mater Interfaces 2018, 5(19), 1800750.

[22]   Jia C, Hu W, Zhang Y, Teng C, Chen Z, Liu J. Facile assembly of a graphitic carbon nitride film at an air/water interface for photoelectrochemical NADH regeneration. Inorg Chem Front 2020, 7(13), 2434–42.

[23]   Cai Z, Song Z, Guo L. Thermo- and photoresponsive actuators with freestanding carbon nitride films. ACS Appl Mater Interface 2019, 11(13), 12770–76.

[24]   Xu J, Thomas JKB, Chabanne L, Neher D, Antonietti M, Shalom M. Liquid-based growth of polymeric carbon nitride layers and their use in a mesostructured polymer solar cell with Voc exceeding 1 V. J Am Chem Soc 2014, 136(39), 13486–89.

[25]   Zhao T, Zhou Q, Lv Y, Han D, Wu K, Zhao L, et al. Ultrafast condensation of carbon nitride on electrodes with exceptional boosted photocurrent and electrochemiluminescence. Angew Chem Int Ed 2020, 59(3), 1139–43.

[26] Zhao Y, Liu Z, Chu W, Song L, Zhang Z, Yu D, et al. Large-scale synthesis of nitrogen-rich carbon nitride microfibers by using graphitic carbon nitride as precursor. Adv Mater 2008, 20(9), 1777–81.

[27] Liu J, Huang J, Dontosova D, Antonietti M. Facile synthesis of carbon nitride micro-/nanoclusters with photocatalytic activity for hydrogen evolution. RSC Adv 2013, 3(45), 22988–93.

[28] Zhang Y, Pan S, Zhang Y, Su S, Zhang X, Liu J, et al. Biomimetic high-flux proton pump constructed with asymmetric polymeric carbon nitride membrane. Nano Res 2022, 16(1), 18–24.

[29] Jia F, Zhang Y, Hu W, Lv M, Jia C, Liu J. In-situ construction of superhydrophilic g-C_3N_4 Film by vapor-assisted confined deposition for photocatalysis. Front Mater 2019, 6, 52.

[30] Fujishima A, Honda K. Electrochemical photolysis of water at a seminconductor electrode. Nature 1972, 238(5358), 37–38.

[31] Grätzel M. Photoelectrochemical cells. Nature 2001, 414(6861), 338–44.

[32] Liu J. NextGen Speaks. Science 2014, 343(6166), 26.

[33] Qin J, Barrio J, Peng G, Tzadikov J, Abisdris L, Volokh M, et al. Direct growth of uniform carbon nitride layers with extended optical absorption towards efficient water-splitting photoanodes. Nat Commun 2020, 11(1), 4701.

[34] Peng G, Albero J, Garcia H, Shalom M. A water-splitting carbon nitride photoelectrochemical cell with efficient charge separation and remarkably low onset potential. Angew Chem Int Ed 2018, 57(48), 15807–11.

[35] Zhang J, Zou Y, Eickelmann S, Njel C, Heil T, Ronneberger S, et al. Laser-driven growth of structurally defined transition metal oxide nanocrystals on carbon nitride photoelectrodes in milliseconds. Nat Commun 2021, 12(1), 3224.

[36] Chen Z, Wang H, Xu J, Liu J. Surface engineering of carbon nitride electrode by molecular cobalt Species and their photoelectrochemical application. Chem Asian J 2018, 13(12), 1539–43.

[37] Han YY, Lu XL, Tang SF, Yin XP, Wei ZW, Lu TB. Metal-free 2D/2D heterojunction of graphitic carbon nitride/graphdiyne for improving the hole mobility of graphitic carbon nitride. Adv Energy Mater 2018, 8(16), 1702992.

[38] Peng G, Volokh M, Tzadikov J, Sun J, Shalom M. Carbon nitride/reduced graphene oxide film with enhanced electron diffusion length: An efficient photo-electrochemical cell for hydrogen generation. Adv Energy Mater 2018, 8(23), 1800566.

[39] Zheng Y, Jiao Y, Zhu Y, Li LH, Han Y, Chen Y, et al. Hydrogen evolution by a metal-free electrocatalyst. Nat Commun 2014, 5(1), 3783.

[40] Lee SH, Choi DS, Kuk SK, Park CB. Photobiocatalysis: Activating redox enzymes by direct or indirect transfer of photoinduced electrons. Angew Chem Int Ed 2018, 57(27), 7958–85.

[41] Liu J, Antonietti M. Bio-inspired NADH regeneration by carbon nitride photocatalysis using diatom templates. Energy Environ Sci 2013, 6(5), 1486–93.

[42] Liu W, Hu W, Yang L, Liu J. Single cobalt atom anchored on carbon nitride with well-defined active sites for photo-enzyme catalysis. Nano Energy 2020, 73, 104750.

[43] Zhang Y, Zhao Y, Li R, Liu J. Bioinspired NADH regeneration based on conjugated photocatalytic systems. Sol RRL 2021, 5(2), 2000339.

[44] Son EJ, Lee SH, Kuk SK, Pesic M, Choi DS, Ko JW, et al. Carbon nanotube-graphitic carbon nitride hybrid films for flavoenzyme-catalyzed photoelectrochemical cells. Adv Funct Mater 2018, 28(24), 1705232.

[45] Xu K, Chatzitakis A, Backe PH, Ruan Q, Tang J, Rise F, et al. In situ cofactor regeneration enables selective CO_2 reduction in a stable and efficient enzymatic photoelectrochemical cell. Appl Catal: B 2021, 296, 120349.

[46] Zhang Y, Huang X, Li J, Lin G, Liu W, Chen Z, et al. Iron-doping accelerating NADH oxidation over carbon nitride. Chem Res Chin Univ 2020, 36(6), 1076–820.

[47] Yang C, Li R, Zhang KA, Lin W, Landfester K, Wang X. Heterogeneous photoredox flow chemistry for the scalable organosynthesis of fine chemicals. Nat Commun 2020, 11(1), 1239.

[48]   Kumru B, Giusto P, Antonietti M. Carbon nitride-coated transparent glass vials as photoinitiators for radical polymerization. J Polym Sci 2022, 60(12), 1827–34.

[49]   Wang Z, Wei W, Shen Y, Liu S, Zhang Y. Carbon nitride-Based Biosensors. Biochemical Sensors: Nanomaterial-Based Biosensing and Application in Honor of the 90th Birthday of Prof Shaojun Dong. World Scientific, 2021, 175–225.

[50]   Zhang X-L, Zheng C, Guo -S-S, Li J, Yang -H-H, Chen G. Turn-on fluorescence sensor for intracellular imaging of glutathione using g-C_3N_4 nanosheet-MnO_2 sandwich nanocomposite. Anal Chem 2014, 86(7), 3426–34.

[51]   Wang Q, Wang W, Lei J, Xu N, Gao F, Ju H. Fluorescence quenching of carbon nitride nanosheet through its interaction with DNA for versatile fluorescence sensing. Anal Chem 2013, 85(24), 12182–88.

[52]   Cheng C, Huang Y, Tian X, Zheng B, Li Y, Yuan H, et al. Electrogenerated chemiluminescence behavior of graphite-like carbon nitride and its application in selective sensing Cu^{2+}. Anal Chem 2012, 84(11), 4754–59.

[53]   She X, Xu H, Xu Y, Yan J, Xia J, Xu L, et al. Exfoliated graphene-like carbon nitride in organic solvents: Enhanced photocatalytic activity and highly selective and sensitive sensor for the detection of trace amounts of Cu^{2+}. J Mater Chem A 2014, 2(8), 2563–70.

[54]   Zhang J, Zhu Z, Di J, Long Y, Li W, Tu Y. A sensitive sensor for trace Hg^{2+} determination based on ultrathin g-C_3N_4 modified glassy carbon electrode. Electrochim Acta 2015, 186, 192–200.

[55]   Cheng C, Huang Y, Wang J, Zheng B, Yuan H, Xiao D. Anodic electrogenerated chemiluminescence behavior of graphite-like carbon nitride and its sensing for Rutin. Anal Chem 2013, 85(5), 2601–05.

[56]   Ji J, Wen J, Shen Y, Lv Y, Chen Y, Liu S, et al. Simultaneous noncovalent modification and exfoliation of 2D carbon nitride for enhanced electrochemiluminescent biosensing. J Am Chem Soc 2017, 139(34), 11698–701.

[57]   Lv Y, Chen S, Shen Y, Ji J, Zhou Q, Liu S, et al. Competitive multiple-mechanism-driven electrochemiluminescent detection of 8-hydroxy-2'-deoxyguanosine. J Am Chem Soc 2018, 140(8), 2801–04.

[58]   Raghu S, Santhosh P, Ramaprabhu S. Nanostructured palladium modified graphitic carbon nitride-high performance room temperature hydrogen sensor. Int J Hydrogen Energy 2016, 41(45), 20779–86.

[59]   Zeng B, Zhang L, Wan X, Song H, Lv Y. Fabrication of α-Fe_2O_3/g-C_3N_4 composites for cataluminescence sensing of H_2S. Sens Actuators: B 2015, 211, 370–76.

[60]   O'regan B, Grätzel M. A low-cost, high-efficiency solar cell based on dye-sensitized colloidal TiO_2 films. Nature 1991, 353(6346), 737–40.

[61]   Ye M, Wen X, Wang M, Iocozzia J, Zhang N, Lin C, et al. Recent advances in dye-sensitized solar cells: From photoanodes, sensitizers and electrolytes to counter electrodes. Mater Today 2015, 18(3), 155–62.

[62]   Xu J, Wang G, Fan J, Liu B, Cao S, Yu J. g-C_3N_4 modified TiO_2 nanosheets with enhanced photoelectric conversion efficiency in dye-sensitized solar cells. J Power Sources 2015, 274, 77–84.

[63]   Wang G, Zhang J, Kuang S, Zhang W. Enhanced electrocatalytic performance of a porous g-C_3N_4/graphene composite as a counter electrode for dye-sensitized solar cells. Chem Eur J 2016, 22(33), 11763–69.

[64]   Yuan Z, Tang R, Zhang Y, Yin L. Enhanced photovoltaic performance of dye-sensitized solar cells based on Co_9S_8 nanotube array counter electrode and TiO_2/g-C_3N_4 heterostructure nanosheet photoanode. J Alloys Compd 2017, 691, 983–91.

[65]   Wang G, Zhang J, Hou S. g-C_3N_4/conductive carbon black composite as Pt-free counter electrode in dye-sensitized solar cells. Mater Res Bull 2016, 76, 454–58.

[66]   Stoumpos CC, Kanatzidis MG. Halide perovskites: Poor man's high-performance semiconductors. Adv Mater 2016, 28(28), 5778–93.

[67] Kim H-S, Lee C-R, Im J-H, Lee K-B, Moehl T, Marchioro A, et al. Lead iodide perovskite sensitized all-solid-state submicron thin film mesoscopic solar cell with efficiency exceeding 9%. Sci Rep 2012, 2(1), 591.

[68] Jeong J, Kim M, Seo J, Lu H, Ahlawat P, Mishra A, et al. Pseudo-halide anion engineering for α-FAPbI$_3$ perovskite solar cells. Nature 2021, 592(7854), 381–85.

[69] Liu T, Yang T, Ma R, Zhan L, Luo Z, Zhang G, et al. 16% efficiency all-polymer organic solar cells enabled by a finely tuned morphology *via* the design of ternary blend. Joule 2021, 5(4), 914–30.

[70] Zhou L, Xu Y, Yu W, Guo X, Yu S, Zhang J, et al. Ultrathin two-dimensional graphitic carbon nitride as a solution-processed cathode interfacial layer for inverted polymer solar cells. J Mater Chem A 2016, 4(21), 8000–04.

[71] Chen X, Liu Q, Wu Q, Du P, Zhu J, Dai S, et al. Incorporating graphitic carbon nitride (g-C$_3$N$_4$) quantum dots into bulk-heterojunction polymer solar cells leads to efficiency enhancement. Adv Funct Mater 2016, 26(11), 1719–28.

[72] Kumru B, Antonietti M, Schmidt BV. Enhanced dispersibility of graphitic carbon nitride particles in aqueous and organic media *via* a one-pot grafting approach. Langmuir 2017, 33(38), 9897–906.

[73] Chen J, Dong H, Zhang L, Li J, Jia F, Jiao B, et al. Graphitic carbon nitride doped SnO$_2$ enabling efficient perovskite solar cells with PCEs exceeding 22%. J Mater Chem A 2020, 8(5), 2644–53.

[74] Xu J, Shalom M, Piersimoni F, Antonietti M, Brenner TJK. Color-tunable photoluminescence and NIR electroluminescence in carbon nitride thin films and light-emitting diodes. Adv Opt Mater 2015, 3(7), 913–17.

[75] Podjaski F, Kroger J, Lotsch BV. Toward an aqueous solar battery: Direct electrochemical storage of solar energy in carbon nitrides. Adv Mater 2018, 30(9), 1705477.

[76] Zhang Y, Pan Q, Chai G, Liang M, Dong G, Zhang Q, et al. Synthesis and luminescence mechanism of multicolor-emitting g-C$_3$N$_4$ nanopowders by low temperature thermal condensation of melamine. Sci Rep 2013, 3(1), 1943.

[77] Liu Z, Wang C, Zhu Z, Lou Q, Shen C, Chen Y, et al. Wafer-scale growth of two-dimensional graphitic carbon nitride films. Matter 2021, 4(5), 1625–38.

[78] Wang W, Cui J, Sun Z, Xie L, Mu X, Huang L, et al. Direct atomic-scale structure and electric field imaging of triazine-based crystalline carbon nitride. Adv Mater 2021, 33(48), 2106359.

[79] Kessler FK, Zheng Y, Schwarz D, Merschjann C, Schnick W, Wang X, et al. Functional carbon nitride materials-design strategies for electrochemical devices. Nat Rev Mater 2017, 2(6), 17030.

[80] Wang S, He T, Chen P, Du A, Ostrikov K, Huang W, et al. In situ formation of oxygen vacancies achieving near-complete charge separation in planar BiVO$_4$ photoanodes. Adv Mater 2020, 32(26), 2001385.

[81] Fu HJ, Moreno-Hernandez IA, Buabthong P, Papadantonakis KM, Brunschwig BS, Lewis NS. Enhanced stability of silicon for photoelectrochemical water oxidation through self-healing enabled by an alkaline protective electrolyte. Energy Environ Sci 2020, 13(11), 4132–41.

Paolo Giusto*

Chapter 12
Carbon nitride thin films as a high refractive index optical material

12.1 Introduction

Since ancient times, brilliance and brightness of gemstones, in particular diamonds, were recognized as symbol of wealth and power. Pliny, the Elder, wrote about diamonds already in 77 A.D. in his work *Naturalis historia:* "The substance that possesses the greatest value, not only among the precious stones, but of all human possessions, is diamond (adamas) . . . Indeed, its hardness is beyond all expression, while at the same time it quite sets fire at defiance and is incapable of being heated." Over the course of about 2000 years, the reputation of diamond has not changed. However, beyond their purely aesthetical value, the brilliance and brightness of diamonds have attracted the attention of the scientific community for applications in optics and photonics. The beautiful reflections that occur at the facets of diamonds are due to its very high refractive index and transparency in the visible range, with highly polished surfaces cut. Recently, high refractive index materials have attracted significant attention for their potential applications in quantum computing, lenses, optical fibers, and more. However, most materials possessing these properties are inorganic materials and obtained through toxic or expensive production processes. Therefore, the synthesis of homogeneous and transparent thin film polymer materials with high refractive index is required for applications in optics, photonics and beyond.

Herein, the synthesis of a polymer thin film with high refractive index and transparency in the visible range is presented, namely polymeric carbon nitride (pCN), by means of chemical vapor deposition (CVD). This method enables the synthesis of flat and homogeneous heptazine-based pCN thin films using as precursors nitrogen-rich small molecules, such as melamine, which are sublimated and subsequently thermally condensed over a target substrate surface. The thin films prepared using this method exhibit high optical quality, making them suitable for use as optical materials where the constraints for flatness and homogeneity are usually high to avoid undesired scattering effects.

The refractive index of pCN is in the same range of diamond in the visible spectrum, making pCN the polymeric material with the highest refractive index and high transparency in the visible range reported so far. However, the 2D structure of pCN causes anisotropy of the refractive index, meaning that the refractive index values are different along different directions within the material; a property that is not present in diamond

---
*Corresponding author: Paolo Giusto, Colloid Chemistry Department, Max Planck Institute of Colloids and Interfaces, 14476 Potsdam, Germany, e-mail: paolo.giusto@mpikg.mpg.de

https://doi.org/10.1515/9783110746976-012

due to its highly symmetrical cubic structure. While pCN is not (yet) associated with power and wealth like diamonds are, this polymer material is eventually expected to receive significant attention for applications in optics and photonics, especially where there are requirements for high flatness, refractive index, and transparency in the visible range.

12.2 Fundamentals of light-matter interaction

This section provides an overview of the fundamental concepts necessary to understand the relation between optical functions and physical observables. For a more detailed description and comprehensive treatment of the optical functions, we invite readers to consult more specialized books [1]. In this context, light is treated as an electromagnetic wave that propagates in vacuum, where speed (c) is defined as

$$c = \sqrt{\frac{1}{\mu_0 \varepsilon_0}} \tag{12.1}$$

where μ_0 is the magnetic permeability of vacuum and ε_0 is the dielectric constant of vacuum. In a medium, the speed of light changes depending on the medium's characteristics μ_m and ε_m. Therefore, we can define the change of the speed of light in vacuum with respect to a medium as the ratio between these two velocities, usually referred to as the refractive index (n) of the medium:

$$\frac{c}{v} = \sqrt{\frac{\mu_m \varepsilon_m}{\mu_0 \varepsilon_0}} = \sqrt{\mu_r \varepsilon_r} = n \tag{12.2}$$

However, since light is constituted by an oscillating electrical (and a magnetic) field with time, the response of a material to an electromagnetic stimulus is depicted, considering a real part (in case losses occur, an imaginary part) of the dielectric constant. The real part (ε_1) is responsible for the propagation phenomena of the electromagnetic wave through the medium, whereas the imaginary one (ε_2) for the losses or attenuation through the medium. Therefore, we can write the complex dielectric constant ($\tilde{\varepsilon}$) as

$$\tilde{\varepsilon} = \varepsilon_1 + i\varepsilon_2 \tag{12.3}$$

and recalling the previous definition of the refractive index (eq. (12.2)), assuming non-magnetic materials ($\mu_r = 1$), we can introduce the complex refractive index ($\tilde{n}$):

$$\tilde{n}^2 = \varepsilon = \varepsilon_1 + i\varepsilon_2 \tag{12.4}$$

with $i = \sqrt{-1}$

$$\tilde{n} = n + ik \tag{12.5}$$

and

$$\varepsilon_1 = n^2 - k^2 \tag{12.6}$$

$$\varepsilon_2 = 2nk \tag{12.7}$$

where n and k are collectively called the optical constants of the material. In particular, the real part of $\tilde{n}$ is usually referred to as the refractive index (n) and the imaginary part (k) as the extinction coefficient. Analogous to what is mentioned previously for the dielectric constant, n is responsible for the propagation of electromagnetic waves through the material, whereas k is for the absorption phenomena. The extinction coefficient vanishes for lossless media, meaning that the material is transparent in the range of wavelength considered. In layman terms, n is responsible for the shining reflections occurring at the diamond's surfaces, whilst it is transparent in the visible range ($k = 0$). However, k is responsible for the green color of the leaves, where the chlorophyll absorbs the red and blue spectrum of the visible and, therefore, they are perceived as green. It is worth reminding that ε_1, ε_2, n and k are all frequency-dependent (and therefore, wavelength-dependent).

These quantities can be correlated to dimensionless physical observables, such as normal incidence reflectivity (R). Assuming a slab of infinite thickness and in contact with air ($n_{air} = 1$), then:

$$R = \left|\frac{1-\tilde{n}}{1+\tilde{n}}\right|^2 = \frac{(1-n)^2 + k^2}{(1+n)^2 + k^2} \tag{12.8}$$

and for materials without losses (transparent, $k \to 0$) it can be further simplified to

$$R = \left(\frac{1-n}{1+n}\right)^2 \tag{12.9}$$

with $0 \geq R \geq 1$. When light impinges a surface, it can be absorbed, reflected or transmitted (we neglect scattering effects for the purpose of this chapter). The sum of all these observables must therefore be equal to unity:

$$R + T + A = 1 \tag{12.10}$$

where T and A are the fractions of light that are transmitted and absorbed. For transparent materials ($A = 0$), it is simple to retrieve T. However, semiconductors (such as pCN) are, by definition, materials where the band gap – the minimum energy required for an electronic transition between the valence band and the conduction band – has values higher than zero and lower than 4 eV [2].

When a material is absorbing light, the impinging ray intensity (I_0) decreases along the material's thickness (z):

$$I(z) = I_0 e^{-az} \tag{12.11}$$

where $I(z)$ is the intensity of light travelling a distance z and a is called the absorption coefficient (cm^{-1}), which is a function of the wavelength. It is worth noting that the abovementioned equation, namely the Lambert-Beer equation, is a simplified model for infinitesimally thin layer of a homogeneous medium. Therefore, we can define the transmittance as

$$T = \frac{I(z)}{I_0} = e^{-az} \tag{12.12}$$

and the absorbance (A):

$$A = -\log_{10} T \tag{12.13}$$

so, the absorbance can be directly correlated to the absorption coefficient:

$$A = \log_{10}(e)az \tag{12.14}$$

as previously mentioned, the imaginary part of the refractive index (k) is responsible for the optical losses. Therefore, we can see from the last equation that it must be related to the absorption coefficient:

$$a = \frac{4\pi k}{\lambda} \tag{12.15}$$

where λ is the wavelength (nm). An important remark here: the author recommends paying attention when measuring absorbance or transmittance, to not confuse intensity losses due to reflectance (especially with high refractive index media) with absorbance. For example, if the impinging light has a wavelength at which the medium has $a = 0$, then the "apparent losses" are likely due to reflectance (or other effects not treated in this chapter, like scattering). In this way, we have correlated the optical functions, n and k, to common observables such as transmittance, reflectance, and absorbance (schematically shown in Fig. 12.1). From this simple series of equations, it is already clear to the reader, the importance of the optical functions. For instance, the higher the refractive index, the higher is the reflectivity, and the higher the k, the more intense is the absorption, when all other parameters are kept constant. To avoid confusion, we report here the definitions of reflectance and reflectivity from the International Commission on illumination: Reflectivity (of a material) is the reflectance of a layer of the material of such a thickness that there is no change of reflectance with increase in thickness; reflectance is the ratio of the reflected radiant or luminous flux to the incident flux in the given conditions. In some fields, the ending –ivity is added to intrinsic, or bulk properties of materials. The ending –ance is reserved for the extensive or extrinsic properties of a fixed quantity of substance, for example, a portion of the substance having a certain length or thickness [3].

Reflectance   Absorbance   Transmittance

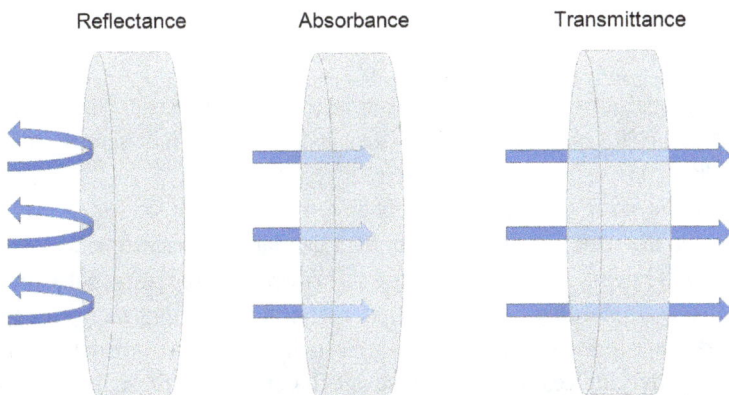

Fig. 12.1: Schematic representation of reflectance, absorbance and transmittance processes.

The absorption coefficient function is a fundamental property for the determination of the band gap energy in semiconductors by means of the Tauc method [4]. The method enables depicting the energy of the band gap in a graphical manner by plotting

$$(ahv)^{1/x} = m(hv - E_g) \tag{12.16}$$

where h is the Planck's constant, v is the light frequency (Hz), E_g is the band gap energy (eV), and m is a proportionality constant [4]. The value of the exponent (x) depends on the nature of the electronic transition, as follows:

- $x = 1/2$ for direct allowed transitions;
- $x = 3/2$ for direct forbidden transitions;
- $x = 2$ for indirect allowed transitions;
- $x = 3$ for indirect forbidden transitions.

In this way, the band gap energy, E_g, can be extrapolated from the region where the y-axis values, $(ahv)^{1/x}$ increase in a linear manner. By extrapolating $(ahv)^{1/x} = 0$, the energy value obtained is the band gap energy. It is worth mentioning that the linear region should be carefully chosen and not confused with defect absorptions states that are near the band edge, such as the "Urbach Tail" [4] (We remind that it is always possible to convert wavelength to energy by means of $E = hc/\lambda \approx 1240/\lambda$).

As previously mentioned, the optical functions are wavelength-dependent. Therefore, the characterization of optical materials relies also on parameters that can describe the materials' optical dispersion, that is, the variation of the refractive index with wavelength. In the visible range and in the absence of absorption, the refractive index usually decreases with wavelength. Optical dispersion is of high importance for the development of optical devices, such as lenses, optical fibers and dispersive prisms; it is usually defined in terms of Abbe number (V_D) for the visible range:

$$V_D = \frac{n_D - 1}{n_F - n_C} \tag{12.17}$$

where n_D, n_F, n_C are the refractive indices of the material at the Fraunhofer D (589.3 nm), F (486.1 nm) and C lines (656.3 nm) respectively [5]. Low values of Abbe number are typical of high optical dispersion materials and vice versa [6].

Optical functions, however, do not depend only on the frequency of the electromagnetic wave, but also on the structural properties of the material. For anisotropic materials, which are materials having different property values along different directions, the refractive index changes according to the direction. Classical linear amorphous polymers (unless orientation is induced) and diamond are usually considered isotropic. This means that the refractive indices do not depend on the structural direction. Two-dimensional materials, such as pCN, boron nitride and transition metal dichalcogenides are anisotropic materials, for which properties, such as optical functions, are direction-dependent. In these cases, we can define the in-plane and out-of-plane optical functions, where the in-plane describes the optical function of the material along the 2D plane, and out-of-plane is the one occurring between the layers. The difference between the in-plane and out-of-plane optical functions is typically called birefringence for refractive index, and dichroism for the extinction coefficient. The evaluation of the anisotropic optical functions is of paramount importance for the development of highly efficient and compact optical devices for light manipulation, including polarizers, wave plates, multilayer mirrors, and phase-matching elements [7].

The concept introduced herein is of fundamental importance for understanding the optical properties of materials for the following discussion on pCN thin films. For a more detailed discussion, a wide range of books on the optics of solid-state materials are available to provide greater insights into the optical functions and light-matter interactions [2].

12.3 Introduction to carbon nitride films and challenges in the synthesis of optical-quality thin films

Carbon nitride materials have recently gained attention and have been widely exploited for their metal-free photocatalytic activity under visible light irradiation. In addition to photocatalysis, polymeric carbon nitride (pCN) has more recently received increasing attention in various applications, such as energy storage [8], optoelectronics [9], polymer/hydrogel synthesis [10], photoelectrochemistry, emulsion stabilizer [11] and water disinfection [12], to name a few. However, in many cases, the preparation of pCN as thin films is required. Thin films are defined as objects where one dimension, that is, the thickness, is much smaller than the other two dimensions that are usually continuous.

In this chapter, the term "thin film" refers to coatings with a thickness in the range of the wavelength of the visible spectrum or lower (ca. 1–1,000 nm) [13]. The processing of pCN thin films is still difficult due to the cross-linked nature of the polymer and the dispersive forces (usually referred to as π–π stacking) between the layers, which makes the material insoluble and even hardly dispersible. As a result, high-quality homogeneous pCN thin films cannot be prepared by common thin films methods, such as spin coating. Therefore, a method for the deposition of highly homogeneous thin film is required to implement this material in optical applications.

In this context, the pioneering works of Liu and Bian, who reported vapor-based methods for the synthesis of pCN thin films for the first time, led to new and exciting opportunities. In the first case, Liu et al. developed the so-called microcontact-printing method, where they infiltrated a precursor (cyanamide) ink into an anodic aluminum oxide membrane placed between two substrates (glass or fluorine-doped tin oxide) and heated it up to 550 °C to deposit a pCN thin film on the substrate surface (Fig. 12.2a). These thin films were also free-standing, after soaking the samples in water, and their thickness was successfully controlled by means of the ink precursor concentration [14]. In the second method, called thermal vapor condensation, Bian et al. used a crucible as a container for the precursor (melamine), capped with the target substrate (Fig. 12.2b). Analogously to the previous case, also here, the pCN thin films could be easily peeled-off from the substrate, and the thickness tuned by means of the precursor amount [15]. Over the years, many other noteworthy methods have been developed and can be found in the extensive literature reviews available [16]. Some of these methods are inexpensive and only require a muffle furnace to convert the precursor in pCN. However, the flatness and homogeneity of pCN thin films are relatively low, which have hindered their application in optics and photonics, despite being widely used in photoelectrochemical cells. Recently, Aida's group proposed an innovative technique that uses a it as model tubular reactor. The test tube contained the precursor (guanidinium carbonate) at the bottom, a target substrate at the top and it was capped with an aluminum foil lid. The tube is then inserted in a muffle furnace with a controlled heating program for the deposition (Fig. 10.1a and b). They called this method "vapor deposition polymerization" [17]. In our group, we further exploited this method using different precursors to create pCN thin film membranes, boron carbon nitride, and more materials with tunable optical properties [18, 19]. Although these vapor methods enable synthesis of thin films with tunable thickness and relatively low roughness, the obtained quality is still not high enough for optical applications. A summary of the most recent synthetic methods for the preparation pCN thin films is reported in [16] and [20].

More recently, Giusto et al. developed an innovative method based on chemical vapor deposition (CVD) to further improve the quality of pCN thin films and pave the way for their application in optical devices. In this method, the CVD consist of two separate heating zones, which enable a tight control of the temperature at the precursor and at the substrate (Fig. 12.3b). Furthermore, the low pressure enables the homogeneous deposition, avoiding the formation of solid clusters in the vapor phase that would

a)

b)

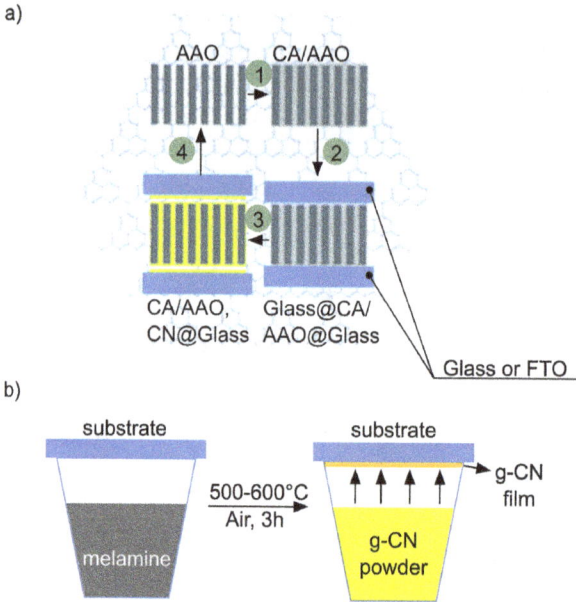

Fig. 12.2: Schematic representation of methods for the synthesis of pCN thin film. (a) The microcontact printing. (b) Thermal vapor condensation (reproduced from reference [20]).

otherwise result in the formation of island-like films with low-homogeneity. Other important considerations include the choice of the precursor and the substrate for the preparation of the thin films. Melamine, for instance, was chosen as a convenient precursor for the preparation of pCN, as it sublimate quantitatively in vacuum at a relatively low temperature (300 °C) without leaving any non-volatile residue. The substrate in the second oven is heated directly to the pCN synthesis temperature (550 °C) before the precursor is sublimated. It is worth noting that in addition to the possible effects of epitaxy on the synthesis of the thin films (which is beyond the scope of this chapter), the substrate must be stable at the process temperature. The characterization of the optical properties of pCN thin films requires the use of highly transparent and highly reflective substrates. For this purpose, silicon and fused silica substrates were chosen, which are stable and inert at the process temperatures. After the deposition process is completed, the substrates appear homogeneously coated with a yellow transparent film of pCN (Fig. 12.3c and e) with a strong blue fluorescence when illuminated with UV light (Fig. 12.3d and f).

Compared to previous methods, this technique offers several advantages, including high homogeneity and flatness of the samples (roughness <1 nm) even on curved edges. The homogeneity is so high that when the samples are analyzed by scanning electron microscopy (SEM), it is difficult to say whether the film is present

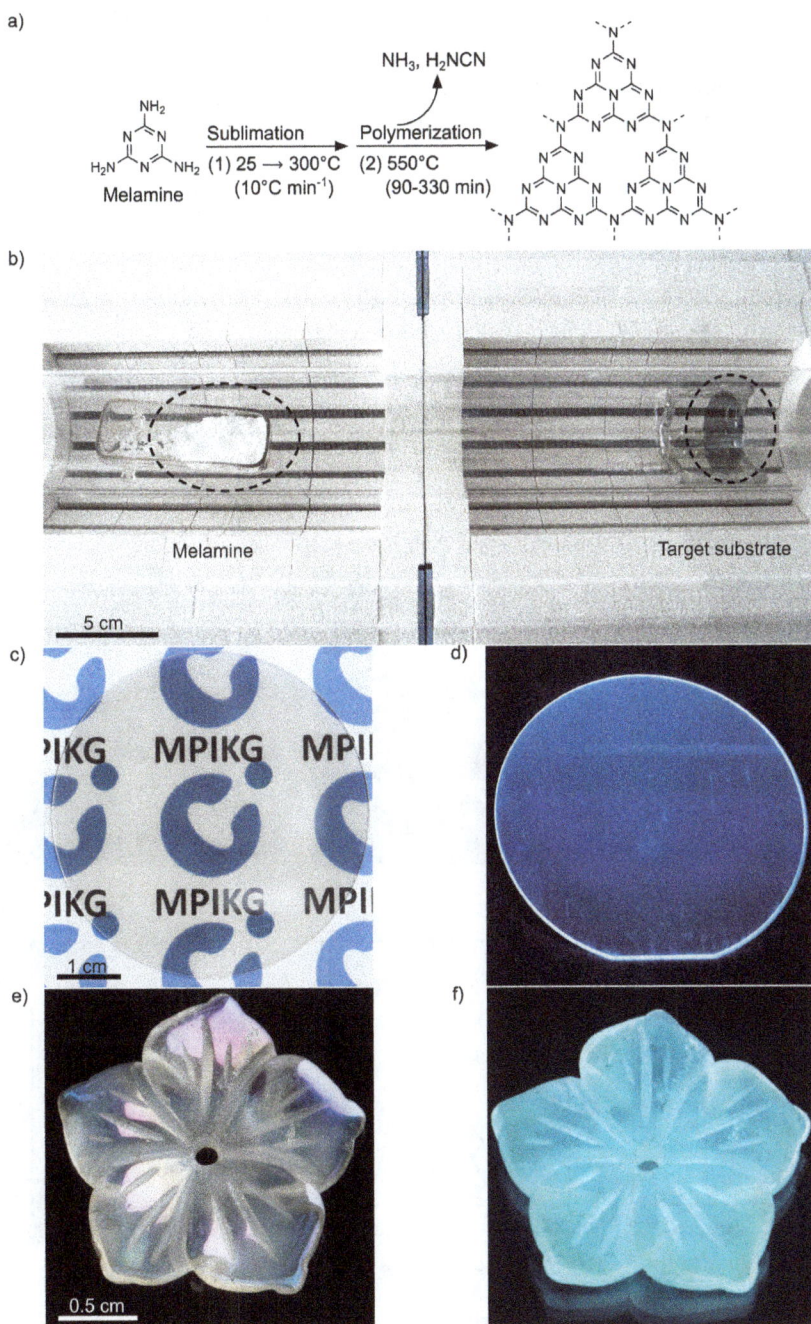

Fig. 12.3: Chemical vapor deposition of pCN thin films. (a) Reaction conditions. (b) The CVD setup used for the preparation. (c) Fused silica substrate after the deposition of pCN thin film. (d) Under UV illumination. (e) A flower-shaped quartz used as a substrate. (f) Under UV illumination [21].

or not (Fig. 12.4a). In such cases, energy-dispersive X-ray spectroscopy (EDX) mapping at a cut edge, where a defect is created on purpose by the cut, enables to clearly distinguish the film from the substrate (Fig. 12.4b–e). The area where the film has been removed by the cut shows a significantly higher intensity of the element present in the substrate (silicon, in this case), whereas in the area where the film is still present, the signals of carbon and nitrogen are more intense and homogeneous. It is important to note that in this case, EDX analysis is used only as a qualitative method for determining the elemental composition of the pCN thin films. The presence of a layered structure, with an interlayer distance of 0.32 nm is observed in transmission electron microscopy (TEM) images, calculated by means of the fast Fourier transform (FFT) obtained from the image, which is consistent with previous reports on pCN materials [14]. The chemical composition and bonding scheme are evaluated by means of X-ray photoelectron spectroscopy (XPS) and electron energy loss spectroscopy (EELS). The XPS and EELS of the thin films reveal the typical features of pCN materials and are comparable to that of pCN bulk materials (Fig. 12.4g-i).

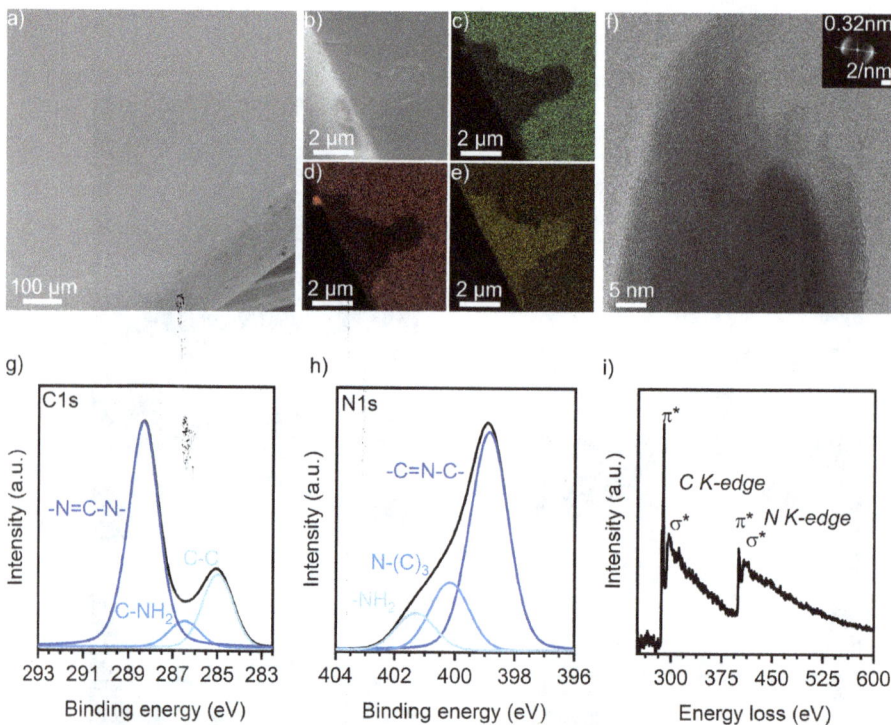

Fig. 12.4: Characterization of pCN thin films from chemical vapor deposition. (a) SEM image at low magnification of a pCN-coated fused silica substrate. (b) SEM image at an edge where part of the film was removed when preparing the sample, and the EDX maps of (c) carbon, (d) nitrogen and (e) silicon. (f) TEM and FFT (insert), (f) C1s and (g) N1s XPS spectra. (h) EELS spectrum of pCN thin films (reproduced with permission from reference [21]).

The relative C/N ratio of 0.71, as calculated from XPS, is very close to the ideal 0.75. Moreover, the EELS spectrum shows a very intense $\pi-\pi^*$ peaks for carbon and nitrogen, speaking of a high degree of conjugation achieved in the material. The Fourier-transform infrared (FTIR) spectrum reveals the presence of stretching modes of amino terminal groups (3,400–3,000 cm^{-1}) and the characteristic heptazine breathing mode (804 cm^{-1}), in good agreement with the XPS analysis. Other techniques require the availability of a very large number of samples to be irreversibly damaged for the purpose, such as solid state nuclear magnetic resonance (ssNMR) and more. Among those, combustion elemental analysis was performed in order to confirm the C/N ratio obtained by means of XPS, which in this case was 0.74. Further details can be found in reference [21].

12.4 Carbon nitride thin films as a high refractive index material

Optical properties, such as photoluminescence and its lifetime, band gap, and transient absorption, to name a few, of bulk pCN are widely reported, especially with regard to applications in photo(electro)catalysis. However, the absence of a reproducible method to obtain highly homogeneous pCN thin films hindered the quantitative experimental determination of the optical functions. The determination of these values will pave the way for introducing pCN thin films in a wide range of optoelectronic and optical devices, such as waveguides, nanoantennas, optical transistors and more [21, 22].

pCN is well known for its typical yellowish appearance due to the absorption edge, which lies in the visible range, typically between 400 and 450 nm [23]. In order to characterize the optical functions, it is, therefore, important to evaluate the reflectance and transmittance spectra, which give qualitative information on the thickness scale, assuming that the optical functions are constant among the samples. As exemplarily shown in Fig. 12.5, the transmittance (black line) and reflectance (blue line) spectra were performed at normal and at 7° incidence, respectively, on a set of samples with increasing thickness (a to d). As previously mentioned, the Lambert-Beer law shows that transmittance decreases exponentially with film thickness for wavelengths below the expected band gap (400–450 nm). The transmittance spectrum in Fig. 12.5a shows a minimum at 305 nm and high transmittance in the visible range, similar to the bare-fused silica substrate (dotted lines). The reflectance spectrum shows also low values, indicating a very low thickness achieved. In the spectra shown in Fig. 12.5b–d, the transmittance in the UV range is very low and similar across all the samples. In these cases, the analysis of the transparent region, that is, above 450 nm will provide further preliminary information. It is worth noting that reflectance and transmittance spectra are perfectly mirrored due to the low scattering of the sample surfaces, confirming the high surface flatness achieved by this method. The decreased transmittance of these samples is not attributable to additional absorption phenomena caused by changes in the electronic structure of the material. It can be easily calculated that in these cases, the sum of the transmittance and reflectance is almost equal to unity and, therefore, optical absorption is null.

The increased reflectance in the visible range for samples C and D is due to optical interference. Optical interference occurs as a result of multiple reflections at the air-film and film-substrate interfaces, where the light is partially reflected and transmitted multiple times (Fig 12.5e). This effect is commonly observed in everyday life, such as the colors arising from soap bubbles (soap does not absorb in the visible range too) or oil layers on water. The optical interference causes the arising of (interference) fringes that are reflectance maxima due to the structural and optical constraints, that is, the thickness and the refractive index mismatch between contiguous phases. Interference fringes require that the film thickness is much lower than the light coherence length:

$$z \ll \frac{\lambda^2}{2\pi\Delta\lambda} \tag{12.18}$$

where $\Delta\lambda$ is the spectral bandwidth, nm. Moreover, the interference extrema, such as the maxima of reflectance or the minima in transmittance, are closely related to the refractive index and thickness of the sample. At normal incidence:

$$z = \frac{m\lambda_j}{4n_j} \tag{12.19}$$

where m is an integer, namely the interference order (a natural positive number), λ_j and n_j are the wavelength and the correspondent refractive index of the film, respectively, at the interference fringe. Since the interference order is usually unknown, the following equation is used to qualitatively determine the thickness as the relative distance between two interference fringes:

$$z = \frac{\Delta j}{4\left(v_{j+\Delta j}n_{j+\Delta j} - v_j n_j\right)} \tag{12.20}$$

with v being the light wavenumber (cm^{-1}). In this way, we could qualitatively define the following thickness ranking between the samples: A < B < C < D. Additionally, the high visibility of the fringes, which depends on the mismatch of the refractive indices of contiguous phases, reveals that pCN has a high refractive index [24]. It is worth noting that the interference occurring in the transmittance spectra above ca. 900 nm is due to the substrate, as it can also be seen on the bare substrate spectra (dotted black line).

To obtain and model the optical functions, pCN thin films were characterized by variable angle spectroscopic ellipsometry (VASE), that enables the modeling of the real and imaginary parts of the complex refractive index over a wide spectral range (Fig. 12.6). Spectroscopic ellipsometry is a non-destructive, non-contact optical technique that is based on the change in the polarization of light reflected from a thin film sample.

a)

b)

c)

d)

e)

Reflected

Transmitted

Incoming

1   2   3

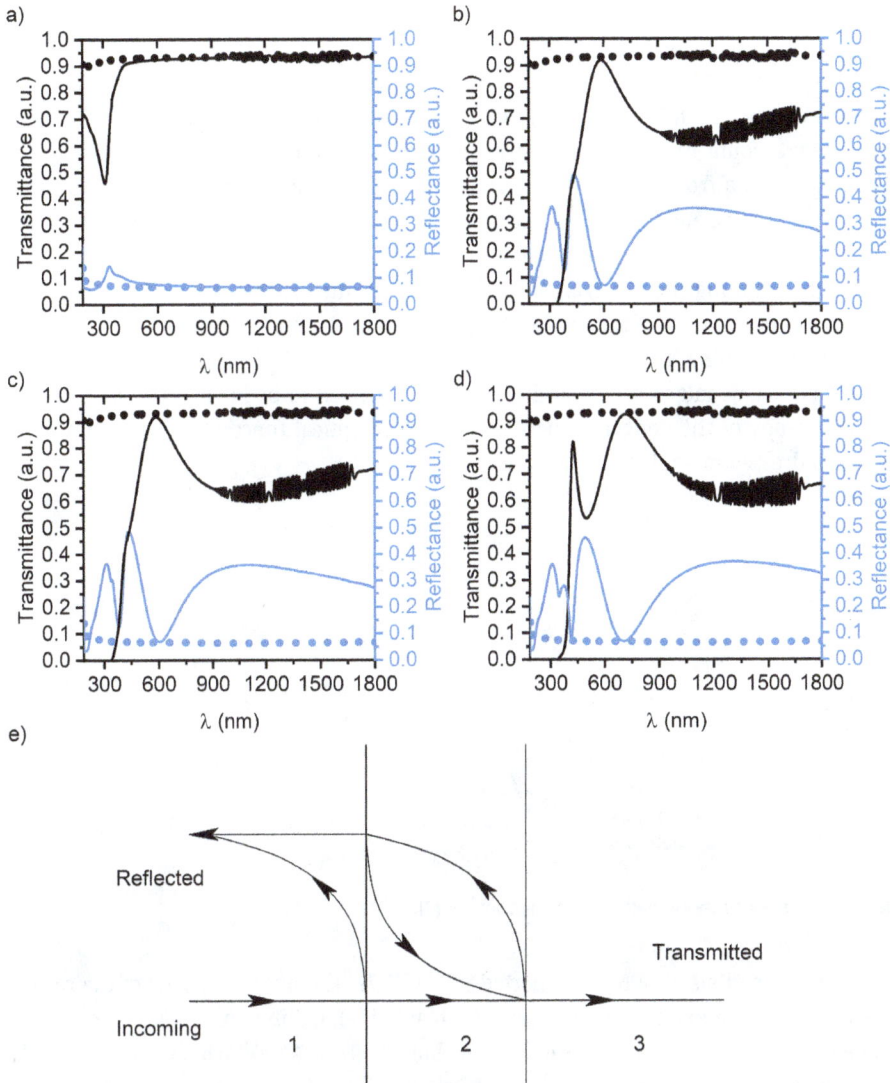

Fig. 12.5: Transmittance (black lines) and reflectance (blue lines) of pCN thin films on fused silica substrates (front thickness). (a) 5 nm, (b) 33 nm, (c) 128 nm and (d) 130 nm. Dotted lines represent the spectra of the bare substrates. (e) Representation of reflectance and transmittance of light at each interface of a finite medium (reproduced with permission from reference [21]).

The setup records the phase shift in the electromagnetic wave from the incident ray and allows for modelling the optical functions [2]. The determination of the optical properties and the models behind them go beyond the scope of this book and the author suggests referring to more specialized physics textbooks for further information on the topic [2]. However, it is important to note that the detected response of ellipsometry measurements is related to the optical constant and thickness of the sample, but can also be correlated to the absorption coefficient, optical anisotropy (which will be described in the next chapter), crystallinity, porosity and more [25]. The data modelling involves regression until the minimum error is achieved by using the ellipsometry results of the samples, while also taking into account reflectance and transmittance spectra. In this way, by recording the ellipsometry data from the same materials but with different thicknesses and the samples' transmittance and reflectance spectra, it is possible to find a unique solution for the optical functions. It is worth mentioning that in this section, we elaborated on the optical functions, assuming the material is isotropic, which means that the material's optical properties are not dependent on the direction within the material. As previously mentioned, pCN thin films are constituted by planes stacked in a 2D fashion and held together by van der Waals interaction, and therefore are expected to be intrinsically anisotropic. The anisotropy of the optical properties in terms of optical functions of pCN thin films will be addressed in a following section.

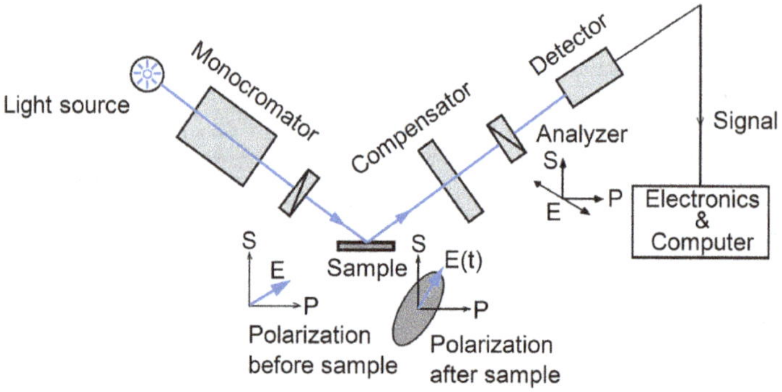

Fig. 12.6: Schematic representation of an ellipsometer [2].

The optical functions of pCN are reported in Fig. 12.7a. The imaginary part of the refractive index, i.e. the extinction coefficient (k, black line), exhibits high values in the UV range where the pCN absorbs, reaching a value of 1.97 at 4.1 eV, which is considerably high when compared to other conjugated polymers [26]. High extinction coefficients in polymeric semiconductors are usually attributed to high density of states, high transition dipole moments, high orientation, and stiffness of the chains [26]. The extinction coefficient function displays a pronounced absorption shoulder at around 3.37 eV which is attributed to π–π* transitions. Additionally, it is possible to retrieve the absorption coefficient function (α) from k. (Fig. 12.7b). As expected, α values are also high, with a maximum value of 8.27×10^5 cm^{-1} at 4.2 eV. Using the Tauc method, the band gap energy value (E_g) of the material can be obtained from the elaboration of the extinction coefficient, which occurs at 2.88 eV or 431 nm (Fig. 12.7c). In the visible range, beyond the band gap energy value, the extinction and absorption coefficients are zero, confirming that the pCN thin films are highly transparent over a large portion of the visible spectrum.

It is worth noting that the optical functions do not provide information on the energy levels, which can be retrieved by means of another spectroscopic characterization, namely ultraviolet photoelectron spectroscopy (UPS) (Fig. 12.7d).

The real part of the refractive index n has values of $n_D = 2.43$ (D stands for the Fraunhofer D line, 589.3 nm) and, further, 2.32 at 1,000 nm, following the typical Sellmeier dispersion trend. The Sellmeier equation and the modeled parameters (Tab. 12.1):

$$n(\lambda) = \left(A + \frac{B\lambda^2}{\lambda^2 - C} + \frac{D\lambda^2}{\lambda^2 - E} \right)^{0.5}$$

(12.21)

Tab. 12.1: Sellmeier parameters.

Parameters	
A	1.545254
B	2.891273
C	2.226221×10^3 μm^2
D	6.917342×10^{-1}
E	3.296790×10^3 μm^2

The optical functions of pCN reveal that this material possesses a very high intrinsic refractive index, comparable to diamond ($n_D = 2.43$) in the visible spectral range, with a high transparency [27]. This value is much higher than that of common polymers with transparency in the visible range, which usually lies between 1.33 and 1.7, seldom exceeding 1.8 in the case of ad hoc synthesized high refractive index polymers (HRIP) [28]. Semiconductor polymers, such as poly(arylenevinylene)s and poly(fluorene)s, have typical values of n in the range 1.6–2.1; however, with a strong absorption in the visible spectrum [29, 30]. High-n materials with high transparency in the visible range, such as pCN, can find application in organic photonic devices for enhanced light–matter interaction, usually dominated by inorganic materials, for example, by TiO$_2$ ($n_D \approx 2.5$–2.6) [31]. The synthesis of high refractive index materials has recently attracted much attention for use in optoelectronic devices such as microlenses for charge coupled devices (CCD), complementary metal–oxide–semiconductor (CMOS) image sensor (CIS) and highly efficient light-emitting diode (LED) encapsulants [28]. The optical functions enables to derive the optical dispersion, usually described in terms of the Abbe number, of the pCN thin films. Eventually, pCN reveals a relatively low Abbe number (9.5), which speaks for a high dispersion of the refractive index in the visible range, as compared to diamond (55.3); however, it is in the range of other widely used high refractive index materials such as titania (rutile, 9.9) [33].

The origin of such high refractive index can be attributed to the pCN chemical structure, which constituted of tri-s-triazine units, and its high density. Indeed, the refractive index of polymers can be calculated by means of the Lorentz–Lorenz equation,

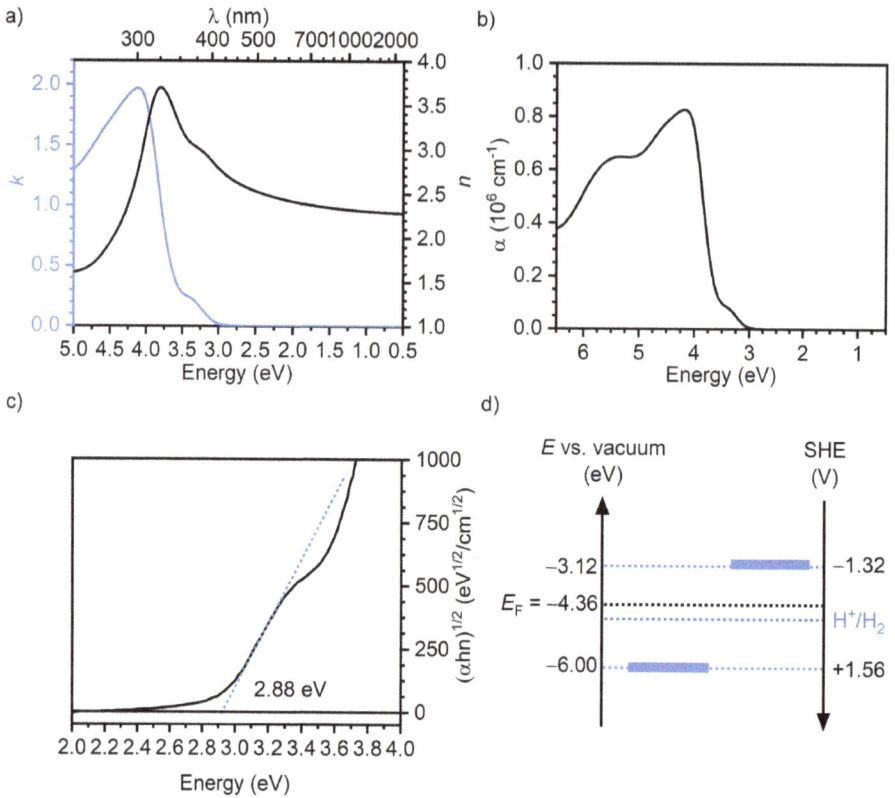

Fig. 12.7: (a) Optical functions (*n* in black, *k* in blue). (b) Absorption coefficient of pCN thin films. (c) Tauc plot indicating the band gap energy. (d) Energy levels obtained from the elaboration of the optical functions and UPS spectra (reproduced with permission from reference [21]).

which defines the refractive index as a function of the repeating unit molecular weight (M_w) and linear molecular polarizability (a_m), the material's density (ρ), or in a simpler manner, as the ratio of molar refraction ($[R]$) and molecular volume of the polymer repeating unit (V_m) [28]:

$$\frac{n^2 - 1}{n^2 + 2} = \frac{4\pi\, a_m \rho N_A}{3} \frac{1}{M_w} = \frac{[R]}{V_m} \tag{12.22}$$

where N_A is the Avogadro's number. Solving the equation for n leads to:

$$n = \sqrt{\frac{1 + 8\pi a_m \rho N_A/3M_w}{1 - 4\pi a_m \rho N_A/3M_w}} = \sqrt{\frac{1 + 2[R]/V_m}{1 - [R]/V_m}} \tag{12.23}$$

from the equation, one can see that higher $[R]$ values lead to higher refractive indices. On the one hand, the tri-s-triazine units possess high $[R]$ values due to the aromatic -C=N-C- groups (4.1) that are circa 2.5 times than the values reported for -C=C- bonds

(1.73) [32]. On the other hand, precise determination of the gravimetric density of thin film materials is challenging. Empirically, materials with higher density are usually stiffer and harder. Therefore, qualitatively evaluating these mechanical properties can provide insights, further supporting the high refractive index found for pCN thin films. Indeed, pCN films were measured by nanoindentation, revealing a Young's modulus and hardness of 36.5 and 2.2 GPa, respectively [21]. Notably, these values are higher than those of hard polymer materials, and in particular, the Young's modulus is even higher than that of glassy carbon, the values of which are reported between 20 and 30 GPa. Diamond films are still much harder; however, for a material synthesized at relatively low temperatures, pCN thin films not only provide outstanding optical properties, but also offer a protective coating against mechanical damage.

Another empirical method, known as the Moss relation, enables calculating the refractive index of a semiconductor material from its band gap energy value, assuming that all the energy levels are scaled by a factor ε_{EFF}^{-2}, where ε_{EFF} is the effective dielectric function, as [34]

$$n^4 E_g = 95 \text{ eV} \tag{12.24}$$

The elaboration of the optical function enables reliable measurement of the thickness of pCN thin films. By increasing the amount of melamine precursor, it is possible to increase the thickness of the films and vice versa, while keeping other conditions of the process the same.

Therefore, the evaluation of the optical functions provides valuable information about the properties of semiconductor materials, such as pCN and boron carbon nitrides [18, 35]. Eventually, we can expect that, based on these findings, a more general application of carbon-nitride-based materials for thin-films-based devices is expected in optics, beyond photocatalysis, optoelectronics, and photonics. In the following section, we will explore the in-plane and out-of-plane optical properties of pCN materials, which will provide additional information on the material structure and confirm the anisotropic behavior of this exciting material.

12.5 Optical anisotropy of carbon nitride thin films

In the previous chapter, we explored the optical properties of pCN thin films, with a particular focus on the refractive index and extinction coefficient. As mentioned, the optical functions presented were obtained assuming the material to be isotropic. However, since pCN thin films are constituted of planes stacked in a 2D manner, it is expected that the material possesses different properties along and between the planes. Indeed, anisotropy is defined as the condition of having different property values along different directions. Materials such as graphite and boron nitride possess directional properties such as mechanical, optical, electrical and thermal. For example, the thermal conductivity of a single layer of boron nitride was reported to be approximately 157 times larger in-plane than between planes [36].

However, the study of anisotropic properties of pCN materials is still in its early stages. One of the pioneering works in this direction, conducted by Noda et al., showed that the out-of-plane electrical conductivity is about two orders of magnitude higher than the in-plane conductivity, providing important information on the pCN electrical conduction mechanism [37]. Recently, Arazoe et al. investigated the pCN actuation in response to humidity variation, revealing the anisotropic structure of the pCN by means of polarized Fourier-transform infrared spectroscopy [17]. Therefore, one can also expect that the optical functions will possess a direction-dependent behavior. Building on these findings and previous examples of 2D materials, our group has further investigated the anisotropy of the optical functions. Birefringence (Δn), that is, the difference between the in-plane and out-of-plane refractive indices, is a fundamental property for light manipulation. It causes the incoming light beam to split into two different rays in uniaxial optical materials, as shown in Fig. 12.8 for calcite crystals.

Fig. 12.8: Birefringence in calcite crystals (reproduced from: https://www.itp.uni-hannover.de/fileadmin/itp/emeritus/zawischa/static_html/kalcal.html).

The performance of optical components based on anisotropic materials depends on the birefringence of the constituting materials. Large Δn values enable smaller and more efficient devices [7]. Transition metal dichalcogenides recently received a lot of attention due to their giant birefringence (≈ 1.5) in the infrared range where no optical absorption occurs, only few materials with high birefringence and high transparency in the visible range have been reported so far. For instance, hexagonal boron nitride exhibits a strong negative birefringence, that is, higher out-of-plane refractive index, while possessing a large band gap (≈ 6 eV) [38]. High refractive index materials, such as diamond, do not present optical anisotropy due to their well-defined cubic lattice structure. Diamond birefringence was detected in stress- or strain-induced diamond (or inclusions of impurities) due to the distortion of the cubic lattice [39].

In the previous section, we observed that pCN thin films exhibit high refractive index with high transparency, homogeneity, and low roughness (ca. 1 nm) in the visible range. The data obtained can be exploited to determine the expected optical anisotropy. The experimental results are expressed in terms of extinction coefficient (k, grey line Fig. 12.9a), ordinary and extraordinary refractive indices (n_o, n_{eo}, black and blue lines in Fig. 12.9a, respectively), and birefringence (Δn), versus energy (Fig. 12.9b). The pCN thin film displays a strong birefringence in the visible range, with a high ordinary refractive index ($n_{o,D}$) of 2.54, which accounts for the in-plane (a, b plane) refractive index, and a comparably low extraordinary refractive index ($n_{eo,D}$, perpendicular to the a, b plane) of 1.83. The calculated birefringence Δn_D of 0.71 ($n_{o,D} - n_{eo,D}$, Fig. 12.9b), which represents the difference between the in-plane and out-of-plane refractive indices, is very high, making pCN thin films suitable for photonic and optoelectronic devices that exploit large optical anisotropy, such as photodetectors, quantum computers, visible light communication, polarizers, microphotonic polarization converters, sensors, and more [7, 40, 41]. It is worth noting that the isotropic model refractive index is not the average refractive index; indeed, as clearly stated by Campoy-Quiles et al.: "*the refractive index deduced using an isotropic model does not correspond to an average refractive index*" [42]. Therefore, the in-plane refractive index is more similar to the isotropic model used in the previous section, and not similar to the average of the in-plane and out-of-plane refractive indices.

As expected, the optical birefringence of pCN increases close to the band gap (2.88 eV, 431 nm) and decreases toward the NIR, with $\Delta n_{1,000\ nm}$ of 0.66 (Fig. 12.9), as a a result of the increased oscillator strength, which occurs closer to the band gap. Furthermore, the strong birefringence in pCN thin films speaks of a high degree of order attained in the material. The pCN thin films did not show any dichroism, that is, the difference between the in-plane and out-of-plane extinction coefficients. However, an anisotropic behavior is found for the imaginary part of the dielectric function (ε_2). Interestingly, the fitting of the ellipsometric data reveals that the pCN layers are almost perfectly stacked on top of each other, but with a tilt angle: the Euler angle theta (θ) that describes the relative tilting with respect to the vertical axis reveals that the units are shifted with respect to the normal by about $-25°$ ($+180° > \theta > -180°$) [43]. These results suggest the presence of a layered structure where the pCN layers grow tilted with respect to the vertical axis, as exemplarily represented in Fig. 12.9.

Eventually, the evaluation of the optical functions provides information beyond the optical properties themselves. Indeed, the interpretation of the Euler angle provides additional information about the structure and the tilting among the layers, which is of paramount importance in different fields, especially, in energy storage and membranes [44]. The high birefringence in the visible range, with high transparency is, furthermore, of high importance for optical and photonic devices [22, 45].

a)

b)

c)

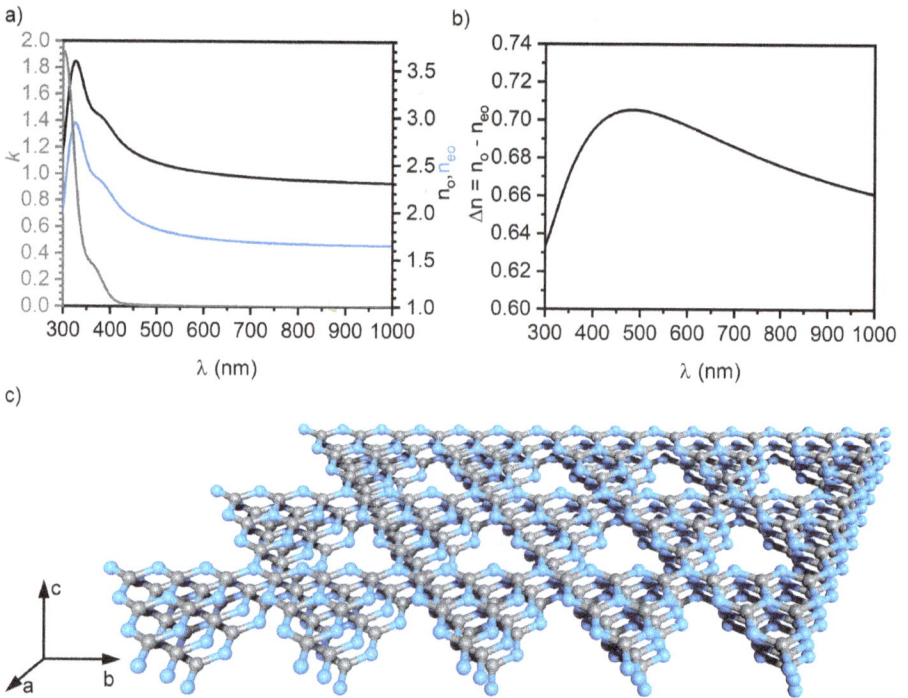

Fig. 12.9: Optical anisotropy of pCN thin films. (a) Ideal chemical structure of a plane of pCN. (b) Anisotropic optical functions (k in grey, n_o in black, and n_{eo} in blue). (c) Birefringence of pCN thin films. (d) Representation of the tilted stack of pCN layers, as derived by ellipsometry [40].

12.6 Summary and conclusions

In this chapter, an innovative method for synthesizing pCN thin films using chemical vapor deposition was shown. The method offers several advantages with respect to previous methods as it enables the control and deposition of flat and homogeneous pCN thin films over large surfaces, regardless of the shape of the substrate shape, on flat substrates, flower-shaped, rough, and even bulky materials. This method allows for the precise control of thickness by means of the precursor amount without reducing the samples' quality making it suitable for a wide range of applications, including photocatalysis, energy storage, and optics. The latter, in particular, has strict requirements regarding flatness and homogeneity, which are met when preparing pCN thin films *via* chemical vapor deposition. The method shown in this chapter can be further applied in the synthesis of other carbon-based materials, whether they are insulators, semiconductors or conductors. The careful understanding of the deposition process and the precursor(s) properties will enable the design of new materials and applications by means of chemical vapor deposition process. Optical characterization, such as transmittance, reflectance and ellipsometry, reveals that pCN possesses a high refractive index with high transparency in the visible range. This is even in the range of diamond, which is widely known for its brilliance and high refractive index, with high transparency in the visible spectrum. Furthermore, a wide set of tool and equations to design and characterize optical materials are provided that can be of help to the reader beyond the scope of this chapter.

Additionally, the layered structure of pCN materials infers anisotropic optical properties in the material. Exploiting the optical functions derived by means of the isotropic model, we explored the in-plane and out-of-plane optical functions, showing that pCN possess high birefringence and n_D values of 2.54 and 1.83, respectively. The strong optical anisotropy and the high values of the in-plane index are of great interest for the development of nanophotonic devices for light manipulation, quantum computing, and sensors, among others.

Eventually, considering the developed techniques for doping, control of the optical band gap, energy levels values, and the other methods widely reported for tuning the chemical and the physical properties of pCN, combined with the possibility to make homogeneous pCN thin films over large areas, it will pave the way for a more general understanding and exploitation of pCN materials. Specifically, adhesion at the substrate surfaces can be exploited for developing photocatalytically active reactor walls for microfluidic devices, heterojunctions and patterned surfaces to exploit the properties of this exciting material [46].

References

[1]   Born M, Wolf E. Principles of optics: Electromagnetic theory of propagation, interference and diffraction of light. Elsevier, 2013.
[2]   Cardona M, Peter YY. Fundamentals of semiconductors. Springer, 2005.
[3]   McCluney R. Radiometry and Photometry. In: Meyers RA, editor. Encyclopedia of Physical Science and Technology. 3rd edition. New York, Academic Press, 2003, pp. 731–58.
[4]   Viezbicke BD, Patel S, Davis BE, Birnie III DP. Evaluation of the Tauc method for optical absorption edge determination: ZnO thin films as a model system. Phys Status Solidi B 2015, 252(8), 1700–10.
[5]   Hearnshaw JB. The analysis of starlight: One hundred and fifty years of astronomical spectroscopy. CUP Archive, 1990.
[6]   Choi JH, Shi F, Margaryan A. Refractive index and low dispersion properties of new fluorophosphate glasses highly doped with rare-earth ions. J Mater Res 2005, 20(1), 264–70.
[7]   Ermolaev G, Grudinin D, Stebunov Y, Voronin K, Kravets V, Duan J, et al. Giant optical anisotropy in transition metal dichalcogenides for next-generation photonics. Nat Commun 2021, 12(1), 1–8.
[8]   Schutjajew K, Giusto P, Härk E, Oschatz M. Preparation of hard carbon/carbon nitride nanocomposites by chemical vapor deposition to reveal the impact of open and closed porosity on sodium storage. Carbon 2021, 185, 697–708.
[9]   Cruz D, Garcia Cerrillo J, Kumru B, Li N, Dario Perea J, Schmidt BV, et al. Influence of thiazole-modified carbon nitride nanosheets with feasible electronic properties on inverted perovskite solar cells. J Am Chem Soc 2019, 141(31), 12322–28.

[10] Kumru B, Shalom M, Antonietti M, Schmidt BV. Reinforced hydrogels *via* carbon nitride initiated polymerization. Macromolecules 2017, 50(5), 1862–69.

[11] Yandrapalli N, Robinson T, Antonietti M, Kumru B. Graphitic carbon nitride stabilizers meet microfluidics: From stable emulsions to photoinduced synthesis of hollow polymer spheres. Small 2020, 16(32), 2001180.

[12] Zhang Y, Su S, Zhang Y, Zhang X, Giusto P, Huang X, et al. Visible-light-driven photocatalytic water disinfection toward *Escherichia coli* by nanowired g-C_3N_4 film. Front Nanotechnol 2021, 3, 35.

[13] McNaught AD, McNaught AD. Compendium of chemical terminology. Blackwell Science Oxford, 1997.

[14] Liu J, Wang H, Chen ZP, Moehwald H, Fiechter S, van de Krol R, et al. Microcontact-printing-assisted access of graphitic carbon nitride films with favorable textures toward photoelectrochemical application. Adv Mater 2015, 27(4), 712–18.

[15] Bian J, Li Q, Huang C, Li J, Guo Y, Zaw M, et al. Thermal vapor condensation of uniform graphitic carbon nitride films with remarkable photocurrent density for photoelectrochemical applications. Nano Energy 2015, 15, 353–61.

[16] Jia C, Yang L, Zhang Y, Zhang X, Xiao K, Xu J, et al. Graphitic carbon nitride films: Emerging paradigm for versatile applications. ACS Appl Mater Interfaces 2020, 12(48), 53571–91.

[17] Arazoe H, Miyajima D, Akaike K, Araoka F, Sato E, Hikima T, et al. An autonomous actuator driven by fluctuations in ambient humidity. Nat Mat 2016, 15(10), 1084.

[18] Giusto P, Arazoe H, Cruz D, Lova P, Heil T, Aida T, et al. Boron carbon nitride thin films: From disordered to ordered conjugated ternary materials. J Am Chem Soc 2020, 142(49), 20883–91.

[19] Giusto P, Cruz D, Rodriguez Y, Rothe R, Tarakina NV. Red carbon thin film: A carbon–oxygen semiconductor with tunable properties by amine vapors and its carbonization toward carbon thin films. Adv Mater Interfaces 2022, 9(21), 2200834.

[20] Xiong W, Huang F, Zhang R-Q. Recent developments in carbon nitride based films for photoelectrochemical water splitting. Sustainable Energy Fuels 2020, 4, 485–503.

[21] Giusto P, Cruz D, Heil T, Arazoe H, Lova P, Aida T, et al. Shine bright like a diamond: New Light on an old polymeric semiconductor. Adv Mater 2020, 32(10), 1908140.

[22] Kuznetsov AI, Miroshnichenko AE, Brongersma ML, Kivshar YS, Luk'yanchuk B. Optically resonant dielectric nanostructures. Science 2016, 354(6314), aag2472.

[23] Bian J, Li J, Kalytchuk S, Wang Y, Li Q, Lau TC, et al. Efficient emission facilitated by multiple energy level transitions in uniform graphitic carbon nitride films deposited by thermal vapor condensation. ChemPhysChem 2015, 16(5), 954–59.

[24] Swanepoel R. Determination of the thickness and optical constants of amorphous silicon. J Phys E: Sci Instrum 1983, 16(12), 1214.

[25] Woollam JA, Co. I. Guide to Using WVASE 32: Spectroscopic ellipsometry data acquisition and analysis software. J. A. Woollam Company, Incorporated, 2008.

[26] Vezie MS, Few S, Meager I, Pieridou G, Dörling B, Ashraf RS, et al. Exploring the origin of high optical absorption in conjugated polymers. Nat Mat 2016, 15(7), 746.

[27] Turri G, Webster S, Chen Y, Wickham B, Bennett A, Bass M. Index of refraction from the near-ultraviolet to the near-infrared from a single crystal microwave-assisted CVD diamond. Opt Mater Express 2017, 7(3), 855–59.

[28] Higashihara T, Ueda M. Recent progress in high refractive index polymers. Macromolecules 2015, 48(7), 1915–29.

[29] Mathy A, Ueberhofen K, Schenk R, Gregorius H, Garay R, Müllen K, et al. Third-harmonic-generation spectroscopy of poly (p-phenylenevinylene): A comparison with oligomers and scaling laws for conjugated polymers. Phys Rev B 1996, 53(8), 4367.

[30] Campoy-Quiles M, Heliotis G, Xia R, Ariu M, Pintani M, Etchegoin P, et al. Ellipsometric characterization of the optical constants of polyfluorene gain media. Adv Funct Mater 2005, 15(6), 925–33.

[31] DeVore JR. Refractive indices of rutile and sphalerite. J Opt Soc Am 1951, 41(6), 416–19.

[32] You N-H, Higashihara T, Oishi Y, Ando S, Ueda M. Highly refractive poly (phenylene thioether) containing triazine unit. Macromolecules 2010, 43(10), 4613–15.

[33] Weatherspoon MR, Cai Y, Crne M, Srinivasarao M, Sandhage KH. 3D Rutile titania-based structures with morpho butterfly wing scale morphologies. Angew Chem Int Ed 2008, 47(41), 7921–23.

[34] Ravindra N, Ganapathy P, Choi J. Energy gap–refractive index relations in semiconductors–An overview. Infrared Phys Technol 2007, 50(1), 21–29.

[35] Giusto P, Cruz D, Heil T, Tarakina N, Patrini M, Antonietti M. Chemical vapor deposition of highly conjugated, transparent boron carbon nitride thin films. Adv Sci 2021, 8(17), 2101602.

[36] Roy S, Zhang X, Puthirath AB, Meiyazhagan A, Bhattacharyya S, Rahman MM, et al. Structure, properties and applications of two-dimensional hexagonal boron nitride. Adv Mater 2021, 33(44), 2101589.

[37] Noda Y, Merschjann C, Tarábek J, Amsalem P, Koch N, Bojdys MJ. Directional charge transport in layered two-dimensional triazine-based graphitic carbon nitride. Angew Chem 2019, 131(28), 9494–98.

[38] Segura A, Artús L, Cuscó R, Taniguchi T, Cassabois G, Gil B. Natural optical anisotropy of h-BN: Highest giant birefringence in a bulk crystal through the mid-infrared to ultraviolet range. Phys Rev Mater 2018, 2(2), 024001.

[39] Howell D. Strain-induced birefringence in natural diamond: A review. Eur J Mineral 2012, 24(4), 575–85.

[40] Giusto P, Kumru B, Cruz D, Antonietti M. Optical anisotropy of carbon nitride thin films and photografted polystyrene brushes. Adv Opt Mater 2019, 10, 2101965.

[41] Wang J, Kühne J, Karamanos T, Rockstuhl C, Maier SA, Tittl A. All-dielectric crescent metasurface sensor driven by bound states in the continuum. Adv Funct Mater 2021, 31(46), 2104652.

[42] Campoy-Quiles M, Etchegoin PG, Bradley DDC. On the optical anisotropy of conjugated polymer thin films. Phys Rev B 2005, 72(4), 045209.

[43] Schmidt D, Booso B, Hofmann T, Schubert E, Sarangan A, Schubert M. Monoclinic optical constants, birefringence, and dichroism of slanted titanium nanocolumns determined by generalized ellipsometry. Appl Phys Lett 2009, 94(1), 011914.

[44] Sakaushi K, Antonietti M. Carbon- and nitrogen-based organic frameworks. Acc Chem Res 2015, 48(6), 1591–600.

[45] Tittl A, Leitis A, Liu M, Yesilkoy F, Choi D-Y, Neshev DN, et al. Imaging-based molecular barcoding with pixelated dielectric metasurfaces. Science 2018, 360(6393), 1105–09.

[46] Mazzanti S, Manfredi G, Barker AJ, Antonietti M, Savateev A, Giusto P. Carbon nitride thin films as all-in-one technology for photocatalysis. ACS Catal 2021, 11(17), 11109–16.

Kai Xiao* and Lei Jiang

Chapter 13
Carbon nitride-based artificial light-driven ion pumps

13.1 Introduction

Graphitic carbon nitride (abbreviated as g-C_3N_4) is a well-known two-dimensional conjugated polymeric semiconductor and has broad applications in photocatalysis and related fields (Fig. 13.1a). The history of g-C_3N_4 as a metal-free organic semiconductor catalyst could only be dated back to 2006 [1, 2]. In the past years, we have witnessed a large number of publications, especially after the discovery of its capability for visible-light-driven photocatalytic hydrogen evolution in 2009 [3]. Due to its appropriate band gap (2.7 eV) and band edge positions, abundance of C and N elements that it is composed of, low cost and nontoxicity, g-C_3N_4 has also been applied in many other areas, such as in electrocatalysis, sensing, membrane filtration and other areas [4–7]. The construction of a light-driven ion pump is one of its emerging applications.

The artificial light-driven ion pump is a technical counterpart of biological light-driven ion pumps, which are able to pump ions against concentration gradients by consuming light energy. In biological systems, light-driven ion pumps are assemblies of integral membrane proteins, which can move ions against a concentration gradient to create a membrane potential, thus converting sunlight energy directly into an osmotic potential (Fig. 13.1b). Biological light-driven ion pumps are structures widely existing in some archaea, such as *Halobacterium halobium* [8]. By absorbing sunlight energy and converting it into excitons, these microorganisms can pump protons across the membrane, generating an osmotic and charge imbalance, which, in turn, powers the synthesis of adenosine triphosphate (ATP) [9, 10]. From a technological angle, the generation of an electrochemical gradient is the precondition for photoelectrical energy conversion by ion pumps. The concept of light-triggered electrochemical gradients and photoelectric conversion are therefore linked.

Scientists have long been inspired by the sophisticated functions of naturally occurring ion pumps and have created a wide range of biomimetic ion pumps using synthetic compounds and artificial nanostructures. "Chemical ion pumps" were developed first by inserting redox shuttles into a lipid bilayer or by integrating a photoredox-active

*Corresponding author: Kai Xiao, Department of Biomedical Engineering, Southern University of Science and Technology (SUSTech), 518055 Shenzhen, P.R. China, e-mail: xiaok3@sustech.edu.cn
Lei Jiang, Key Laboratory of Bio-inspired Smart Interfacial Science and Technology of Ministry of Education, School of Chemistry, Beihang University, 100191 Beijing, P.R. China

https://doi.org/10.1515/9783110746976-013

molecular switch into a membrane material, as shown in Figure 13.1c [11, 12]. Therein, C-P-Q represents the carotene–porphyrin–naphthoquinone molecular triade. Upon photon absorption, C-P-Q gives a charge separated state ($C^{\bullet+}$-P-$Q^{\bullet-}$) with a positive charge localized at carotenoid moiety and negative charge at the naphthoquinone (Step 1). Q_s represents a lipophilic quinone. Reduction of Q_s to the radical anion ($Q_s^{\bullet-}$) occurs near the external bilayer–water interface via electron transfer from the naphthoquinone radical anion (Step 2). In this step, $C^{\bullet+}$-P-$Q^{\bullet-}$ is oxidized to $C^{\bullet+}$-P-Q. Protonation of $Q_s^{\bullet-}$ near the external aqueous interface gives semiquinone radical $HQ_s^{\bullet}$ (Step 3). $HQ_s^{\bullet}$ delivers the electron and proton, either via diffusion or self-exchange reaction with other Q_s molecules, to the oxidation site localized at the carotenoid radical cation near the inner membrane surface (Step 4). Oxidation of $HQ_s^{\bullet}$ by $C^{\bullet+}$-P-Q (formation of transient protonated lipophilic quinone specie HQ_s^{+}, Step 5) followed by release of H^+ into intraliposomal volume recovers C-P-Q and Q_s (Step 6), and generates the proton-motive force – higher concentration of H^+ inside the liposome compared to the outer volume. However, these "chemical ion pumps" with good biocompatibility can only drive specific ions (proton or calcium ions) by lipophilic H^+- or Ca^{2+}-binding shuttle molecules and have limited efficiency. For example, artificial photosynthetic reaction centers can only create a several- to dozen-fold concentration gradient with an overall quantum yield of only 0.4% – 1% [11, 12].

With the development of nanotechnology, it is now possible to fabricate nanometer-sized physical orifices equipped with an ion pump function, a so-called "physical ion pump" (Fig. 13.1e). In contrast to "chemical ion pumps", which are based on the proton-coupled electron transfer process, the "physical ion pump" can drive ions against a concentration gradient by light–induced surface charge redistributions [13–15]. However, the "physical ion pump" has poor biocompatibility and the concentration gradients are still low, only several-fold, because of the weak transmembrane driving force created by light illumination.

Despite all important advances achieved in this area of research, synthetic ion-to-gradient conversion is still, by far, more ineffective when compared to biological systems, which can pump ions even against steep concentration gradient to create a membrane potential directly by conversion of light energy (Fig. 13.1d). However, it is clear that such artificial light driven ion pumps, preferably, superior to natural and applicable in a wider range of chemical conditions, i.e., solvent, temperature and salinity, would have a myriad of potential applications.

a)

b)

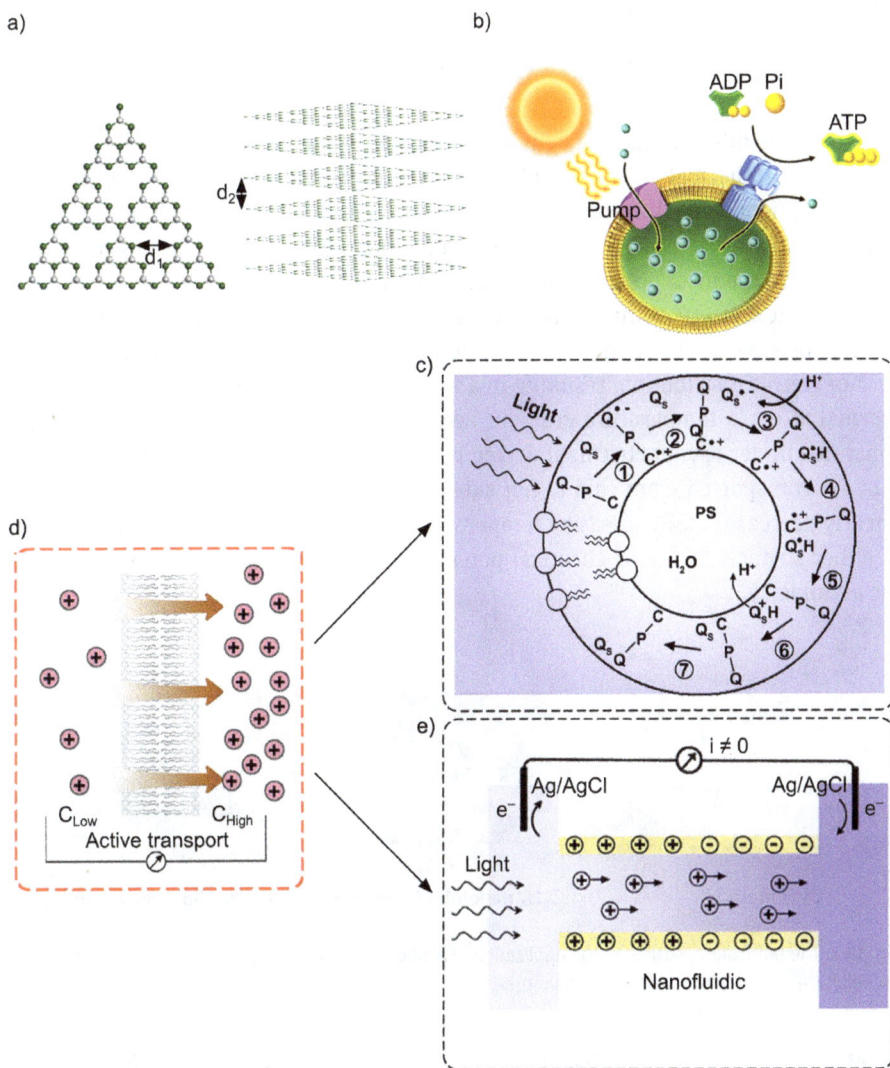

c)

d)

e)

Fig. 13.1: (a) A fragment of perfectly condensed heptazine-based g-C_3N_4 features characteristic in-plane periodicity of heptazine (C_6N_7) units and AAA type π–π stacking. The d_1 and d_2 represent specific intrinsic pores created by layers and the spacing in-between layers, respectively. (b) Light-driven ion pumps play a critical role for solar energy harvesting exploited by several archaea species. Electrochemical gradients are created by the light-induced ion pumping process and are used to power various biological processes. ADP, adenosine diphosphate; Pi, phosphate; ATP, adenosine triphosphate. (c) Artificial chemical ion pump. (d) Schematic of active ion transport (ion pump). (e) Artificial physical ion pump.

13.2 Ion pump based on single-phase symmetric carbon nitride nanotube

To date, three different mechanisms, including photochemical [13, 16], photoelectrical [17] and photothermal effects [18, 19], have been proposed to explain light-driven ion transport phenomena. In general, photochemical and photoelectrical effects break the symmetry of surface charge distribution, while photothermal effect results in the dissymmetry of chemical potential along the nanofluidic device. For photochemical and photoelectrical effects, either a photoisomerization reaction or photoinduced electron transfer process occurring in the nanofluidic device alters its surface properties, e.g., surface charge distribution, resulting in a unidirectional ion transport. For the photothermal effect, a temperature gradient across the nanofluidic device may easily be generated under asymmetric light-driven heating; then, ions with a concurrent water flux are transported from cold to hot sides because the chemical potential of water linearly decreases with increasing temperature. All in all, the asymmetric distribution of either surface charge or chemical potential will drive ions movement to balance the asymmetry.

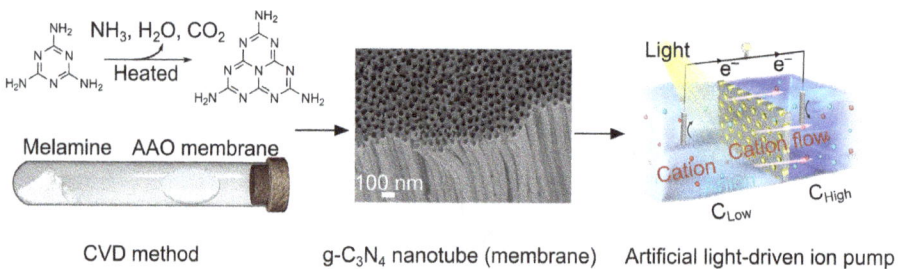

Fig. 13.2: The fabrication, structure and application of artificial light-driven ion pump based on single-phase symmetric carbon nitride nanotube.

Based on the photoelectric effect, recently, Xiao et al. [20] described a universal light-induced ion pump system by fabricating single-phase symmetric carbon nitride nanotube (Fig. 13.2). The synthesis uses melamine as starting material in a typical vapor deposition polymerization (VDP) process and an anodic aluminum oxide (AAO) with a pore diameter of 100 nm as a substrate. The CNNM (carbon nitride nanotube membrane) inner diameter can be well controlled from 0 nm (complete filling, i.e., a nanorod is formed inside the AAO membrane channel) to ~90 nm, by changing the amount of melamine used in the VDP process. Figure 13.2 shows typical SEM image of CNNM, in which the tube has an external diameter ~90 nm and inner diameter ~30 nm. The ion pump properties were measured in home-made electrolyte cells. The carbon nitride nanotube membrane was symmetrically placed in contact with two KCl solutions differing by a factor of 100 in concentration (C_H = 0.01 M; C_L = 0.0001 M) and initially illuminated from the low concentration side.

Figure 13.3a and b show the typical current–time (I–T) and voltage–time (V–T) characteristic of CNNM measured in dark and under simulated solar illumination of 143 mW cm^{-2}. Without illumination, a positive zero-volt current (a current without applying external bias voltage) and negative open-circuit voltage are generated because of selective ions diffusion driven by concentration gradient [21]. Throughout illumination, the zero-volt current decreases from about 0.1 to −0.3 μA, and the open-circuit voltage increases from −0.1 to 0.35 V. This change of current and voltage indicates that the direction of ionic movement is reversed inside the nanotube under illumination, that is, ions are moving against the concentration gradient, which can be further confirmed by inductively coupled plasma mass spectrometry (ICP-MS can be used to detect the concentration change in high concentration side (C_L)). Calculations suggest that the current change is translated into the ability of a single nanotube to pump ~1,500 ions per second against 100-fold concentration gradient, an order of magnitude smaller than the bacteriorhodopsin proton pump or halorhodopsin Cl ion pump [10], but an unprecedented breakthrough for artificial ions pump. In addition, the CNNM-based ion pump shows stable and fully repeatable instant response to illumination, which is ascribed to the fast separation of electrons and holes in CNNM under light irradiation [22].

Further measurements show that the ion pump "power" is closely connected to the light power density. The dependence can be confirmed by the direction of ionic current in Fig. 13.3c. Only at high illumination density (>100 mW cm^{-2}), the CNNM can pump ions transport against 100-fold concentration gradient. Furthermore, the CNNM can pump ions even against 5,000-fold concentration gradient with a light illumination of 380 mW cm^{-2} (Fig. 13.3d), an efficiency that has not been realized before by artificial ion pumps [23], but is comparable to the biological ion pump [10]. The CNNM-based pump system also shows obvious relationship to the light wavelength (Fig. 13.3e). To different monochromatic light with the same power density (112.5 mW cm^{-2}), high energy blue light has the strongest influence on the CNNM "power" to pump ions, while low energy yellow light has almost no effect. These results are consistent with the light absorbance of CNNM. More importantly, the artificial pump system is universal and does not differentiate electrolytes including acid, saline and alkali solutions. In this way, it is clearly an extension of the biological ions pump.

The surface charge redistribution of CNNM due to the photoinduced separation of electrons and holes is thought to be the key in explaining the ion pumping phenomenon. As illustrated in Fig. 13.4a, the CNNM in the initial state is negatively charged because of the incomplete polymerization or condensation, which results in electron-rich –imide groups, a fraction of which is deprotonated. Under these conditions, the collected ionic current is positive due to the cation diffusion from the high concentration side to the low concentration side, that is the concentration gradient induced potential difference (V_{CG}) across the cation-selective CNNM is the only source of ionic current (Fig. 13.4b). Meanwhile, Ag/AgCl electrode will undergo a chemical reaction

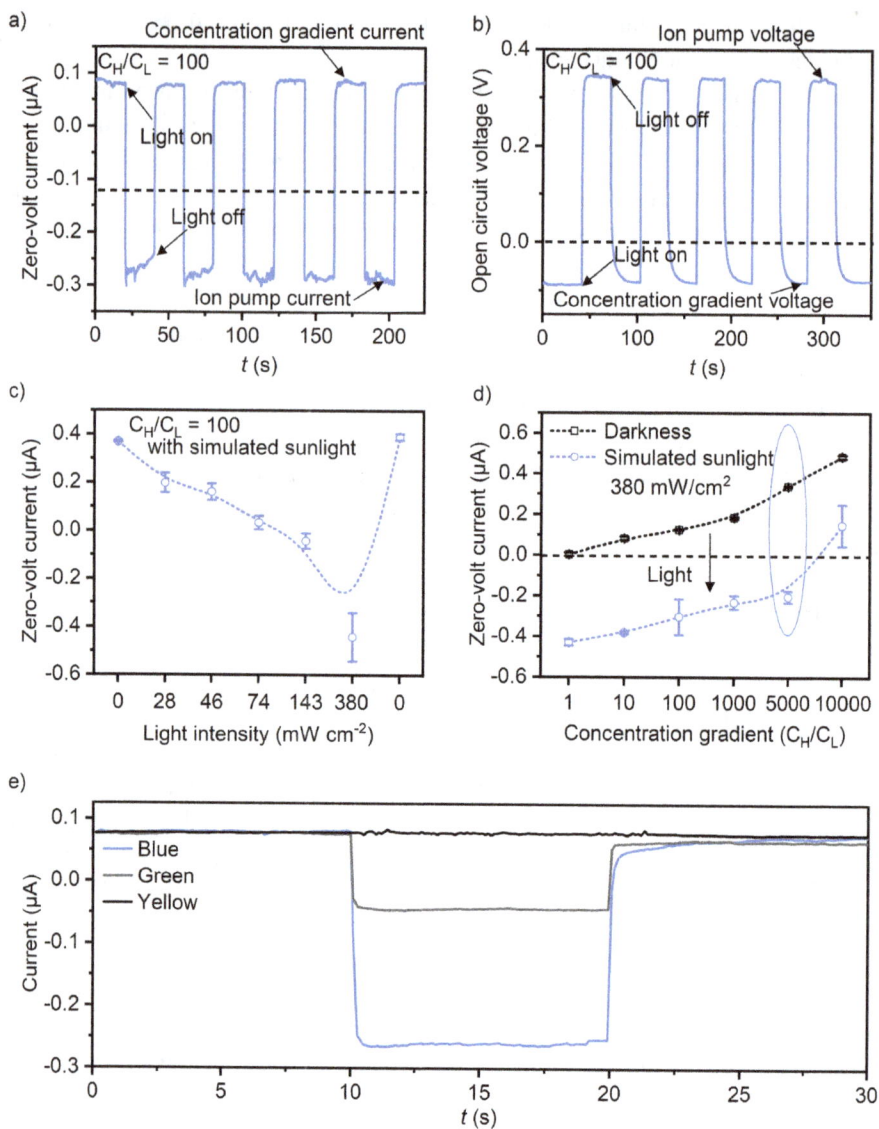

Fig. 13.3: (a) Measured cyclic constant zero-volt current with the alternating periods of illumination and dark at 100-fold KCl concentration gradient. (b) Measured open-circuit voltage across the CNNM, before and after illumination at 100-fold KCl concentration gradient. (c) Zero-volt current as a function of light density from 0 to 380 mW cm^{-2}. Only the light density stronger than 74 mW cm^{-2} can change the direction of ionic current at 100-fold KCl concentration gradient. For 0 mW cm^{-2}, the ionic current was recorded twice to illustrate stability of CNNM. (d) Zero-volt current as a function of concentration gradient from 1-fold to 10,000-fold. The CNNM-based ion pump can realize "uphill" ions transport process at up to 5,000-fold concentration gradient. (e) Zero-volt current as a function of incident light wavelength (blue: 405 nm; green: 515 nm; yellow: 590 nm) at 100-fold KCl concentration gradient. The ionic current is consistent with the light absorbance by carbon nitride material.

a)

b)

Without light irradiation

c)

CNNM molecular structure

Solution   Carbon nitride

d)

Without light irradiation

e)

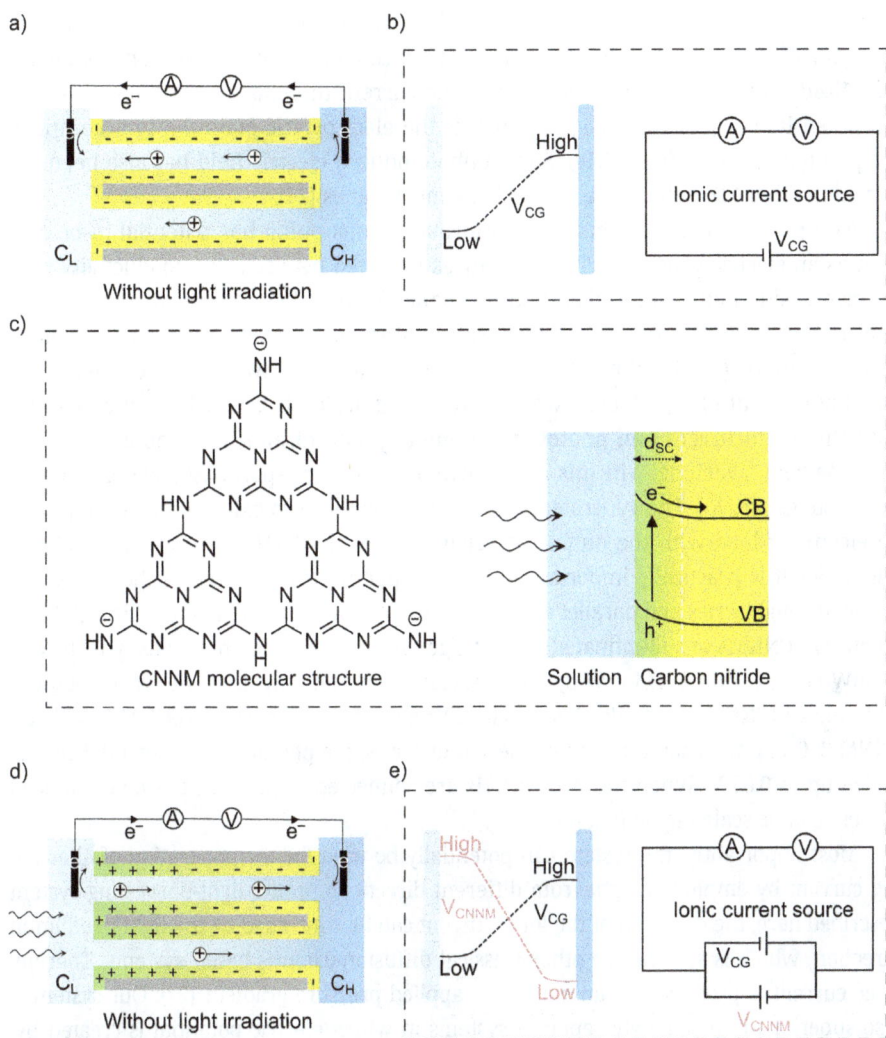

Fig. 13.4: (a) Schematic of surface charge distribution on the nanotube, before illumination. Under such conditions, low-density negative surface charge is homogeneously distributed on nanotube. (b) Equivalent circuit of the device shown in panel (a), the concentration gradient potential (V_{CG}) is the only ionic current source. (c) Schematic representation of the negatively charged molecular structure and photoinduced separation of electrons and holes in carbon nitride. (d) Schematic of surface charge distribution on the nanotube, after illumination. Under such conditions, the separation of electrons and holes results in the heterogeneous charge distribution. (e) Equivalent circuit of the device shown in panel (d), transmembrane potential (V_{CNNM}) provides a reverse potential to V_{CG}.

with chloride anions to guarantee the charge balance of electrolyte. When illuminated from the C_L side, the electrons separate from the holes and move into the bulk of carbon nitride or the unilluminated side (Fig. 13.4c) [22], resulting in the positively charged surface in the illuminated side, while the unilluminated side is still negatively charged. That is the origin of asymmetric surface charge distribution across CNNM (Fig. 13.4d). Under these conditions, the ionic current direction is determined by the net potential ($V_{net} = V_{CG} - V_{CNNM}$), which is the electrostatic potential to accomplish the pumping process (Fig. 13.4e). Such a photoinduced electric field has also been observed in other 2D materials, e.g., graphene membranes [24].

As a proof of concept, this high-performance ions pump has potential to be also used as an electric generator. The CNNM was mounted between two conductivity cells (termed A side and B side) filled with the same 0.001 M KCl electrolyte. A light with power density of 380 mW cm^{-2} from A side can produce a stable open circuit voltage up to 550 mV (Fig. 13.5a). When the photoelectrochemical cell was short-circuited, a stable photocurrent of 2.4 μA cm^{-2} was recorded (Fig. 13.5b). It is worth mentioning that both the photocurrent and photovoltage build up and disappear without hysteresis after the light switching, which is superior to other ion transport-based energy conversion systems [25, 26]. The generated power can be supplied to external circuit to power an electronic load with the output power density of the CNNM of up to 1.2 mW m^{-2} (Fig. 13.5c). It is practically important that the output of CNNM can be further scaled up simply through series and parallel connections of multiple devices. As shown in Fig. 13.5d, when two CNNMs are illuminated by two light sources of different intensity (CNNM 1: 46 mW cm^{-2}; CNNM 2: 143 mW cm^{-2}) separately, they can only generate weak open circuit voltage (CNNM 1: 0.23 V; CNNM 2: 0.42 V) and zero-volt current (CNNM 1: 0.31 μA; CNNM 2: 0.42 μA). When they are connected in series, the photoinduced potential can be scaled up to 0.65 V. When the two CNNMs are connected in parallel, the photoinduced current can be scaled up to 0.71 μA.

Most importantly, the system can potentially be used for the generation of alternating current by employing light from different directions. In the light-harvesting system described here, the direction of the ionic current can be controlled by the light irradiation direction, which is superior to other classical diffusion-osmosis-based systems. Therein, ionic current is produced by an externally applied pressure gradient [27]. Our system is also superior to other newly reported systems in which electric potential is created by, for example, gradient of temperature [28] or concentration of electrolyte [26, 29, 30]. As shown in Fig. 13.5e, positive current arising from illumination B side (46 mW cm^{-2}) changes its direction when A side is illuminated instead (143 mW cm^{-2}). Meanwhile, simultaneous illumination of both sides (A + B) generates a partially mutually compensating ionic current. In this particular system, different photocurrent patterns can be generated when different light sequences are applied – ion flux follows the light.

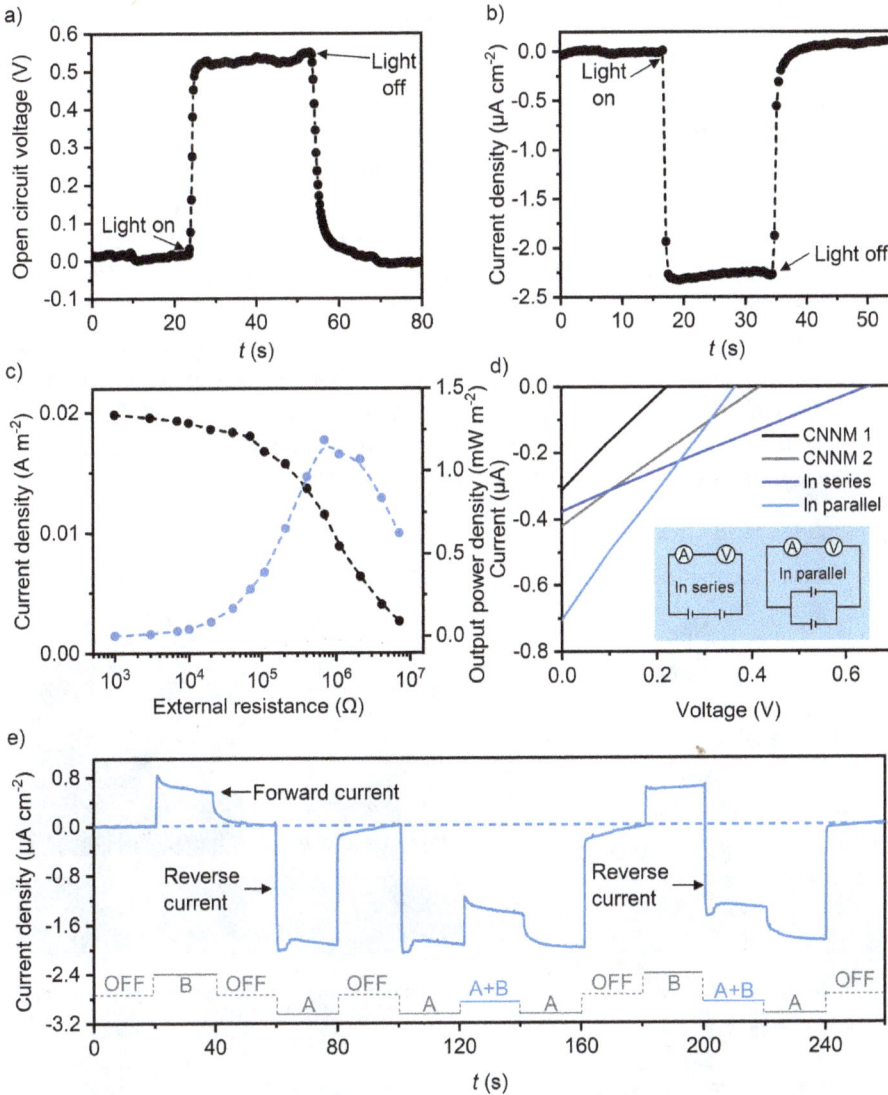

Fig. 13.5: (a,b) Open-circuit voltage (a) and current density (b) generated by light-induced ions transport. (c) The generated power can be supplied to the external circuit and power an electronic load. The output power density reaches its peak value of 1.2 mW m^{-2} at the external resistance of ~400 kΩ. (d) Current–voltage curves of two individual CNNM and their series and parallel connections. Inset: circuit diagram. CNNM 1 illuminated with simulated solar light 46 mW cm^{-2}; CNNM 2 – 143 mW cm^{-2}. (e) Light with different power densities (A side: 143 mW cm^{-2}; B side: 46 mW cm^{-2}) from different directions resulted in alternating current. CNNM assembled in a series.

13.3 Ion pumps based on single-phase asymmetric carbon nitride nanotubes

Breaking symmetry on the nanoscale often results in unexpected phenomena. To implement an asymmetric structure in carbon nitride nanotube, the fabrication process must be slightly changed. Xiao et al. [31] reported that an asymmetric carbon nitride nanotube membrane (ACNNM) can be fabricated by the same VDP method with CNNM, when placing the AAO substrate perpendicular to the tube. Asymmetric nanotubes are expected to be generated if the substrate is positioned in the chamber as a partition wall, which separates the polymerization chamber into two areas: chamber 1 and chamber 2 (Fig. 13.6a). Chamber 1 will have a high carbon nitride vapour concentration (C_{High}), while chamber 2 will have a lower concentration (C_{Low}) because of the direction of N_2 flow (which is a gas carrier in the VDP process). SEM and TEM images clearly confirm the asymmetric structure of the membrane (Fig. 13.6b and c).

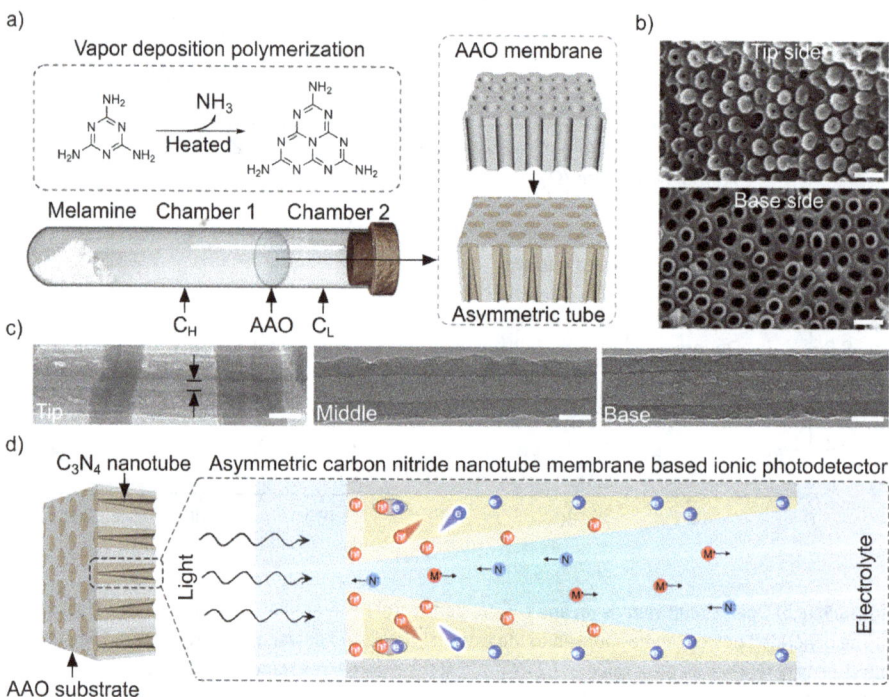

Fig. 13.6: (a) The fabrication process of ACNNM. (b) Cross sections of ACNNM at the tip and base sides, scale bar 200 nm. (c) TEM images of ACNNM at the tip, middle, and base sides, scale bar 50 nm. (d) Schematic diagram of ACNNM used in this work and the mechanism of light-induced ions transport.

Similar to CNNM, ACNNM shows light-driven ion transport (ion pump) property, and the authors use it to construct an ionic photodetector. It is known that the working principle of conventional electron transport photodetectors is based on the separation of electrons and holes in the semiconductor generated by the incident photons [32]. The operation principle of an ionic photodetector is based on separation of the electrolyte ions (cation and anion) induced by an asymmetric surface charge distribution on a semiconductor surface. Indeed, light is first converted into separated charges located along a gradient, which then, however, induce a flux of mobile ions for charge compensation. The liquid character can guarantee reproducible and stable contacts, good biocompatibility and structural plasticity. In this work, ACNNM is homogeneously negatively charged in its initial state [33], but changes to a state with an asymmetric surface charge distribution when irradiated with light (Fig. 13.6d). Then, the mobile ions in the electrolyte move to balance the asymmetric surface charge, which can be described as a light-induced ionic current.

Based on these carefully designed ACNNM, the authors examined the figures of merit of the ionic photodetector, following the "6S" principles: high spectrum selectivity, high signal-to-noise ratio, high sensitivity, fast response speed, high stability and self-power property. Figure 13.7a presents the spectrum selectivity of ACNNM-based ionic photodetectors. The time-dependent photocurrents were measured at –0.5 V bias potential with same light power density (50 mW cm^{-2}), but different wavelengths (blue light: 405 nm; green light: 515 nm; yellow light: 590 nm). It can be seen that the ionic photodetector device exhibits the strongest responsivity to high energy blue light, while a weaker responsivity is found for low energy yellow light. The obvious different responsivity to different light energy follows the absorbance of the semiconducting ACNNM well.

The ACNNM-based ionic photodetector also exhibits a high signal-to-noise ratio. Figure 13.7b shows the dark current and photocurrent at 50 mW cm^{-2} blue light irradiation and 0 V bias. In general, the dark current at 0 V bias is less than 0.02 nA, defined as "off" state (which is maybe much smaller, but is limited by the sensitivity of the used instrument), while a photodetector under illumination is defined as "on" state, which is about 0.1 µA. The current ratio between on and off states (on/off ratio) reveals the sensitivity of the photodetector to certain irradiation. For the ACNNM-based ionic photodetector, the on/off ratio can reach to 5,000, which is high compared with conventional photodetectors [34].

Figure 13.7c depicts a representative set of the time-dependent currents under light illumination with different incident power at –0.5 V bias potential. Following light illumination, the current gain shows a clear dependence on power of incident light. The photocurrent is calculated as

$$I_{pc} = I_{light} - I_{dark} \qquad (13.1)$$

where I_{light} and I_{dark} are the ionic current under light illumination and dark, respectively.

The responsivity is given by

$$R = I_{pc}/P \qquad (13.2)$$

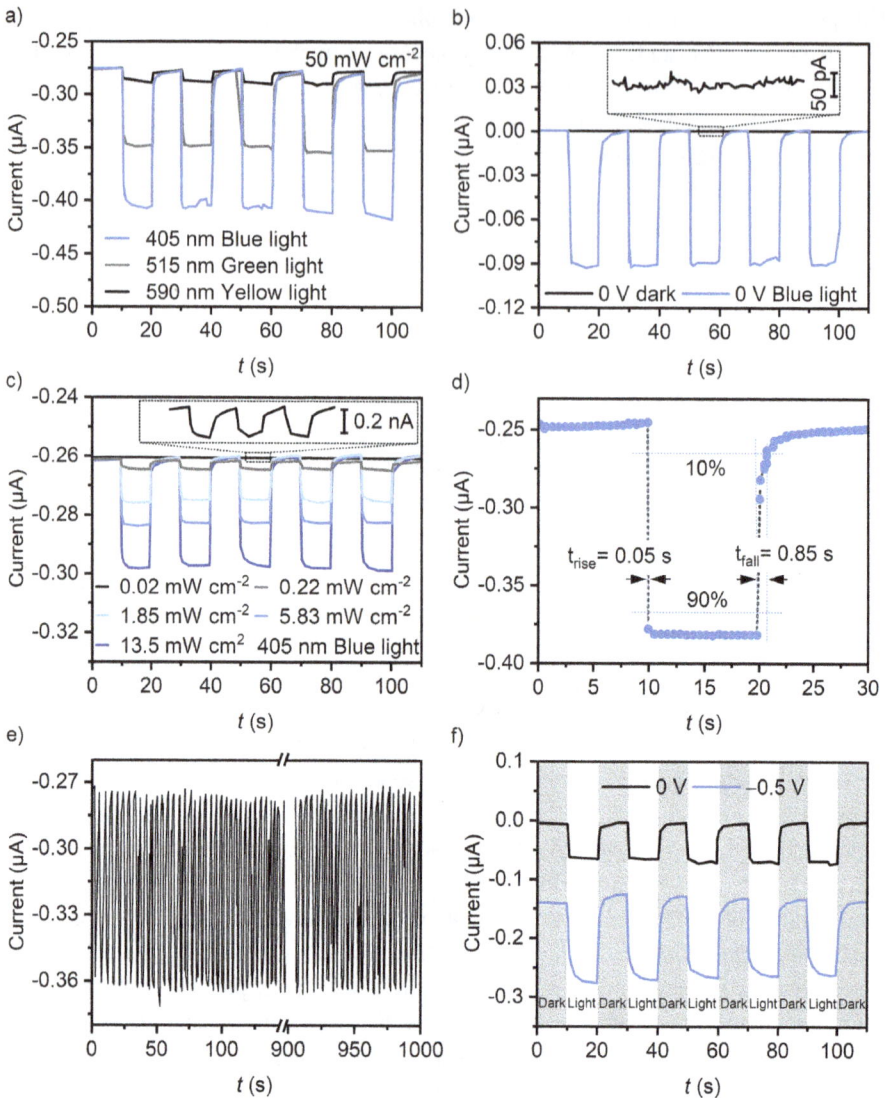

Fig. 13.7: (a) Photocurrent responses of the ionic photodetector at different light illuminations (blue, green, yellow, power density: 50 mW cm^{-2}) show high spectrum selectivity at −0.5 V bias. (b) Photocurrents of the ionic photodetector measured in dark and under light illumination (50 mW cm^{-2} blue light at 0 V bias) show a high signal-to-noise ratio. (c) Time-dependent photocurrents of ACNNM-based ionic photodetector as a function of incident light power show high sensitivity (blue light at −0.5 V bias). (d) The measured photocurrent responses indicate a rise time of less than 0.05 s and a fall time of less than 0.85 s, indicating a fast sensing speed. (e) Photoresponse of the ionic photodetector for 330 cycles (1,000 s) illustrating the stability. (f) Photocurrent response of the ionic photodetector at different bias (black: 0 V; blue: −0.5 V). The ionic photodetector still works without external power source (0 V).

where P is the incident light power density. Within the incident power range of 0.04 to 15 mW cm^{-2}, the photodetector shows a linear response with photocurrent from 1 nA to 0.1 µA, while the responsivity decreases from 30 to 3 µA W^{-1}. We have to state that the responsivity of our ionic photodetector has no advantage compared with other photodetectors, but has much room to improve by increasing ion transport rate, e.g., by increasing electrolyte temperature or going to quantum tunneling fluid.

Fast response to optical signals, which is here coupled to the charge (represented by electrons or ions) transport and collection, is critical for optoelectronic devices [35]. Figure 13.7d shows the temporal photocurrent response of the ACNNM-based ionic photodetector. The measured switching times for the rise (current increasing from 10% to 90% of the peak value) and fall (current decreasing from 90% to 10% of the peak value) of the photocurrent are 0.05 s and 0.85 s, respectively, indicating a rather fast response speed. It is an obvious disadvantage that due to the involved liquid transport of ions, the response time of ionic photodetector must be much slower than that of an electronic photodetector, which is in the range of several nanoseconds. However, it is fast enough to meet the requirements of most of the photoelectric devices and could be further improved by increasing ion transport speed. The fact that the decrease of the photocurrent is significantly longer than the increase is due to the relaxation of the photo-induced surface charges within the semiconductor, which are obviously rather long-living. In addition, the ionic photodetector is a four-step signal conversion process: light signal is converted into surface-localized photocharges, which create an ionic current to be finally read out in an electronic signal. We predict that the ionic photodetector should also be faster and has a higher responsivity when it is used in some nanometer or micrometer-sized ion readout device, just like synapses and neurons [36].

Stability is one of the most important characteristics of optoelectronic devices. Figure 13.7e shows the response of photocurrent to optical pulses at a time interval of about 3 s. It is found that the dynamic photo response of the ionic photodetector remains stable and reproducible at least on a 1,000 s timescale. For about 330 cycles, the photocurrent quickly increases as soon as the light is turned on and then drops to the original value when the light is turned off, and this process experiences no obvious changes. In addition, some devices were retested after several weeks and displayed exactly the same response, which points at high stability of the photodetector even under environmental ageing.

The ability to operate without any power supply is also highly appealing, especially in wet and biological environments. The time-dependent current curves of the ACNNM device under 50 mW cm^{-2} blue light illumination (Fig. 13.7f) showed similar photoelectric response performance at bias potential of –0.5 V and without bias. This suggests that the ionic photodetector can work without external power (0 V) as well as with power supply (–0.5 V). Meanwhile, it is also noticed that the photoinduced current gain at 0 V bias is about 0.075 µA, which is a bit smaller than it is at –0.5 V bias (0.14 µA). This is due to the fact that the light generates charge pairs in the semiconductor nanotubes, while the applied potential bias when applied in the correct direction only improves charge separation. In other words, the semiconductor nanotube acts as a photovoltaic cell driving its own ability to sense light.

13.4 Ion pumps based on double-phase g-C$_3$N$_4$/TiO$_2$ nanotube

Both in symmetric and asymmetric carbon nitride nanotubes, the directional ions transport is based on the effective photoinduced separation of electrons and holes. Therefore, an effective method to improve performance in ion transport is to realize effective electron–hole separation by constructing semiconductor heterojunction nanotube membranes. Xiao et al. [37] described that TiO$_2$/g-C$_3$N$_4$ semiconductor heterojunction nanotube membrane can realize an improved ion transport properties compared to carbon nitride nanotube.

The heterojunction of the semiconductor nanotubes can be fabricated by two deposition steps (Fig. 13.8a). In the first step, amorphous TiO$_2$ nanotubes with various wall thicknesses were fabricated by the atomic-layer deposition (ALD) method, using an AAO membrane with pore diameters of about 100 nm as the substrate. Then, the amorphous TiO$_2$ nanotubes were crystallized by thermal annealing at 500 °C for 2 h. In the second step, the anatase TiO$_2$ nanotubes were coated with a 10 nm layer g-C$_3$N$_4$ by chemical vapor deposition (CVD). In this way, a TiO$_2$/g-C$_3$N$_4$ heterojunction nanotube was fabricated (Fig. 13.8b). The light-driven ion transport can also be easily reversed by changing the position of g-C$_3$N$_4$ and TiO$_2$ (Fig. 13.8a and 13.8c), and a second g-C$_3$N$_4$/TiO$_2$ nanotube membrane by the same deposition methods but with reverse order was made (Fig. 13.8c). In these, g-C$_3$N$_4$ is placed in the inner layer and TiO$_2$ in the outer layer. Figure 13.8d shows a typical TEM image of the g-C$_3$N$_4$/TiO$_2$ heterojunction nanotube and the enlarged cross section clearly shows that the wall is composited by an inner g-C$_3$N$_4$ layer and an outer TiO$_2$ layer, each of which has a thickness about 7 nm.

The authors then studied the pumping properties of this TiO$_2$/g-C$_3$N$_4$ heterojunction nanotube membrane as an example and its ion pump properties. Figure 13.9a shows the cycle-constant zero-volt current across the nanotube membrane upon illumination with simulated solar light at 300 mW cm^{-2}. Without illumination, the zero-volt current is negligible, while it increased to about 9 μA cm^{-2} upon illumination, indicating that light provides an external driving force for ion movement. Calculations suggest that the measured current translates into the ability of a single nanotube to transport actively ~ 5,500 ions per second. It is an unprecedented breakthrough for artificial light-driven ion transport systems and much closer to the performance of bacteriorhodopsin sodium pump [38] or halorhodopsin Cl ion pump [39]. The directional photo-driven ion transport phenomenon can be directly confirmed by the change of ion concentration in the two cells and monitored in real-time by employing a scanning ion-selective electrode technique (SIET). In

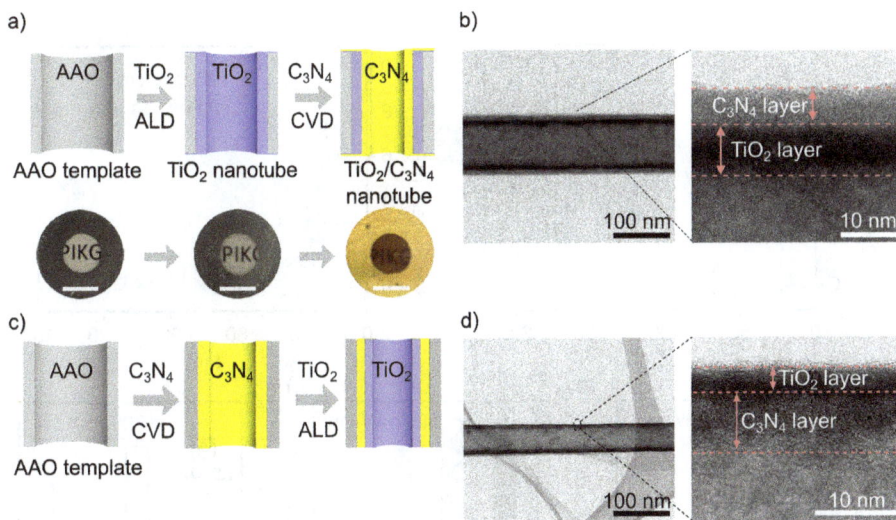

Fig. 13.8: (a) Fabrication process of TiO_2/g-C_3N_4 heterojunction nanotubes including two steps. Step 1: TiO_2-layer deposition by ALD; Step 2: g-C_3N_4 layer deposition by CVD, scale bar 0.5 cm. (b) TEM image of single TiO_2/ g-C_3N_4 nanotube and enlarged area next to the wall surface. (c) Fabrication process of the g-C_3N_4/TiO_2 heterojunction nanotube and TEM image of a single g-C_3N_4/TiO_2 nanotube and enlarged area next to the wall surface. Step 1: g-C_3N_4 layer deposition by CVD; Step 2: TiO_2 layer deposition by ALD.

addition, the membrane shows an instantaneous stable and fully repeatable response to illumination. The ionic current remains stable even at extended illumination time. Further measurements show that the ion transport is closely connected to the illumination power density. The dependence is confirmed by the ionic current measurements shown in Fig. 13.9b. With the decrease of power density from 300 to 54 mW cm^{-2}, the ionic current decreases gradually from 9 to 0.8 μA cm^{-2}. It is worth mentioning that the photo-induced voltage correlates positively with light power density, while it is only dozens of millivolt and much smaller than that recorded through pure g-C_3N_4 nanotube membrane [20].

In this work, the effect of wall thickness of TiO_2 nanotube is also studied. With the increase of wall thickness from 5 nm to 15 nm, the ionic current decreases from about 9 to 2.5 μA cm^{-2} (Fig. 13.9c). This could be ascribed to a less efficient photochemical charge separation and more interfacial recombination of electrons and holes, when a thicker layer of TiO_2 is deposited. The light-driven ion transport system also shows an obvious dependence on the light wavelength (Fig. 13.9d). When applying various monochromatic light sources with the same power density (300 mW cm^{-2}), white light and high-energy blue light generate a comparable and high driving force for ion transport, while low-energy green and yellow light show a much weaker effect. This is consistent with the light absorbance of the g-C_3N_4 layer exposed to light [3]. In general, the isoelectric point of g-C_3N_4 fabricated from different precursors is in

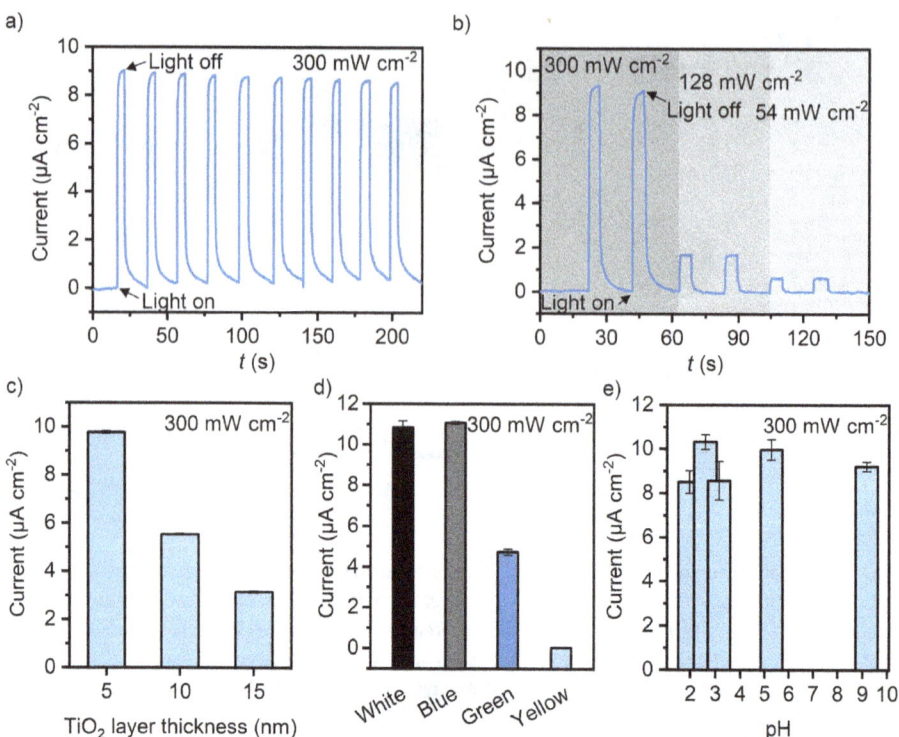

Fig. 13.9: (a) Measured cyclic constant zero-volt current with the alternating illumination and dark at 0.1 M KCl concentration. (b) Zero-volt current as a function of light density of 54, 128 and 300 mW cm^{-2}. (c) Zero-volt current as a function of TiO$_2$ layer thickness. (d) Zero-volt current as a function of monochromatic light (blue: 405 nm; green: 515 nm; yellow: 590 nm) obtained at the same power density of 300 mW cm^{-2}. The magnitude of ionic current is consistent with the light absorbance of the outer g-C$_3$N$_4$ layer. (e) Zero-volt current as a function of pH. Error bars in panels (c–e) represent the standard deviations of five independent experiments.

the range of 3.5 to 5 [40], while the light-driven ion transport system is universal and works constantly in a wider pH value range from 1.9 to 9.5 (Fig. 13.9e). In strong alkaline solution with a pH value of 12.5, it shows a different phenomenon, since the surface of g-C$_3$N$_4$ photo-corrodes under such conditions. Meanwhile, the ionic current shows a positive correlation with electrolyte concentration.

Redistribution of surface charge in heterojunction nanotube triggered by the photo-induced separation of electrons and holes is thought to be a key step to progress with light-driven ion transport phenomena. As illustrated in Fig. 13.10a, the initial TiO$_2$/g-C$_3$N$_4$ nanotube is symmetrically weakly negatively charged due of the acidic nature of the inner g-C$_3$N$_4$ layer. Under these conditions, there is no ionic current in the external circuit. When illuminated from one side of the H-cell, the surface charge density in the irradiated side of the TiO$_2$/g-C$_3$N$_4$ nanotube increases due to the built-in potential in the heterojunction, resulting from band bending. The built-in potential drives movement of

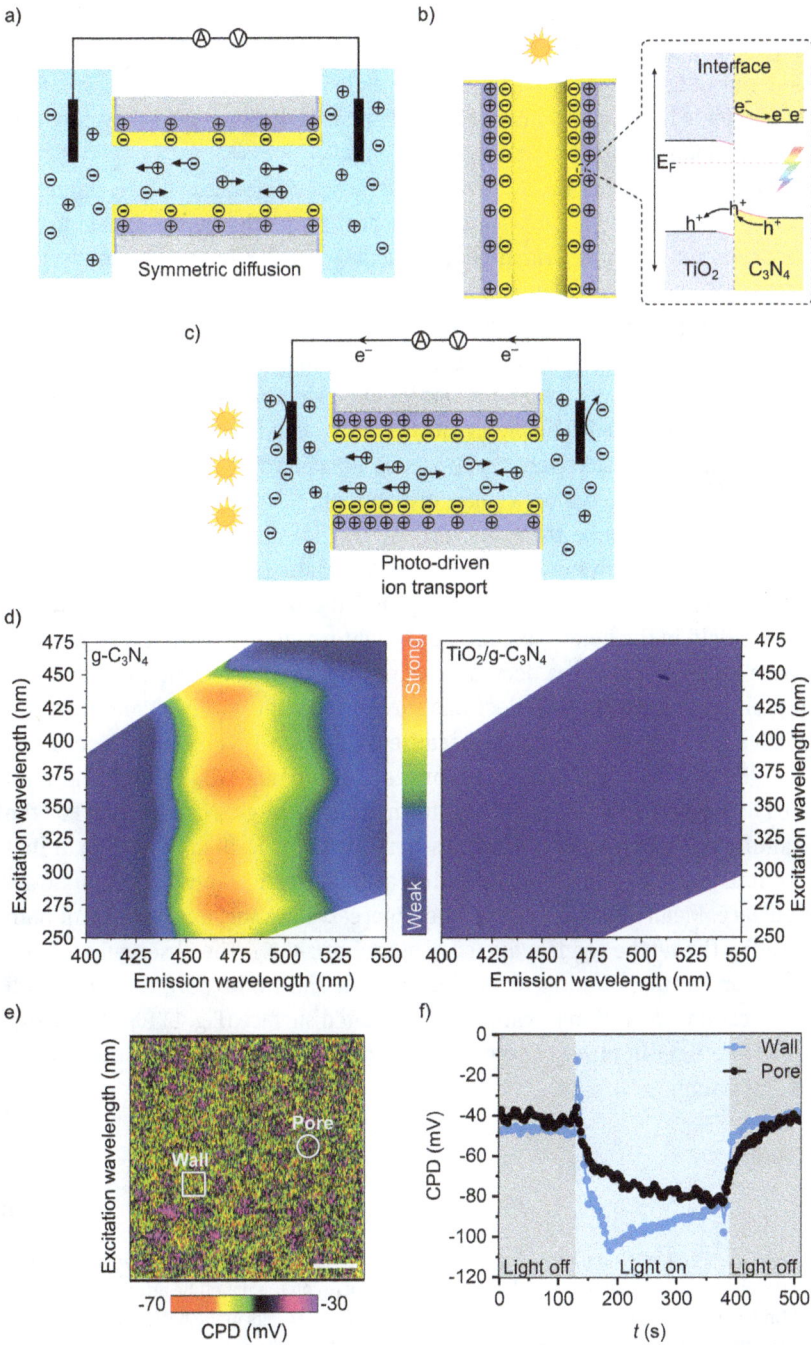

Fig. 13.10: (a) Schematic of the surface charge distribution on the nanotube before illumination. In this state, low-density negative charge is homogeneous distributed over the nanotube. (b) Light-induced

the photogenerated holes from the g-C$_3$N$_4$ layer to the TiO$_2$ layer (Fig. 13.10b). As a result, a positively charged TiO$_2$ layer and a negatively charged g-C$_3$N$_4$ layer are formed. As asymmetric negative surface charge is created, cations move from the unilluminated side to the illuminated side, while anions move in opposite direction (Fig. 13.10c). In this way, a light-driven ion transport system is established.

The work discussed earlier has shown that already a single-phase g-C$_3$N$_4$ nanotube exhibits similar but weaker light-driven ion transport properties, which is due to less efficient separation of photogenerated charge carriers [20, 31]. In the present system, the nanoscopic TiO$_2$/g-C$_3$N$_4$ heterojunction structure provides two different phases for each charge. The proposed mechanism is further confirmed by fluorescent mapping. As shown in Fig. 13.10d, the fluorescent mappings of g-C$_3$N$_4$ nanotube system (left) exhibits strong signals, which represent recombination of photo-generated electrons and holes. However, the fluorescence is negligible in TiO$_2$/g-C$_3$N$_4$ system (right), which means that radiative recombination is effectively suppressed in the heterojunction structure.

In order to directly observe the charge distribution, the authors mapped the surface potential by Kelvin probe force microscopy (KPFM) under unilateral illumination. As for the TiO$_2$/g-C$_3$N$_4$ system, surface potential of the pore area exceeds that of the wall area by 10 mV (Fig. 13.10e), indicating the band bending and generation of a built-in electric field in the heterojunction. After light irradiation, both the surface potential of pore area and wall area decrease clearly, giving solid evidence of the directional transport of photogenerated electrons (to the outer g-C$_3$N$_4$ layer) and holes (to the inner TiO$_2$ layer) (Fig. 13.10f). This is because g-C$_3$N$_4$ and TiO$_2$ have different band gaps, and g-C$_3$N$_4$ is much more attractive for electrons, while TiO$_2$ is much more attractive for holes when a heterojunction is formed between g-C$_3$N$_4$ and TiO$_2$. For the g-C$_3$N$_4$ nanotube, the potential of the area assigned to pores is 40 mV lower than that of the wall, indicating upward band bending and electrons capture by the g-C$_3$N$_4$ nanotube surface. Meanwhile, light irradiation increases the surface potential of both areas, indicating the n-type semiconductor property of g-C$_3$N$_4$. The observed accumulation of electrons at the illuminated side of the wall and in the pore coincides with the measured cations migration toward the illuminated surface (Fig. 13.10c). Thus, this indicates that the light-induced charge redistribution is responsible for the driving force of ion migration.

Fig. 13.10 (continued)
separation of electrons and holes in g-C$_3$N$_4$. The holes are transferred from g-C$_3$N$_4$ to TiO$_2$, while electrons in opposite direction. (c) Schematic of the surface charge distribution on the nanotube after unilateral illumination. Under such conditions, separation of electrons and holes results in the heterogeneous (gradient) distribution of the negative charge. (d) Fluorescent mappings of g-C$_3$N$_4$ (left) and TiO$_2$/g-C$_3$N$_4$ (right) nanotube membranes. (e) KPFM image of the TiO$_2$/g-C$_3$N$_4$ nanotube membrane. Scale bar, 200 nm. (f) Evolution of surface potential of the TiO$_2$/g-C$_3$N$_4$ nanotube in pore area and wall area upon illumination with light.

13.5 Summary

In recent years, techniques were developed to fabricate carbon nitrides with precise nanostructure, for example, by the CVD method for constructing g-C_3N_4 membranes and nanotubes. In this chapter, we showed that such carbon nitride nanotubes can be used to construct ion pumps and ionic photodetectors based on the universal strategy that ion transport can be coupled along interfaces and pores with photoinduced movement of carriers. Undoubtedly, the artificial light-driven ion pump is an exciting tool based on carbon nitride ordered structures. We firmly believe that the efforts devoted to carbon nitride-based ion pumps will likely pay off in the very near future, for example, in the frontiers of sustainable energy, environment and biosensing. One latest example along those lines is to construct confined flow catalytic system to realize enhanced organic photocatalysis [41]. Despite encouraging results already, further improvements in composition and structure of the organic semiconductor, of deposition techniques and alternative fabrication techniques and, of course, membrane layouts might bring such systems to levels of being able to challenge natural ion pumps and bio osmotic electricity generation.

References

[1]   Goettmann F, Fischer A, Antonietti M, Thomas A. Chemical synthesis of mesoporous carbon nitrides using hard templates and their use as a metal-free catalyst for Friedel-Crafts reaction of benzene. Angew Chem Int Ed Engl 2006, 45(27), 4467–71.

[2]   Vinu A. Two-dimensional hexagonally-ordered mesoporous carbon nitrides with tunable pore diameter, surface area and nitrogen content. Adv Funct Mater 2008, 18(5), 816–27.

[3]   Wang X, Maeda K, Thomas A, Takanabe K, Xin G, Carlsson JM, et al. A metal-free polymeric photocatalyst for hydrogen production from water under visible light. Nat Mater 2009, 8(1), 76–80.

[4]   Ong WJ, Tan LL, Ng YH, Yong ST, Chai SP. Graphitic carbon nitride (g-C_3N_4)-based photocatalysts for artificial photosynthesis and environmental remediation: Are we a step closer to achieving sustainability? Chem Rev 2016, 116(12), 7159–329.

[5]   Lakhi KS, Park DH, Al-Bahily K, Cha W, Viswanathan B, Choy JH, et al. Mesoporous carbon nitrides: synthesis, functionalization, and applications. Chem Soc Rev 2017, 46(1), 72–101.

[6]   Liu J, Wang H, Antonietti M. Graphitic carbon nitride "reloaded": emerging applications beyond (photo)catalysis. Chem Soc Rev 2016, 45(8), 2308–26.

[7]   Arazoe H, Miyajima D, Akaike K, Araoka F, Sato E, Hikima T, et al. An autonomous actuator driven by fluctuations in ambient humidity. Nat Mater 2016, 15(10), 1084–89.

[8]   El-Sayed MA. On the molecular mechanisms of the solar to electric energy conversion by the other photosynthetic system in nature, bacteriorhodopsin. Acc Chem Res 1992, 25(7), 279–86.

[9]   Tributsch H. Light driven proton pumps. Ionics 2000, 6(3-4), 161–71.

[10] Gadsby DC. Ion channels versus ion pumps: the principal difference, in principle. Nat Rev Mol Cell Biol 2009, 10(5), 344–52.

[11] Steinberg-Yfrach G, Rigaud J-L, Durantini EN, Moore AL, Gust D, Moore TA. Light-driven production of ATP catalysed by F_0F_1-ATP synthase in an artificial photosynthetic membrane. Nature 1998, 392(6675), 479–82.

[12] Gust D, Moore TA, Moore AL. Mimicking photosynthetic solar energy transduction. Acc Chem Res 2001, 34(1), 40–48.

[13] Zhang Z, Kong X-Y, Xie G, Li P, Xiao K, Wen L, et al. "Uphill" cation transport: A bioinspired photo-driven ion pump. Sci Adv 2016, 2(10), e1600689.

[14] Yang J, Liu P, He X, Hou J, Feng Y, Huang Z, et al. Photodriven active ion transport through a Janus microporous membrane. Angew Chem Int Ed 2020, 59(15), 6244–48.

[15] Jiang Y, Ma W, Qiao Y, Xue Y, Lu J, Gao J, et al. Metal–organic framework membrane nanopores as biomimetic photoresponsive ion channels and photodriven ion pumps. Angew Chem 2020, 59, 12795.

[16] Xiao K, Kong X-Y, Zhang Z, Xie G, Wen L, Jiang L. Construction and application of photoresponsive smart nanochannels. J Photochem Photobiol C 2016, 26, 31–47.

[17] Yang J, Hu X, Kong X, Jia P, Ji D, Quan D, et al. Photo-induced ultrafast active ion transport through graphene oxide membranes. Nat Commun 2019, 10(1), 1171.

[18] Hong S, Zou G, Kim H, Huang D, Wang P, Alshareef HN. Photothermoelectric response of $Ti_3C_2T_x$ mxene confined ion channels. ACS Nano 2020, 14(7), 9042–49.

[19] Lao J, Lv R, Gao J, Wang A, Wu J, Luo J. Aqueous stable Ti_3C_2 Mxene membrane with fast and photoswitchable nanofluidic transport. ACS Nano 2018, 12(12), 12464–71.

[20] Xiao K, Chen L, Chen R, Heil T, Lemus SDC, Fan F, et al. Artificial light-driven ion pump for photoelectric energy conversion. Nat Commun 2019, 10(1), 74.

[21] Siwy Z, Fuliński A. Fabrication of a synthetic nanopore ion pump. Phys Rev Lett 2002, 89(19), 198103.

[22] Dittrich T, Fiechter S, Thomas A. Surface photovoltage spectroscopy of carbon nitride powder. Appl Phys Lett 2011, 99(8), 084105.

[23] Xiao K, Schmidt OG. Light-driven ion transport in nanofluidic devices: photochemical, photoelectric, and photothermal effects. CCS Chem 2021, 2938–49.

[24] Lozada-Hidalgo M, Zhang S, Hu S, Kravets VG, Rodriguez FJ, Berdyugin A, et al. Giant photoeffect in proton transport through graphene membranes. Nat Nanotechnol 2018, 13(4), 300–03.

[25] Zhang Q, Xiao T, Yan N, Liu Z, Zhai J, Diao X. Alternating current output from a photosynthesis-inspired photoelectrochemical cell. Nano Energy 2016, 28, 188–94.

[26] Xie X, Crespo GA, Mistlberger G, Bakker E. Photocurrent generation based on a light-driven proton pump in an artificial liquid membrane. Nat Chem 2014, 6(3), 202–07.

[27] van der Heyden FHJ, Bonthuis DJ, Stein D, Meyer C, Dekker C. Power generation by pressure-driven transport of ions in nanofluidic channels. Nano Lett 2007, 7(4), 1022–25.

[28] Xie G, Li P, Zhang Z, Xiao K, Kong X-Y, Wen L, et al. Skin-inspired low-grade heat energy harvesting using directed ionic flow through conical nanochannels. Adv Energy Mater 2018, 1800459.

[29] Zhang Z, Sui X, Li P, Xie G, Kong X-Y, Xiao K, et al. Ultrathin and ion-selective Janus membranes for high-performance osmotic energy conversion. J Am Chem Soc 2017, 139(26), 8905–14.

[30] Wen L, Hou X, Tian Y, Zhai J, Jiang L. Bio-inspired photoelectric conversion based on smart-gating nanochannels. Adv Funct Mater 2010, 20(16), 2636–42.

[31] Xiao K, Tu B, Chen L, Heil T, Wen L, Jiang L, et al. Photo-driven ion transport for a photodetector based on an asymmetric carbon nitride nanotube membrane. Angew Chem Int Ed 2019, 58(36), 12574–79.

[32] Bao C, Yang J, Bai S, Xu W, Yan Z, Xu Q, et al. High performance and stable all-inorganic metal halide perovskite-based photodetectors for optical communication applications. Adv Mater 2018, e1803422.

[33] Xiao K, Giusto P, Wen L, Jiang L, Antonietti M. Nanofluidic ion transport and energy conversion through ultrathin free-standing polymeric carbon nitride membranes. Angew Chem Int Ed Engl 2018, 57(32), 10123–26.

[34] Brar VW, Sherrott MC, Jariwala D. Emerging photonic architectures in two-dimensional opto-electronics. Chem Soc Rev 2018, 47(17), 6824–44.

[35] Wang H, Zhang C, Chan W, Tiwari S, Rana F. Ultrafast response of monolayer molybdenum disulfide photodetectors. Nat Commun 2015, 6, 8831.

[36] Song YA, Melik R, Rabie AN, Ibrahim AM, Moses D, Tan A, et al. Electrochemical activation and inhibition of neuromuscular systems through modulation of ion concentrations with ion-selective membranes. Nat Mater 2011, 10(12), 980–86.

[37] Xiao K, Giusto P, Chen F, Chen R, Heil T, Cao S, et al. Light-driven directional ion transport for enhanced osmotic energy harvesting. Natl Sci Rev 2021, 8, nwaa23.

[38] Inoue K, Ono H, Abe-Yoshizumi R, Yoshizawa S, Ito H, Kogure K, et al. A light-driven sodium ion pump in marine bacteria. Nat Commun 2013, 4, 1678.

[39] Bamberg E, Tittor J, Oesterhelt D. Light-driven proton or chloride pumping by halorhodopsin. Proc Natl Acad Sci USA 1993, 90(2), 639–43.

[40] Zhu B, Xia P, Ho W, Yu J. Isoelectric point and adsorption activity of porous g-C_3N_4. Appl Surf Sci 2015, 344, 188–95.

[41] Zou Y, Xiao K, Qin Q, Shi J-W, Heil T, Markushyna Y, et al. Enhanced organic photocatalysis in confined flow through a carbon nitride nanotube membrane with conversions in the millisecond regime. ACS Nano 2021, 15(4), 6551–61.

Markus Antonietti*, Oleksandr Savateev and Xinchen Wang

Chapter 14
Looking into the crystal ball of a sustainable future chemistry with carbon nitride

After going through the earlier contributions and many others of 20 years of intensified carbon nitride research (with more than 38,000 papers published), we may state that, indeed, a door has been opened in the walls of Science. Many of the reported experiments indeed changed our chemical view on what materials composed of only covalently linked lightweight elements can do. As the elder graphene or carbon nanotubes, the carbon nitride family is composed of only covalent sp^2–bonds, bringing not only the highest Young's modulus and mechanical performance, but also electric conductivity. Contrary to graphene, the introduction of nitrogen changes the electronic structure to open up a band-gap – medium band gap semiconductors instead of a semimetal are obtained.

It was stated even at the beginning of this book that carbon nitrides are easy to synthesize, sustainable, economically feasible, and exerted a certain appeal over 200 years to diverse generations of chemists, including Friedrich Wöhler or Linus Pauling. We guess the appeal came from symmetry, structure and composition. This early structural appeal left the functional properties of carbon nitrides intact, and research was being focused mostly on structure. This is our explanation as to why carbon nitride is one of the oldest polymer structures ever reported in the scientific literature, while at the same time, the 2D material dimension and electronic properties began to be explored only 20 years ago.

To sum up the many returns of carbon nitrides to application and the development of science the final wrap up is as follows:

- As an electrocatalyst, it is very active in a variety of applications, first described in hydrogen evolution [1]. With appropriately adjusted band positions, it has the potential to even replace platinum.

Acknowledgments: Editors express their gratitude to Dr. Lu Peng for providing carbon nitride powders that were used to create the image for the cover page, and to Dr. Paolo Giusto and Dr. Vitaliy Shvalagin for proofreading the manuscript.

*Corresponding author: Markus Antonietti, Colloid Chemistry Department, Max Planck Institute of Colloids and Interfaces, Am Muehlenberg 1, 14476 Potsdam, Germany,
e-mail: markus.antonietti@mpikg.mpg.de
Oleksandr Savateev, Colloid Chemistry Department, Max Planck Institute of Colloids and Interfaces, Am Muehlenberg 1, 14476 Potsdam, Germany
Xinchen Wang, State Key Laboratory of Photocatalysis on Energy and Environment, College of Chemistry, Fuzhou University, Fuzhou 350116, P.R. China

https://doi.org/10.1515/9783110746976-014

- As photocatalyst, it not only generates hydrogen [2] and oxygen [3] with record efficiencies, it is even able to do full artificial photosynthesis (oxygen liberation and carbon dioxide reduction) with only one catalytic system, without any cocatalysts applied [4].
- In organic chemistry, carbon nitride species can match the reactivity of Ir-based low molecular weight photoactive complexes, while being heterogeneous and affordable, at the same time [5]. In chemical space, more and more reactions are discovered, with at least 19 new reactions described, some of them even peculiar for carbon nitrides [6].
- For robotics and autonomous microsystems, solar batteries [7], light-driven carbon nitride pumps [8] and also autonomous microswimmers [9] were described. The simplicity of many of these systems can be strictly related to the multifunctionality of carbon nitride [10].
- As catalytic, noninnocent supports, carbon nitrides and their relatives are among the most efficient supports for single-atom and small nanoparticle catalysis [11, 12]. For instance, Fe_1 loaded on Na-PHI is able to catalyze alkane-monooxidation by reversibly stabilizing the corresponding $Fe(IV) = O$ oxo-species [13].
- For optics, the very high index of refraction that even matches diamond could be instrumentalized [14]. Carbon nitride films can be homogeneously deposited by CVD, while structuring is possible by laser writing or e-beam lithography [15]. First metaoptic devices have been presented [16].
- Carbon nitride coatings are, at the same time, mechanically strong and hard, with a Young's modulus of 36.5 GPa and hardness of 2.2 GPa. For an essentially porous material, these are extreme values.
- Use of carbon nitride in even advanced perovskite cells as buffer or electron collection layer improved the overall efficiency by another 2% [17]. Such improvements go through all fields of energy conversion or molecular sensing, as described in a whole chapter above.

This is quite an unexpectedly rich harvest, and it is also a sign how digitalization and the internet have driven international exchange and accelerated science, when compared to earlier days.

This chapter is, however, also an outlook, and we will try a prediction of possible futures, and what experts see and expect to occur in the next years. Predictions into the future are, of course, notoriously wrong and more blurred the further the vision goes. Nevertheless, such predictions might be stimulating or even just entertaining, and they certainly serve as indications that the pipeline of carbon nitride research is not empty, but still more rapidly filed by an increasing number of involved scientists.

Some of the bets on the possible future development are listed as follows:

1) Carbon nitride is just a first example of a practically endless field of covalent C, N, O materials with reduced dimensionality and electronic conjugation. Even within the

carbon–nitrogen composition space, scientists already reported on C_3N [18], C_2N [19, 20], $C_2N_2(NH)$ [21] as "carbon subnitrides" as well as on the nitrogen-rich side on $C_3N_{5.4}$ [22] and C_3N_7 [23], and all of these systems bring new properties, such as band positions and band gaps, and also ferromagnetism and special ion binding properties (Fig. 14.1) [24].

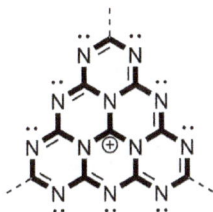

Fig. 14.1: A structure of alternative "carbon nitride."

2) First experiments already indicate that the C-N-space can be profitably expanded to the ternary systems C-N-B [25], C-N-P [26] or even C-N-O [27]. It was already reported that this adds chemical functionality and electronic properties, for instance, moving the valance band to even more positive values. Such potential oxidants with a new deep oxidation chemistry were formerly only accessed by metal oxides with very positive metal centers. A courageous prediction based on that is that this might reduce our dependency on rare metal even more than carbon nitride already did. Here is a vision to replace any noble metal chemistry by similar electrons placed in covalent structures.

3) It is still unclear if "carbon oxides" without nitrogen can play the same role as carbon nitrides, but again there are, indeed, endless solid "carbonoxides," not only the easily recognized CO and CO_2. The so-called "red carbon" is only one of the many possible species (Fig. 14.2) [28]. The problem of these structures is that CO and CO_2 are good leavings groups, but the leftover carbon nanostructures are interesting as such and potentially very electron-rich.

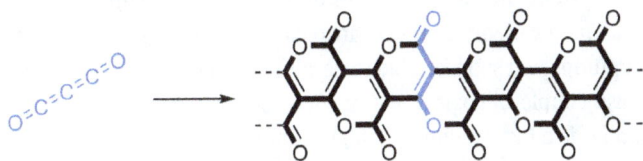

Fig. 14.2: Synthesis of "carbon oxide" composed of conjugated ladder of polypyrone ribbons by polymerization of carbon suboxide (C_3O_2).

4) The optical properties of then transparent carbon nitride films were just recently instrumentalized, but allow production of metamaterials in the visible spectrum. Condensed carbon nitride, in spite of being highly porous, now is close to matching diamond already, and it is an exciting question how to further increase index of refraction – either by chemical modification of the backbone to increase electronic polarizability or by

supramolecular chemistry filling the pores with another more contributing compound. Carbon nitride inclusion compounds, as such, will be exciting also beyond optics.

5) We know from measurements on poly(heptazine imides), as also from supramolecular model structures that the structural pore channels are filled with water, which cannot be removed even under harsh conditions, but it is mobile and allows flux through the system. Such highly bound water and the involved hydrogen bridges (imide groups are H-bridge donors, while triazines are H-bridge acceptors) create a new type of pore water pool with different boiling point, pH value, electrochemical window, chemical solubility and so on. One might learn a lot about bound water in biological structures too, from such model experiments.

6) Heterogeneous organocatalysis and photocatalysis with carbon nitride derivatives (without and with light) are some of the most active areas within the carbon nitride field. The availability of new structures with different oxidation and reduction potential will certainly expand chemistry beyond current metalorganic species, and especially for deep oxidation reactions (more positive than +2.0 V), there is a lost world to discover.

7) Dark photochemistry [29] and also recent papers on multisite collective proton-electron transfer indicate that reductions in water can now be performed (as with biological catalysts) by electron-proton pairs [30]. The advantage is that depending on the structure, such proton electron pairs can have a higher reduction potential than H_2 or noble metal hydrides as such. In addition, for conjugated multisite systems, the reduction potential is getting higher with electron loading, as the conduction band is slowly filled up from the bottom to higher values, until electron-electron repulsion gets too high. Thus, we might dream about really unusual reduction reactions, currently only accessible by chemical reductants, such as aluminum hydrides. A similar effect must also exist on the oxidation side, i.e., oxidation by hole–hydroxy couples, but this, to our knowledge, has not been described up to now.

8) Chemical nano- and microsystems for active devices and small scale engineering as actuators and pumps, or solar batteries, are also very inspiring and full of ideas for non-chemists. Photo-osmotic ion pumps will soon be able to desalinate water, separate diverse ions from each other and also couple biological and artificial systems on a cellular/liposomal level. Small-scale robotics is a big upcoming field, but should be described by more qualified engineering experts.

9) For semiconductors, there is a productive material properties level above the scale of molecules, which is the special coupling to another semiconductor. With different doping, this creates diodic or transistor junctions, while with only two metals or two semiconductors, this creates Schottky heterojunctions with spontaneous charge transfer over the interface. This highly polar/dipolar state is also very effective in chemistry [31] and hard to create with molecules.

10) Finally, carbon nitrides must be made available on larger scales to enable engineers and physicists to touch the materials. Synthesis is, in principle, very simple, but there is a lot of know-how involved; this is presumably to be addressed by start-ups or medium scale companies. Current commodity industry will only enter, e.g., the use as a catalyst, once diverse versions are commercially available.

At the very end, we should not forget to mention that carbon nitride also hits the "zeitgeist" of the modern age of mythos and sustainability. Born in flames even from simple molecules such as urea, it is not only Pharaoh's Serpent, but rather a phoenix that has risen from the ashes. As such, the material is not only born in a sustainable fashion but also enables sustainable processes as artificial photosynthesis, photocatalysis, metal-free electrochemistry and organocatalysis. On top of it, it functionally cross-couples light and ions pumping, or light and electron transfer, even down to the micrometer scale.

It is thus a true molecular structure of the twenty-first century!

References

[1] Zheng Y, Jiao Y, Zhu Y, Li LH, Han Y, Chen Y, et al. Hydrogen evolution by a metal-free electrocatalyst. Nat Commun 2014, 5(1), 3783.

[2] Zhang G, Lin L, Li G, Zhang Y, Savateev A, Zafeiratos S, et al. Ionothermal synthesis of triazine–heptazine-based copolymers with apparent quantum yields of 60 % at 420 nm for solar hydrogen production from "Sea water". Angew Chem Int Ed 2018, 57(30), 9372–6.

[3] Yang X, Tang H, Xu J, Antonietti M, Shalom M. Silver phosphate/graphitic carbon nitride as an efficient photocatalytic tandem system for oxygen evolution. Chem Sus Chem 2015, 8(8), 1350–8.

[4] Xia P, Antonietti M, Zhu B, Heil T, Yu J, Cao S. Designing defective crystalline carbon nitride to enable selective CO_2 photoreduction in the gas phase. Adv Funct Mater 2019, 29(15), 1900093.

[5] Savateev A, Ghosh I, König B, Antonietti M. Photoredox catalytic organic transformations using heterogeneous carbon nitrides. Angew Chem Int Ed 2018, 57(49), 15936–47.

[6] Savateev A, Kurpil B, Mishchenko A, Zhang G, Antonietti M. A "waiting" carbon nitride radical anion: A charge storage material and key intermediate in direct C–H thiolation of methylarenes using elemental sulfur as the "S"-source. Chem Sci 2018, 9(14), 3584–91.

[7] Podjaski F, Kröger J, Lotsch BV. Toward an aqueous solar battery: Direct electrochemical storage of solar energy in carbon nitrides. Adv Mater 2018, 30(9), 1705477.

[8] Xiao K, Chen L, Chen R, Heil T, Lemus SDC, Fan F, et al. Artificial light-driven ion pump for photoelectric energy conversion. Nat Commun 2019, 10(1), 74.

[9] Sridhar V, Podjaski F, Kröger J, Jiménez-Solano A, Park B-W, Lotsch BV, et al. Carbon nitride-based light-driven microswimmers with intrinsic photocharging ability. Proc Natl Acad Sci 2020, 117(40), 24748–56.

[10] Podjaski F, Lotsch BV. Optoelectronics meets optoionics: Light storing carbon nitrides and beyond. Adv Energy Mater 2021, 11(4), 2003049.

[11] Zhao M, Feng J, Yang W, Song S, Zhang H. Recent advances in graphitic carbon nitride supported single-atom catalysts for energy conversion. ChemCatChem 2021, 13(5), 1250–70.

[12] Xiao X, Zhang L, Meng H, Jiang B, Fu H. Single metal atom decorated carbon nitride for efficient photocatalysis: Synthesis, structure, and applications. Sol RRL 2021, 5(6), 2000609.

[13] da Silva MAR, Silva IF, Xue Q, Lo BTW, Tarakina NV, Nunes BN, et al. Sustainable oxidation catalysis supported by light: Fe-poly (heptazine imide) as a heterogeneous single-atom photocatalyst. Appl Catal B 2022, 304, 120965.

[14] Giusto P, Cruz D, Heil T, Arazoe H, Lova P, Aida T, et al. Shine bright like a diamond: New light on an old polymeric semiconductor. Adv Mater 2020, 32(10), 1908140.

[15] Zhang J, Zou Y, Eickelmann S, Njel C, Heil T, Ronneberger S, et al. Laser-driven growth of structurally defined transition metal oxide nanocrystals on carbon nitride photoelectrodes in milliseconds. Nat Commun 2021, 12(1), 3224.

[16] Sun L, Hong W, Liu J, Yang M, Lin W, Chen G, et al. Cross-linked graphitic carbon nitride with photonic crystal structure for efficient visible-light-driven photocatalysis. ACS Appl Mater Interfaces 2017, 9(51), 44503–11.

[17] Cruz D, Garcia Cerrillo J, Kumru B, Li N, Dario Perea J, Schmidt BVKJ, et al. Influence of thiazole-modified carbon nitride nanosheets with feasible electronic properties on inverted perovskite solar cells. J Am Chem Soc 2019, 141(31), 12322–8.

[18] Yang S, Li W, Ye C, Wang G, Tian H, Zhu C, et al. C_3N – A 2D crystalline, hole-free, tunable-narrow-bandgap semiconductor with ferromagnetic properties. Adv Mater 2017, 29(16), 1605625.

[19] Tian Z, López-Salas N, Liu C, Liu T, Antonietti M.. C_2N: A class of covalent frameworks with unique properties. Adv Sci 2020, 7(24), 2001767.

[20] Mahmood J, Lee EK, Jung M, Shin D, Jeon I-Y, Jung S-M, et al. Nitrogenated holey two-dimensional structures. Nat Commun 2015, 6(1), 6486.

[21] Horvath-Bordon E, Riedel R, McMillan PF, Kroll P, Miehe G, van Aken PA, et al. High-pressure synthesis of crystalline carbon nitride imide, $C_2N_2(NH)$. Angew Chem Int Ed 2007,46(9),1476–80.

[22] Sathish C, Premkumar S, Chu X, Yu X, Breese MBH, Al-Abri M, et al. Microporous carbon nitride $(C_3N_{5.4})$ with tetrazine based molecular structure for efficient adsorption of CO_2 and water. Angew Chem Int Ed 2021, 60(39), 21242–9.

[23] Kim IY, Kim S, Premkumar S, Yang J-H, Umapathy S, Vinu A. Thermodynamically stable mesoporous C_3N_7 and C_3N_6 with ordered structure and their excellent performance for oxygen reduction reaction. Small 2020, 16(12), 1903572.

[24] López-Salas N, Albero J. C_xN_y: New carbon nitride organic photocatalysts. Front Mater 2021, 8, 772200.

[25] Giusto P, Arazoe H, Cruz D, Lova P, Heil T, Aida T, et al. Boron carbon nitride thin films: From disordered to ordered conjugated ternary materials. J Am Chem Soc 2020, 142(49), 20883–91.

[26] Liu B, Ye L, Wang R, Yang J, Zhang Y, Guan R, et al. Phosphorus-doped graphitic carbon nitride nanotubes with amino-rich surface for efficient CO_2 capture, enhanced photocatalytic activity, and product selectivity. ACS Appl Mater Interfaces 2018, 10(4), 4001–9.

[27] Battula VR, Kumar S, Chauhan DK, Samanta S, Kailasam K. A true oxygen-linked heptazine based polymer for efficient hydrogen evolution. Appl Catal B 2019, 244, 313–9.

[28] Odziomek M, Giusto P, Kossmann J, Tarakina NV, Heske J, Rivadeneira SM, et al. "Red Carbon": A rediscovered covalent crystalline semiconductor. Adv Mater 2022, 2206405.

[29] Lau VW-h, Klose D, Kasap H, Podjaski F, Pignié M-c, Reisner E, et al. Dark photocatalysis: Storage of solar energy in carbon nitride for time-delayed hydrogen generation. Angew Chem Int Ed 2017, 56(2), 510–4.

[30] Mazzanti S, Schritt C, ten Brummelhuis K, Antonietti M, Savateev A. Multisite. PCET with photocharged carbon nitride in dark. Exploration 2021, 1(3), 20210063.

[31] Cao S, Low J, Yu J, Jaroniec M. Polymeric photocatalysts based on graphitic carbon nitride. Adv Mater 2015, 27(13), 2150–76.

Index

https://doi.org/10.1515/9783110746976-015

www.ingramcontent.com/pod-product-compliance
Lightning Source LLC
Chambersburg PA
CBHW080708220326
41598CB00033B/5342